Management and Ecology of Lake and Reservoir Fisheries

Management and Ecology of Lake and Reservoir Fisheries

EDITED BY

I. G. COWX

Hull International Fisheries Institute
University of Hull, UK

Fishing News Books
An imprint of Blackwell Science

Blackwell Science

A Division of Blackwell Science Ltd
Editorial Offices:
Osney Mead, Oxford OX2 0EL
25 John Street, London WC1N 2BS
23 Ainslie Place, Edinburgh EH3 6AJ
350 Main Street, Malden
MA 02148 5018, USA
54 University Street, Carlton
Victoria 3053, Australia
10, rue Casimir Delavigne
75006 Paris, France

Other Editorial Offices:

Blackwell Wissenschafts-Verlag GmbH
Kurfürstendamm 57
10707 Berlin, Germany

Blackwell Science KK
MG Kodenmacho Building
7-10 Kodenmacho Nihombashi
Chuo-ku, Tokyo 104, Japan

Iowa State University Press
A Blackwell Science Company
2121 S. State Avenue
Ames, Iowa 50014-8300, USA

First published 2002

Set in Times and produced by
Gray Publishing, Tunbridge Wells, Kent

DISTRIBUTORS
Marston Book Services Ltd
PO Box 269
Abingdon
Oxon OX14 4YN
(*Orders:* Tel: 01865 206206
Fax: 01865 721205
Telex: 83355 MEDBOK G)

USA and Canada
Iowa State University Press
A Blackwell Science Company
2121 S. State Avenue
Ames, Iowa 50014-8300
(*Orders:* Tel: 800-862-6657
Fax: 515-292-3348
Web: www.isupress.com
email: orders@isupress.com)

Australia
Blackwell Science Pty Ltd
54 University Street
Carlton, Victoria 3053
(*Orders:* Tel: 03 9347 0300
Fax: 03 9347 5001)

A catalogue record for this title is available from the British Library

ISBN 0-85238-283-9

Library of Congress
Cataloging-in-Publication Data
Management and ecology of lake and reservoir fisheries/edited by I.G. Cowx.
p. cm.
Proceedings of a meeting held in Hull, UK in April 2000.
Includes bibliographical references (p.).
ISBN 0-85238-283-9
1. Fishery management–Congresses. 2. Freshwater fishes–Ecology–Congresses. I. Cowx, I. G. (Ian G.)

SH328.M33 2001
333.95′6–dc21 2001043018

For further information on Fishing News Books, visit our website:
www.blacksci.co.uk/fnb/

Contents

II Anthropogenic activities/rehabilitation and mitigation

III Management

Preface

During a meeting of the European Inland Fisheries Advisory Commission (EIFAC) in Portugal in June 1998, the need to understand fully the key issues regarding the management and ecology of lake and reservoir fisheries was recognised because of their importance to society in general. In view of this importance it was felt that current knowledge of the status and management of lake and reservoir fisheries warranted further discussion and dissemination.

To this end, the International Fisheries Institute at the University of Hull, in co-operation with the EIFAC, organised a symposium and workshop on *Management and Ecology of Lake and Reservoir Fisheries* which took place in Hull, UK in April 2000. The objectives of the symposium were:

- to exchange knowledge among countries and to present, in reviews, an appraisal of the status of lake and reservoir fisheries, assessment methodology, constraints on their development, issues and options regarding their management, and the benefits and problems associated with these activities;
- to identify constraints and gaps in our knowledge that affect the application of fisheries management policy in temperate and tropical lake and reservoir fisheries;
- to recommend and promote action to improve the management of lake and reservoir fisheries;
- to provide guidelines for the policy formulation, planning methodology and evaluation of future activities.

The main conclusions of the symposium and workshop are outlined below and in the selected chapters which make up these proceedings. It is hoped they will stimulate fisheries scientists, managers and academics to collaborate in further research to improve our understanding of lake and reservoir fisheries and promote further research and collaboration to maintain and enhance these important resources worldwide.

Stock assessment for management purposes

It was recognised that lacustrine fisheries are formed in diverse types of habitats from small ponds to large lakes and reservoirs. Consequently, there is a need for an array of methods to provide adequate assessment of stocks in those habitats. These vary from environmental correlation methods, through sampling gears and romote methods (e.g. hydroacoustics) to catch assessment surveys (catch effort methods). It was recognised that there is no perfect sampling technique and a combination of methods is required to gain the desired precision in stock biomass. In this respect, there is a need to define the objectives of the assessment activity to ensure the appropriate precision in information is gained. In many instances it was questioned whether stock assessment was necessary as trends in species composition, catches etc., can provide adequate information for management at minimal costs.

The major constraints on stock assessment were inadequate development in gear technology, poor data analytical methods, weak understanding of the limitation of stock assessment procedures, and lack of resources.

Anthropogenic activities/rehabilitation and mitigation

It was recognised that issues affecting lake and reservoir fisheries are either related to the fishery itself (e.g. overexploitation, inadequate recruitment or maintenance of biodiversity), the ecosystem or watershed (e.g. pollution, nutrient enrichment, hypolimnetic anoxia), or the human dimension. It is possible:

- to identify and prioritise problems impacting upon lake and reservoir fisheries and aquatic communities;
- but it is less easy to identify and carry out appropriate technical solutions to these problems, mainly because of the scale of action required.

The mechanisms for rehabilitation or improvement of lakes and reservoir fisheries are in their infancy and often restricted to stock enhancement procedures (targeting the fishery problems), or simple regulation of pollution and land use changes.

From the human dimension, problems arise from fisher dissatisfaction, poor catch quality, excessive fishing effort, access problems and multiple resource user conflicts. Most of the factors causing problems for fish communities are outside the control of the fisheries sector. Those involved in fisheries must therefore broaden and strengthen their cause, by interacting and making alliances with other interested parties, in seeking to limit damage to aquatic ecosystems, and promote rehabilitation and enhancement activities.

Management issues

Several key management issues were raised which link to other outputs of the symposium.

- Whilst stock assessment may not be required *per se* there is a need to improve fisheries statistical monitoring procedures to provide baseline information on exploitation levels.
- There is a need to improve communication linkages between fisheries managers, scientists and those utilising the resource. This can be the first step in management of the resources which appears to be the most desirable way to manage large-scale lacustrine fisheries.
- The profile of lake fisheries needs raising in general, and particularly where there is a multiple array of resource users who are often in conflict or potential conflict. To support this action there is a growing awareness of the importance for economic

and social valuation of fisheries to ensure they are well represented in all development activities. It is recommended that priority be given to developing and promoting economic valuation of inland fisheries.

- If water and aquatic resources are to be exploited on a sustainable basis in the future, concerted effort is needed to resolve the conflicts between user groups. Where possible, this must be based on sound scientific evidence, close liaison between user groups, full cost-benefit analysis and transparency in the decision-making process. Where scientific information is not available this should not prevent decisions being made, but the precautionary approach should be adopted. If resolution of conflict is to be successful it must involve cross education of all user groups, recognition of stockholder participation and needs, and be probably implemented at the local community level.
- There is also the need for robust methods for prioritising demands for the water and aquatic resources of lake and reservoirs that balance human requirements against protection of the environment and biodiversity.

The production of these proceedings has involved considerable effort by a number of people. In particular, thanks must go to the following for their contribution to workshops held prior to the main symposium and support in reviewing the chapters for the proceedings: U. Amarasinghe, M. Aprahamian, S.S. De Silva, D. Gerdeaux, C.R. Goldspink, J. Harvey, P. Hickley, S. Hughes, A.T. Ibbotson, G. Peirson, M.R. Perrow, M. Petrere Jr, D. Tweddle, T. Vehenan, R.L.J. Wazenböck, R. Welcomme and I.J. Winfield. I would like to thank Jon Harvey and Emma Doy for their considerable assistance in the running of the symposium and Julia Cowx for the production of these proceedings. Finally, I would like to thank the many international funding agencies and organisations for their financial support, thus ensuring truly international coverage of the issues and the success of the symposium.

Ian G. Cowx
University of Hull International Fisheries Institute

Section I
Stock assessment for management purposes

Chapter 1
Hydroacoustic fish stock assessment in lakes in England and Wales

S. HUGHES* and J. HATELEY
National Coarse Fisheries Centre, Environment Agency, Kidderminster, Worcestershire, UK

Abstract

Conventional survey techniques alone often cannot provide adequate and timely information on fish communities in smaller lakes. Mobile horizontal and vertical hydroacoustic methods and complimentary fish capture techniques can provide information to meet a wide range of requirements and are suited to the monitoring contexts and constraints peculiar to England and Wales. Consistent collection of hydroacoustic data allows nationally comparable results. The basis for a nationally applied, consistent strategy for data collection in lakes is outlined.

Keywords: fish, hydroacoustic, lakes, sampling strategy.

1.1 Introduction

Inland fishery managers, environmental regulators and aquatic conservationists often require information about fish communities and stocks to discharge adequately their responsibilities. This information must be reliable, timely, cost effective and adequate for its purpose, and based upon appropriate, sound and defensible data (Hickley & Aprahamian 2000).

This chapter examines the role of hydroacoustic techniques in monitoring lake fish communities in England and Wales. The basis of a nationally applied, consistent strategy for data collection is outlined.

1.2 Operating context

1.2.1 *Freshwater angling*

Freshwater recreational fishing is one of the most popular participant sports in England and Wales, with an estimated 2.9 million anglers, which equates to approximately 3.5% of the population. Of these, 2.3 million are coarse anglers, and the majority (52%) fish mostly on stillwaters (National Rivers Authority 1995a). There is evidence from

*Correspondence: National Coarse Fisheries Centre, Environment Agency, Hoo Farm Industrial Estate, Worcester Road, Kidderminster DY11 7RA, UK (email: simon.hughes@environment-agency.gov.uk).

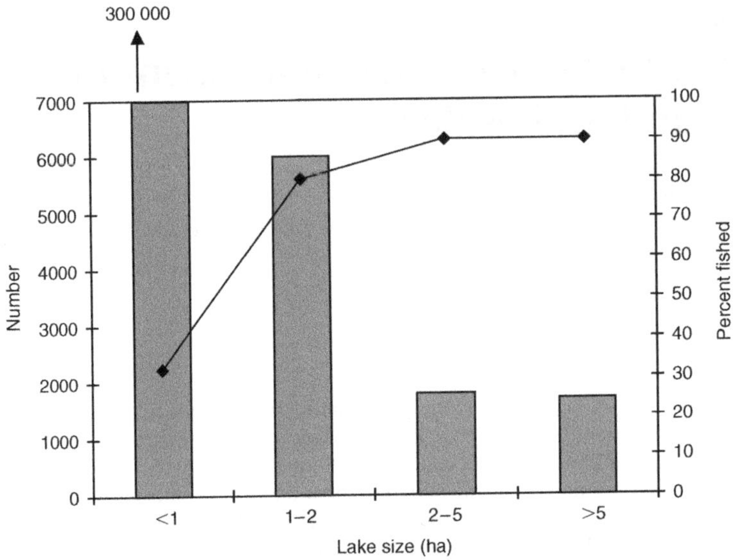

Figure 1.1 Estimated number (bar) and percentage use (◆) of different sizes of still waters in England and Wales

angling groups to suggest that the national trend is towards increased numbers fishing in stillwaters.

In England and Wales the total number of stillwaters is estimated at over 310 000, which includes over 300 000 ponds of <1 ha surface area, and 1700 lakes >5 ha of which only about 180 lakes are >50 ha (Fig. 1.1). The majority of fisheries are based on smaller lakes (0.05–15 ha). The Environment Agency receives many requests from fishery owners and managers to use non-intrusive hydroacoustic methods to provide quick assessments of fish stock abundance, distribution and relative size structure. They often have the objective of maximising natural recruitment to support catch and release recreational fisheries, and the information is either used to inform subsequent management programmes, or evaluate outcomes from past management intervention.

1.2.2 *Managing the water environment*

The EC Directive Establishing a Framework for Community Action in the Field of Water Policy (2000/60/EC), otherwise known as the Water Framework Directive (WFD), came into force in September 2000. The Directive is concerned with the assessment of ecological quality (encompassing both chemistry and biology) to establish the status of water bodies, and to assess any changes in the status resulting from programmes of management, regulation or mitigation measures.

Frequent assessment is required for all bodies of water that are at risk of failing to meet environmental objectives, and for water bodies into which effluent is discharged. Monitoring must include parameters of biological quality if this is the element most sensitive to the pressures to which the water is subject. In addition, special investigation is required where the reason for any exceedances is unknown; to ascertain reasons for failure to meet environmental objectives and to determine the magnitude and impacts of accidental pollution. The change of emphasis from water quality criteria to wider environmental objectives places a much greater demand on biological (fish, invertebrates and flora) information.

1.2.3 *Rare and threatened fish species*

The EC Directive on the Conservation of Natural Habitats and of Flora and Fauna (92/43/EEC) requires member States to undertake surveillance of listed species of community interest, to maintain their favourable conservation status. Two whitefish species that appear under Annex V of the Directive ('species whose taking in the wild and exploitation may be subject to management measures') inhabit a small number of lakes in England and Wales. Vendace, *Coregonus albula* L., are only found in Bassenthwaite and Derwentwater, where they are threatened by eutrophication and habitat destruction. *Coregonus lavaretus* L. populations (known as schelly in England and gwyniad in Wales) face similar pressures, notably in Brotherswater, Ullswater and Llyn Tegid.

Both species are protected under Schedule 5 of the UK Wildlife and Countryside Act 1981, thereby eliminating angler catch data and making large-scale destructive sampling unacceptable. In view of this, and the size and depth of the waters concerned, mobile hydroacoustic methods remain the only practicable approaches available to monitor the status of these endangered stocks (National Rivers Authority 1995b).

Highly localised fish populations that are currently not afforded primary legislative protection also deserve monitoring effort. For example, Arctic charr, *Salvelinus alpinus* (L.), found in a small number of Cumbrian and Welsh lakes and self-sustaining stillwater brown trout, *Salmo trutta* L., populations. In these cases, hydroacoustic surveys would supplement population data obtained from the exploited stocks by conventional methods.

1.2.4 *Operating constraints*

In coarse fisheries in England and Wales, a very strong catch–release tradition exists, and many anglers take great pride in their ability to return the majority of their catch unharmed to the water. As a result, many of the data collection methods conventionally used in lake fish sampling elsewhere (e.g. gillnets, traps, trawls, longlines) that involve a high degree of fishing mortality would be perceived as unacceptable for routine sampling in many situations.

Partly as a consequence of these constraints (albeit with a few notable exceptions), historically there has been a legacy of hands off monitoring of many lake fisheries and fish communities. Accordingly, increasing emphasis is placed on fish stock and community assessment techniques built around non-destructive hydroacoustic methods.

1.3 Development and application of hydroacoustic methods

In an international context, applications of hydroacoustic techniques in freshwater lakes have matured rapidly, without many of the difficulties faced by workers in the marine field (MacLennan & Holliday 1996). Many of these developments have concentrated principally upon effectively copying mobile, vertical beaming, marine applications of hydroacoustic techniques, for example, in assessing shoaling fish species in large, deep lakes (Bagenal, Dahm, Lindem & Tuunainen 1982; Dahm, Hartmann, Lindem & Löffler 1985; Hartmann, Dahm, Dawson, Doering, Jurgensen, Lindem, Löffler, Raemhild & Volzke 1987). Lake applications like these are rare in England and Wales.

In England and Wales much of the information required is from small, shallow, eutrophic lakes supporting species rich communities of coarse fish that are often associated with physical boundaries such as the lake bed and surface or water plants. Unfortunately all of these factors present problems for conventional hydroacoustic techniques and equipment.

Development of lake applications in England and Wales has mirrored experience elsewhere (e.g. Goldspink 1990; Baroudy & Elliott 1993). The past inadequacies and limitations of equipment designed for marine applications have been largely overcome. Mobile, horizontal beaming methods suitable for use with multi-species fish communities in small, shallow lakes have been developed successfully (Ken-ichi, Zhu & Taizo 1986; Kubecka, Duncan & Butterworth 1992; Kubecka, Duncan, Duncan, Sinclair & Butterworth 1994; Gorin, Borisenko & Pavlov 1997; Duncan & Kubecka 1998; Kubecka & Wittingerova 1998).

1.4 A consistent strategy for hydroacoustic sampling

The carefully managed application of hydroacoustic methods to coarse fish monitoring has great potential to deliver comprehensive, high quality results quickly. A feature of fish community assessments by these methods remains the high variability of the results from successive surveys, due to a number of causes including sampling design. A strategy for the consistent collection of hydroacoustic information has been developed to reduce this variability and to enable nationally comparable fishery monitoring results for lakes. This will increase the ability to observe and understand processes occurring in these systems, and lead to increased confidence in the interpretation of fisheries problems.

1.4.1 *Beam orientation*

Hydroacoustic methods for sampling fish in very deep waters are well established (MacLennan & Simmonds 1992), and have become more common in shallow waters (Kubecka *et al.* 1994). Horizontal beaming can allow a much greater water volume to be sampled in shallow waters than vertical beaming.

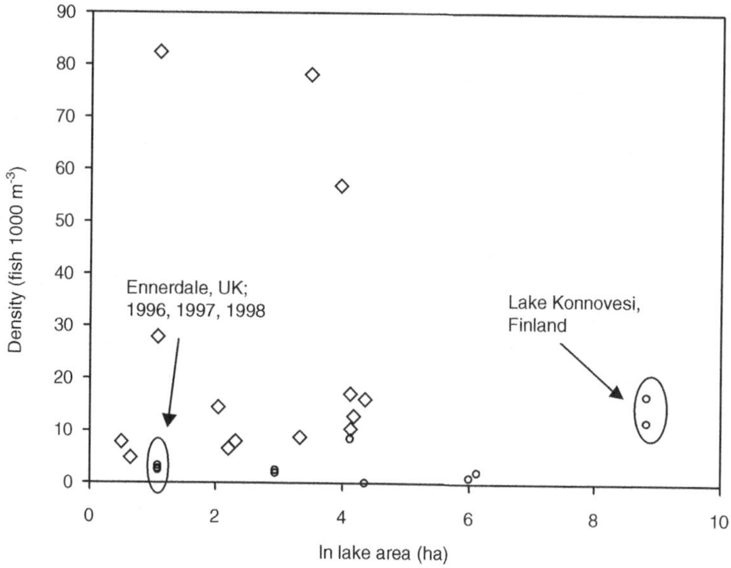

Figure 1.2 Vertical (○) and horizontal (◊) hydroacoustic fish stock assessment from lakes in the north-west England and Wales, and Lake Konnovesi, Finland (after Bagenal *et al.* 1982)

Few studies, however, have tackled adequately the problems of assessing fish stocks in water bodies which have intermediate depths (10–100 m) and extensive shallow littoral waters. Those that have (Gauthier, Boisclair & Legendre 1997; Kubecka & Wittingerova 1998) demonstrated that using vertical beaming led to a significant underestimation of fish stocks in stillwaters of intermediate depth inhabited by coarse fish. A combination of mobile horizontal and mobile vertical sounding was found to be a much more reliable tool in assessing coarse fish distribution and abundance in these environments. Results (Fig. 1.2) from mobile horizontal and mobile vertical beaming in lakes in the north-west England and Lake Konnovesi, Finland (Bagenal *et al.* 1982) demonstrated that the estimated abundance is generally higher when horizontal beaming is used.

Lakes deeper than 10–15 m should be sampled using a combination of vertical beaming in deep water and horizontal beaming in shallower zones. Lakes shallower than 15 m are generally more suited to horizontal beaming alone, however, vertical beaming alone can be valuable for survey planning and population assessments of species occupying deep water zones. Only transducers with relatively narrow beam angles (4–6°) with minimal side lobes should be used in horizontal surveys.

1.4.2 *Temporal aspects*

Successful sonar surveys require that most fish are not tightly shoaled or closely associated with boundaries (lake bed, surface or other physical habitats), so that they can more readily be resolved as single targets. Even in shallow waters, fish behaviour can alter the vertical distribution of individuals, and a number of environmental factors can alter fish association with boundaries as well as their shoaling behaviour. Each of these can influence the results of an hydroacoustic survey and should be considered and exploited in developing a sampling strategy.

Continuous annual studies of hydroacoustic estimates of fish abundance in temperate lakes and rivers have identified major seasonal changes in fish behaviour (e.g. National Rivers Authority 1993). Abundance obtained during summer months was greater than in winter months, when fish were more tightly shoaled near the bottom (Appenzeller & Legget 1992; Kubecka 1996).

Variability in results on a day-to-day or week-to-week scale identified considerable diel migrations in a variety of freshwater fish species (Unger & Brandt 1989; National Rivers Authority 1993; Winfield, Fletcher & Cubby 1993; Jurvelius & Sammalkorpi 1995; Comeau & Boisclair 1998; Kubecka & Duncan 1998). These results suggest that hydroacoustic stock surveys in temperate lakes should be carried out at night, when fish are aggregated less densely and larger fish were less likely to be associated with the lake bed, and consequently more likely to be detected. Furthermore, detailed diel hydroacoustic surveys showed that night time light levels related to moon phase affected estimates of fish abundance, sometimes by up to 50% (Luecke & Wurtsbaugh 1993), and could account for an eight-fold increase in the number of fish performing horizontal migrations (Gaudreau & Boisclair 2000).

These studies strongly support the conclusion that to ensure consistent and comparable hydroacoustic fish population estimates, the timing of survey fieldwork should be considered carefully and standardised. Hydroacoustic survey fieldwork should therefore be carried out between about 1 h after sunset and about 1 h before sunrise, from June to October, avoiding periods of full moon.

1.4.3 *Reducing interference*

In shallow water, the effectiveness of horizontal scanning sonars is limited chiefly by hydroacoustic reverberation from, for example, near surface air bubbles, surface waves, dense aggregations of plankton, macrophytes and rough bottom sediments. This boundary reverberation 'noise' can obscure even the largest size fish targets (Trevorrow 1998).

Entrained bubbles caused by boat wakes and wave action can mask even large fish targets, while reverberation from rough sediments can be such that small fish targets associated with them cannot be detected (Trevorrow 1998). The authors observed similar phenomena where small targets near the surface were obscured by reverberation from water surface 'roughness' caused by heavy rainfall. Reverberation from small non-fish targets such as plankton can act to mask small targets even at relatively short range.

To a degree, these limitations can be accounted for by careful beam steering and restriction of the maximum useable range (Kubecka *et al.* 1994). It is, however, important that population surveys are not compromised by restricting ranges and avoiding the boundaries, with which fish often associate. Survey timing and site selection can therefore be critically important to the success of an hydroacoustic fish population survey. Conditions with wind speeds <10 knots, no rainfall and where other boat traffic is absent are ideal. Rough sediments, large stands of macrophytes and dense plankton aggregations should be avoided, but this may not be practicable in many circumstances. Survey vessels must be sufficiently small to adequately manoeuvre in smaller lakes, but stable enough to allow effective beam steering.

1.4.4 *Spatial aspects*

Fish in lakes often have non-random, patchy or clumped distributions (Baroudy & Elliott 1993). Guidance on a minimum transect length for mobile surveys is necessary to ensure sufficient water volume is covered to provide a reliable stock assessment.

Current survey designs establish lake transect length from

$$\text{Transect length (km)} = 3 \times (\text{lake area (ha)})^{0.5}. \qquad \text{(Aglen 1989)}$$

Whilst more suited to meeting the requirements of marine and large lake surveys, this method for determining transect length can be used to establish a rule of thumb minimum level of coverage, which may be increased as local conditions dictate. Survey replication or coverage in smaller systems can be altered easily and quickly with little disruption to the overall programme.

1.4.5 *Operators, equipment and post-processing*

Proper application of the technique is technically and physically challenging. Highly skilled and knowledgeable practitioners are required for all stages of data collection, analysis and reporting. Data collection carried out at night during summer months can prove extremely arduous and disruptive to those involved.

To ensure consistently high quality results, the method should be used under the supervision of experienced practitioners. The quality and variability of results can be compromised by the use of inexperienced operators, and inappropriate equipment at the wrong sampling times and locations.

The degree of variability or bias in results related to the use of different equipment, deployment techniques and post-processing practices remains undocumented. A consistent strategy would aim to standardise these as far as possible, however, that may overly constrain application of the technique. These potential sources of bias warrant further investigation.

Hydroacoustic techniques, like all sampling methods, have a number of limitations (e.g. in species identification), consequently a good approach is to use other complimentary techniques to provide information that hydroacoustics cannot.

1.5 A future role for hydroacoustics

The WFD in Europe is driving the development of fish community status classification schemes that utilise hydroacoustic assessments of fish stocks. Successful application of these tools will enhance the value of hydroacoustic methods for assessing ecological status and ensure an increasing demand for their use.

At the same time, increased participation in and demand for lake fisheries have driven the development of new techniques for recreational fishery management, many of which aim to optimise or enhance natural recruitment to establish good quality catch and release coarse fisheries. Fishery managers will be able to predict likely fishery performance of different stock densities (Environment Agency 2001), and there is increasing demand for the quick and unobtrusive assessments that hydroacoustic methods can provide.

Increasing acceptance of results from hydroacoustic surveys will stimulate demand for further information. Cost effectiveness will remain high and the unobtrusive nature of data collection is consistent with conservation objectives of the Western world. These increasing demands are likely to continue to encourage sonar manufacturers to develop equipment and operating systems specifically designed for shallow water applications.

Acknowledgement

The authors gratefully acknowledge the assistance of colleagues who have contributed data, information and opinion. The views expressed are entirely those of the authors.

References

Aglen A. (1989) *Reliability of acoustic fish abundance estimates.* Dr Scientifique Thesis, Department of Fisheries Biology, University of Bergen, Norway.

Appenzeller A.R. & Legget W.C. (1992) Bias in hydroacoustic estimates of fish abundance due to acoustic shadowing: evidence from day–night surveys of vertically migrating fish. *Canadian Journal of Fisheries and Aquatic Sciences* **49**, 2179–2189.

Bagenal T.B., Dahm E., Lindem T. & Tuunainen P. (1982) EIFAC experiments on pelagic fish stock assessment by acoustic methods in Lake Konnovesi, Finland. Rome: FAO, *EIFAC Occasional Paper* **14**, 16 pp.

Baroudy E. & Elliott J.M. (1993) The effect of large-scale spatial variation of pelagic fish on hydroacoustic estimates of their population density in Windermere (Northwest England). *Ecology of Freshwater Fish* **2**, 160–166.

Comeau S. & Boisclair D. (1998) Day-to-day variation in fish horizontal migration and its potential consequence on estimates of trophic interactions in lakes. *Fisheries Research* **35**, 75–81.

Dahm E., Hartmann J., Lindem T. & Löffler H. (1985) EIFAC experiments on pelagic fish stock assessment by acoustic methods in Lake Constance. Rome: FAO, *EIFAC Occasional Paper* **15**, 14 pp.

Duncan A. & Kubecka J. (1998) Editorial. *Fisheries Research* **35**, 149–150.

Environment Agency (2001) Stocking densities in stillwater coarse fisheries. R&D Technical Report W2(99)01. Bristol: Environment Agency, 88 pp.

Gaudreau N. & Boisclair D. (2000) Influence of moon phase on acoustic estimates of the abundance of fish performing daily horizontal migration in a small oligotrophic lake. *Canadian Journal of Fisheries and Aquatic Sciences* **57**, 581–590.

Gauthier S., Boisclair D. & Legendre P. (1997) Evaluation of a variable angle scanning method to estimate relative abundance and distribution of fish using a single-beam echosounder in shallow lakes. *Journal of Fish Biology* **50**, 208–221.

Goldspink C.R. (1990) The distribution and abundance of young (I+–II+) perch – *Perca fluviatilis* L., in a deep eutrophic lake, England. *Journal of Fish Biology* **36**, 439–447.

Gorin A.N., Borisenko E.S. & Pavlov D.S. (1997) Assessment of the spatial distribution and abundance of fish in Lake Quistacocha, Peruvian Amazonia, with the use of parametric sonar. *Journal of Ichthyology* **37**, 511–522.

Hartmann J., Dahm E., Dawson J., Doering P., Jurgensen S., Lindem T., Löffler H., Raemhild G. & Volzke V. (1987) EIFAC experiments on pelagic fish stock assessment by acoustic methods in Lake Tegel. Rome: FAO, *EIFAC Occasional Paper* **17**, 26 pp.

Hickley P. & Aprahamian M. (2000) Fisheries science and the managerial imperative. In I.G. Cowx (ed.) *Management and Ecology of River Fisheries.* Oxford: Fishing News Books, Blackwell Science, pp. 346–360.

Jurvelius J. & Sammalkorpi I. (1995) Hydroacoustic monitoring of the distribution, density and the mass-removal of pelagic fish in a eutrophic lake. *Hydrobiologia* **316**, 33–41.

Ken-ichi T., Zhu Z.-R. & Taizo M. (1986) Studies on acoustic estimation of fish populations in Lake Donghu. *Acta hydrobiologica sinica/Shuisheng Shengwu Xuebao. Wuhan* **10**, 311–326.

Kubecka J. (1996) Use of horizontal dual-beam sonar for fish surveys in shallow waters. In I.G. Cowx (ed.) *Stock Assessment in Inland Fisheries.* Oxford: Fishing News Books, Blackwell Science, pp. 165–178.

Kubecka J. & Duncan A. (1998) Diurnal changes of fish behaviour in a lowland river monitored by a dual-beam echosounder. *Fisheries Research* **35**, 55–63.

Kubecka J., Duncan A. & Butterworth A. (1992) Echo counting or echo integration for fish biomass assessment in shallow waters. In M. Weydert (ed.) *European Conference on Underwater Acoustics.* London: Elsevier, pp. 129–132.

Kubecka J., Duncan A., Duncan W., Sinclair D. & Butterworth A. (1994) Brown trout populations of three Scottish lochs estimated by horizontal sonar and multimesh gillnets. *Fisheries Research* **20**, 29–48.

Kubecka J. & Wittingerova M. (1998) Horizontal beaming as a crucial component of acoustic fish stock assessment in freshwater reservoirs. *Fisheries Research* **35**, 99–106.

Luecke C. & Wurtsbaugh W.A. (1993) Effects of moonlight and daylight on hydroacoustic estimates of pelagic fish abundance. *Transactions of the American Fisheries Society* **122**, 112–120.

MacLennan D.N. & Holliday D.V. (1996) Fisheries and plankton acoustics: past, present and future. *ICES Journal of Marine Science* **53**, 513–516.

MacLennan D.N. & Simmonds E.J. (1992) *Fisheries Acoustics.* London: Chapman and Hall, 325 pp.

National Rivers Authority (1993) Hydroacoustic methods of Fish Surveys, R&D Note 196. Bristol: National Rivers Authority, 163 pp.

National Rivers Authority (1995a) Socio-economic review of angling 1994, R&D Note 385. Bristol: National Rivers Authority, 32 pp.

National Rivers Authority (1995b) Echo-sounding studies of *Coregonus albula* and *Coregonus laveratus* in England and Wales, with particular reference to dissolved oxygen levels. R&D Note 425. Bristol: National Rivers Authority, 74 pp.

Trevorrow M.V. (1998) Boundary scattering limitations to fish detection in shallow waters. *Fisheries Research* **35**, 127–135.

Unger P.A. & Brandt S.B. (1989) Seasonal and diel changes in sampling conditions for acoustic surveys of fish stocks in small lakes. *Fisheries Research* **7**, 353–366.

Winfield I.J., Fletcher J.M. & Cubby P.R. (1993) Confirmation of the presence of schelly, *Coregonus lavaretus*, in Brotherswater, UK. *Journal of Fish Biology* **42**, 621–622.

Chapter 2
The use of a split-beam echosounder for fisheries assessment in reservoirs: necessity for validation

N. CADIC*, P. IRZ and C. ARGILLIER
Cemagref, Ressources Ichtyologiques en Plans d'eau Research Unit, Montpellier, France

Abstract

The validity of echosounding in reservoirs was tested by comparing the output from echosounding (with vertical split-beam echosounder) with gillnetting and dam draining in five French reservoirs (100–1000 ha). Fish biomass estimates were highly variable (dependent, among other things, on the user's choices), and the output has little value because of the relatively large contribution of a few big fish. Vertical fish distributions provided by echocounting computation gave valuable insight into diel distribution of fish and zooplankton in relation to depth. Echocounting proved an efficient complementary tool to gillnetting in behavioural studies. One of the most important outputs was the size distribution of fish (target strength, in dB, for acoustic traces, and weight for real fish, in g), especially on reservoirs with a mean depth >10 m.

Keywords: biomass, fish behaviour, hydroacoustics, size distribution, target strength.

2.1 Introduction

Hydroacoustic assessment of fish populations in European freshwaters is poorly developed compared with marine environments, particularly with respect to the quantification of fish biomass. There is recognition (Appenzeller 1997; Comeau & Boisclair 1998; Kubecka & Duncan 1998; Lyons 1998) of the need to improve baseline data or generate valid methods in acoustics, such as exist for electric fishing (used in rivers) or gillnets (for lakes).

This chapter describes an echocounting method, tested for 3 years in five reservoirs, to assess the vertical distribution of fish, fish stock biomass and fish size distribution (target strength (TS), in dB). These outputs were assessed for accuracy by comparison size with data collected simultaneously using gillnets and total counts (census) during dam draining.

*Correspondence: Ressources Ichtyologiques en Plans d'eau Research Unit, 361, rue J.F. Breton, BP 5095, 34033 Montpellier Cedex, France (email: nadou.cadic@cemagref.fr).

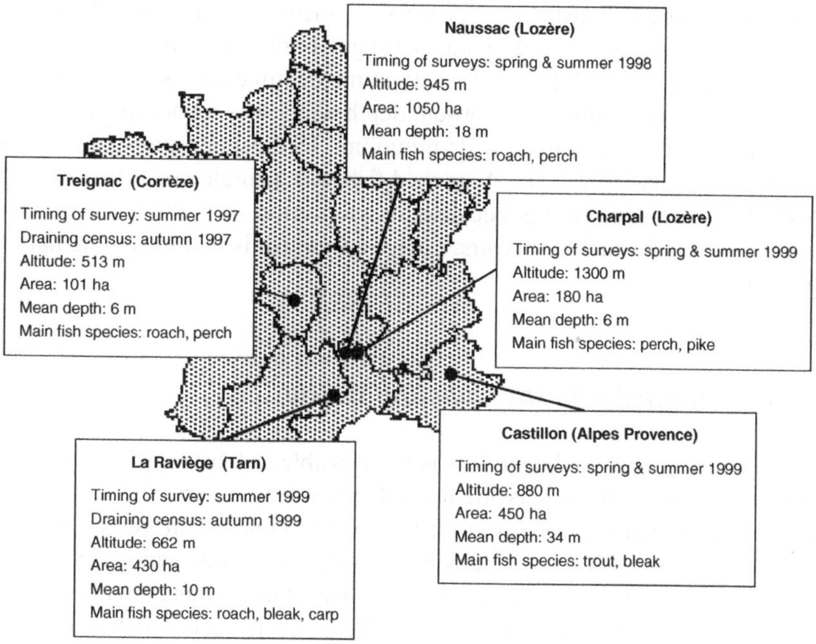

Figure 2.1 Location of study reservoirs in France

2.2 Materials and methods

2.2.1 *Study locations*

Between 1997 and 1999, five reservoirs were studied (Fig. 2.1). Three (Charpal, Naussac and Castillon) were studied twice in the same year (once in spring and again in summer). The other two reservoirs were studied with respect to total population counts (census) during dam draining (Treignac and La Raviège).

2.2.2 *Acoustic sampling and echocounting*

During each field trip, five acoustic surveys were conducted (three in the daytime, two at night, within a 4-day period). Each acoustic survey involved navigating a continuous path over the entire reservoir, at a speed of 1.5 m s^{-1}, with a 70-kHz EY500 split-beam echosounder of 11.3° beamwidth, operating vertically. Because of the large choice of settings for the equipment and thus, potential for a big variation in interpretation a single setting for all reservoirs was chosen at the beginning of the project in 1997. Also, for data computation, only echocounting was used because previous studies (Guillard 1991; Kubecka, Duncan & Butterworth 1992; Simrad 1995; MacLennan & Menz 1996) indicated large differences in interpretation using the echointegration computation mode, due to the user's choice of setting.

Counting was carried out using post-processing software Simrad EP 5.3 (trace tracking analysis menu) with the following settings: minimum number of detections required to form a tracked fish = 1; maximum centimetres between two detection = 60 cm (only traces inside this 60-cm depth window were accepted); maximum number of missing pings per track = 1; minimum TS detection threshold = −65 dB (or −60 dB in Treignac reservoir). A tracked fish was a single target (scattered fish). This method gave reliable results because the proportion of multiple targets (fish in shoals) was very low in such reservoirs. Each tracked fish is defined by the maximum value of its TS.

2.2.3 *Post-stratified acoustic sampling system*

Due to the cone shape of the beam, it was not possible to have equal sampling effort in each unit volume of the reservoir. It is, therefore, necessary to use a weighting system to allow comparison of samples: the total volume of the reservoir is three-dimensionally (3-D) divided into elementary sampled volume units (ESVU). Each ESVU is a cube with 1 m depth and an area of about 2.5 ha (rectangle with sides of one hundredth of a minute longitude or latitude). Thus, each tracked fish was weighted with the inverse of the sampling rate of the ESVU that contains it. The sampling rate could be expressed in linear metres (if covered path is taken into account), or cubic metres (if insonified cone volume is taken into account). The lake-scale assessment of the detected fish community was obtained by adding these weighted tracked fish.

2.2.4 *Acoustic data computations*

To fit the vertical distribution of the targets, all the pings with traces were displayed on the *X*-axis. On the *Y*-axis, several series were edited: depth to the bottom, depth of the tracked fish (the fish being divided into three sizes: small, medium and large). This display of the acoustic data is simple to obtain, but does not allow weighting of the targets in terms of their detection probability related to their depth. The surface layer (depth <2.5 m) was, therefore, not taken into account.

To estimate biomass, various relationships were used to convert the TS value into weight. Figure 2.2 shows these relationships: some were obtained by marine studies (Love 1977; Gerlotto 1987) and the others (reservoir names in quotation marks) were the results of graphical fittings of the TS distribution on the weight distribution of the gillnetted fish (Cadic, Irz & Argillier 1998) for each reservoir.

An equivalent weight was allocated to each tracked fish (Table 2.1 shows the calculations used to estimate biomass). A minimum TS threshold value was fitted, depending on the reservoir and on its fish communities (in this case −42 dB), to reduce confusion between small scattered fish and other types of traces (bubbles, zooplankton, suspended organic matter). Several relationships were used (general or particular to each reservoir). By summing the equivalent weights, the total detected biomass was estimated.

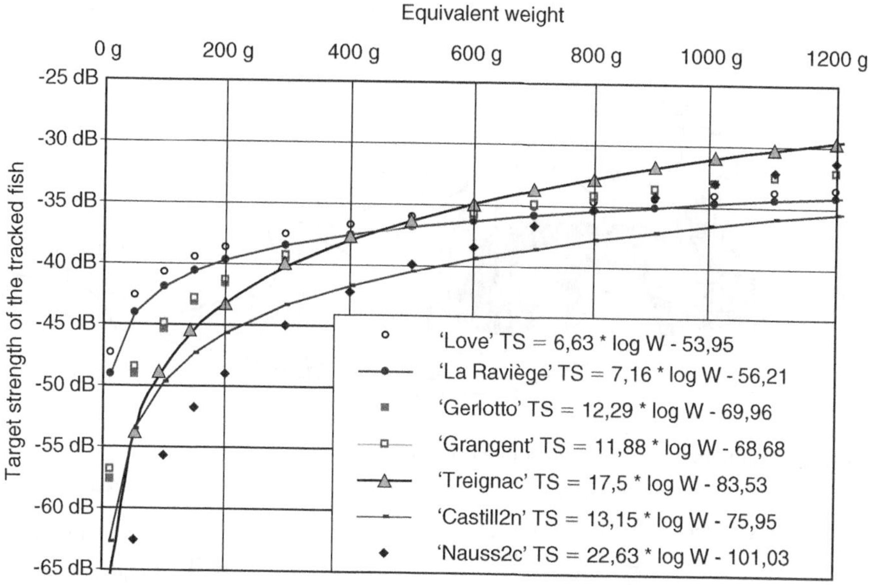

Figure 2.2 Real weight *v.* acoustic size relationships employed in the study

Table 2.1 Calculations to convert echocounting results into total biomass: examples with different TS *v.* weight relationships on two reservoirs (Treignac and Naussac)

Reservoir	Sampling rate (% of volume)	TS *v.* weight (*W*) relation used	Number of tracked fish (>−42 dB)	Total biomass detected in sampled volume (kg)	Total reservoir's estimated biomass (t)
Treignac	1.5	$TS = 12.29 \log W - 69.96$	92	54	3.3
Naussac	1.6	$TS = 12.29 \log W - 69.96$	578	249	15.8
Treignac	1.5	$TS = 17.50 \log W - 83.53$	92	46	2.9
Naussac	1.6	$TS = 22.63 \log W - 101.03$	578	358	22.6

Comparison of acoustic results on fish size distribution with gillnet results, histograms of TS distribution were displayed. For easy comparison between different reservoirs and between catch and echocounting data, a common mean relationship between the arithmetic TS scale (in dB) and the logarithmic weight scale (W, g) was chosen: $TS = 13.8 \log_{10}(W) - 70.5$ (Fig. 2.2). TS class interval is 2 dB, with the TS scale beginning at −59 dB, and weight scale beginning at 18 g. Therefore, information on small-sized fish including recruits, of particular importance in fish population studies, was not available. However, this minimum threshold does not limit the possibility of comparison with gillnet catch because they are selectivity against such small-sized fish (weight under 15 g in study gillnets) (Irz 1998).

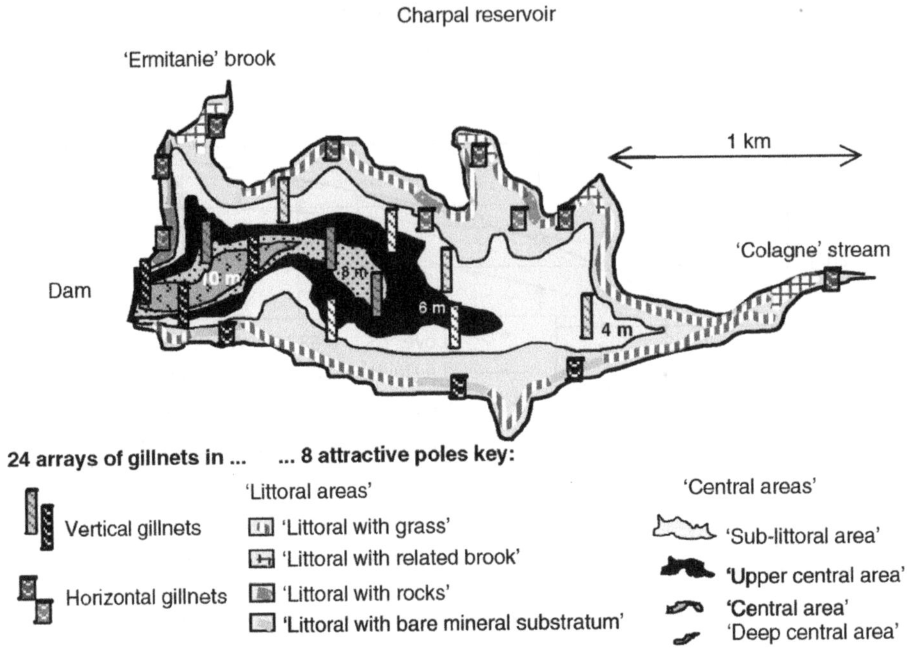

Figure 2.3 Map illustrating one example (Charpal reservoir) of gillnet sampling: locations of gillnets depends on the characterised 'attractive poles' of the reservoir

2.2.5 *Gillnet sampling*

Fish sampling was carried out using multimesh gillnets with knot-to-knot mesh sizes of 10, 15, 20, 30, 40, 50, 60 and 70 mm. The sampling design was stratified by habitat or attractive poles (Degiorgi 1994) (Fig. 2.3). Each habitat was sampled three times within a 4-day period. The lake-wide assessment of the fish community was obtained by weighting the catch of each habitat with respect to its surface area.

2.2.6 *Draining operation and fish census method*

Only two reservoirs were drained. A metal grating was placed downstream of the dam to collect fish during the draining. About 50 fish-pickers were required during peak drainage flow. They briefly sorted out the fish into eight different containers (Table 2.2), depending on species and size. All containers were weighed and samples were detailed by fish biologists to verify both the species composition and the total weight (in the sampled containers, each fish was individually measured: total length, mm; weight, g). Some errors were evident in such a census as a result of the relatively simple sampling process carried out both up- and downstream of the dam: in particular, small fish (weight under about 15 g) and eels *Anguilla anguilla* (L.) swim through the grid.

Table 2.2 Sorting out of collected fish during dam draining census

No.	Sorting types	Main species collected
1	Small Cyprinidae	*Rutilus rutilus, Alburnus alburnus*
2	Large Cyprinidae	*Cyprinus carpio, Tinca tinca, Abrimis brama*
3	Salmonidae	*Salmo trutta*
4	Small carnivorous	(without Salmonidae)
5	Large carnivorous	(without Salmonidae) *Perca fluviatilis, Esox lucius, Stizostedion luciperca*
6	'Pest species'	*Lepomis gibbosus, Abramis brama*
7	Eel	*Anguilla anguilla*
8	Dead fish	All species

Certain species, particularly rheophilous species, such as chub *Leuciscus cephalus* (L.), or dace *Leuciscus leuciscus* (L.) swim into the tributaries. These errors were, however, considered to have little effect on the total biomass or on the assessment of the species relative abundance.

2.3 Results

Before calibrating the acoustic output against the gillnet catch or the census results, interpretation of the different acoustic outputs with the user's choices and variations in fish behaviour (diel migration) were assessed.

2.3.1 *Importance of the users' choice on the acoustic outputs*

One of the most important settings during acoustic data computation is the minimum threshold of the TS value. A low threshold could provide information on juvenile fish distribution but might mix up small fish with bubbles or large zooplankton (Guillard, Boet, Gerdeaux & Roux 1994; Richeux 1996). The comparison of vertical distributions obtained with different values of the TS minimum threshold (Fig. 2.4) showed the importance of this parameter on the outputs.

The example displayed in Fig. 2.5 highlights the dependence of biomass assessment upon both the TS range and the TS–weight conversion formulae. The comparison reveals a three-fold variation in the estimate. This partially results from the importance of a limited number of large fish in the overall biomass estimate. It was confirmed (Fig. 2.6) by gillnet catches that a few (17 large fish over 1 kg, 2% by number) dominated the biomass represented 25 kg, 38% by weight).

Another important users' choice is the unit used for the weighting method. The effect of two post-processing weighting methods was investigated by comparing the TS distributions obtained (Fig. 2.7). It appears that the weighting influences the biomass estimate, but the TS distribution remains stable.

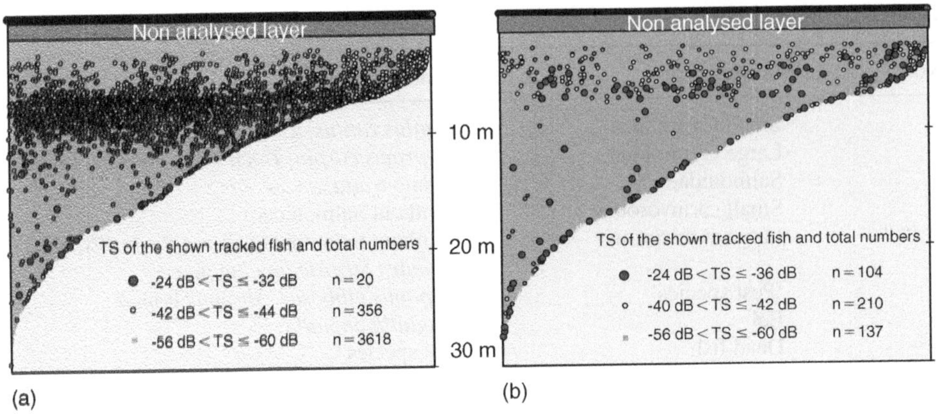

Figure 2.4 Vertical distribution of tracked fish (La Raviège reservoir) with two different displays: (a) with low TS values (where sediment's bubbles, zooplankton and alevin could be merged), (b) with upper TS values

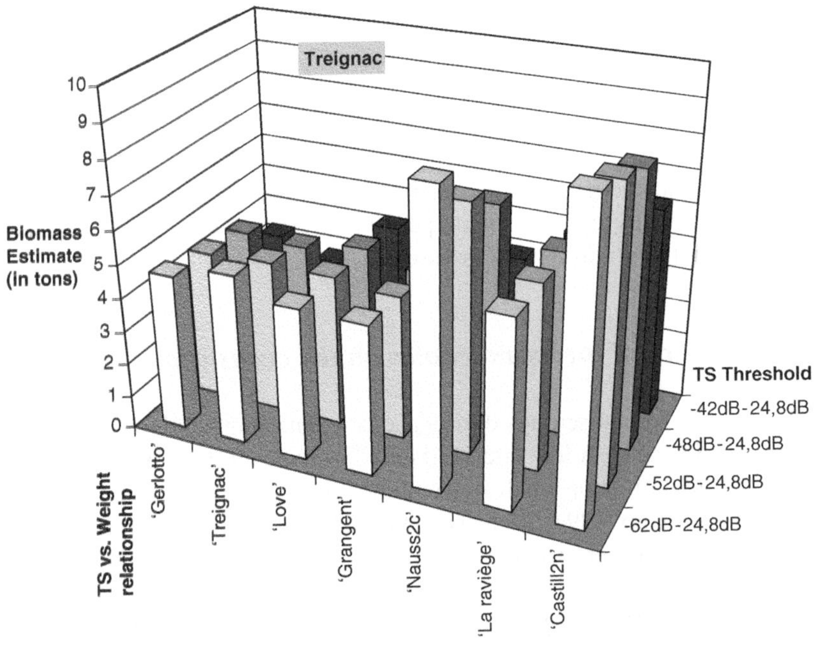

Figure 2.5 Different biomass estimates on Treignac reservoir, depending on various TS thresholds and TS *v.* weight relationships

2.3.2 *Importance of the fish diel migrations on the acoustic results*

Vertical tracked fish distributions provide valuable data on the diel depth migration of fish as well as plankton (Fig. 2.4). Echocounting, therefore, is an efficient complementary tool to gillnetting (set for 24 h in this study) in behavioural studies. However, due to

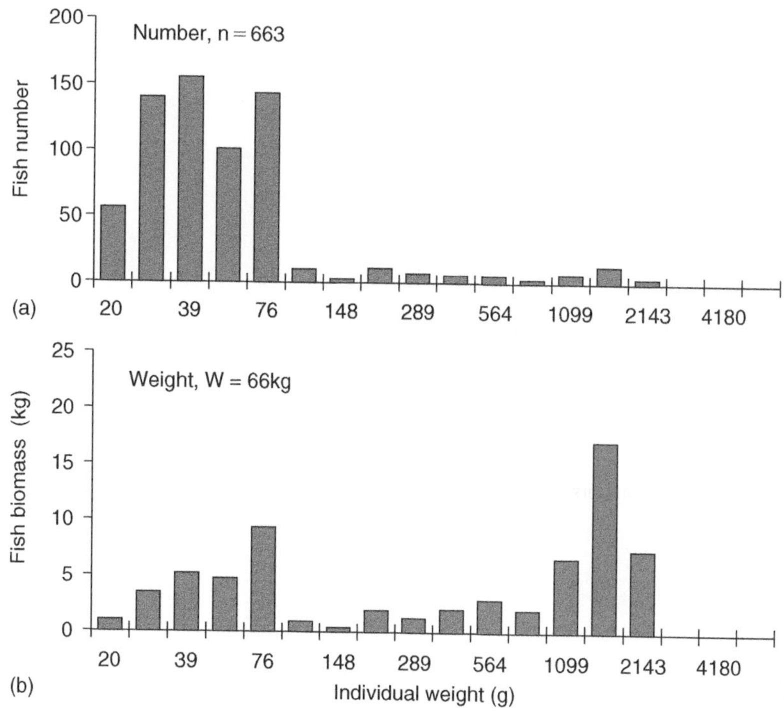

Figures 2.6 Fish caught by gillnets in La Raviège reservoir divided into 17 weight classes: (a) numerical distribution; (b) biomass distribution

fish behaviour, such vertical outputs do not allow comparison between seasons or between different lakes.

For biomass estimate, many authors (Gerlotto 1987; Barange & Hampton 1994; Appenzeller 1997; Lyons 1998) recommended night acoustic surveys, because fish are more scattered and active, which is convenient for echocounting. However, in this study (Fig. 2.8) the TS distributions proved to be rather stable between day and night, which suggested that the vertical migration patterns were similar for all fish sizes, but the fluctuations in target detectability strongly affect the biomass estimates. This highlights the point that it is not always beneficial to limit data acquisition to night-time because the phototactic behaviour of some species leads them to deeper layers at daylight, reducing the biases related to the blind zone (from surface to 2.5 m in depth).

2.3.3 *Importance of the fish seasonal variations on the acoustic results*

The seasonal variability in the three types of outputs was studied on the three lakes that were surveyed twice, in spring and summer, during the same year. Vertical distribution or biomass estimate were not edited, because of their high sensitivity to daily variations

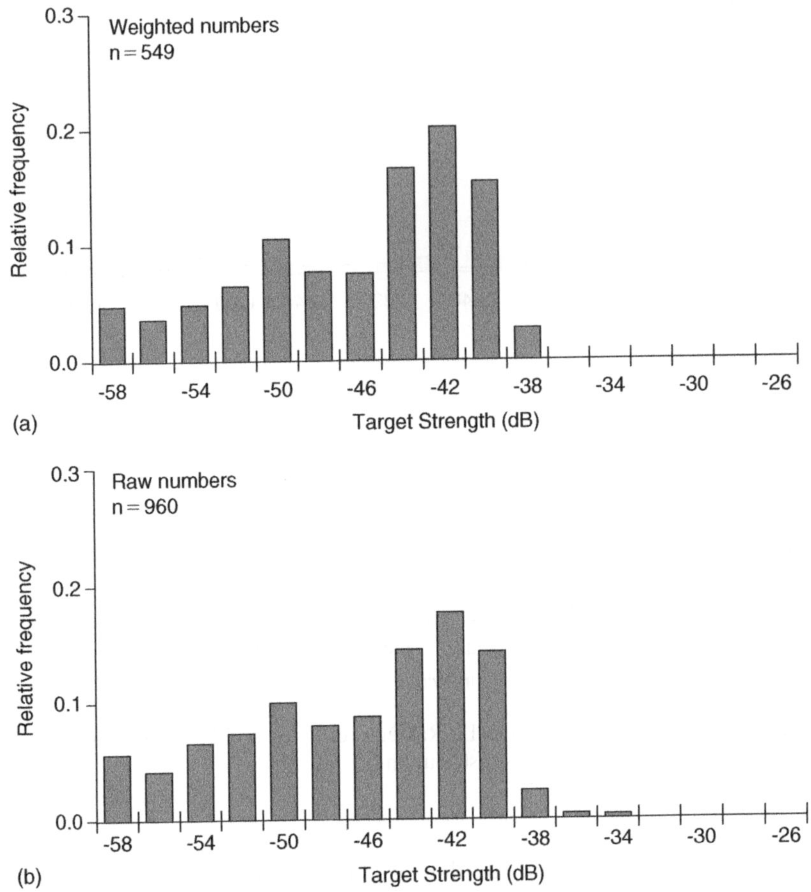

Figures 2.7 Target size (dB) distribution of the Charpal reservoir with: (a) weighted numbers (the post-processing allows stratification of data by elementary volume) or (b) raw numbers (without weighting)

of fish behaviour. The TS distributions for these three lakes (Fig. 2.9) in springtime and in summer showed a relative stable pattern which was little affected by the season, whereas the total number of fish detected was highly variable. This seems to confirm the previous assumption that vertical migration patterns are similar for all fish sizes. Thus, different vertical distributions in summer or in spring can strongly affect acoustic outputs such as the biomass estimate or vertical distribution, even if the size distribution patterns stay similar. Nevertheless, the effect of seasonal variation on the TS distribution was more marked in Charpal reservoir, which was the shallowest of the three reservoirs studied (mean depth = 6 m, compared to Castillon = 34 m, and Naussac = 18 m). This reservoir showed another characteristic: the total number of the fish detected was higher in the spring (despite the low water temperature) than in the summer, possibly because of the presence of an anoxic layer on the bottom during spring forcing fish to migrate to the surface layer and thus, making them more difficult to detect.

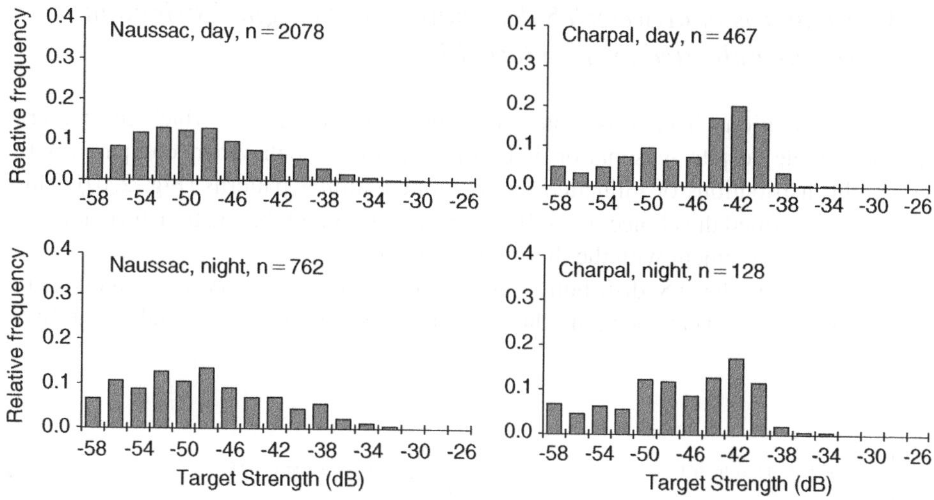

Figure 2.8 Target size (dB) distribution by day and by night in Naussac reservoir in September 1998 and Charpal reservoir in June 1999

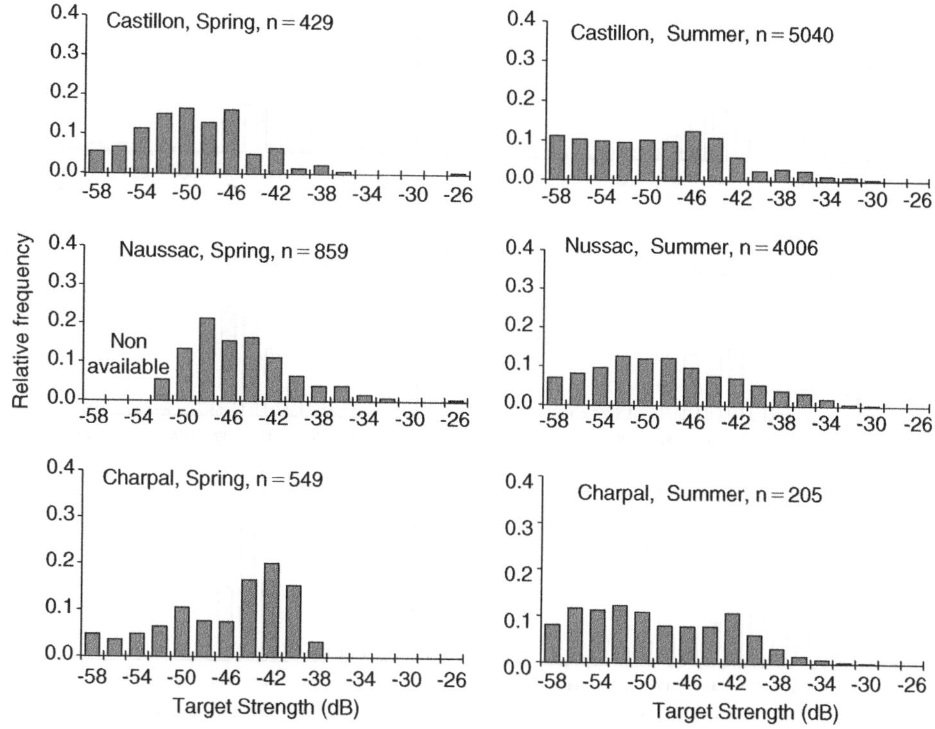

Figure 2.9 Target size (dB) distributions in the Castillon, Naussac and Charpal reservoirs during spring and summer

2.3.4 *Comparison between TS distribution and weight distribution (of the gillnetted or inventoried fish)*

The TS distributions from Castillon and Naussac reservoirs (in which the pattern remained stable despite seasonal or diel variations in fish behaviour) compared with the weight distributions of gillnetted fish exhibited similar patterns (Fig. 2.10). This similarity confirmed the choice of the fitting between TS and the log transformation of fish weight, in agreement with the theory of acoustics.

Comparison of the TS distribution from the other two reservoirs, which were sampled only once a year (negating the possibility to display the seasonal variability)

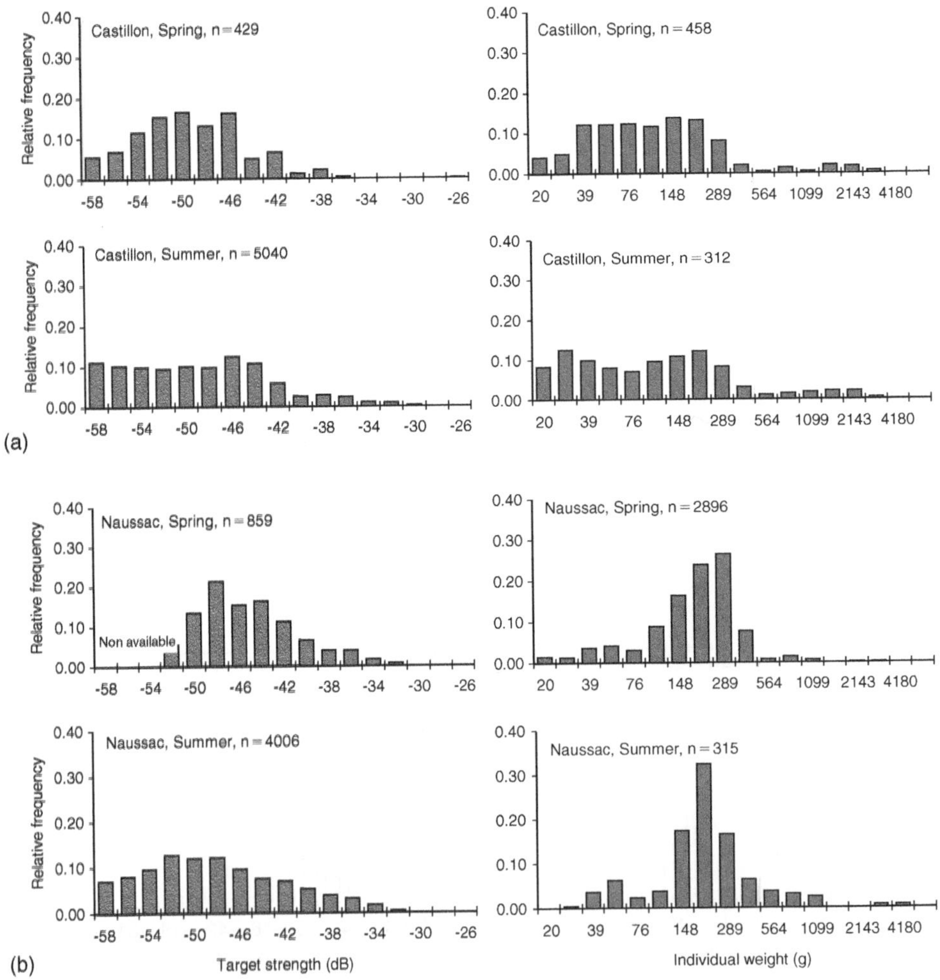

Figure 2.10 Comparison between target size (TS, dB) distribution and fish real weight (g) in Castillon (a) and Naussac (b) reservoirs during spring and summer

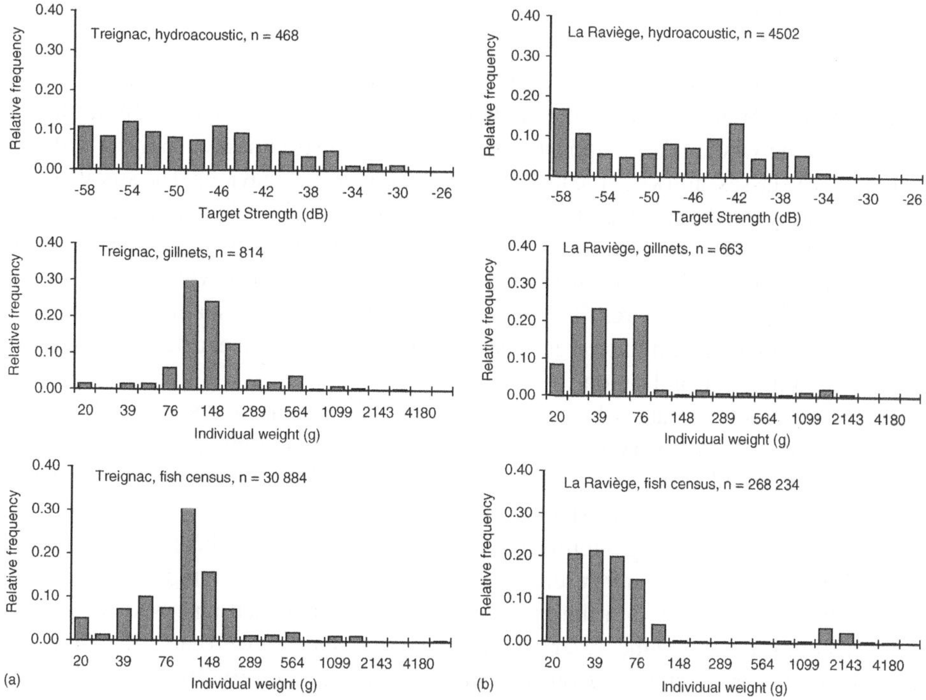

Figure 2.11 Comparison between target size (TS, dB) distribution and weight (g) distribution of gillnetted and total census by draining in Treignac (a) and La Raviège (b) reservoirs

against the distributions obtained from a total count of the fish community when the reservoirs were drained were less complementary (Fig. 2.11). For Treignac reservoir, this could be explained by (MacLennan & Menz 1996) TS values being more uniformly distributed, due to the stochastic nature of TS, which is highly variable even for the same species and size of fish. For La Raviège reservoir outputs, the dissimilarity could not be explained, but may be due to the high relative abundance of large carp (about 2.5 kg) in this reservoir (shown both by gillnet and census results). The TS distribution pattern also seemed characteristic, with two different modes (−58 and −42 dB) which were even present in the weight distributions, but tailing out from the previous TS histogram. It is possible that the TS of the carp was lower than the relationship TS *v.* weight predicted because this species (Comeau & Boisclair 1998) appears in lateral aspect when swimming or digging on the bottom.

When all the data available are considered, they provide a mechanism to aid interpretation of the acoustic histograms, which are particularly uniformly distributed with less pronounced modes than samples caught by gill netting or inventoried by draining operations. Furthermore, this uniformly distributed pattern was particularly prominent in the shallow water reservoirs (i.e. Treignac, with the mean depth = 6 m, Charpal = 6 m and La Raviège = 10 m), in which the blind layer (depth below 2.5 m), not taken into account, is an important factor.

2.4 Discussion

Echocounting provides many useful results, which can be combined with other sources of information. When checking the reliability of three of these outputs by a global approach to the whole lake, it appears that the vertical distribution, displayed in three different size classes, allows a true understanding of the fish behaviour. The study of the fish community via the real weight distribution with all species taken together, and via the TS distribution, is another method of characterising, for a given season, the fish community condition. The TS distribution is, however, an indicator which is easy to set up, stable and gives reliable results, but needs, particularly for the shallow water reservoirs, to be compared with the weight distributions shown by gillnet catch results or by census (i.e. top-down calibration).

For the biomass estimate, which is favoured by fisheries managers, it is quite difficult to use either when a very small number of large fish represents proportionally the main part of the total biomass, or when the TS and the real weight distributions are not similar (i.e. Charpal, which is a shallow water reservoir).

As there is now a growing interest in the use of the vertical split-beam echosounder, and its limits in shallow water reservoirs, there is need for further studies to evaluate the variability inherent within the methods and the acoustic settings used by means of replications on the same reservoir, over several successive years. In addition, echointegration-processing of acoustic data from the same reservoirs should also be carried out to get a more accurate quantitative estimate of biomass. For shallow water reservoirs, new acoustic devices are needed which allow scanning of the total water column, including the surface layer (i.e. multibeam sonar, or elliptical transducers used horizontally).

Acknowledgements

We thank Sarah Ashmore (Griffith University, Brisbane, Australia) and Madeleine Sagliocco (Cemagref) for assistance with translation, EDF (Electricité de France) manager of the French hydropower reservoirs, Francois Gerlotto and Jean Guillard for assistance during Echospace group activities.

References

Appenzeller A.R. (1997) Ist der Fischbestand im Ammersee zu hoch? Akustiche Fischbestandsaufnahmen in bayerischen Voralpenseen. *Fischer und Teichwirt* **5**, 200–206.

Barange M. & Hampton I. (1994) Influence of trawling on *in situ* estimates of Cape horse mackerel (*Trachurus trachurus capensis*) target strength. *Journal of Marine Sciences* **51**, 121–126.

Cadic N., Irz P. & Argillier C. (1998) A comparison between dam draining stocktaking, gillnet sampling and acoustic sampling, for fish biomass assessment on a reservoir. In A. Allippi & G.B. Cannelli (eds) *Proceedings on the Fourth European Conference on Underwater Acoustics*, Vol. 1. Rome: CNR-IDAC Publications, pp. 261–266.

Comeau S. & Boisclair D. (1998) Day-to-day variations in fish horizontal migration and its potential consequence on estimates of trophic interactions in lakes. *Fisheries Research* **35**(1–2), 75–81.

Degiorgi F. (1994) *Etude de l'organisation spatiale de l'ichtyofaune lacustre: prospection multi-saisonnière de 6 plans d'eau de l'Est de la France à l'aide de filets verticaux*. Thèse de Doctorat Sci., Besançon: Université de Franche-Comté, 208 pp. (in French).

Gerlotto F. (1987) Medicion de las caracteristicas austicas de algunos pecès tropicales. *Memoria Société Ciensas Naturales La Salle, Venezuela,* **XLVII**, pp. 127–128, 149–167. (in French).

Guillard J. (1991) *Estimation des stocks pisciaires lacustres par echointegration: problèmes méthodologiques*. Thèse Doctorat Sci., Lyon: Université Cl. Bernard, 156 pp. (in French).

Guillard J., Boet P., Gerdeaux D. & Roux P. (1994) Application of mobile acoustic techniques fish surveys in shallow water: the River Seine. *Regulated Rivers: Research and Management* **9**, 121–126.

Irz P. (1998) Etude comparative de deux stratégies d'échantillonnage piscicole en milieu lacustre. *Mémoire de DEA*, Toulouse III: Université Paul Sabatier, 35 pp. (in French).

Kubecka J., Duncan A. & Butterworth A.J. (1992) Echocounting or echointegration for fish biomass assessment in shallow waters. In M. Weydert (ed.) *Proceedings of the European Conference on Underwater Acoustics*. London: Elsevier, pp. 129–132.

Kubecka J. & Duncan A. (1998) Acoustic size vs. real size relationships for common species of riverine fish. *Fisheries Research* **35**(1–2), 115–125.

Love R.H. (1977) Target Strength of an individual fish at any aspect. *Journal of the Acoustics Society of America* **62**, 1397–1403.

Lyons J. (1998) A hydroacoustic assessment of fish stocks in the River Trent, England. *Fisheries Research* **35**(1–2), 83–90.

MacLennan D.N. & Menz A. (1996) Interpretation of *in situ* target-strength data. *ICES Journal of Marine Science, Fisheries and Plankton Acoustics* **53**, 233–236.

Richeux C. (1996) *Les populations ichtyologiques de trois retenues du Massif Central. Application des techniques d'échosondage pour l'évaluation des stocks et l'étude de la répartition spatio-temporelle des biocénoses pisciaires*. Thèse de Doctorat, Toulouse: Université P. Sabatier, 194 pp. (in French).

Simrad (1995) Operation. Simrad EY500 Portable Scientific Echosounder Instruction Manual P3402E/A., 81 pp.

Chapter 3
Estimation of the fish yield potential of lakes in north-east Germany

U. BRÄMICK*
Institute of Inland Fisheries Potsdam-Sacrow, Gross Glienicke, Germany

Abstract

Between 1997 and 1999, the fish yield potential for 204 lakes in north-east Germany was assessed on the basis of total phosphorus concentration during then spring turnover. Yield potentials ranged from 20 to 93 kg ha^{-1} year^{-1} for different lake categories. Highest yield were predicted for polymictic hypertrophic lakes. To evaluate the quality of the estimation, yield potentials were compared to yield statistics taking into consideration changes in fishing effort. On average, calculated yield potentials were superior to statistical data by 53–83%. Possible reasons for these deviations are discussed.

Keywords: fish yield, hypertrophic lakes, phosphorus.

3.1 Introduction

The state of Brandenburg in the lowlands of north-east Germany is characterised by more than 3000 lakes. The majority of these waters are state owned and the fishing rights are mainly rented by commercial fishermen. Estimation of the fish yield potential of these lakes in an objective way is of considerable interest for a number of reasons. First, for the implementation of appropriate management strategies and sustainable exploitation of fish stocks, basic knowledge about the natural productivity of still waters is essential. Furthermore, in cases of fish kills due to pollution, an objective basis for compensation is required. Finally, the monetary value of fishing rights is based on the potential natural fish yield.

A number of different approaches have been developed to estimate either fish biomass, fish production or fish yield (catch) in lakes and reservoirs (for review see Downing, Plante & Lalonde 1990; Knoesche & Barthelmes 1998). Predictors like benthos biomass (Alm 1923; Wundsch 1936) or morphometric characters (Ryder 1965) have been widely applied, but they suffer from problems of poor correlation (Hanson & Leggett 1982; Barthelmes 1988; Downing *et al.* 1990; Downing & Plante 1993; Knoesche & Barthelmes 1998). Instead, it has been suggested that total phosphorus is a promising predictor for fish biomass, yield and production. Because the Phosphorus–Primary Production–Fish

*Correspondence: Institute of Inland Fisheries Potsdam-Sacrow, Jägerhof, 14476 Gross Glienicke, Germany (email: uwe.braemick@ifb-potsdam.de).

Yield concept (P–PP–F) of Knoesche & Barthelmes (1998) was developed in the same region of this study, it was chosen for estimating the fish yield potential in Brandenburg lakes.

3.2 Materials and methods

Between 1997 and 1999, fish yield potential was estimated for a total of 204 lakes in north-east Germany. Study lakes ranged from 1.1 to 1207 ha in area, 0.5 to 50 m in depth and mesotrophic to hypertrophic in primary productivity.

All lakes were fished once in spring or autumn mainly by electric fishing and gill netting with multimesh net fleets. In some lakes, additional or exclusive fish sampling was carried out by trap nets and commercial gill nets with different mesh sizes. Total phosphorus concentration (TP, $\mu g\ L^{-1}$) was determined in samples drawn from epilimnic water layers during the spring turnover between the beginning of March and mid-April. Possible stratification of water bodies was excluded by vertical temperature and oxygen profiles. In addition, the following parameters were collected: Secchi disc transparency (SDT, cm), pH and conductivity (μS). For dimictic lakes, bathymetric maps were provided.

Fish yield potential was estimated according to the P–PP–F model of Knoesche & Barthelmes (1998) with minor adaptations. Primary productivity (PP, $g\ cm^{-2}$) was calculated from total phosphorus concentration (TP, $\mu g\ L^{-1}$). For dimictic lakes, the equation of Koschel, Haubold, Kasprzak, Kuchler, Proft & Ronneberger (1981), i.e. $PP = 148 \log TP - 39.6$, was used. This relation was derived from north-east German hardwater lakes with protracted stratification regimes during summer and winter, and reflects the nutrient cycle in such water bodies. Due to a frequent nutrient turnover in polymictic lakes, spring TP values in shallow lakes are not representative for the PP model of Koschel *et al.* TP values of those lakes were corrected using data sets from German lakes (German States Association for Water 1998). This revealed the following relation between TP concentrations in dimictic and polymictic lakes of a similar trophic index and thus a comparable value of PP: $TP_{dimictic} = 1.48198\ TP^{1.2278}_{polymictic}$. Using this equation, TP concentrations of polymictic lakes were corrected to values of a dimictic lake with the same trophic index and could thus be used to calculate PP.

Total phosphorus was converted into fish yield potential (FYP, $kg\ ha^{-1}$) following Bulon & Vinberg (1981) as: $FYP = (1.8 \pm 0.9)\ PP/10$. Subsequently, FYP was corrected for dependence on geomorphometric and fish population characters for each lake. As proposed in Knoesche & Barthelmes (1998), the share of the hypolimnetic area was calculated from a bathymetric map (after Ventz 1974) and used for FYP adjustment in cyprinid lakes.

Bream, *Abramis brama* (L.), roach, *Rutilus rutilus* (L.), and perch, *Perca fluviatilis* L., are the most abundant species in north-east German lakes. These species were graded as an indicator of stunted growth to allow adjustment of FYP due to a reduced share of energy being available for somatic growth (Knoesche & Barthelmes 1998). Interviews with commercial fishermen and experimental fish samples were used for grading.

Because predatory fish species like pikeperch, *Stizostedion lucioperca* (L.), or eel, *Anguilla anguilla* L., are among the most valuable species in German freshwater fisheries, the share of predatory fish was calculated separately as 30% of the total fish yield potential.

Commercial inland fisheries in the region of the former East Germany have undergone a decline in fishing effort in the last 10 years, induced by a rapidly changing economic environment. In contrast, the PP–FYP relationship of Bulon & Vinberg (1981) was validated under conditions of intense fishing without market limitations. Therefore FYPs calculated were adjusted to the current average fishing pressure for comparison with recent fisheries statistics. Fishermen were interviewed to determine the decline in fishing effort.

Commercial fish catch statistics from 59 lakes were used to compare calculated FYPs and actual yields. The variance of these data was analysed independently for measured and corrected TP values, lake area, lake depth, conductivity, pH during summer, Secchi disc transparency, extent of hypolimnic area and degree of stunting in selected fish species.

3.3 Results

Lakes examined hosted between 5 and 18 species, while 29 species were found in total. The most widespread species were roach, perch, eel, pike, *Esox lucius* L., and rudd, *Scardinus erythrophthalmus* (L.), which were found in more than 80% of the lakes examined. Stunted growth in *A. brama*, *R. rutilus* and *P. fluviatilis* led to strong yield deductions in polymictic and small lakes. In dimictic lakes, stunting increased with higher nutrient loads (Fig. 3.1).

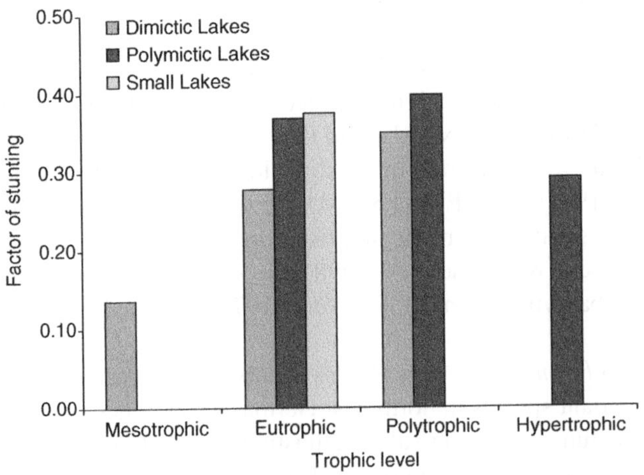

Figure 3.1 Stunted growth of selected fish species in different lake categories

Total phosphorus concentrations in lakes during spring turnover varied between 5 and 2195 $\mu g\,L^{-1}$, with 90% of the lakes showing values between 20 and 500 $\mu g\,L^{-1}$ (median value 66 $\mu g\,L^{-1}$). Secchi disc transparency was measured in a range of 0.3–6.5 m, and summer pH and conductivity values of 5.1–10.1 and 38–1304 μS were recorded.

Fish yield predictions according to the modified P–PP–F procedure were in the range 5–135 $kg\,ha^{-1}\,year^{-1}$. When lakes were grouped according to their TP value (trophic status) and stratification regime, annual mean yield potentials of 20.9–91.3 $kg\,ha^{-1}$ were estimated (Table 3.1).

Due to the estimation procedure, calculated FYPs increased with increasing trophic levels both in dimictic and polymictic lakes. With the exception of dimictic, polytrophic lakes, FYPs estimated differed significantly between trophic classes ($P \leqslant 0.01$). At the same time, polymictic lakes showed a tendency towards higher yield potentials compared to dimictic lakes, which was statistically significant in eutrophic conditions. All small lakes (<5 ha in area) were classified as eutrophic with FYP values between dimictic and polymictic lakes of the same trophic class.

Commercial fishermen indicated, that fishing effort in the Brandenburg lakes has generally declined over the past 10 years, with considerable differences between gear types used (Fig. 3.2). While fyke and gill net use has been stable over the past 20 years, seine netting and electric fishing usage has dropped markedly. These two gears are currently used less than half as much as 10 years ago. Summarising all gear types, current fishing effort is about 70% of that recorded in the 1980s. Consequently, FYPs were reduced by 30% in response to the decline in fishing effort, compared to conditions of the Bulon and Vinberg relationship (Table 3.2).

When applying the P–PP–F estimation procedure, 46 out of 59 FYPs calculated were superior to yield statistics. The median deviation of the FYP value from yield statistics was higher in dimictic than in polymictic lakes (Table 3.2). In extreme cases, FYPs exceeded statistics 5–7 times, while underestimations remained comparably moderate. When the data were used for TP-based yield estimation procedures of other authors (Hanson & Leggett 1982; Downing *et al.* 1990), median deviations from yield statistics increased. Contrary to the P–PP–F model, these procedures led to a substantial underestimation of yield statistics in most lakes (Table 3.2).

The variation in yield statistic data is significantly influenced by lake area, extent of hypolimnic area, stunted growth and Secchi disc transparency (ANOVA, $P < 0.05$). Neither measured nor corrected TP values were detected as significant causes of variance.

Table 3.1 Estimated fish yield potentials ($kg\,ha^{-1}$) in different lake categories (small lakes are defined as lakes with <5 ha in area)

Lake category	Dimictic (n = 72)	Polymictic (n = 114)	Small (n = 18)
Mesotrophic (n = 32)	20.9^{a}		
Eutrophic (n = 116)	29.1^{b}	48.7^{c}	40.3^{b,c}
Polytrophic (n = 49)	40.2^{b,c,d}	59.6^{d}	
Hypertrophic (n = 7)		91.3^{e}	

Values marked with different letters are significantly different at $P < 0.01$ (t-test).

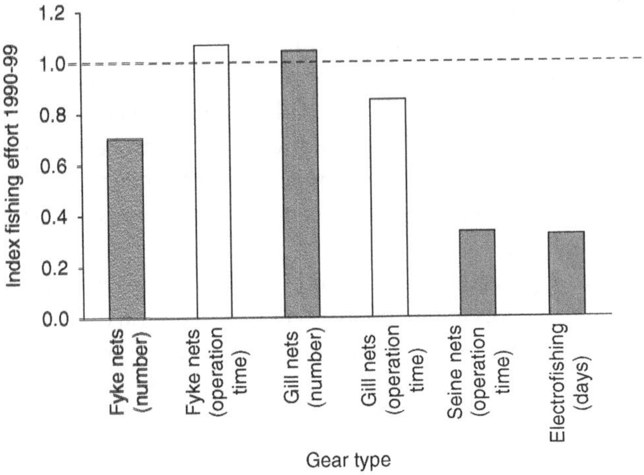

Figure 3.2 Current fishing effort adjustment in Brandenburg lakes (period 1990–1999) for the most common gear types compared to times when the Bulon–Vinberg relationship was established (period 1980–1989 = index 1; $n = 67$ lakes)

Table 3.2 Deviations (%) between calculated FYPs and data from yield statistics between 1991 and 1999 ($n = 59$ lakes). FYPs are calculated after Knoesche & Barthelmes (1998) (P–PP–F), Hanson & Leggett (1982) and Downing *et al.* (1990)

Estimation procedure	Median of deviation (%)	Maximum overestimation (%)	Maximum underestimation (%)
P–PP–F (Dimictic lakes)	83.7	533.8 ($n = 18$)	−20.1 ($n = 3$)
P–PP–F (Polymictic lakes)	53.1	766.1 ($n = 28$)	−48.2 ($n = 10$)
Hanson & Leggett (1982)	−159.1	429.8 ($n = 14$)	−2027.0 ($n = 45$)
Downing *et al.* (1990)*	−147.0	258.0 ($n = 14$)	−1289.4 ($n = 45$)

*Yield potentials were assumed to be 30% of the calculated fish production value.

3.4 Discussion

Lake trophy has been shown to be the most meaningful parameter to predict fish biomass, fish production and fish yield, respectively, in lakes of the northern and southern hemisphere (Ryder 1965; Oglesby 1977; Liang, Melack & Wang 1981; Jones & Hoyer 1982; Quiros 1990; Downing & Plante 1993). In this context, total phosphorus has been highlighted as an appropriate starting point for estimates (Bulon & Vinberg 1981; Hanson & Leggett 1982; Downing *et al.* 1990; Quiros 1991). Besides its significant correlation with fish biomass or yield, TP is much less expensive to measure, and more representative, than for instance chlorophyll *a* or primary production. For these reasons, fish yield potential of 204 lakes in the lowlands of north-east Germany were assessed using the P–PP–F procedure of Knoesche and Barthelmes (1998) with two adjustments:

the correction of measured TP values in polymictic lakes according to the German States Association for Water (1998) and the adjustment of the FYP calculated to the current fishing effort.

Estimated FYPs varied between 5 and 135 kg ha^{-1} $year^{-1}$ with median values of 26, 55 and 40 kg ha^{-1} $year^{-1}$ in dimictic, polymictic and small lakes, respectively. The increase in FYPs in lakes with a higher trophy is attributed to the TP–PP–FYP correlation of the estimation procedure but is in agreement with results of other TP-based procedures (Hanson & Leggett 1982; Downing *et al.* 1990; Quiros 1991). The out performance of polymictic compared to dimictic lakes is a result of the adaptation of the P–PP–F procedure to the individual features of lakes. Contrary to dimictic lakes, frequent nutrient turnover in shallow lakes is making sedimented nutrients reaccessible to primary producers during the vegetative growth period and therefore lead to a higher productivity compared to stratified waters with a similar spring concentration of TP. This was taken into account by correcting measured TP values in polymictic lakes before PP calculation. In addition, the absence of light and often lack of oxygen associated with cold hypolimnic layers with a low production intensity in dimictic lakes gave a higher predicted fish yield for polymictic lakes. The P–PP–F model considers this factor by correcting FYPs in accordance with the extent of the hypolimnic lake area. As a measure for the share of energy used for somatic growth, grading classes of selected indicator species were analysed from catch statistics as well as from experimental samplings. The results indicate, that the rate of stunted individual growth increased with lake trophy in dimictic lakes, but was generally higher in shallow water bodies. Although stunting led to substantial FYP deductions in polymictic lakes it did not decrease the yield expectations to the level of dimictic lakes.

To evaluate the quality of the FYPs calculated, commercial yield statistics from the last decade were provided as a standard for comparison. On average, yield potentials were substantially higher than statistical data available for 59 lakes, even after a reduction of FYPs by 30% due to a currently lower fishing effort.

Reasons for these differences are numerous.

(1) Statistics of commercial fisheries do not include sport-fishing catches, which would have increased the statistical data into the region of calculated potentials if these catches were assumed to be around 10–15 kg ha^{-1}.
(2) The accuracy of commercial yield statistics is variable. This view is supported by the lack of any correlation between TP and yield statistics. Coincidently, yield statistic data were found to vary significantly with hypolimnic lake area as well as the rate of stunting, which supports the choice of these factors for the adaptation of the estimation procedure to individual lake features.
(3) Changes in fishing effort due to altered economic conditions, market preferences or alternative sources of income may generate different actual yields. In their entirety these drawbacks of yield statistics underline the necessity for an objective procedure to assess the yield potential of lakes independently of statistical data.

If TP data are used to calculated yield potentials according to models of other authors (Hanson & Leggett 1982; Downing *et al.* 1990), results are generally inferior to P–PP–F estimates and deviations from statistical data increase. The empirically-derived correlations of TP with fish yield (Hanson & Leggett 1982) and fish production (Downing *et al.* 1990) were at least partially based exclusively on catch statistics of anglers and therefore do not seem to fit the conditions of commercially exploited fish populations in north-east German lakes. For example, Hanson & Leggett (1982) based their model on fish yield values in a range of 0.4–42 kg ha^{-1}, while the current statistical data cover a range of 2.5–99.8 kg ha^{-1}.

The study demonstrated, that the P–PP–F procedure is a practical way of calculating the yield potential of lakes. To improve the estimation procedure, future research should concentrate on the assessment of PP in polymictic lakes, the adjustment of the PP–FYP relationship of Bulon & Vinberg (1981) to local conditions and a more precise figure of the energy conversion in stunted fish populations.

Acknowledgements

I am indebted to Robert Frenzel for assistance during data sampling. Thanks are also due to R. Knoesche, D. Barthelmes and R. Lemcke for critical discussions of the P–PP–F procedure and its modification.

References

Alm G. (1923) Über die Prinzipien der quantitativen Bodenfaunistik und ihre Bedeutung für die Fischerei. *Verhandlungen-Internationale Vereinigung für theoretische und angewandte Limnologie* **1**, 168–180.

Barthelmes D. (1988) Fish predation and resource reaction: biomanipulation background data from fisheries research. *Limnologica* **19**, 51–59.

Bulon W.W. & Vinberg G.G. (1981) The dependence between primary production and fish production of water bodies. In *Basis for the Study of Aquatic Ecosystems*. Leningrad: 5–9 pp. (in Russian).

Downing J.A. & Plante C. (1993) Production of fish populations in lakes. *Canadian Journal of Fisheries and Aquatic Sciences* **50**, 110–120.

Downing J.A., Plante C. & Lalonde S. (1990) Fish production correlated with primary productivity, not the morphoedaphic index. *Canadian Journal of Fisheries and Aquatic Sciences* **47**, 1929–1936.

German States Association for Water (1998) Assessment of water bodies – stagnant waters. Schwerin, 74 pp. (in German).

Hanson J.M. & Leggett W.C. (1982) Empirical prediction of fish biomass and yield. *Canadian Journal of Fisheries and Aquatic Sciences* **39**, 257–263.

Jones J.R. & Hoyer M.V. (1982) Sportfish harvest predicted by summer chlorophyll a concentration in midwestern lakes and reservoirs. *Transactions of the American Fisheries Society* **11**, 176–179.

Knoesche R. & Barthelmes D. (1998) A new approach to estimate lake fisheries yield from limnological basic parameters and first results. *Limnologica* **28**, 133–144.

Koschel R., Haubold G., Kasprzak P., Kuchler L., Proft G. & Ronneberger D. (1981) Eine limnologische Zustandsanalyse des Feldberger Haussees. *Acta hydrochimical Hydrobiologie* **9**, 255–279 (in German).

Liang Y., Melack J.M. & Wang J. (1981) Primary production and fish yield in Chinese ponds and lakes. *Transaction of the American Fisheries Society* **110**, 346–350.

Oglesby R. (1977) Relationship of fish yield to lake phytoplankton standing crop, production, and morphoedaphic factors. *Journal Fisheries Research Board of Canada* **34**, 2271–2279.

Quiros R. (1990) Predictors of relative fish biomass in lakes and reservoirs of Argentina. *Canadian Journal of Fisheries and Aquatic Sciences* **47**, 928–939.

Quiros R. (1991) Empirical relationships between nutrients, phyto- and zooplankton and relative fish biomass in lakes and reservoirs of Argentina. *Verhandlungen-Internationale Vereinigung für theoretische und angewandte Limnologie* **24**, 1198–1206.

Ryder R.A. (1965) A method for estimating the potential fish production of north-temperate lakes. *Transaction of the American Fisheries Society* **94**, 214–218.

Ventz D. (1974) *Die Einflussnahme von Umgebungsfaktoren und morphometrischen Faktoren auf den Stoffhaushalt von Seen*. PhD Thesis, Dresden University.

Wundsch H.H. (1936) Die Arbeitsmethoden der Fischereibiologie. In E. Abderhalden (ed.) *Handbuch biologischer Arbeitsmethoden, Abt. IX, Teil 2/II.* Berlin und Wien: pp. 853–1208.

Chapter 4
Effects of area and location on pikeperch yields in Finnish lakes

J. LAPPALAINEN* and T. MALINEN
Department of Limnology and Environmental Protection, University of Helsinki, Finland

Abstract

A postal survey was carried out in 1998 to evaluate the state of pikeperch *Stizostedion lucioperca* (L.) stocks in Finnish lakes. Annual median yield of pikeperch was 0.25 kg ha^{-1} year^{-1} (min−max 0.01−5.82 kg ha^{-1} year^{-1}, n = 67) in surveyed lakes. In 61.2% of the lakes, yields were <0.5 kg ha^{-1} year^{-1} and in 79.1%, <1.0 kg ha^{-1} year^{-1}. A positive relationship was found between the yields and lake area, and a negative relationship between yields and latitude (°N). In the latter analysis, the effects of lake area were removed from yields. Yield per recruit modelling was applied to six sites from southern (26°E, 61.5°N) to northern (28°E, 69°N) Finland. This showed that both recruitment and growth decreased northwards, as did yields. The reduction in recruitment had greater effect on yields than growth.

Keywords: Finland, lake, latitude, postal survey, pikeperch, yield.

4.1 Introduction

Pikeperch, *Stizostedion lucioperca* (L.), is among the most valuable fish species in commercial and recreational fisheries in Finland (Finnish Game and Fisheries Research Institute 1999, 2000). Based on a recent lake survey, pikeperch were found to occur naturally in 650 lakes in southern and central Finland, but it has been introduced to a further 1600 lakes (Lappalainen & Tammi 1999; Tammi, Lappalainen, Mannio, Rask & Vuorenmaa 1999). Despite the large number of stockings, studies dealing with the success of these introductions have only recently started (Ruuhijärvi, Salminen & Nurmio 1996).

Large variations in year-class strength and annual yield are characteristic of pikeperch populations (Lappalainen & Lehtonen 1995; Lappalainen, Erm & Lehtonen 1995). For example, in Pärnu Bay yields varied between 75 and >300 t between 1959 and 1992. Several factors can potentially affect fish yield in lakes. These include fishing effort, water quality, temperature and fish species present (Ranta & Lindström 1989; 1990; 1998; Ranta, Lindström & Rask 1992a; Ranta, Lindström & Salojärvi 1992b; Lappalainen *et al.* 1995). In addition it was hypothesised that the yields should decrease with increasing latitude because growth of pikeperch is strongly temperature-

*Correspondence: Department of Limnology and Environmental Protection, P.O. Box 27, FIN-00014 University of Helsinki, Finland (email: Jyrki.T.Lappalainen@Helsinki.Fi).

dependent and the mortality during the first winter is size-dependent (Kirjasniemi & Valtonen 1997; Lappalainen, Erm, Kjellman & Lehtonen 2000).

The present chapter examines the influence of lake area and location (latitude) on pikeperch yields in Finnish lakes. The study was based on a postal survey completed in 1998. The survey aimed to clarify the origin of pikeperch stocks, whether stocked or natural, number of stockings and stocked juveniles, fishing regulations in pikeperch lakes and the occurrence and abundance of other fish species. This chapter, however, only considers questions concerning the yields and importance of pikeperch.

4.2 Materials and methods

4.2.1 *Postal survey*

A postal questionnaire was sent to 760 heads of the fishing associations in 1998. The addresses were obtained from the Employment and Economical Development Centre for Central Finland. The first mailing was followed up by a second action to all those who had not responded after 4 weeks. Because some of the addresses were incorrect only five counties and 415 lakes were included in this study. The number of returns was 261 and 68 during the first and second mailings, respectively. Thus the total return rate was 79%. Note that the number of answers to each question is always lower than the maximum return rate. Data on the location (coordinates) of each lake and its area were also obtained from the Employment and Economical Development Centre for Central Finland.

The relationship between fish yield and lake area was tested by regression analysis. The residuals from this relationship were correlated (Spearman) with the location of lakes (minutes in latitude to one decimal place). Annual yields (kg ha^{-1}) were also related to the importance of pikeperch in each lake estimated by the heads of the fishing associations.

4.2.2 *Location and temperature*

Yield per recruit (Y/R) modelling was used to estimate the effects of recruitment and growth on yields in six sites from southern (26°E, 61.5°N) to northern (28°E, 69°N) Finland (Fig. 4.1). These six sites were the same as those used by Lappalainen & Lehtonen (1997).

Water temperature were estimated from daily air temperatures, which were generated for the six study sites with the so called 'weather generator' for the present climate (Carter, Posch & Tuomenvirta 1995) based on three variables: precipitation, air temperature and cloud cover. This works by simulating the time series of daily precipitation, which is the independent variable in the procedure. Daily temperatures and cloud cover values are then correlated with the occurrence of wet and dry days. These generated variables have similar statistical properties to the climatic variables observed from 1961 to 1990 at each site.

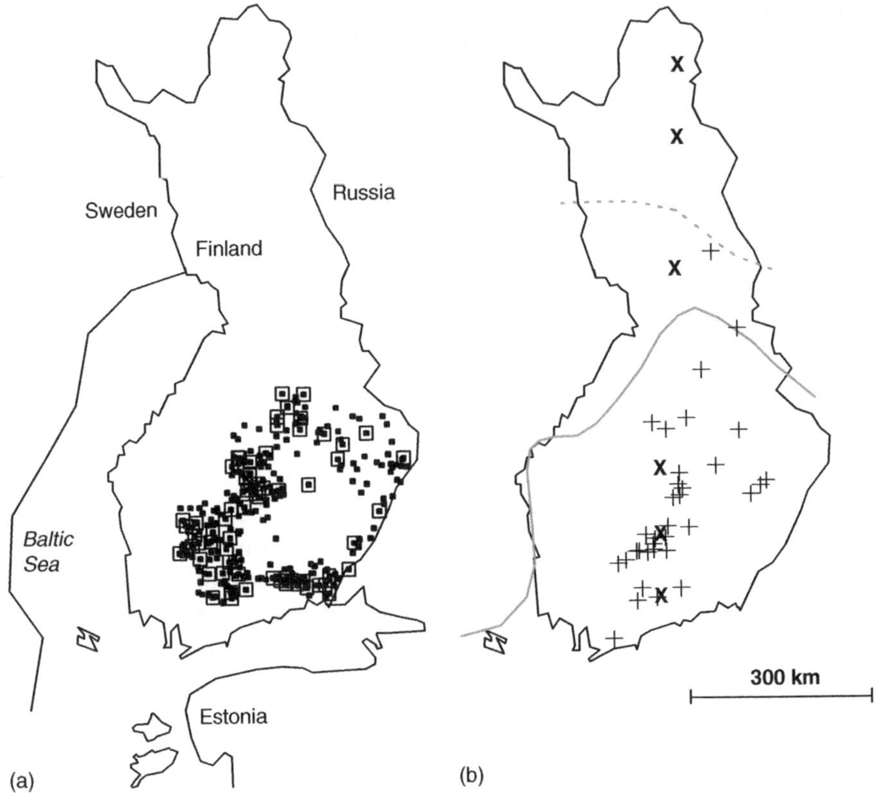

Figure 4.1 Locations of surveyed lakes and six sites for Y/R modelling. (a) Squares denote returned postal survey and open squares the lakes in which the yields also were estimated. (b) X denotes the six sites where the Y/R modelling was evaluated, and the + sign marks lakes in which the growth data of pikeperch were available. Northern distribution range of pikeperch is marked with solid line. The area between solid and broken line denotes the region where only two or three pikeperch lakes exist

Baseline climate data for 100 years were generated for the six study sites. Within the surveyed lake area the mean annual air temperature decreased from 4.5°C to 2°C from south to north, respectively (Atlas of Finland 1987), and summer (air temperature over 10°C) was approximately 3 weeks longer, 120 days *v.* 100 days, in southern Finland compared with central Finland (Atlas of Finland 1987). These differences in air temperature are also reflected in water temperatures in lakes (Kuusisto & Lemmelä 1976).

Pikeperch do not occur in the two northernmost sites (Fig. 4.1), probably because the climate is too cold (Lappalainen & Lehtonen 1997). The generated daily air temperature scenarios were then applied to estimate the corresponding changes in the daily surface water temperatures. This estimation was based on observations that the mean air temperature was 1.57°C lower than water temperature at sites south of 65°N during May and October and 1.86°C lower at sites north of 65°N during June and October (E. Kuusisto, unpublished data).

Day degrees over 15°C were used because Lappalainen *et al.* (2000) found this threshold better explained growth of age-0 pikeperch during the first summer. Annual water temperature sums were calculated from 1 May until 31 October.

To include the effects of first winter mortality in recruitment, the proportion of under 6 cm TL juveniles was calculated based on water temperature, as this is the lowest length at which age-0 pikeperch are able to survive through the first winter (Kirjasniemi & Valtonen 1997; Lappalainen *et al.* 2000). Lappalainen *et al.* (2000) showed that the number of age-0 pikeperch <6 cm depends on water temperature (*WT*) measured as degree-days over 15°C. The proportion was estimated by

$$\text{Proportion} = \exp(1.6 - 0.0185\,WT)\,[(1 + \exp(1.6 - 0.0185\,WT))]^{-1}. \qquad (4.1)$$

4.2.3 *Growth*

Pikeperch length at age was based on published or unpublished data in 35 Finnish lakes (Kokko, Kaijomaa & Kokko 1984; Heikinheimo-Schmid & Huusko 1987; Salo 1988; Hakaste 1992a, b; Ruuhijärvi *et al.* 1996; Sarell & Nyberg 1998; Keskinen, Marjomäki, Valkeajärvi, Salonen & Helminen 1999; Sutela, Huusko, Hyvrärinen & Pursiainen 1995; Vinni 1999). Because only few old pikeperch were aged in some lakes, the actual lengths could not be used directly to estimate the growth at each of the six sites. Therefore the growth was estimated as follows. First the length at ages 2 and 4 were estimated in relation to latitude. Then the relationships between length at age 4 and lengths at ages 5–14 were estimated from the combined dataset including all study lakes based on 10 regression analyses. Thus the lengths at ages 5–14 were based only on the observed growth pattern regardless to the location of the lakes. Altogether these procedures yielded 12 different lengths at age estimates for each of the six sites, two length estimates based on relationships between length and latitude (ages 2 and 4), and 10 length estimates based on regressions (ages 5–14). Finally, equation (4.2) was fitted to lengths at ages 2 and 4–14 at each of the six sites:

$$y = \mathrm{a}\ln(x) + \mathrm{b}, \qquad (4.2)$$

where y is length at age (cm) and x is age from 2 and 4–14. These six models were used to estimate growth at each of the six sites. Equation (4.2) was used here because it fitted to the length data better than the more general used von Bertalanffy growth equation.

4.2.4 *Yield per recruit model and scenarios*

The effects of estimated first winter mortality and growth on yields were examined using the age- and size-structured yield per recruit-model (Buijse, Pet, van Densen, Machiels & Rabbinge 1992). The model computes average yield per 1000 age-0 pikeperch based on growth rate, fishing and natural mortality rates and a selection factor which produces gradual recruitment to the fishery. In the present model pikeperch was assumed to be harvested only with gill nets.

The selection of gill nets was computed as in Buijse *et al.* (1992), but the left-hand side of the selection curve was taken from the curve generated by the 45-mm gill net (bar length) and selectivity was assumed to be unity after the peak of the curve. This should be a realistic description for the average selectivity of pikeperch in Finland, because in most pikeperch lakes the minimum allowable mesh size is 40–50 mm, although larger mesh sizes are also used (e.g. Keskinen *et al.* 1999; Finnish Game and Fisheries Research Institute 2000). In addition, because of the high fishing mortality (Lehtonen 1983; Salonen, Helminen & Sarvala 1996), the right-hand side of the selection curve only has a marginal effect on the results.

The standard deviation of length was set to be constant contrary to Buijse *et al.* (1992), because there was no sign of dispersion in length distribution with age in the largest dataset from Lake Vesijärvi (J. Ruuhijärvi, unpublished data). Natural and fishing mortality values were set according to literature (Lehtonen 1983; Salonen *et al.* 1996).

Three different scenarios were analysed with the Y/R model. First, growth was allowed to change as found in the six sites, but the recruitment (i.e. winter mortality) was kept constant as found in the southernmost site (26°E, 61.5°N). In the second scenario the recruitment was allowed to change while the growth was kept constant, also as found for the southernmost site. In the last Y/R model both the recruitment and growth were allowed to change.

4.3 Results

4.3.1 *Yields, lake area and location*

Median pikeperch yields were 0.25 kg ha^{-1}, with a range from 0.01 to 5.82 kg ha^{-1} (min – max, $n = 67$) in the surveyed lakes (Fig. 4.2). In 61.2% of the lakes, yields were less than 0.5 kg ha^{-1} and in 79.1% less than 1.0 kg ha^{-1} (Fig. 4.2). Yields correlated with lake area (both log-transformed; $r^2 = 0.31$, $P < 0.0001$, d.f. = 66; Fig. 4.3(a)). When the effect of lake area was removed yield diminished towards the north ($r_s = -0.31$, $P < 0.05$, $n = 67$; Fig. 4.3(b)). This corresponds to a decline of 0.6 kg ha^{-1} to 0.07 kg ha^{-1} within the surveyed area.

The heads of the fishing associations evaluated the pikeperch catches to be very important in 17 lakes, important in 50 lakes, usual in 73 lakes and not important in 85 lakes (total $n = 225$). The average yield was 2.70 kg ha^{-1} in lakes, considered very important by the heads (SE = 0.65, min = 0.69, max = 5.82, $n = 9$). The same averages for the three other classes important, usual and not important were 1.1 kg ha^{-1} (SE = 0.32, min = 0.01, max = 3.65, $n = 16$), 0.31 kg ha^{-1} (SE = 0.07, min = 0.01, max = 1.11, $n = 27$) and 0.10 kg ha^{-1} (SE = 0.03, min = 0.01, max = 0.38, $n = 15$), respectively. All four classes differ significantly from each other (Mann–Whitney *U*-test).

The decrease in yield towards the north suggests that either the most important pikeperch lakes are situated in southern Finland, or that the least important lakes are situated in northern Finland. The latter may not be detected here because the survey

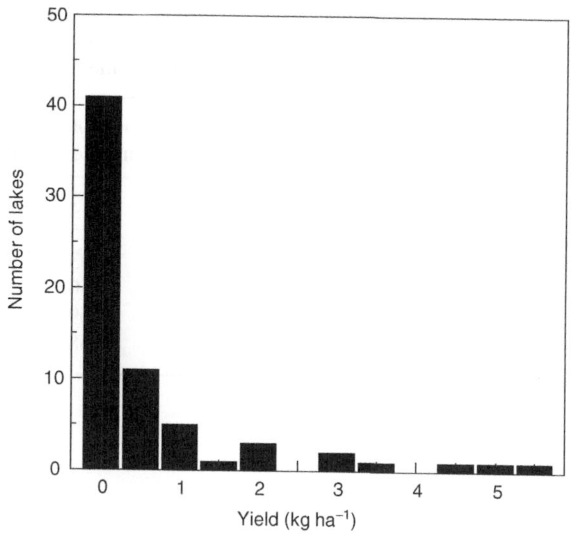

Figure 4.2 Pikeperch yields in 67 Finnish lakes

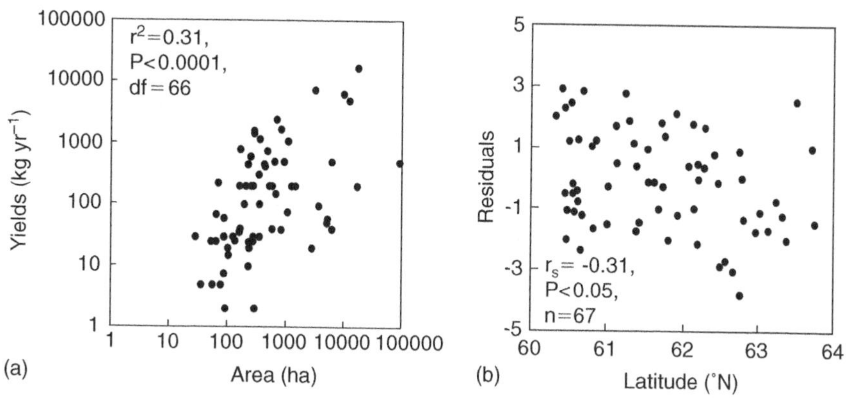

Figure 4.3 (a) Relationship between annual yield (kg) and lake area (ha); (b) relationship between residuals from (a) and latitude (°N)

was targeted mainly on pikeperch lakes, and so the northernmost lakes were not included in the survey, because they have no pikeperch stocks. When the very important lakes were classified as 1 and the other three classes to 0, the logistic regression analysis showed that there is a negative relationships between the importance of lakes and latitude, so the most important pikeperch lakes are, indeed, situated mainly in the southern Finland ($P = 0.02$). The other two possible classifications (very important and important = 1, usual and not important = 0; very important, important and usual = 1, not important = 0) against latitude were non-significant.

Table 4.1 Estimated total mortality of age-0 pikeperch during the first winter based on degree-days over 15°C in each site. Both the water temperature and proportion are based on simulation for 100 years. In the parenthesis is the minimum and maximum value

Site	Water temperature as degree-days (>15°C)	Proportion of <6.0 cm juveniles in autumn (estimated total mortality)
26°E, 61.5°N	187 (87.2–317.7)	13.7 (1.4–50.0)
26°E, 62.5°N	146.8 (59.7–278.7)	25.0 (2.8–62.5)
26°E, 63.5°N	124 (71.3–219.8)	33.6 (7.9–57.3)
28°E, 66°N	87 (27.6–169)	50.1 (18.1–75.1)
28°E, 68°N	43.3 (8.6–102.1)	69.3 (43.2–81.1)
28°E, 69°N	45.2 (11.7–113.9)	68.5 (37.9–80.2)

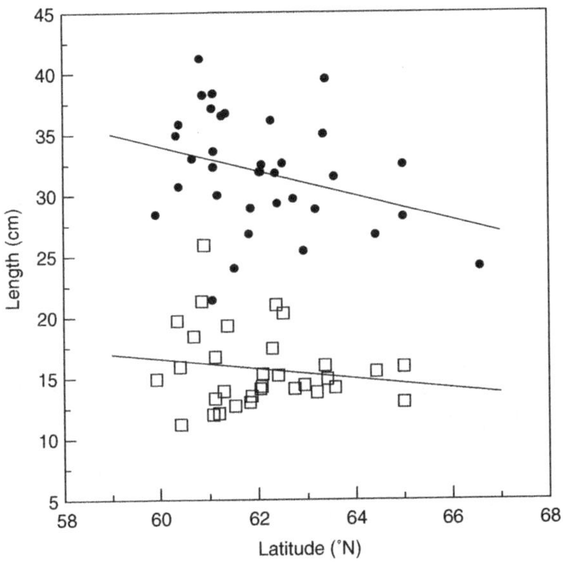

Figure 4.4 The relationships between length at age 2 (open squares) and age 4 (dots) with latitude

4.3.2 *Y/R modelling*

The proportion of age-0 pikeperch <6 cm, i.e. those to die during the first winter, increased towards north as the temperature decreased (Table 4.1). Lengths at both ages 2 and 4 reduced towards the north (Fig. 4.4). All three scenarios showed that the yields decreased with increasing latitude. The reduction in growth had less effects on yields than the corresponding reduction in recruitment (Fig. 4.5).

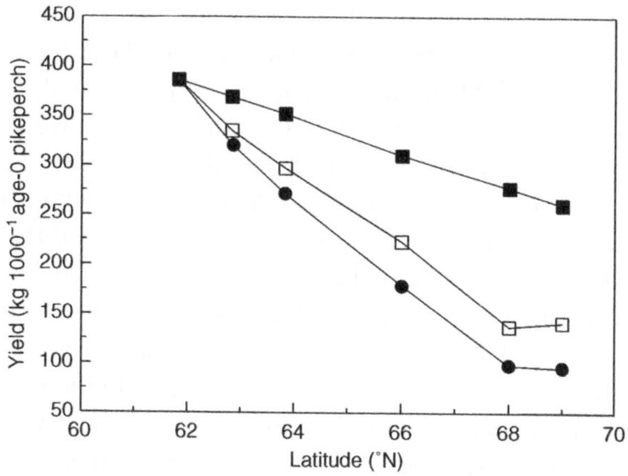

Figure 4.5 Estimated yields in six sites in Y/R modelling. Line with filled squares = growth decline and constant recruitment (winter mortality of 15% in southern site, see Table 4.2), line with open squares = estimated recruitment and constant growth (in southern site), and line with solid circles = estimated recruitment and growth at each six sites

4.4 Discussion

Pikeperch stockings has increased greatly in the 1990s. Almost 10 million, age-0 juveniles are stocked annually (e.g. Ruuhijärvi *et al.* 1996). According to the heads of the fishing associations the two main objectives of stocking are to enhance the stocks in lakes or to introduce it as a new species to a lake (Lappalainen, Lehtonen, Tammi & Lappalainen 1998). At least three studies have shown a positive correlation between pikeperch stockings and subsequent yields (Ruuhijärvi *et al.* 1996; Sutela & Hyvärinen 1998; Salminen, Ruuhijärvi & Ilmarinen 1999). In the present study the possible effects of stocking on yields were not evaluated. However, within the five counties surveyed there seems to be almost equal annual stockings. Spearman correlation between the latitude and the sum of stockings in 1988–1994 was non-significant in the 67 lakes where yield was estimated. This suggests that the negative relationship between yields and latitude may have not be affected by pikeperch stockings, but this needs more thorough analysis.

The present yield data were based on a postal survey. The obvious advantage of a postal survey is that large areas can be covered at relatively low cost. Another advantage is that the yields were estimated by persons familiar with their lakes. Furthermore, the yields were estimated at the same time, so the differences between lakes due to timing should be minimal.

Pikeperch is mainly fished with gill nets. As the minimum legal size of pikeperch is 37 cm in total length, the most typical gill nets used are 40–55 mm in bar length (Keskinen *et al.* 1999; Finnish Game and Fisheries Research Institute 2000). In recent years recreational fisheries have taken more catches using spinning rods and trolling gears (Finnish Game and Fisheries Research Institute 2000).

Several studies have shown that the annual yields of pikeperch in certain locations are highly variable (van Densen, Gazemier, Dekker & Ouderlaar 1990; Lappalainen *et al.* 1995). Here focus was on yields in different lakes, and so annual variations were not considered. However, the yields obtained (0–6 kg ha^{-1} $year^{-1}$) were comparable with other lakes in Finland. According to Sutela & Hyvärinen (1998), the maximum yield of pikeperch was 1.3 kg ha^{-1} $year^{-1}$ in Lake Mikitänjärvi, which is located within the latitudes 64–65°N. In that lake stocking may have increased the yields (Sutela & Hyvärinen 1998). In central Finland the yield levels varied from 0 to 3.8 kg ha^{-1} $year^{-1}$ in 14 lakes (Keskinen *et al.* 1999). Similar values of 0.1–4.1 kg ha^{-1} $year^{-1}$ were given by Ranta *et al.* (1992b), who also estimated the effects of fishing effort and water quality on yields and found water quality important. In their study water quality was analysed as two principal components, and the component which explained the yields was negatively correlated with pH, alkalinity and conductivity. In central and southern Europe the maximum annual yields are higher, but they also show the typical fluctuations from near 0–10 kg ha^{-1} $year^{-1}$ (Mikulski 1964; Biro 1990; van Densen *et al.* 1990).

Lehtonen, Hansson & Winkler (1996) also showed that the length of 4-year-old pikeperch decreased towards the north, but the study was based on data mainly from the Baltic Sea and adjacent lakes, whereas the present data were from Finnish lakes. According to Keskinen *et al.* (1999) the length of 4-year-old pikeperch correlated positively with total phosphorus and negatively with lake area and lake depth in 34 lakes in central Finland. Total phosphorus seemed to explain the length at age 4 well up to 20 $\mu g\,L^{-1}$, while greater phosphorus levels seemed to have no effects on length (Keskinen *et al.* 1999). Thus, a part of the growth fall off in this study might also be due to the overall decrease in phosphorus levels in lakes towards the north (Tammi *et al.* 1999). Unfortunately no data were available to tease out the possible effects of phosphorus on lengths. The Y/R modelling also showed that the decrease in recruitment

Table 4.2 Summary of variables and parameter values used in Y/R modelling

	Latitude						
Variable	61.5	62.5	63.5	66	68	69	References
First winter mortality (%)	13.7	25.0	33.6	50.1	69.3	68.5	
a (in equation (4.2))	27.87	27.10	26.33	24.40	22.86	22.08	
b (in equation (4.2))	4.87	4.48	4.08	3.09	2.30	1.90	
Standard deviation of length distribution (cm)	2.22	2.22	2.22	2.22	2.22	2.22	Ruuhijärvi, unpublished data
M $year^{-1}$ (age-group 1)	0.2	0.2	0.2	0.2	0.2	0.2	
M $year^{-1}$ (age-groups 2–14)	0.15	0.15	0.15	0.15	0.15	0.15	Lehtonen (1983)
F $year^{-1}$ in completely recruited age-groups	1	1	1	1	1	1	Salonen *et al.* (1996)

had more effect on yields that the corresponding decrease in growth. Even though pikeperch grows more slowly in the north, low natural mortality cannot reduce the numbers as much as first winter mortality. In the southernmost sites, pikeperch recruits to the gill net fishery (45 mm bar length) at age 4 but not until 6 years in the northernmost site where it occurs naturally (site 28°E, 66°N).

There are possible weaknesses in the current Y/R modelling. The parameter values for Y/R modelling were obtained from published articles, and the natural and fishing mortalities were based on studies from southern Finland (see Table 4.2). Furthermore it was suggested that all catches were taken with similar gill nets. Although more realistic modelling could be achieved with more appropriate parameters, these had to be very different to those used here to alter the conclusions drawn, which is considered unlikely.

Acknowledgements

This study was financed by the Finnish Cultural Foundation and Emil Aaltonen Foundation (J.L.). We are grateful to our colleagues, M. Vinni, J. Ruuhijärvi, K. Nyberg and H. Peltonen, who gave us the unpublished growth data of pikeperch.

References

Atlas of Finland (1987) *Climate*. National Board of Survey & Geographical Society of Finland. Appendix 131, 32 pp.

Biro P. (1990) Population parameters and yield-per-recruit estimates for pikeperch (*Stizostedion lucioperca* L.) in Lake Balaton, Hungary. In W.L.T. van Densen, B. Steinmetz & R.H. Hughes (eds) *Management of Freshwater Fisheries*. Wageningen: Pudoc, pp. 248–261.

Buijse A.D., Pet J.S., van Densen W.L.T., Machiels M.A.M. & Rabbinge R. (1992) A size- and age-structured simulation model for evaluating management strategies in a multispecies gill net fisheries. *Fisheries Research* **13**, 95–117.

Carter T., Posch M. & Tuomenvirta H. (1995) SILMUSCEN and CLIGEN User's Guide; guidelines for construction of climatic scenarios and use of a stochastic weather generator in the Finnish Research Programme on Climate Change (SILMU). *Publications of the Academy of Finland* 5/95, 62 pp.

Finnish Game and Fisheries Research Institute (1999) Commercial Inland Fishery 1998. *SVT Agriculture, Forestry and Fishery 1999*, 13, 13 pp.

Finnish Game and Fisheries Research Institute (2000) Recreational Fishing 1998. *SVT Agriculture, Forestry and Fishery 2000*, 1, 27 pp.

Hakaste T. (1992a) Vanajaveden reitin yläosan kalataloudellinen yhteistarkkailu: kalansaaliit 1990, saaliskirjanpito 1989 ja 1990, ahven-, kuha-, lahna- ja sulkavakantojen tila saalisnäytteiden perusteella vuosina 1985–1990. *Kokemäenjoen vesistön Vesiensuojeluyhdistys r.y.* Julkaisu 256, 105 pp. (in Finnish).

Hakaste T. (1992b) Vanajaveden reitin alaosan Kalataloustarkkailu vv. 1989 ja 1990. *Kokemäenjoen vesistön Vesiensuojeluyhdistys r.y.* Julkaisu 266, 133 pp. (in Finnish).

Heikinheimo-Schmid O. & Huusko A. (1987) Kemijärven kalatalouden nykytila ja ehdotukset hoitotoimenpiteiksi. *Riista- ja kalatalouden tutkimuslaitos. Monistettuja julkaisuja* 69, 212 pp. (in Finnish).

Kirjasniemi M. & Valtonen T. (1997) Winter mortality of young-of-the-year pikeperch *(Stizostedion lucioperca). Ecology of Freshwater Fish* **6**, 155–160.

Keskinen T., Marjomäki T. J., Valkeajärvi P., Salonen S. & Helminen H. (1999) Kuhakantojen hoito Keski-Suomessa. Nykytila ja kehittämissuunnitelma. *Kala- ja riistahallinnon julkaisuja nro* 37, 46 pp. (in Finnish).

Kokko H., Kaijomaa V.-M. & Kokko T. (1984) Pohjois-Karjalan läänin alueellinen kalataloussuunnittelu. Osa V. Koekalastusten sekä kasvu- ja ikäjakaumien taulukot. Joensuun yliopisto. *Karjalan tutkimuslaitoksen monisteita.* No. 4/1984, 60 pp. (in Finnish).

Kuusisto E. & Lemmelä R. (1976) On the surface water temperature of lakes in Finland. *Aqua Fennica* 1976, 18–25.

Lappalainen J. & Lehtonen H. (1995) Year-class strength of pikeperch (*Stizostedion lucioperca* L.) in relation to environmental factors in a shallow Baltic bay. *Annales Zoologici Fennici* **32**, 411–419.

Lappalainen J. & Lehtonen H. (1997) Temperature habitats for freshwater fishes in a warming climate. *Boreal Environmental Research* **2**, 69–84.

Lappalainen J. & Tammi J. (1999) Kuhajärvien lukumäärä tarkentuu. *Suomen Kalastuslehti* 2/1999, 16 (in Finnish).

Lappalainen J., Erm V. & Lehtonen H. (1995) Pikeperch, *Stizostedion lucioperca* (L.), catch in relation to juvenile density and water temperature in Pärnu Bay, Estonia. *Fisheries Management and Ecology* **2**, 113–120.

Lappalainen J., Lehtonen H., Tammi J. & Lappalainen S. (1998) 'Kuhakantakartoitus 1998' – kyselyn alustavia tuloksia. *Suomen Kalastuslehti* 8/1998, 6–10 (in Finnish).

Lappalainen J., Erm V., Kjellman J. & Lehtonen H. (2000) Size-dependent winter mortality of age-0 pikeperch (*Stizostedion lucioperca*) in Pärnu Bay, the Baltic Sea. *Canadian Journal of Fisheries and Aquatic Sciences* **57**, 451–458.

Lehtonen H. (1983) Stocks of pike-perch and their management in the Archipelago Sea and the Gulf of Finland. *Finnish Fisheries Research* **5**, 1–16.

Lehtonen H., Hansson S. & Winkler H. (1996) Biology and exploitation of pikeperch, *Stizostedion lucioperca* (L.), in the Baltic Sea area. *Annales Zoologici Fennici* **32**, 525–535.

Mikulski J.St. (1964) Some biological features of perchpike lakes. *Verhandlungen der Internationalen Vereinigung für Limnologie* **15**, 151–157.

Ranta E. & Lindström K. (1989) Prediction of lake-specific fish yield. *Fisheries Research* **8**, 113–128.

Ranta E. & Lindström K. (1990) Water quality versus other determinants of species-specific yield of fish in northern Finnish lakes. *Fisheries Research* **8**, 367–379.

Ranta E. & Lindström K. (1998) Fish yield versus variation in water quality in the lakes of Kuusamo, northern Finland. *Annales Zoologici Fennici* **35**, 95–106.

Ranta E., Lindström K. & Rask M. (1992a) Fish catch and water quality in small lakes. *Fisheries Research* **13**, 1–7.

Ranta E., Lindström K. & Salojärvi K. (1992b) Water quality, fishing effort and fish yield in lakes. *Fisheries Research* **15**, 105–119.

Ruuhijärvi J., Salminen M. & Nurmio T. (1996) Releases of pikeperch (*Stizostedion lucioperca* (L.)) fingerlings in lakes with no established pikeperch stock. *Annales Zoologici Fennici* **33**, 553–567.

Salminen M., Ruuhijärvi J. & Ilmarinen P. (1999) Tuki-istutukset yli kaksinkertaistivat Lohjanjärven kuhasaaliin. *Suomen Kalastuslehti* 1/1999, 34–38. (in Finnish).

Salo H. (1988) Kuhan kalastus ja saalisvarat Vanajanselällä vuonna 1985. *Hämeen Kalastuspiirin Kalastustoimisto. Tiedoitus* **4**, Osa A, 1–55. (in Finnish).

Salonen S., Helminen H. & Sarvala J. (1996) Feasibility of controlling coarse fish populations through pikeperch (*Stizostedion lucioperca*) stocking in Lake Köyliönjärvi, SW Finland. *Annales Zoologici Fennici* **33**, 451–457.

Sarell J. & Nyberg K. (1998) Reis- ja Vuohtojärvellä tehtyjen rapu- ja kuhakompensaatioiden tuloksellisuus. *Alueelliset ympäristöjulkaisut* 101, 89 pp. (in Finnish).

Sutela T. & Hyvärinen P. (1998) Oulujoen vesistön kuhakantojen tila ja hoidon kehittämismahdollisuudet. *Kala- ja riistaraportteja nro* 113, 18 pp. + 5 appendixes (in Finnish).
Sutela T., Huusko A., Hyvärinen P. & Pursiainen M. (1995) Kuha Oulujoen vesistöalueella. Riistan- ja kalantutkimus. *Kala- ja riistaraportteja nro* 36, 26 pp. (in Finnish).
Tammi J., Lappalainen A., Mannio J., Rask M. & Vuorenmaa J. (1999) Effects of eutrophication on fish and fisheries in Finnish lakes: a survey based on random sampling. *Fisheries Management and Ecology* **6**, 173–186.
van Densen W.L.T., Gazemier W.G., Dekker W. & Ouderlaar H.G.J. (1990) Management of the fish stocks in Lake IJssel, The Netherlands. In W.L.T. van Densen, B. Steinmetz & R.H. Hughes (eds) *Management of Freshwater Fisheries*. Wageningen: Pudoc, pp. 313–327.
Vinni M. (1999) Kalojen kasvu ja ravinto Gennarbynlahdessa. Department of Limnology and Environmental Protection, University of Helsinki. MS thesis. 46 pp. +3 appendixes (in Finnish).

Chapter 5
Fish stock assessment of Lake Schulen, Flanders: a comparison between 1988 and 1999

I. SIMOENS*, J. BREINE, H. VERREYCKEN and C. BELPAIRE
Institute for Forestry and Game Management, Groenendaal-Hoeilaart, Belgium

Abstract

Lake Schulen, a shallow man-made lake (90 ha) in Flanders, Belgium, was excavated in the 1970s, as a flood storage reservoir for the neighbouring River Demer (Scheldt basin). In the autumns of 1988 and 1999 extensive fish stock assessments were carried out. To avoid selectivity of fishing gear, four fishing techniques (electric fishing, gill netting, fyke netting and seine netting) were used. The fish biomass was assessed by a multiple-mark–recapture method. Twenty-three fish species were captured in both surveys but five species found in 1988 were not caught in the recent survey, when five new species were collected. Important observations in 1999 were the occurrence of the non-indigenous Asian topmouth gudgeon, *Pseudorasbora parva* (Schlegel), which was reported for the first time in Flanders in 1992 and has now invaded most lakes and rivers, and the presence of 0+ specimens of the European catfish, *Silurus glanis* L., indicating the reproduction of this species in the lake. Shifts in the fish population structure and recruitment were observed not only between the two sampling periods, but also between different sampling zones in the lake. Based on the fish community, the biotic integrity of the lake was assessed with a multi-metric Index of Biotic Integrity developed for standing waters.

Keywords: Belgium, fish, stock assessment, Flanders, Index of Biotic Integrity.

5.1 Introduction

Extreme variation in abundance and/or species composition over space and time is a feature of fish communities in (larger) lakes, which necessitates assessment for lake management (Martin-Smith 1998). Of all the biological components of lakes, fish are the most difficult and time-consuming to sample as they are diverse and do not necessarily reflect local conditions in large lakes. To overcome the biases associated with sampling fishes in large lakes, the use of different gear types is recommended (EPA 1998). The trends in the fish population and community structure obtained from stock assessment exercise can indicate when and where action has to be taken. In temperate European shallow lakes, fish communities may be valuable for recreational fisheries

*Correspondence: Institute for Forestry and Game Management, Duboislaan 14, B-1560 Groenendaal-Hoeilaart, Belgium (email: Ilse.Simoens@lin.vlaanderen.be).

(Perrow, Jowitt & Gonzalez 1996). However, in Flanders (Belgium) few possibilities for lake angling exist. One exception is Lake Schulen, one of the larger lakes in Flanders, and has been the subject of intensive study to understand the ecosystem dynamics in an effort to help managers in the development of stock enhancement activities.

The objective of this chapter is to determine the changes in the fish community structure between 1988 (Belpaire, Verreyeken, Van Vlasselaer & Ollevier 1989) and 1999, and assess the implications for management of the fish stocks of the lake.

5.2 Materials and methods

Lake Schulen (N50°57′30″ E05°08′50″), a shallow man-made lake in Flanders (Belgium), was excavated in 1974 to act as a flood storage reservoir for the neighbouring River Demer (Scheldt basin). It has a surface area of 90 ha and an average depth of 5 m (about 4.5 million m^3 water volume). The lake has a sandy substratum with little vegetation. The lake was originally surveyed in 1988 when it was divided into five zones (A–E) (Fig. 5.1) covering a total of 42 ha, representing all the different habitats in the lake. In 1988 fish were collected over a 35-day period, between September and November. In 1999 fishing took place over 18 days in September and October. In both surveys the same four fishing techniques were used (electric fishing, gill netting, fyke netting and seine netting). The data were supplemented with information from anglers' questionnaires.

Electric fishing was carried out from a boat using two hand-held anodes powered from a 5-kW generator with an adjustable output voltage of 300–500 V (DC) and a

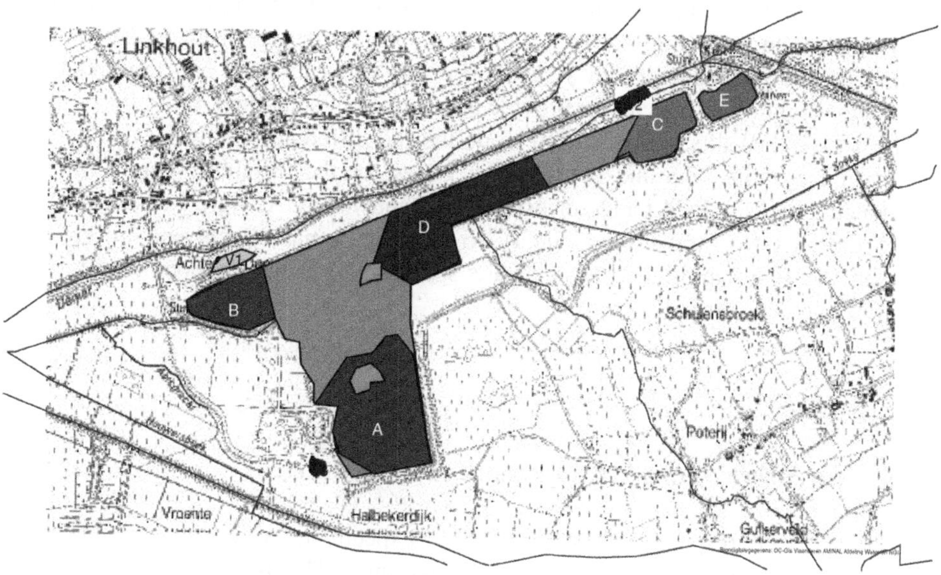

Figure 5.1 The different zones in Lake Schulen

Table 5.1 Number of samples of fish collected by each fishing method in each zone in 1988 and 1999

	1988				1999			
Zone	Electric fishing	Fyke nets	Gill nets	Seine nets	Electric fishing	Fyke nets	Gill nets	Seine nets
A	2	4	2	2	4	3	2	7
B	3	3	2	2	4	3	3	4
C	4	2	3	1	4	2	2	4
D	4	11	8	5	7	3	6	10
E	4	3	1	0	3	2	4	0

pulse frequency of 480 Hz. The fishing period was set at 1 h for each sample. The gill nets were 1.5 × 30 m and had knot-to-knot mesh sizes of 45, 50, 60 and 70 mm. For each sample, four nets were set for 2 h in the pelagic area of the target zone. The fykes were double fykes with each part 5 m long and a first hoop of 90 cm diameter; both parts were connected by a 12-m wing. They were placed randomly in groups of four in a particular zone and fished for 2 days. The seine nets were of variable size (from 30 up to 100 m long and 5 m deep). Details about the sampling frequency for the different fishing gears in each zone in 1988 and 1999 are given in Table 5.1.

Fish caught were identified, counted, and individually weighed (nearest g) and measured (total length (TL) up to 1 mm). Fish longer than 8 cm TL were marked by cutting part of the left pectoral fin and by freeze branding with liquid nitrogen (−196°C). For each zone a different mark was used. All fish captured were returned to their zone of origin.

Where possible, biomass (kg ha^{-1}) was calculated using a multiple-mark–recapture method (Robson & Spangler 1978). For each species for which marked fish were recaptured, the abundance (with a standard error) was estimated and multiplied by the mean weight of the species, to provide an estimate of standing crop.

The Kruskall–Wallis test was used to define the differences between fish assemblages in the entire lake and between the zones for the different years. Principal component analysis (PCA) and correspondence analysis (CA) were executed on the whole data set (log transformed data), to examine zonal effects and differences between the two sampling periods. Rare species (<1% of the population) were removed to reduce bias.

5.3 Results

5.3.1 *Inventory of the fish species of Lake Schulen*

In 1988 and 1999 respectively, 8483 and 10 693 specimens were caught (Table 5.2). Both surveys found 23 species, dominated by cyprinids. The fish species belong to the various trophic levels (piscivores, planktivores, invertivores and omnivores).

Table 5.2 Number of fish and proportion of 0+ fish in total catch (representing recruitment) collected in Lake Schulen in 1988 and 1999

	1988		1999	
Species	Total fish	% 0+ of catch species	Total fish	% 0+ of catch species
Abramis brama	336	11.9	524	25.5
Alburnus alburnus	3		1	
*Ameiurus nebulosus**	781	0.7	232	2.2
Anguilla anguilla	407		1722	
Barbatula barbatula	–		3	
Blicca bjoerkna	5		516	
Carassius auratus gibelio	7		270	18.3
Carassius carassius	4		–	
Cyprinus carpio	22		17	
Cobitis taenia	1		–	
Esox lucius	325	70.0	71	0.2
Gasterosteus aculeatus	4		–	
Gobio gobio	15		29	
Gymnocephalus cernuus	194	47.7	446	34.7
Lepomis gibbosus	123	1.8	222	0.6
Leuciscus cephalus	–		1	
Leuciscus idus	43	4.8	223	12.2
Misgurnus fossilis	–		1	
Perca fluviatilis	2401	30.6	1571	44.6
Pseudorasbora parva	–		104	0.5
Pungitius pungitius	6		–	
Rhodeus sericeus amarus	1		1	
Rutilus rutilus	3263	15.4	4385	49.2
Scardinius erythrophtalmus	241	13.3	71	43.0
Silurus glanis	–		8	100
Stizostedion lucioperca	102	16.7	214	95.2
Tinca tinca	198	72.2	61	6.9
Umbra pygmaea	1		–	
Total species	23		23	

*The black bullhead (*Ameiurus melas*) is probably also present in Lake Schulen. Due to the difficult determination in the field between the brown and the black bullhead, both species were treated as *Ameiurus nebolosus*.

Five species found in 1988 were not caught in 1999, whilst five new species were recorded in 1999. The newly recorded species in 1999 were topmouth gudgeon, *Pseudorasbora parva* (Schlegel), chub, *Leuciscus cephalus* (L.), stone-loach, *Barbatula barbatula* (L.), weatherfish, *Misgurnus fossilis* (L.) and wels, *Silurus glanis* L.

Umbra pygmea (DeKay) (striped or eastern mudminnow), *Carassius carassius* (L.) (crucian carp), *Cobitis taenia* (L.) (spined loach), *Pungitius pungitius* (L.) (ten-spined stickleback) and *Gasterosteus aculeatus* L. (three-spined stickleback) were five species present in 1988 but not caught in 1999. Only <10 specimens of these species were caught in 1988, so it is not clear whether these fish species are now extinct from the lake or whether they were just not caught in the 1999 survey. There is, however, no obvious reason why these species should in particular have disappeared since 1988.

In 1988 anglers caught five species only (perch, *Perca fluviatilis* L., roach, *Rutilus rutilus* (L.), bream, *Abramis brama* (L.), brown bullhead, *Ameiurus nebulosus* (Le Sueur) and pikeperch, *Stizostedion lucioperca* (L.)). Fourteen species (subtract brown bullhead and add silver bream *Blicca bjoerkna* (L.), gibel carp, *Carassius auratus gibelio* (L.), common carp, *Cyprinus carpio* L., crucian carp, eel, *Anguilla anguilla* (L.), rudd, *Scardinius erythrophtalmus* (L.), ruffe, *Gymnocephalus cernuus* L., pike *Esox lucius* L., ide, *Leuciscus idus* (L.), and tench, *Tinca tinca* (L.)) were captured in 1999 by anglers. These differences can largely be explained by the higher response by anglers to the questionnaire in 1999 compared with 1988. All of these species were also caught in the fish surveys.

5.3.2 *Species composition and population structure in 1988 and 1999*

Substantial differences in the species composition of catches were found between 1999 and 1988 (Table 5.2). The proportion of ruffe, bream and eel in the total catch was much higher in 1999, while the number of roach, perch and brown bullhead declined markedly. On both occasions, however, roach was the most abundant species.

A strong decline in relative abundance of pike (3.6–0.5%) and perch (27.7–13.1%) was recorded, but the contribution of pikeperch remained relatively stable (1.4–1.8%). The proportion of 0+ fish in the total catch per species (Table 5.2) revealed a shift in recruitment in the piscivorous trophic level from pike (70–0.2%) to pikeperch (strong increase from 16.7% to 95.2%).

A considerable increase in the average weight of pike was found in 1999 (915 g) compared with 1988 (257 g). Beside low recruitment, this could also be explained by enhanced cannibalism due to reduced habitat complexity (less macrophytes) and consequently loss of refuges (Grimm & Backx 1990).

The proportion of the bream and silver bream increased from 1988 to 1999 whilst roach decreased, although the contribution of 0+ roach in 1999 was higher than in 1988.

5.3.3 *Fish species distribution between different zones (1988–1999)*

Significant differences (PCA; $P < 0.01$) were observed in the fish assemblages between 1988 and 1999 (Fig. 5.2(a)) (48% of the variance was explained by the two first factors). The second axis in Fig. 5.2(b) explained the differences among the zones where two groups A–B–D and E–C can be distinguished.

CA on the same matrix, indicating the time and zone effects on the fish community, confirmed the differences (54% of the variance explained by the first two axes), but

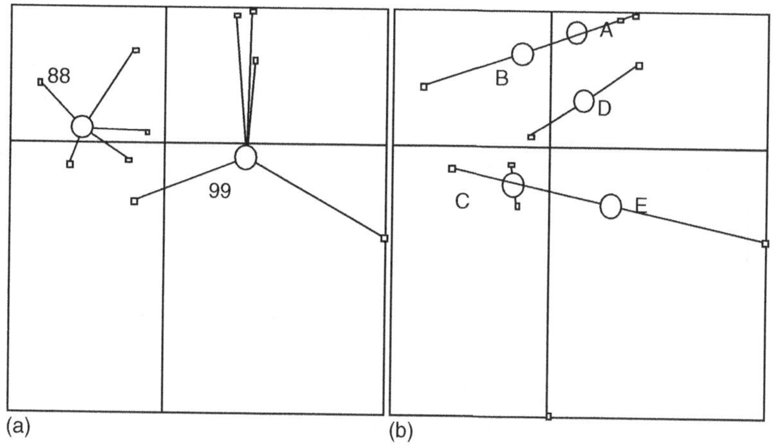

Figure 5.2 (a) and (b) PCA scatter plot for whole 1988 + 1999 data set, excluding rare (<1% of total catch) species

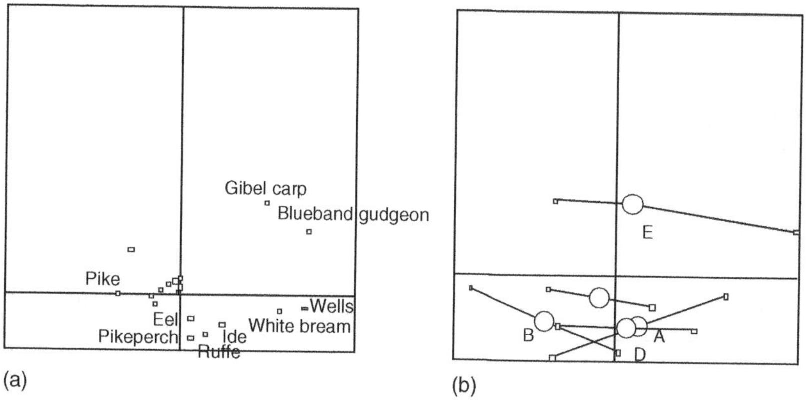

Figure 5.3 (a) and (b) FCA scatterplot on log transformed data (log $x + 1$) (x = % of total population in zones A–E) 1988 and 1999 excluding rare (<1%) species

zone E was distinct from the other zones (Fig. 5.3(b)). Zone E differs due to the higher density of topmouth gudgeon and gibel carp (Fig. 5.3(a)).The position of zone E is atypical since it receives water directly from the River Demer and acts as a sediment trap because of the narrow inlet to the rest of the lake.

When considering data from 1988 and 1999 separately, PCA revealed a difference between zones (Fig. 5.4(a) and (b)). In 1988 the zones were more heterogeneous than in 1999. In 1988 the abundance of brown bullhead in zone E accounted for its position in the plot. Zone C is differentiated from the other zones due to the presence of gibel

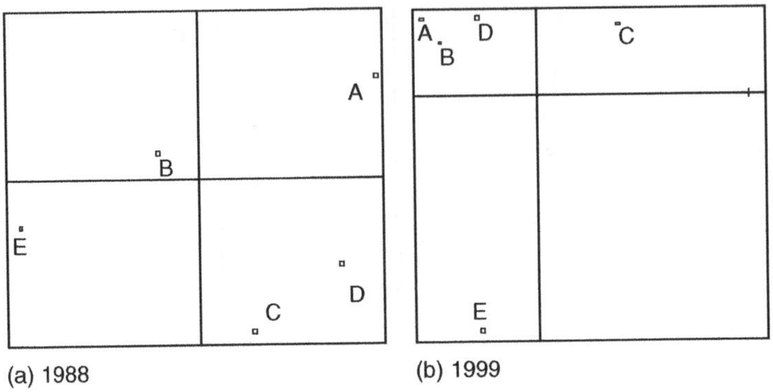

Figure 5.4 (a) and (b) nPCA scatterplot on log transformed data (log $x + 1$) (x = % of total population in zones A–E) 1988 and 1999 respectively excluding rare (<1%) species

Table 5.3 Recaptures of different species of fish in relation to zone of release

Species	Recaptured in same zone	Recaptured in different zone
Brown bullhead	7	24
Eel	44	4
Gibel carp	1	0
Ide	10	0
Perch	18	3
Pike	4	0
Pumpkinseed	5	0
Roach	4	0
Ruffe	1	0
Rudd	1	0

carp and bream. In 1999 zone E was distinct mainly because of the high density of *P. parva*, gudgeon and silver bream. Pike and roach densities explained the position of zone C in 1999.

From the recapture data (Table 5.3) most species appeared sedentary, except for bullhead, which was more actively moving and showed no home range specificity.

5.3.4 *Fish biomass assessment in Lake Schulen*

The standing stock (kg ha^{-1}) of each species for 1988 and 1999 (Table 5.4) was only estimated when there was at least one marked fish recaptured. These estimates should be treated as minimum estimates because few fish were recaptured, however, the shifts in biomass of each species between the two sampling periods reflect trends in

Table 5.4 Biomass ± standard error ($kg\,ha^{-1}$) of different fish species in Lake Schulen in 1988 and 1999

Species	1988	1999
Bream	167.8 ± 99.8	
Brown bullhead	5.4 ± 16.0	1.8 ± 15.1
Eel	45.0 ± 70.3	77.6 ± 15.0
Gibel carp		6.2 ± 98.8
Ide	0.4 ± 42.3	0.9 ± 29.4
Perch	278.0 ± 11.3	3.1 ± 10.4
Pike	4.7 ± 13.2	6.4 ± 54.2
Pikeperch	26.7 ± 98.4	
Pumpkinseed	0.3 ± 33.0	0.1 ± 40.3
Roach	160.5 ± 19.4	125.1 ± 56.0
Ruffe	1.1 ± 56.9	3.2 ± 57.3
Rudd	6.5 ± 36.0	0.8 ± 48.6
Tench	0.6 ± 67.4	
Total	447.1	225.2

abundance. The large difference in the total biomass can be explained by the lack of an estimate of bream in 1999. Bream were abundant in the catches but no marked specimen was captured.

5.3.5 *Biotic integrity*

The establishment of an Index of Biotic Integrity (IBI) for standing waters in Europe is relatively new. Belpaire, Smolders, Vanden Auweele, Ercken, Breine, Van Thuyne & Ollevier (2000) recently described a method for standing waters in Flanders. This index was used to calculate the integrity class for both sampling periods 1988 and 1999 (Table 5.5). The index showed no major shifts in scoring between the two periods, with the lake exhibiting a reasonable biotic integrity. This suggests that the lake ecosystem is relatively stable, despite a shift in the fish community.

5.4 Discussion

This study found a rich species diversity in Lake Schulen. Important shifts in the fish community structure occurred between 1988 and 1999. New species (of which the topmouth gudgeon and the wels were the most conspicuous) were caught while apparently other species disappeared.

The small Asian topmouth gudgeon was accidentally transported to Europe in the early 1960s with Chinese carps (first recorded in Romania in 1961). Over the last 40 years it has spread all over Europe and it was first recorded, in Flanders in 1992, in the

Table 5.5 The IBI score for each metric for the lake in 1988 and 1999

	1988		1999	
Parameter	Value	Score	Value	Score
Total number of species	16	5	23	5
Mean tolerance value	2.25	5	2.5	5
Total biomass (kg ha^{-1})	447.05	4	295.3*	5
Weight ratio piscivores/ non-piscivores	0.09	3	0.048	2
Type species	2.3	2	2.3	2
Pike recruitment and biomass (kg ha^{-1})	4.7 (recruitment)	3	6.4	3
Tench recruitment and biomass (kg ha^{-1})	0.69 (recruitment)	3		
Weight percentage of non-native species	7.24	2	21.4	1
Final score		3.38		3.29
Integrity class		4		4
Assessment		Reasonable		Reasonable

*Underestimated.

River Demer basin (De Charleroy & Beyens 1998). It is now common in almost all river and lentic systems of Flanders. The presence of large numbers of 0+ specimens as well as adult fish in the lake indicates that the population has naturalised. Its effect on the indigenous fish population is unclear, however, interaction with native species can be expected and should be monitored.

The appearance of chub and stoneloach in 1999 was based on one specimen of each. This was probably due to improvement in water quality in the River Demer over the last few years, and populations of both species now occur in the river (Breine, Van Thuyne, Belpaire, De Charleroy & Beyens 1999); the specimens in the lake probably entered from the river during flood periods.

Although some references show that the wels was an indigenous fish species in Western Europe and even in Flanders (found in archeological sites until the 12th century; Van Neer & Ervynck 1993), no natural populations of this species remain (Bruylants *et al.* 1989; De Nie 1996). In Lake Schulen a number of large wels were illegally stocked in the early 1990s by anglers and eight juvenile specimens (ranging from 8 to 14 cm) were caught in the 1999 survey. The presence of small specimens of wels is unique in Flanders. There is no indication that these small specimens were stocked suggesting this species is breeding in the lake. This is supported by some stretches in zone A corresponding to the habitat needed for the natural recruitment of wels, i.e. shallow with reed fringes and hanging tree roots (Huet 1970; De Nie 1996).

Only one specimen of *Misgurnus fossilis* was found in the lake. The weatherfish is very rare in Flanders and is protected by the Flemish Fisheries Law. *Misgurnus fossilis* was recorded in 1994 in a brook adjacent to the lake and other specimens were found in several locations in the Demer catchment in 1995 (De Charleroy & Beyens 1998) and 1998 (Breine *et al.* 1999). Due to cryptic habitats (often hiding in the mud) the weatherfish is difficult to catch using the methods applied, therefore the number of specimens present in a water body is often underestimated.

Reasons for the shift in community structure between 1988 and 1999 are varied and complex. In 1991 the lake had almost a complete fringe of emergent vegetation with some submerged species (Schoonjans, Belpaire, Podoor & Van Assche 1991). In 1999 barely any aquatic vegetation, needed for spawning of pike, rudd and tench, was present in the lake. This is probably the main reason for the low recruitment of these species in 1999.

Large roach and bream populations can cause a reduction of aquatic vegetation. In search for benthic food, these animals can increase turbidity by stirring up the sediment, which also enhances the nutrient flow in the water column. This brings unfavourable conditions for the development of submerged macrophytes and stimulates algal blooms (Scheffer 1998).

Persson, Diehl, Johansson, Andersson & Hamrin (1991) studied shifts in fish communities along the productivity gradient of 13 temperate lakes. They observed that a high biomass of percids was caused by different species depending on the productivity gradient of the lake. In mesotrophic systems this was due to a high biomass of perch, as opposed to pikeperch in eutrophic systems. In eutrophic systems, a higher abundance of ruffe was also present. A similar shift took place in Lake Schulen, although the pikeperch population did not increase in biomass but only in numbers. Also ruffe abundance increased in 1999 compared with 1988. This success of pikeperch and ruffe has been related to adaptations of these species to feed under poor light conditions prevalent in eutrophic systems (Ali, Ryder & Anctil 1977). Pikeperch may also take advantage of the presence of ruffe as a food source (Löffler 1998).

The decrease of perch is probably due to the competitive superiority of roach and bream in eutrophic, turbid waters, which is related to their efficient capture of zooplankton. Also green algae is an important food resource for roach (Johansson & Persson 1986) making this species more adaptable to eutrophic waters.

Eel is now very abundant in the lake. The main reasons for the expansion of this population are the yearly restockings with glass eels (4 kg glass eel year^{-1}) and the large ruffe population, which is probably a good food source for eel. The increasing food requirement of the present eel population could explain the decline in abundance of bullhead from 7.6% to 1.9%. The active movement pattern of bullhead in the lake could also be a sign of this lack of food.

The IBI for both sampling periods was nearly identical and classified the lake as having a moderate biotic integrity. Karr (1981) defined this as an ecosystem with a species richness somewhat below expectation (due to loss of intolerant forms), some species with less than optimal abundance or size distribution and with a trophic structure showing some signs of stress.

A multi-disciplinary research programme on the lake (water quality, plankton community, etc.) could provide additional data on the lake ecosystem, however, the present study has already provided much information for management of the lake fisheries.

Acknowledgements

This study was possible thanks to funds from the Provincial Fishery Committee, Limburg (Belgium). Many thanks to all people who assisted in the field with the fish stock assessments and to J. Neyrinck and N. Roset for their help in the statistical analysis.

References

Ali M.A., Ryder R.A. & Anctil M. (1977) Photoreceptors and visual pigments as related to behavioral responses and preferred habitat of perches (*Perca* spp.) and pikeperches (*Stizostedion* spp.). *Journal of the Fisheries Research Board of Canada* **34**, 1475–1480.

Belpaire C., Verreycken H., Van Vlasselaer L. & Ollevier F. (1989) Evaluatie van het visbestand van het Schulensmeer. Studierapport in opdracht van de Administratie voor Ruimtelijke Ordening en Leefmilieu, KULeuven, 105 pp. (in Dutch).

Belpaire C., Smolders R., Vanden Auweele I., Ercken D., Breine J., Van Thuyne G. & Ollevier F. (2000) An index of biotic integrity characterizing fish populations and ecological quality of Flandrian water bodies. *Hydrobiologia* **434**, 17–33.

Breine J.J., Van Thuyne G., Belpaire C., De Charleroy D. & Beyens J. (1999) Het visbestand in de Demer anno 1999. IBW, Groenendaal, 43 pp. + Annexes (in Dutch).

Bruylants B., Vandelannoote A. & Verheyen R.F. (1989) De vissen van onze Vlaamse beken en rivieren (Hun ecologie, verspreiding en bescherming), V.Z.W. WEL, D/1989/4092/1, 272 pp. (in Dutch).

De Charleroy D. & Beyens J.B. (1998) Het visbestand in het Demerbekken. Inventarisatie van de vissoorten en hun verspreiding. Instituut voor Bosbouw en Wildbeheer, Groenendaal. Mededelingen 1998 (2), 103 pp. (in Dutch).

De Nie H.W. (1996) Atlas van de Nederlandse zoetwatervissen. Media Publ.-III, Doetinchem, 151 pp. (in Dutch).

EPA (1998) Lake and reservoir bioassessment and biocriteria. Technical Guidance Document: Office of Wetlands, Oceans and watersheds, Office of Science and Technology, Office of Water, U.S. EPA Washington DC, 841-B-98-007, EPA, 118 pp. + Appendices.

Huet M. (1970) *Traité de Pisciculture*. Bruxelles: Ed. De Wyngaert, 718 pp.

Grimm M.P. & Backx J. (1990) The restoration of shallow eutrofic lakes and the role of northen pike, aquatic vegetation and nutrient concentration. *Hydrobiologia* **200/201**, 557–566.

Johansson L. & Persson L. (1986) The fish community of temperate, eutrophic lakes. In B. Riemann & M. Søndergaard (eds) *Carbon Dynamics of Eutrophic, Temperate Lakes: The Structure and Functions of the Pelagic Environment.* Amsterdam: Elsevier, pp. 237–266.

Karr J.R. (1981) Assessment of biotic integrity using fish communities. *Fisheries* **6**, 21–27.

Löffler H. (1998) Pikeperch, *Stizostedion lucioperca*, in Lake Constance (Bodensee-Obersee): an example of a successful introduction? In I.G. Cowx (ed.) *Stocking and Introduction of Fish.* Oxford: Fishing News Books, Blackwell Science, pp. 201–208.

Martin-Smith K.M. (1998) Temporal variation in fish communities from the upper Segama river, Sabah, Malaysian Borneo. *Polska Archiv fur Hydrobiologie* **45**, 185–200.

Perrow M.R., Jowitt A.J.D. & Gonzalez L.Z. (1996) Sampling fish communities in shallow lowland lakes: point sample electric fishing vs electric fishing within stop-nets. *Fisheries Management and Ecology* **3**, 303–313.

Persson L., Diehl L., Johansson L., Andersson G. & Hamrin S.F. (1991) Shifts in fish communities along the productivity gradient of temperate lakes – patterns and the importance of size-structured interactions. *Journal of Fish Biology* **38**, 281–293.

Robson D.S. & Spangler G.R. (1978) Estimation of population abundance and survival. In S.D. Gerkings (ed.) *Ecology of Freshwater Fish Production*. Oxford: Blackwell Scientific Publications, pp. 27–51.
Scheffer M. (1998) *Ecology of Shallow Lakes*. Lelystad: Riza, London: Chapman and Hall, 287 pp.
Schoonjans L., Belpaire C., Podoor N. & Van Assche J. (1991) De aquatische en oevervegetatie van het Schulensmeer, huidige toestand en aanbevelingen tot herstel. Studierapport in opdracht van het Ministerie van de Vlaamse Gemeenschap Administratie Milieu, Natuurbehoud en Landinrichting, KULeuven, IBW. Wb. V.R.91.04.110 pp. (in Dutch).
Van Neer W. & Ervynck A. (1993) Archeologie en vis. Herlevend verleden 1, Instituut voor het Archeologisch Patrimonium, D/1993/6024/1, 96 pp. (in Dutch).

Chapter 6
Ecology of European whitefish, *Coregonus lavaretus*, in two Austrian lakes in relation to fisheries management and lake productivity

J. WANZENBÖCK*, H. GASSNER, Y. HASSAN, B. LAHNSTEINER and G. HAUSEDER
Institute for Limnology, Austrian Academy of Sciences, Mondsee, Austria

Abstract

Whitefish, *Coregonus lavaretus* L., abundance, age structure and life-history parameters were studied in two Austrian, alpine lakes in relation to exploitation levels and total phosphorus concentration. Lake Traunsee is characterised by a four-fold lower nutrient concentration but higher fishing pressure resulting in low fish density as revealed by hydroacoustics and the lack of older age classes (≥3+). Lake Hallstättersee has higher phosphorus concentrations but lower fishing pressure resulting in higher fish densities. Despite the higher productivity of Lake Hallstättersee, growth of whitefish was poorer, but the population included older age classes; the age at maturity was slightly higher and fecundity was lower compared with Lake Traunsee. The results suggest that fish abundance regulated by fishing pressure has greater influence on growth and life-history strategies than lake productivity inferred by total phosphorus concentration.

Keywords: coregonids, fecundity, growth, Hallstättersee, maturity, stock size, Traunsee.

6.1 Introduction

The interplay between life-history strategies, population dynamics and prediction of fishery yield is central to fisheries ecology. Fishery yield is largely determined by two major structuring forces, i.e. system (lake) productivity (=bottom-up force) and exploitation level (top-down force, i.e. mortality induced by predators and parasites). On the one hand lake primary and secondary productivity determines the amount of food resources available to a particular fish community and their populations, thus setting the limit to maximum stock size. On the other hand exploitation levels exerted by all predators and parasites directly influence stock size and population density, and this indirectly affects the level of possible competition for available resources.

The general positive relationship between lake productivity and fish stock size or abundance is well known (e.g. Peters 1986; Plante & Downing 1993; Knösche &

*Correspondence: Institute for Limnology, Austrian Academy of Sciences, Mondseestrasse 9, A–5310 Mondsee, Austria (e-mail: josef.wanzenboeck@oeaw.ac.at).

Barthelmes 1998). Also for oligotrophic lakes in the northern hemisphere inhabited by coregonids, which are often also the main target species of a fisheries, such a positive relationship is well established (Mills 1985; Kirchofer 1995; Müller & Bia 1998). Many of the lakes in the Alps have undergone human-induced eutrophication during past decades followed by recent oligotrophication. In parallel, fishery yields (mainly of coregonids) have increased with eutrophication and decreased with oligotrophication (Eckmann & Rösch 1998; Müller & Bia 1998). In most cases a correlation between lake productivity, stock size and exploitation level was observed, i.e. larger stocks due to increased lake productivity were also fished more intensively and vice versa when lake productivity declined. In two Austrian lakes (Lake Traunsee and Lake Hallstättersee) lake productivity and exploitation levels have become uncoupled in recent years. Lake Traunsee has undergone a pronounced oligotrophication period with an 80% decline in total phosphorus content, and therefore declining productivity, during the past 20 years. However, exploitation levels by the fishery remained relatively high, resulting in a collapse of the fishery during recent years. Lake Hallstättersee has undergone only moderate oligotrophication (50% decline of total phosphorus content) and exploitation levels by the fishery were adjusted resulting in only a marginal decline in yields. Thus the whitefish population of Lake Traunsee might have responded to considerable lower productivity but at the same time to density-dependent competitive release from conspecifics due to overfishing. The population in Hallstättersee might have adapted to higher lake productivity (four times higher than total phosphorus levels) compared to Lake Traunsee, but also to higher levels of possible intraspecific competition due to presumable higher densities (as inferred from higher fishery yields at lower fishing effort – see below).

The objective of this study was to compare the stocks of whitefish, *Coregonus lavaretus* L., of the two lakes to assess if, and in which way, the populations reacted to the different scenarios with respect to stock size and life-history parameters such as growth, age-at-maturity and fecundity. With such a comparison it should be possible to get deeper insight into the interplay of the two main structuring forces (bottom-up and top-down) affecting lake fish populations.

6.2 Study sites

Lakes Traunsee and Hallstättersee are situated in the River Traun drainage basin in the lake district 'Salzkammergut' east of Salzburg, Austria (Fig. 6.1). Both lakes are typical oligotrophic, deep, alpine lakes with rather low retention times due to the relatively large River Traun flushing the lakes (Table 6.1). The fish communities (Table 6.2) of both lakes are dominated by whitefish, which serve as the main commercial target species. In Lake Traunsee, two growth forms of whitefish are present: a normal growing form and a slow-growing, dwarf form which is normally found at depths >40 m. In Lake Hallstättersee only a normal growing form occurs.

Total phosphorus levels, mean values based on measurements every 5 m from the entire water column at the deepest place at the end of the summer stagnation period, have fallen at Lake Traunsee from a peak in 1979 of 14 to 3 $\mu g\,L^{-1}$ during the past

Figure 6.1 Location of Lake Hallstättersee and Lake Traunsee within the Salzkammergut lake district in Austria

Table 6.1 Limnological characteristics of Lake Hallstättersee and Lake Traunsee

	Hallstättersee	Traunsee
Metres above sea level	508	422
Area (ha)	858	2560
Maximum depth (m)	125	191
Mean depth (m)	65	90
Volume (million m^3)	557	2302
Theoretical water renewal time (year)	0.5	1.0
Mean total phosphorus content ($mg\,m^{-3}$)	13	4
Trophic classification	Oligo-mesotrophic	Oligotrophic

4 years. In Lake Hallstättersee total phosphorus declined from a peak in 1994 of 22 to 11 $\mu g\,L^{-1}$ during recent years (Schwarz & Jagsch 1998).

6.2.1 *Exploitation levels by the fisheries*

Comparison of the exploitation levels of the two lakes revealed that whitefish populations were exclusively targeted by commercial fishermen using gillnets (only in Lake Hallstättersee were whitefish targeted by anglers in 1999). In Lake Hallstättersee the fishery is run by the federal forest agency employing one fisherman who operates three nets a day during the fishing season. The nets are 70 m long and 9 m high (1892 m^2) and with a mesh size of 38 mm (knot to knot). The exploitation level by commercial

Table 6.2 Fish species of Lake Hallstättersee and Lake Traunsee

Group	Species	Common name	Traunsee	Hallstättersee
Coregonidae	*Coregonus lavaretus*	Whitefish	X	X
Salmonidae	*Salvelinus alpinus*	Arctic char	X	X
	Salmo trutta morpha lacustris	European lake trout	X	X
	Oncorhynchus mykiss	Rainbow trout	X	
Gadidae	*Lota lota*	Burbot	X	X
Percidae	*Perca fluviatilis*	Perch	X	X
	Stizostedion lucioperca	Pikeperch	X	
Cyprinidae	*Rutilus rutilus*	Roach	X	X
	Abramis brama	Common bream	X	
	Leuciscus cephalus	Chub	X	X
	Phoxinus phoxinus	Minnow	X	X
	Rutilus frisii meidingeri	Pearlfish	X	
	Chalcalburnus chalcoides	Danubian bleak	X	
	Tinca tinca	Tench	X	
	Cyprinus carpio	Carp	X	
Esocidae	*Esox lucius*	Pike	X	X
Cobitidae	*Barbatula barbatula*	Stoneloach	X	
Cottidae	*Cottus gobio*	Bullhead	X	X
Anguillidae	*Anguilla anguilla*	Eel	X	
Gobiidae	*Proterorhinus marmoratus*	Tubenose goby	X	

fishermen in Lake Traunsee is not known exactly. From the 52 professional fishing rights approximately 15 have been used during recent decades. For coregonids each fisherman is allowed to operate eight nets; four nets of the size 50 × 5 m with 34 mm mesh size which target the normal growing form, and four nets of the size 50 × 4.5 m with 22 mm mesh size targeting the dwarf form. The 34 mm nets can be used everywhere in the lake, but the 22 mm nets have to be set below 8 m water depth. This amounts to 120 nets set in Lake Traunsee every day during the fishing season. Taking into account the differences in lake size, in Lake Hallstättersee there are 2.9 m^2 of netting per ha lake area compared with 11.6 m^2 ha^{-1} in Lake Traunsee. The fishing effort was therefore approximately four times higher at Lake Traunsee and taking into consideration the differences in mesh sizes, this fishing pressure is even higher.

Catch statistics were considered to be reliable for Lake Hallstättersee, and yields were stable over the past 20 years. A slight peak was observed in 1988 (4 years after the peak in total phosphorus) at 12 kg ha^{-1} and since then the annual yield declined slowly to about 9 kg ha^{-1} in 1995 (see Wanzenböck & Jagsch 1988). There is no stocking in Lake Hallstättersee and reproduction is completely natural. Unfortunately there are no long-term and reliable catch statistics available for Lake Traunsee. However, the sporadic notes on fishery yields suggest that catches were high in the early 1980s

(10–14 kg ha^{-1}) but have declined since then until a collapse was reported by the fishermen in recent years despite regular stocking with larvae and fingerlings.

6.3 Methods

6.3.1 *Hydroacoustics*

Echosounding surveys were carried out during the nights starting on 2 April, 16 July and 26 November 1998, 6 April, 11 May, 5 June and 16 September 1999 on Lake Hallstättersee and during the nights starting on 25 March, 22 May, 16 July, 25 September, 28 December 1998 and 18 March, 6 May, 24 June and 26 August 1999 on Lake Traunsee. Night sampling is important because fish schools disperse, leading to a more uniform distribution. A Simrad EY-500 digital split-beam echosounder with a working frequency of 120 kHz was used for the surveys. Nominal beam angle of the elliptical transducer was 4° and 10° at the 3 dB level. Pulse duration was 0.1 ms and ping rate was between 0.3 and 0.7 s ping^{-1}, depending on water depth (maximum 191 m). The system was calibrated with a standard copper sphere. The transducer was mounted on a pole on the side of the boat at a depth of 0.4 m; boat speed was maintained at 6–8 km h^{-1}. To keep track of the transects during the night a differentially-corrected GPS system (Trimble Pathfinder Pro XR, real-time correction via the Austrian Radio network) was used. Hydroacoustic surveys produced echograms for each of the 17 and 11 transects (Traunsee and Hallstättersee, respectively) set perpendicular to the long axis of the lakes. The abundance estimates were carried out using 20 log R TVG (times varied gain) amplification and the sound level threshold was set to −55 dB to omit noise and other unwanted small echoes. Based on total area backscattering strength (total sa) and the dB-distribution of single trace echoes and their sum (trace sa), the software (Simrad EP-500) calculated the number of fish per area and per volume (with reference to the form of the conical sound field of the transducer). Processing of data was limited to depths >4 m due to physical constraints of the echosounder. The split-beam system allows estimates of *in situ* target strength which can be related to fish size and subsequently to fish biomass (MacLennan & Simmonds 1992) using the following procedure. First the proportion of multiple echoes (e.g. from fish schools) was minimised by surveying the lake during night because they distribute evenly during the night (Ptak & Appenzeller 1998). A high proportion of single targets is important because the statistical precision of the measured target strength distribution increases with the number of single targets. The measured target strength signals (dB) were converted to fish total length (cm) using the empirical relation between these variables developed by Love (1971), for data obtained by a 120 kHz system. The resulting size distribution of fish was similar to the size distribution of fish from net catches. The biomass (wet weight in g) was calculated from total length (cm) using empirical length–weight regressions for each lake. Biomass for each transect was then calculated by multiplying fish weight by the number of fish in each 1 dB class. For further details of fish biomass calculations see Gassner & Wanzenböck (1999).

6.3.2 *Catches*

Gillnetting and purse seining were used to verify the information from the echosounder and to collect material for the analysis of age structure and life-history parameters. Gillnetting was carried out using monofilament nets (most of them 50 m long, 3 m high) each with a different mesh size (15, 26, 32, 38, 42, 45, 50 or 60 mm knot to knot). These eight nets were combined to form a net series (1050 m^2). Two such net series were set in the pelagic zone of Lake Traunsee (5 and 10 m depth) for 5 days in October 1999. Nets were checked for fishes twice a day, in the morning and in the evening. One similar net series was set in the pelagic zone of Lake Hallstättersee for one night in October 1999.

The purse seine used was 125 m long, 12 m high and had a minimum mesh size of 6 mm. Several shoots were performed on both lakes between July and December during nights in 1998 and 1999.

All fish captured were measured (total length, nearest 0.5 cm) and weighed (total fresh weight, nearest 1 g). Age was determined from scales taken from the shoulder below the dorsal fin. Maturity and ripening of oocytes was assessed according to Nikolski (in Lagler 1978). Fecundity was determined gravimetrically by counting eggs in weighed subsamples of ripe gonads. Egg diameter was measured from at least 50 unswollen eggs per ripe female using a dissecting microscope.

6.4 Results

The hydroacoustic surveys revealed that stock size was considerably larger in Hallstättersee than Traunsee. However, when abundances and biomasses were analysed with respect to fish length (Fig. 6.2), the differences were more pronounced at catchable sizes of whitefish (>20 cm) compared to smaller, juvenile whitefish (<20 cm). There were approximately four times more fish of catchable sizes in Hallstättersee (158 fish ha^{-1}) compared with Traunsee (40 fish ha^{-1}). When the whole fish stock was estimated hydroacoustically and expressed as average biomass from all surveys the ratio was approximately 1 : 3.7 in Traunsee and Hallstättersee, respectively (i.e. 17.8 kg $ha^{-1} \pm 1.89$ SE, $n = 9$, *v.* 66.4 kg $ha^{-1} \pm 4.47$ SE, $n = 6$). The pronounced differences in stock size from the hydroacoustic surveys were corroborated by CPUE values in the gillnetting campaign. When standardised to total net area and duration of exposure, 329 m^2 of netting had to be set in Lake Traunsee for 24 h to catch one whitefish of any size, compared with only 9 m^2 for Lake Hallstättersee. This converts roughly to a ratio of 1 : 30.

The age and size structure of the whitefish populations (Fig. 6.3) showed that few older (3+ and 4+) and larger fishes were caught at Lake Traunsee despite a much higher (six times) fishing effort. The growth pattern revealed that the normal growing form of whitefish in Lake Traunsee had a slightly larger mean total length in all age groups than fishes in Lake Hallstättersee. In the 1+ and 2+ age groups the differences in mean total length were statistically significant (Mann–Whitney *U*-test, $P < 0.001$). The dwarf form of whitefish in Lake Traunsee grew (by definition) at a much slower rate.

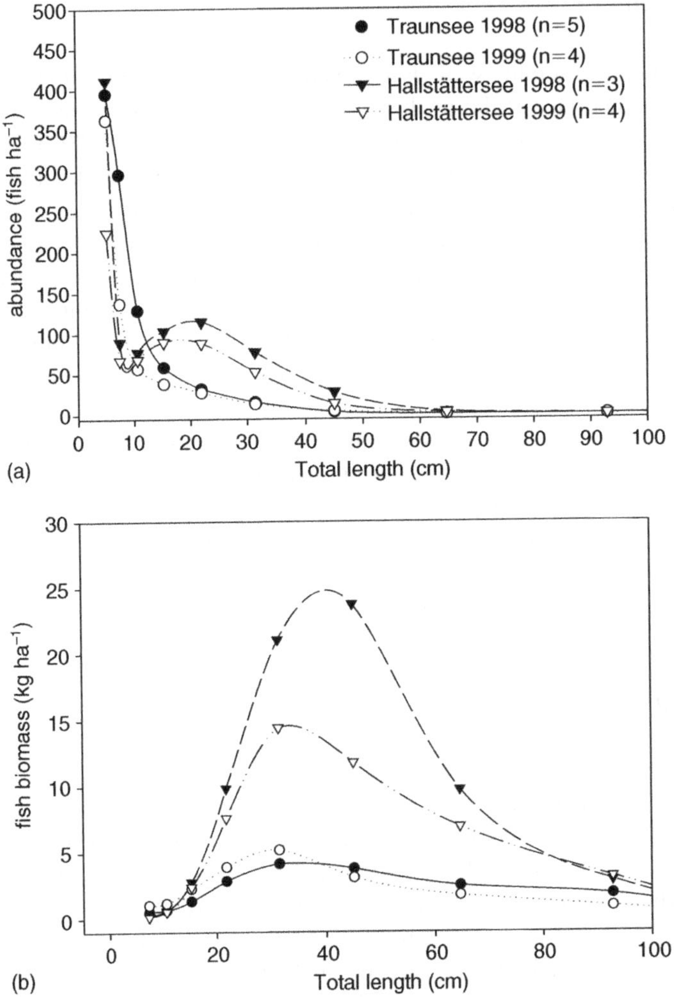

Figure 6.2 Fish abundance (a) and fish biomass (b) *v.* total length according to hydroacoustic observations in two different years

Maturity was reached by most males in autumn of their second year of life (1+): percentages of mature males were 58% at Lake Hallstättersee and 78% at Lake Traunsee. In older males the percentage mature was 90% and 98%, respectively. In females the percentage of mature fish in age group 1+ was 8% and 22% for Hallstättersee and Traunsee, respectively. The age when most females reached sexual maturity was 2+, comprising 90% and 75% for this age group for Hallstättersee and Traunsee, respectively. In older females, maturity was >90% in both lakes.

The interrelationship between total length, age and maturity, together with the mean length of whitefish caught with commercially used gillnets (as depicted in Fig. 6.4), gives further insight into the effects of prevailing management practices. In Hallstättersee, all fish >30 cm (regardless of their age) were mature. Using a mesh size of 38 mm

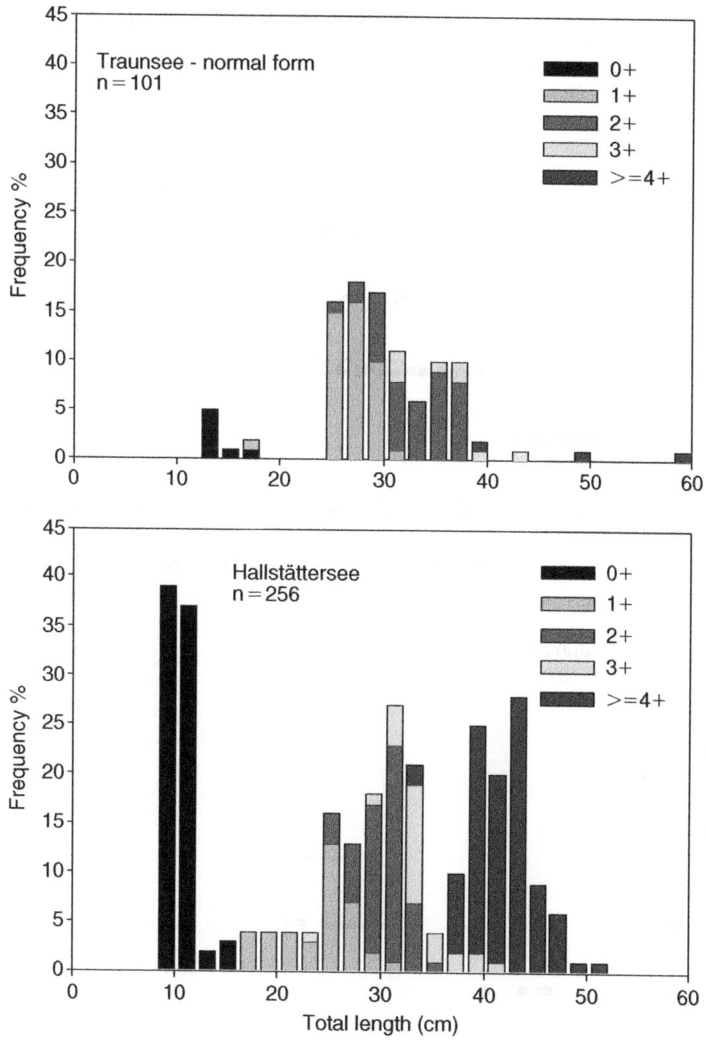

Figure 6.3 Age and size structure of the whitefish caught with gillnets and purse seines (note that fishing effort at Traunsee was much higher than at Hallstättersee, see text)

ensures that only mature fish are exploited. On Lake Traunsee, despite their better growth, a proportion of fish >30 cm were still immature. With the mesh sizes of gillnets used there, young and immature fish are threatened, especially by nets with 22 mm mesh sizes which should target the dwarf form.

Whitefish in Lake Traunsee had significantly higher fecundity over the whole range of fish sizes than at Lake Hallstättersee (*t*-test of fecundity per 100 g body weight, $P < 0.001$, Fig. 6.5). The gonadosomatic indices of females were not significantly different (t-test, $P = 0.602$) and therefore the differences in fecundity must be correlated to different egg sizes. Whitefish females in Lake Traunsee indeed produced eggs with

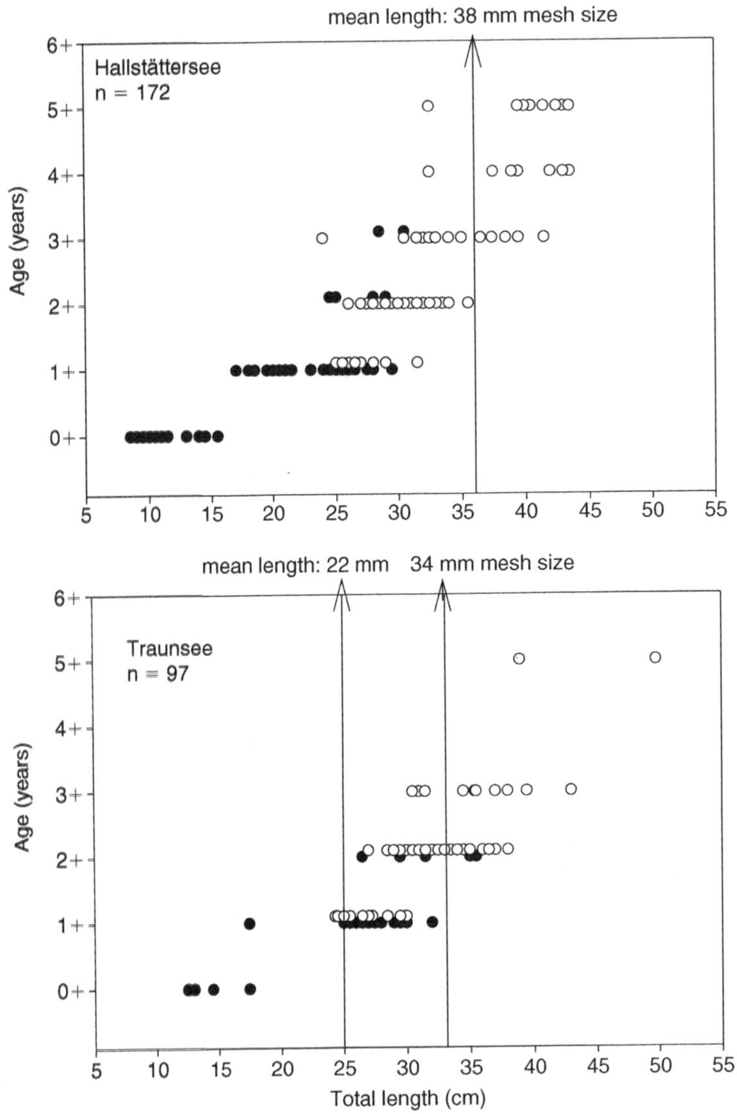

Figure 6.4 Sexually mature fish (○) and immature fish (●) in relation to age and length. The vertical arrows depict the mean length of fish caught with commercial gillnets

significantly lower diameters (*t*-test, $P < 0.001$) than females in Lake Hallstättersee (Fig. 6.6) irrespective of fish size.

6.5 Discussion

Despite the lower phosphorus level of Lake Traunsee, growth of whitefish was higher than Lake Hallstättersee. This contradicts the studies of Kirchhofer (1995) and Müller

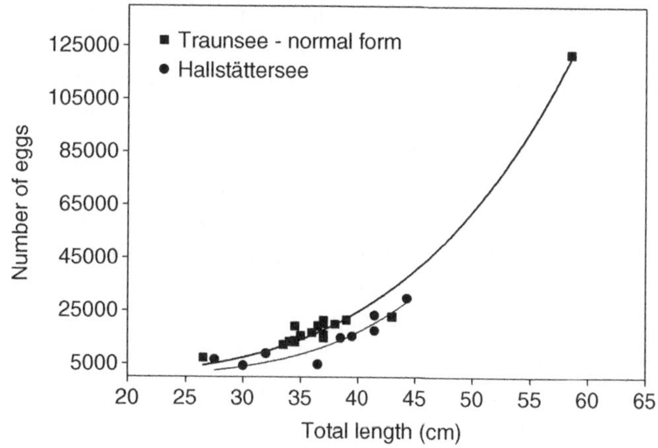

Figure 6.5 Fecundity *v.* length of whitefish from Lake Traunsee (normal growing form only) and from Lake Hallstättersee

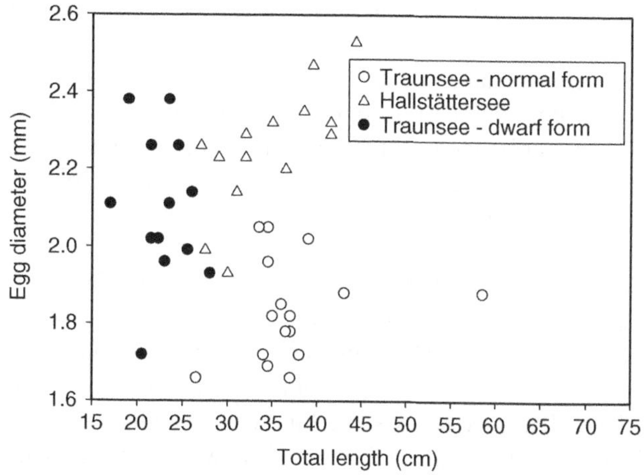

Figure 6.6 Egg diameter *v.* female length for whitefish from Lake Traunsee (normal growing and slow-growing form) and from Lake Hallstättersee

& Bia (1998) who found a direct correlation between growth of whitefish and total phosphorus level. However, this correlation was found only for some year classes and some growth forms studied by Kirchhofer (1995). He found increasing growth in slow-growing forms of Lake Biel (Switzerland) despite a decline in total phosphorus levels. Besides phosphorus concentrations he tried to relate the differences in growth to trends in lake temperature. However, temperature trends were only available for 11–13 years and growth patterns differed between growth forms and year classes even in the same lake. This was in accordance to data from lake fertilisation experiments (Mills 1985; Mills & Chalanchuk 1987) in which no correlation between growth of age 0 fish and fertilisation could be established, whereas a positive correlation was found for older fish.

Irrespective, in none of these studies were data on actual population density available. Nevertheless, Mills, Chalanchuk, Allan & Mohr (1995) observed increased growth and recruitment in experimental lakes after removal of 26% and 40% of the adult fish. This increase in growth could be observed for all age groups at least in some lakes. Bigger sizes of all age classes were also apparent in this study.

The much higher abundance of whitefish in Lake Hallstättersee, in combination with their lower growth, suggests that intraspecific competition for food resources was the key factor. It was assumed that overfishing, especially of larger sizes, in Lake Traunsee might have led to release from intraspecific competition and to higher availability of food to the remaining individuals, despite lower productivity of the lake. A parallel study on zooplankton (C. Jersabek unpublished data) revealed a higher biomass and larger average size of crustaceans in Lake Traunsee compared to Lake Hallstättersee, corroborating this assumption. Overfishing of the Traunsee stock is apparent given the relatively low abundance of larger fish as seen from hydroacoustic data and the lack of fish ⩾3+ in gillnet catches. Overfishing seems to be due to 22-mm nets allowed in water below 8 m deep. This is not enough to protect juvenile individuals of the normal growing form. At depths <40 m only whitefish of the normal growing form were caught in the present survey. The dwarf form was only found below 40 m in this study.

The differences between lakes in the percentage proportional contribution of 0+ whitefish in the total population (Fig. 6.3) were considered to be not representative because gillnets are not effective in catching small fish (Lagler 1978). Most of the 0+ fish were caught by purse seine at Lake Hallstättersee. However, because purse seines sample rather small areas and catches show high variability, it was not considered appropriate to interpret these differences.

The ratio of larger fishes which were catchable in commercial nets (>20 cm), was higher (i.e. 1 : 4 according to hydroacoustic data and 1 : 30 according to CPUE values of gillnets for lakes Traunsee and Hallstättersee, respectively) than the ratio of smaller and younger fish (3–20 cm, i.e.1 : 1.26 according to hydroacoustics). This pattern suggests that in the Lake Traunsee stock some kind of mechanism was compensating, at least partially, for the low abundance of potential spawners. Indeed, a higher fecundity was observed in the Lake Traunsee stock leading to relatively higher larval densities than expected from the abundance of large fish in the lake. The densities of whitefish larvae showed a ratio of 1 : 2 in Lake Traunsee and Lake Hallstättersee, respectively (B. Lahnsteiner, unpublished data).

In conclusion, this study suggests that fish abundance determined by exploitation (top-down force) may be a stronger structuring force for whitefish populations with regard to growth and other life-history parameters than lake phosphorus concentration determining productivity (bottom-up force).

Acknowledgements

The study was financed by the local government of Upper Austria (Project: The influence of SOLVAY-effluents on the ecological integrity of Lake Traunsee). Additional funds came from the project 'Fish communities and the ecological integrity of standing waters' which was financed by the Austrian Federal Ministry of Science and Transport, the Austrian Federal Ministry of Agriculture and Forestry, the local government of Upper Austria, the Austrian Federal Forestry, the consortium Lake Irrsee and the Fishery Union of Salzburg. Many thanks to K. Maier, G. Tischler and many others for their help with extensive fieldwork.

References

Eckmann R. & Rösch R. (1998) Lake Constance fisheries and fish ecology. *Archiv für Hydrobiologie* **53**, 285–301.

Gassner H. & Wanzenböck J. (1999) Hydroakustische Fischbestandserhebungen in vier Salzkammergutseen. *Österreichs Fischerei* **52**, 122–128 (in German).

Kirchofer A. (1995) Growth characteristics of coregonid populations in three lakes with different trophic states and decreasing nutrient concentrations. *Archiv für Hydrobiologie* **46**, 61–70.

Knösche R. & Barthelmes D. (1998) A new approach to estimate lake fisheries yield from limnological basic parameters and first results. *Limnologica* **28**, 133–144.

Lagler K. (1978) Capture, sampling and examination of fishes. In T.B. Bagenal (ed.) *Methods for Assessment of Fish Production in Fresh Water.* IPB Handbook No. 3. Oxford: Blackwell Scientific Publications, pp. 7–47.

Love R.G. (1971) Dorsal aspect target strength of an individual fish at any aspect. *Journal of the Acoustical Society of America* **49**, 816–823.

MacLennan D.N. & Simmonds E.J. (1992) *Fisheries Acoustics.* London: Chapman & Hall, 325 pp.

Mills K.H. (1985) Responses of lake whitefish (*Coregonus clupeaformis*) to fertilization of Lake 226, the experimental lakes area. *Canadian Journal of Fisheries and Aquatic Sciences* **42**, 129–138.

Mills K.H. & Chalanchuk S.M. (1987) Population dynamics of lake whitefish (*Coregonus clupeaformis*) during and after the fertilization of lake 226, the experimental lakes area. *Canadian Journal of Fisheries and Aquatic Sciences* **44**, 55–63.

Mills K.H., Chalanchuk S.M., Allan D.J. & Mohr L.C. (1995) Responses of lake whitefish (*Coregonus clupeaformis*) to exploitation at the Experimental Lakes Area, northwestern Ontario. *Archiv für Hydrobiologie* **46**, 361–368.

Müller R. & Bia M.M. (1998) Adaptive management of whitefish stocks in lakes undergoing re-oligotrophication: the Lake Lucerne example. *Archiv für Hydrobiologie* **50**, 391–399.

Peters R.H. (1986) The role of prediction in limnology. *Limnology and Oceanography* **31**, 1143–1159.

Plante C. & Downing J.A. (1993) Relationship of salmonine production to lake trophic status and temperature. *Canadian Journal of Fisheries and Aquatic Sciences* **50**, 1324–1328.

Ptak J.K. & Appenzeller A.R. (1998) Size, depth and frequency of pelagic Lake Constance whitefish (*Coregonus lavaretus* L.) shoals during the seasons: a hydroacoustic study. *Archiv für Hydrobiologie* **50**, 237–248.

Schwarz K. & Jagsch A. (1998) Gewässerschutzberichte. Lakes of Upper Austria Federal Agency for Water Management, Institute for Freshwater Ecology, Fisheries Biology and Lake Research. Report No. 20 on CD-ROM.

Wanzenböck J. & Jagsch A. (1998) Comparison of larval whitefish densities in lakes with different schemes of larval stocking and fishing practice. *Archiv für Hydrobiologie* **50**, 497–505.

Chapter 7
Fisheries of Lake Victoria: an underwater perspective

R. TUMWEBAZE*
Department of Fisheries, Entebbe, Uganda

A. GETABU
KMFRI, PO Box 1881, Kisumu, Kenya

J. BAYONA
TAFIRI, PO Box 475, Mwanza, Tanzania

D. MACLENNAN
UNECIA Ltd, Lake Victoria Fisheries Research Project, Jinja, Uganda

I.G. COWX
University of Hull International Fisheries Institute, Hull, UK

Abstract

Considerable data exist about the ecology and exploitation patterns of the commercially-exploited fishes of Lake Victoria. However, to date it has not been possible to assess the potential standing stock of these species in the lake. This paper gives the first acoustic abundance estimates of fish stocks of Lake Victoria over two seasons and assesses the variation in distribution in relation to limnological features such as thermal stratification. The pelagic community was found to be dominated by *Rastrineobola argentea* (Pellegrin) and haplochromines, while the semi-pelagic/demersal community was dominated by Nile perch, *Lates niloticus* (L.). The pelagic distribution of fish stocks when the lake is thermally stratified favours use of hydroacoustics but makes bottom trawling ineffective.

Keywords: hydroacoustics, *Lates niloticus*, *Rastrineobola argentea*, stock assessment.

7.1 Introduction

Only two lake-wide fish stock assessment surveys have been undertaken on Lake Victoria. The first was conducted by Graham (1929) between 1927 and 1928; the second by FAO/UNDP in collaboration with the East African Freshwater Fisheries Research Organisation (EAFRO) between 1969 and 1971 (Kudhongania & Cordone 1974). At the time of the first survey, the most important commercial fishery was *Oreochromis esculentus* (Graham) followed by *Oreochromis variabilis* (Boulenger) exploited by gillnetting. The primary target of the survey was to protect the stocks of *O. esculentus* by establishing the impact of different fishing gears and methods on

*Correspondence: Department of Fisheries, Entebbe, Uganda.

commercially-important species. Around the late 1960s the fishery had shifted to smaller species, i.e. haplochromines, and the method used was bottom trawling. The management strategy on Lake Victoria shifted to establishing the magnitude of the stocks of haplochromines and finding ways of utilising them.

Current commercial catches from the lake indicate that Nile perch, *Lates niloticus* (L.), Nile tilapia, *Oreochromis niloticus* (L.), and *Rastrineobola argentea* (Pellegrin) are the most abundant fish taxa. At the time of the last lake-wide survey (1969–1970), these fish were distributed at the bottom, which made stock assessment using bottom trawling justifiable. Since then, there have been changes in the productivity cycle of the lake (Ochumba & Kibaara 1989; Goldschmidt, Witte & Wanink 1993; Hecky 1993; Gophen 1995). These changes have led to a reduction in the dissolved oxygen content of the bottom waters when the lake is thermally stratified, increasing the prominence of fish in the pelagic zone. Bottom trawling alone may therefore not be the most suitable method to assess the magnitude of the stocks in the lake. Hydroacoustics, in conjuction with pelagic trawling, was selected to overcome this problem. The hydroacoustic surveys aim to generate knowledge to aid rational exploitation and management of the stocks and identify optimal fishing effort.

Acoustic methods have the advantage of rapidly generating data. A large volume of water can be surveyed in a shorter time compared to other sampling methods, such as trawling (Mathisen, Tumble, Lemberg & Johnson 1983), and coverage of big areas increased reliability of abundance estimates (Hansson 1993).

Since the first measurement of fish abundance by acoustics in the 1970s (MacLennan & Simmonds 1992), many species have been monitored by this method (Nakken & Ulltang 1983; Thorne 1983; Porteiro, Carrera & Miquel 1996; Bailey, Maravelias & Simmonds 1998). However, there are limitations with acoustic methods. The best conditions required are: single species of fish of uniform density and stable behaviour; and calm weather (Midttun & Nakken 1977). These conditions are rarely met. Acoustic methods require expensive equipment and skilled operators. They also rely on other methods such as trawling for information on organisms responsible for the echo traces.

Fortunately, of the three commercial species *L. niloticus*, *O. niloticus* and *R. argentea* in Lake Victoria, *O. niloticus* is predominantly an inshore species and can be spatially separated from the other two species over much of the lake. *Rastreneobola argentea*, is a pelagic schooling species and can be spatially separated from *L. niloticus*. This simplifies the system and makes it more suitable for use of acoustics.

This chapter gives the first acoustic abundance estimates of fish stocks of Lake Victoria over two seasons and assesses the variation in distribution in relation to limnological features such as thermal stratification.

7.2 Materials and methods

7.2.1 *Hydroacoustic data*

Acoustic and fish data were collected during August 1999 and February 2000 along pre-designed cruise tracks (Fig. 7.1) at a speed of 9–10 knots, using a 120-kHz EY500

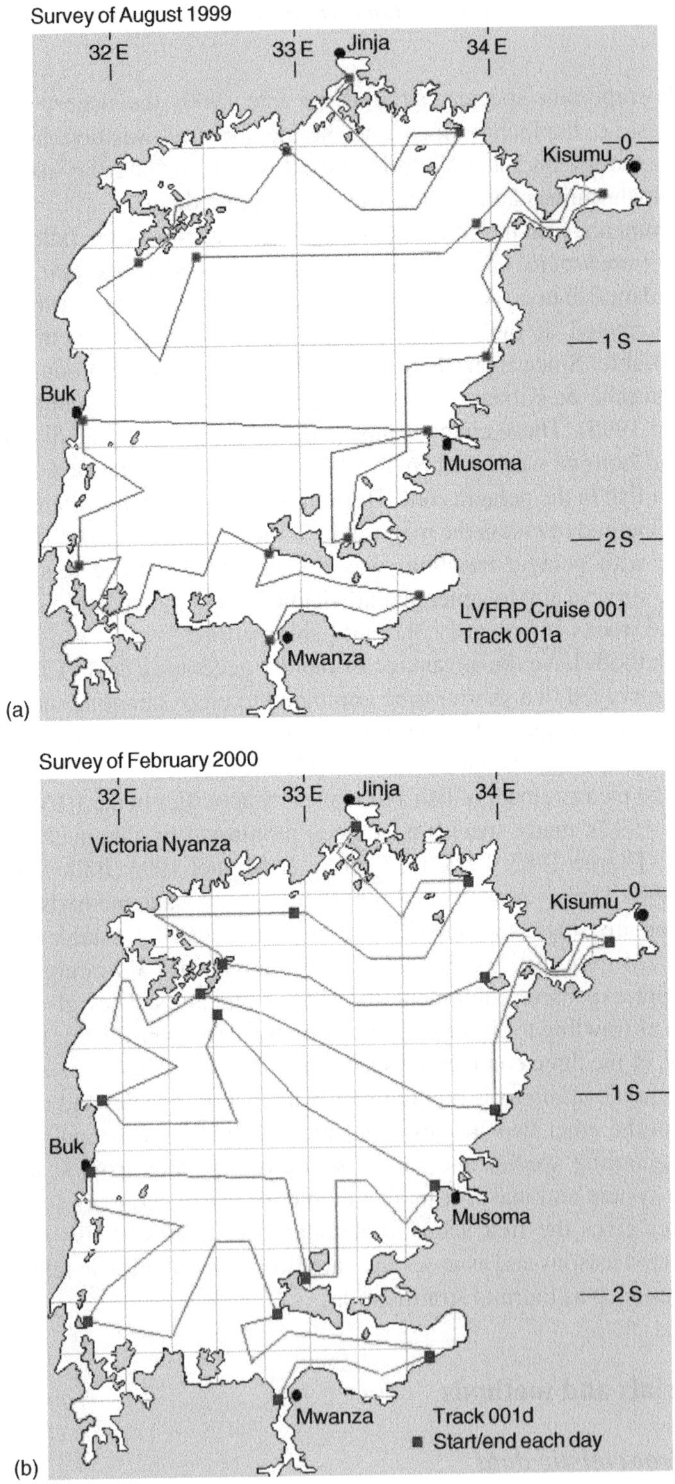

Figure 7.1 Maps showing a pre-planned cruise track for the lake-wide hydroacoustic survey in (a) August 1999 and (b) February 2000

split-beam echosounder of 9° beamwidth. The transducer was hull-mounted at a depth of 1 m and the echo recording started at 5 m below the surface down to 0.2 m above the lake bottom. The unit was calibrated using a 23 mm copper sphere of −40.4 dB before and after the surveys.

Acoustic data were collected mostly during the day but some recording was done in the early hours of the night. A paper echogram indicating the position of echo traces was printed continuously. Integration tables for 10-m depth layers of acoustic data (*Sa* values) were printed every 6 min. The vessel's position at the time the acoustic data were printed was obtained using a global position system (GPS) and recorded in a master log. Details from the master log and integrator tables were transferred to an Excel spreadsheet. At the end of each day, digital echograms were recorded on 600 MB compact disks (CDs) for later analysis.

7.2.2 *Fishing*

Sampling of echo traces previously observed on the echogram was made using a 3.5 m × 3.5 m frame trawl. The cod end was lined with mosquito netting to ensure retention of the smallest organisms. Depth deployment of the frame trawl was monitored using a Furuno CR 24 netsonde connected to a shipboard monitor and a net recorder attached to the frame trawl. Positioning of the frame trawl to the required depth was done by adjusting the boat speed or warp length. Fishing depth ranged from 7 to 40 m in waters between 17 and 70 m deep. Fishing duration was between 10 and 30 min at a speed of about 3 knots. Ship position at the beginning and end of each haul was obtained from a GPS and recorded.

7.3 Results

7.3.1 *General remarks*

The first lake-wide hydroacoustic survey was carried out during the time the lake was not thermally stratified, and fish were distributed down to the bottom (Fig. 7.2(a)). The second survey was carried out when the lake was stratified and fish were distributed off the bottom (Fig. 7.2(b)). For both surveys, distribution of fish in the shallow waters was down to the bottom (Fig. 7.2(c)). All surveys were carried out during the day because the small pelagics were too scattered at night to make any meaningful interpretation.

7.3.2 *Acoustic densities with depth*

The total area backscattering coefficient (*Sa* values), which represents the combined echoes returned by all the insonified targets between the surface and the lakebed were used. Acoustic densities were correlated with the bottom depth (Fig. 7.3). Both surveys show a similar trend with highest densities in waters below 40 m deep.

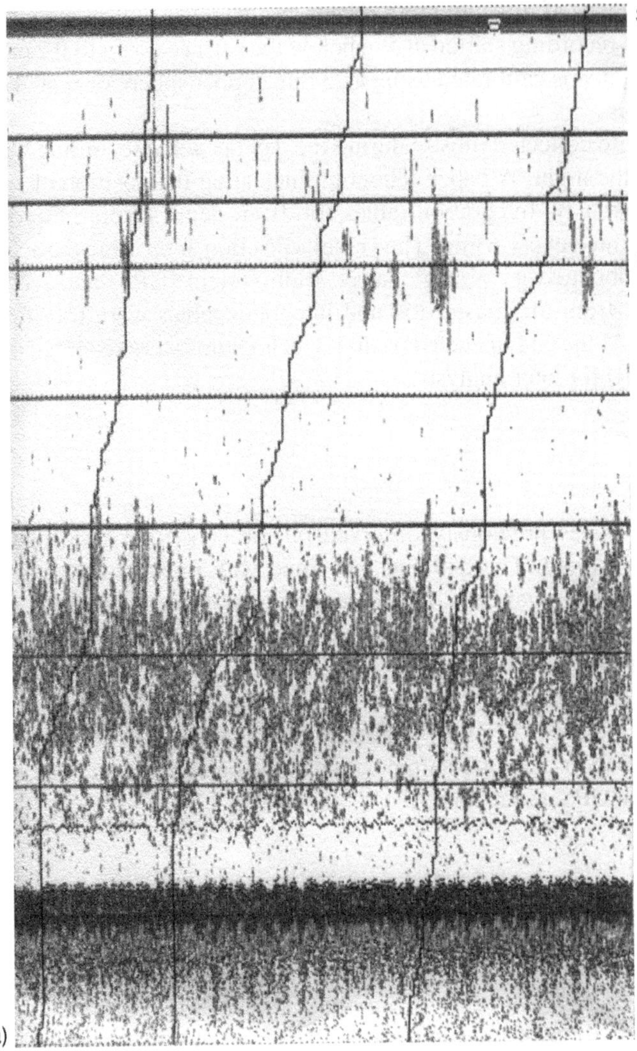

Figure 7.2 Distribution of fish traces in the water column: (a) survey of August 1999 in deep water; (b) survey of February 2000 in deep water; and (c) example of distribution in shallow waters in both surveys

7.3.3 *Geographical distribution of acoustic densities*

Geographical distribution of acoustic densities was broken down into six regions (Fig. 7.4). Region six had the lowest densities for both surveys (Fig. 7.5).

7.3.4. *Size structure*

The size structures for *R. argentea* and haplochromines differed for each region (Fig. 7.6). In particular Region 4 appears to have a greater proportion of small *R. argentea* than

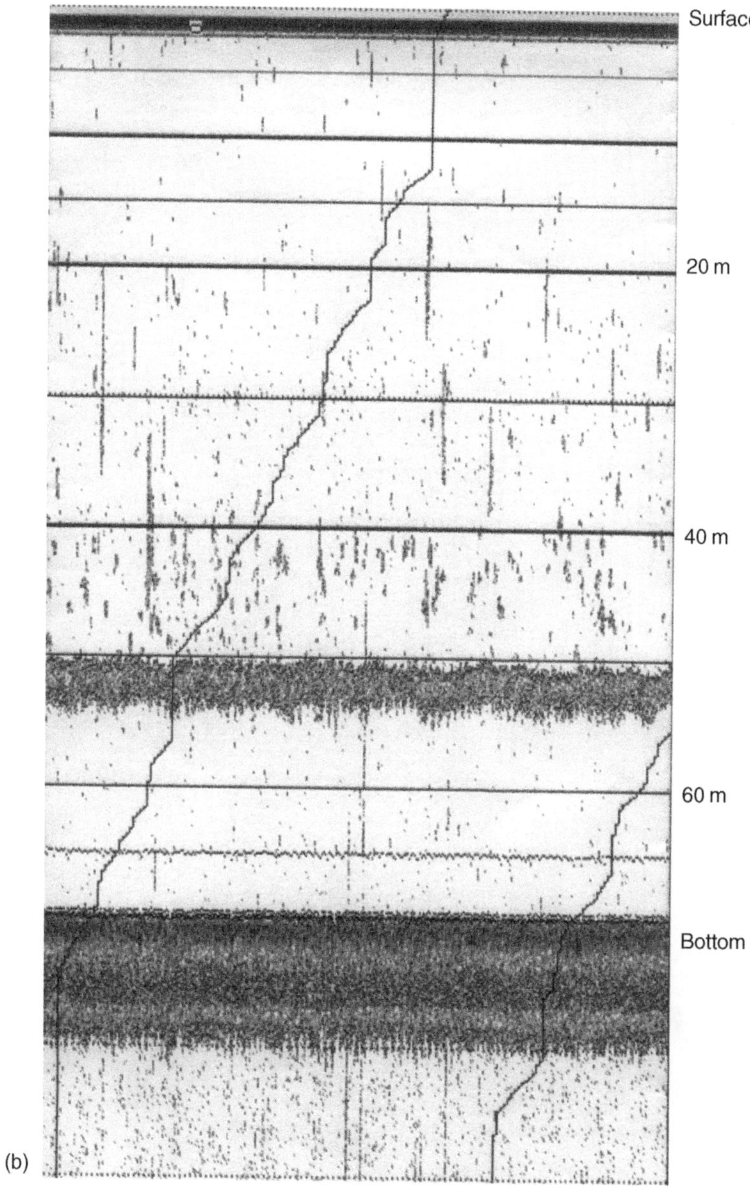

Figure 7.2 Continued

other regions in both surveys (Fig. 7.6(a)). Haplochromines exhibited a wide range of sizes especially in Regions 1, 2 and 4, in both surveys and Region 6 in the August survey. These also certainly represent different species but no identification has been forthcoming to date. Only larger fish were caught in Region 5 and few fish were caught in Regions 5 and 6 in February.

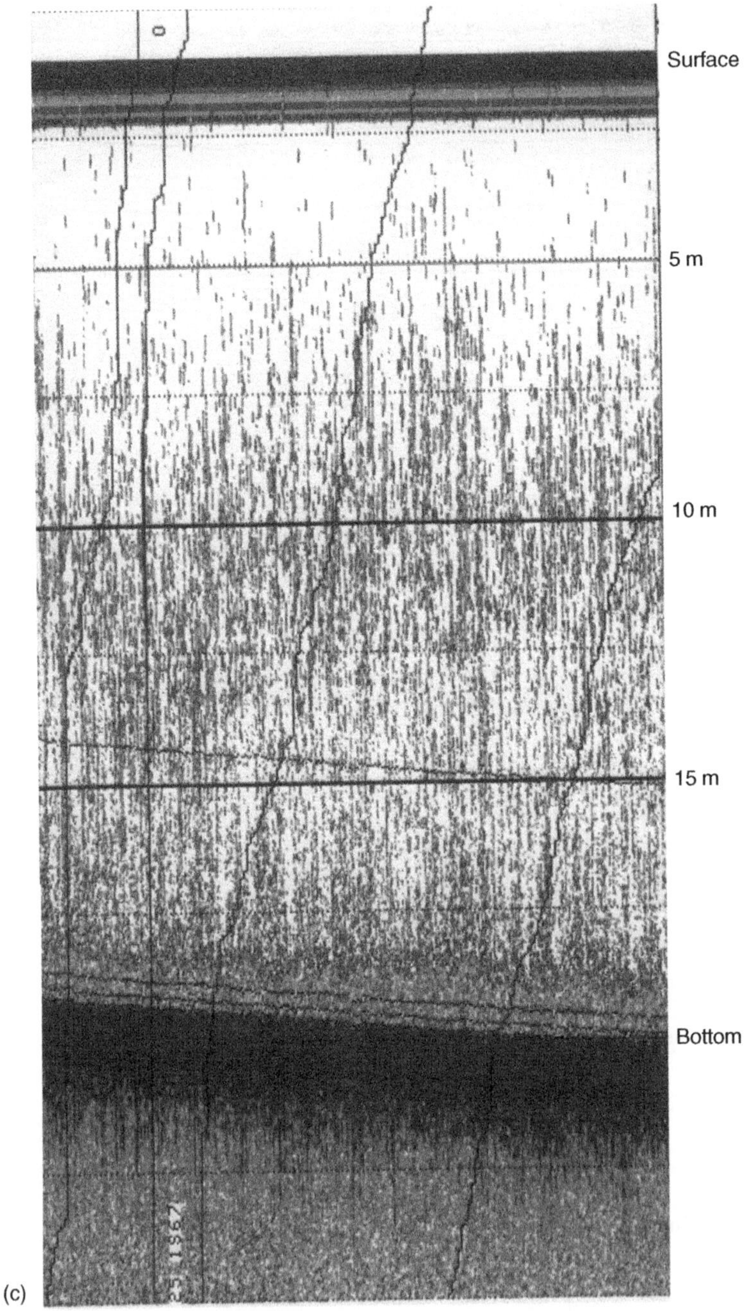

Figure 7.2 Continued

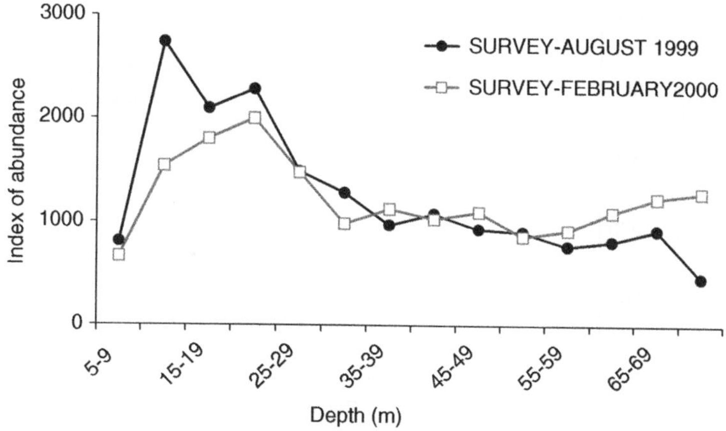

Figure 7.3 Variation in the total area backscattering coefficient with bottom depth (m). Each value is a mean of all measurements over a 5 m depth increment

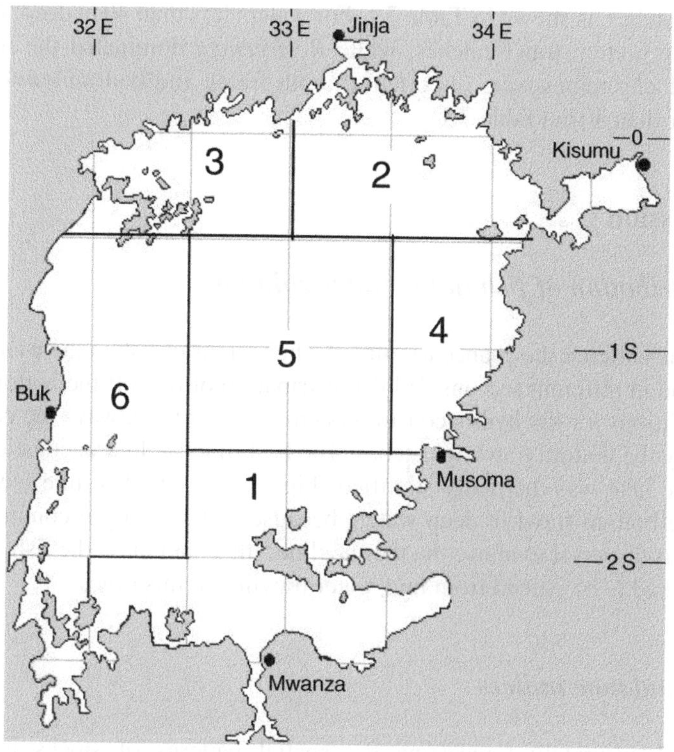

Figure 7.4 Map showing the breakdown of Lake Victoria into the six regions used for analysis

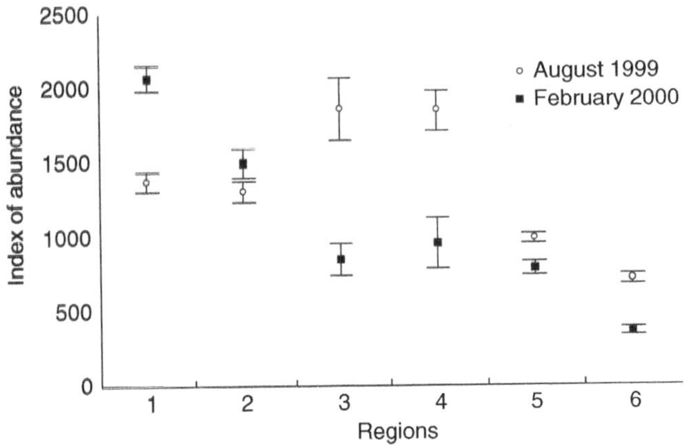

Figure 7.5 Geographical distribution of acoustic densities relating to regions shown in Fig. 7.4

7.3.5 *Size composition*

Species composition for the hydroacoustic survey of February 2000, for bottom and frame trawl catches, is shown in Table 7.1. For waters less than 40 m deep, *L. niloticus* dominated the bottom trawl catches, while *R. argentea* dominated the frame trawl catches. Haplochromines were important in both frame and bottom trawl catches in waters deeper than 40 m (Table 7.1).

7.4 Discussion

7.4.1 *Distribution of fish in the water column*

Hydroacoustics offered the chance to study the behaviour of fish in the water at different depths and in different seasons. Whilst distribution of fish off the bottom provides the best conditions for the hydroacoustic system this was not always the case, as was conformed by the bottom trawling catches. The best time for hydroacoustic surveying was when the lake was thermally stratified (Fig. 7.2(b)). At this time few fish were caught in the bottom trawl in deep waters because of low oxygen conditions which restricted their dispersal to above the thermocline. Information on the fish community composition had to be gained from mid-water trawling at this time.

7.4.2 *Abundance indices*

Mapping and accurately forecasting the spatial patterns of stocks is important for conservation to protect the spawning biomass and juveniles, for profit optimisation by fishing on higher densities of harvestable stock, or for calculating the biomass and

(a) *Rastrineobola argentea*

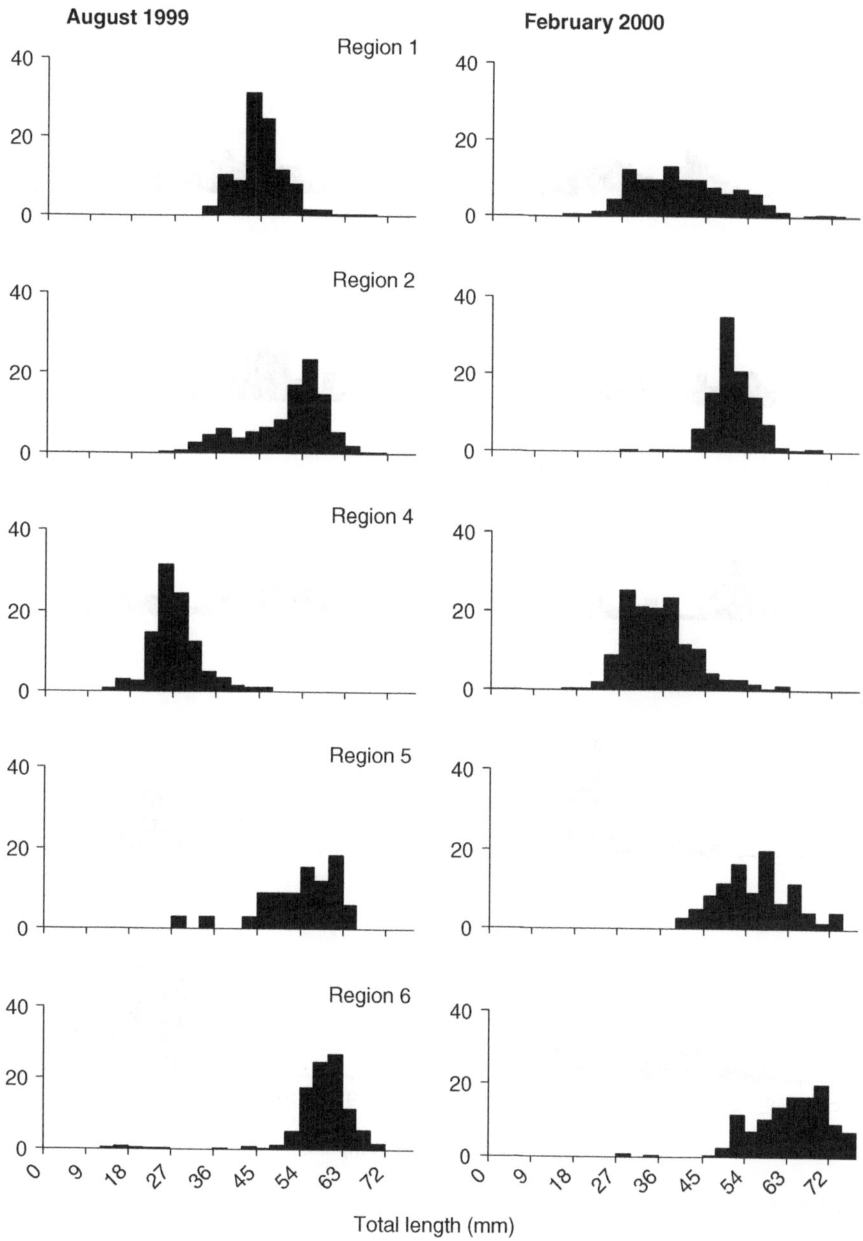

Figure 7.6 Size structure of: (a) *Rastrineobola argentea* and (b) haplochromines in Regions 1–6 shown in Fig. 7.4

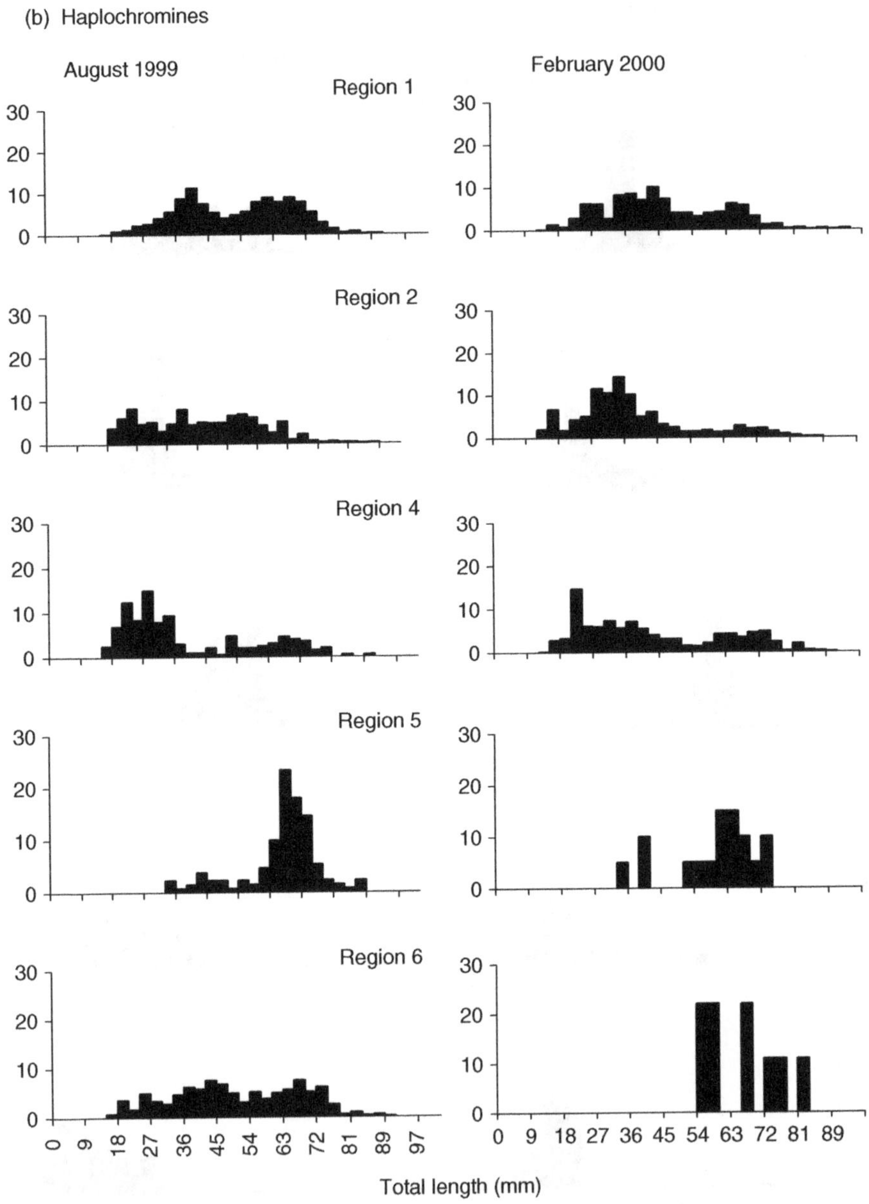

Figure 7.6 (b) Continued

setting confidence limits on the estimates. Results from both surveys indicated a decrease of abundance with depth. Differences in productivity may be related to differences in limnological factors whereby shallow waters have higher nutrient concentrations originating from the inflowing rivers and run-off, which are responsible for production of food resources. Shallow waters also act as breeding and nursery areas for most of the fish. Possible reasons why Region 6 had the lowest densities despite part

Table 7.1 Composition of fish species (% of catch by weight) in bottom trawl and frame trawl catches during the February 2000 hydroacoustic survey

	Depth (m)			
	Bottom trawl		Frame trawl	
	<40	>40	<40	>40
Lates niloticus	95.4	31.3	19.3	34.7
Oreochromis niloticus	2.6		<0.1	
Rastrineobola argentea	1.0	1.6	43.9	26.6
Haplochromines	0.9	60.2	36.8	30.6
Synodontis spp.		6.6		0.2
Barbus spp.		0.2		0.2
Caridina nilotica		0.1		7.1

of its area being along the shore line may be related to limited suitable spawning and nursery areas since the shoreline is more or less straight, strong winds which dominate the area most of the year, or there is a lower input of nutrients because there are fewer rivers on the western side of the lake.

7.4.3 *Size structure*

Regional and seasonal differences in population size structure were identified from the length frequency information collected during both surveys. These differences may be due to horizontal migration movements between inshore and offshore area as fish move to feed or spawn.

7.4.4 *Species composition*

Information from bottom and frame trawl catches suggests that the pelagic community is dominated by *R. argentea* and haplochromines while the semi-pelagic or demersal community is dominated by Nile perch. Below 40 m deep, haplochromines dominate both frame and bottom trawl catches. The frame trawl was only efficient for small fish and therefore the contribution of Nile perch in the pelagic community was probably under-represented.

7.5 Conclusions

- Frame and bottom trawls alone probably do not provide the necessary information required for apportioning the acoustic densities into the taxonomic groups of fish. A mid-water trawl that can be towed at a faster speed is also required.

- Pelagic distribution of the fish especially when the waters are stratified favours use of hydroacoustics.
- From the limited analysis, the results show valuable information on the relative abundance and distribution of fish populations.
- More surveys with similar coverage will be required to establish seasonal changes.

Acknowledgements

We thank the European Union for funding Lake Victoria Fisheries Research Project (Ref: ACP-PPR227), coordinated by Mr M. Van der Knaap. We thank Mr G. Passiotis and the crew of the RV Explorer for their work on the vessel. We thank the directors of FIRRI, KMFRI and TAFIRI for their assistance during the surveys.

References

Bailey M.C., Maravelias C.D. & Simmonds E.J. (1998) Changes in the spatial distribution of autumn spawning herring (*Clupea harengus* L.) delivered from annual acoustic surveys during the period 1984–1996. *ICES Journal of Marine Science* **55**, 545–555.

Goldschmidt T., Witte F. & Wanink J. (1993) Cascading effects of the introduced Nile perch on the detritivorous/phytoplanktivorous species in the sublittoral areas of Lake Victoria. *Conservation Biology* **7**, 686–700.

Gophen M., Ochumba P.O.B. & Kaufman L.S. (1995) Some aspects of perturbation in the structure and biodiversity of the ecosystem of Lake Victoria (East Africa). *Aquatic Living Resources* **8**, 27–41.

Graham M. (1929) The Victoria Nyanza and its fisheries. *A report on the fish survey of Lake Victoria 1927–1928 and appendices*. Crown Agents for the Colonies, London, 255 pp.

Hansson S. (1993) Variation in hydroacoustic abundance of pelagic fish. *Fisheries Research* **16**, 203–222.

Hecky R.E. (1993) The eutrophication of Lake Victoria. *Verhandlungen der Internationale Vereinigung Limnologie* **25**, 39–48.

Kudhongania A.W. & Cordone A.J. (1974) Bathospatial distribution pattern and biomass estimates of major demersal fishes in Lake Victoria. *African Journal of Tropical Hydrobiology and Fisheries* **3**, 15–31.

Maclennan D.N. & Simmonds E.J. (1992) *Fisheries Acoustics*. London: Chapman & Hall, Fish and Fisheries Series 5, 325 pp.

Mathisen R.E., Tumble R.J., Lemberg N.A. & Johnson R.L. (1983) Hydroacoustic indices as input to dynamic management systems of fish stocks. ICES/FAO Symposium of Fisheries Acoustics. *Rome: FAO Fisheries Report* **300**, 223–248.

Midttun L. & Nakken O. (1977) On acoustic estimation studies with echo integrators. *ICES Rapports et Procès-Verbaux des Réunions* **170**, 253–258.

Nakken O. & Ulltang ø. (1983) A comparison of the reliability of acoustic estimates of fish stock abundances and estimates obtained by other assessment methods in the Northeast Atlantic. ICES/FAO Symposium of Fisheries Acoustics. *Rome: FAO Fisheries Report* **300**.

Ochumba P.B.O. & Kibaara D.I. (1989) Observations on blue-green algal blooms in the open waters of Lake Victoria, Kenya. *African Journal of Ecology* **27**, 23–34.

Porteiro C., Carrera P. & Miquel J. (1996) Analysis of Spanish acoustic surveys for sardine, 1991–1993: Abundance estimates and inter-annual variability. *ICES Journal of Marine Science* **53**, 429–433.

Thorne (1983) Application of hydroacoustic assessment techniques to three lakes with contrasting fish distributions. ICES/FAO Symposium of Fisheries Acoustics. *Rome: FAO Fisheries Report* **300**.

Chapter 8
Analysis of exploitation patterns for Nile perch, *Lates niloticus*, in Lake Victoria

O.C. MKUMBO*, C. EZEKIEL, Y.L. BUDEBA
Tanzania Fisheries Research Institute, Mwanza, Tanzania

I.G. COWX
International Fisheries Institute, University of Hull, Hull, UK

Abstract

Since the explosion of Nile perch, *Lates niloticus* (L.), in Lake Victoria in the early 1980s they have sustained a rapidly developing fishery. The species is now the predominant component of the catch but there is evidence of overexploitation of the stocks. This chapter examines trends in the fishery since the early 1980s and presents evidence to show that overexploitation is occurring in certain localities of the lake, especially around urban centres and industrial processing units. Opportunities for the management of the resource are discussed and potential implications for the fishery are outlined.

Keywords: fishery management, overexploitation, recruitment.

8.1 Introduction

The fisheries of Lake Victoria have undergone remarkable changes over the past 30 years. Signs of overfishing were reported as early as 1970 when catch rates for tilapia dropped from 50–100 fish per 50-m long gillnet with 127 mm stretched mesh (Worthington & Worthington 1933) to less than five fish (Ssentongo 1972). *Lates niloticus* (L.) and tilapiines introduced in the late 1950s further altered the fishery. Although this brought undeniable benefits to the economy (Greboval 1990), a number of native fish species disappeared or have decreased to very low levels. The lake's ecosystem and food web have changed and indeed are still in the process of change, thus affecting the fisheries and the lake resources in general (Ogutu-Ohwayo 1990; Witte, Goldschmidt, Wanink, van Oijen, Goudswaard, Witte-Maas & Bouton 1992). Increased pollution and clearing of the peripheral wetlands (Kaufman 1992), which served as fish nursery grounds, may also be seriously affecting the fisheries.

Despite a continuous decline in many endemic food species, such as *Oreochromis esculentus* Graham, *Bagrus docmak* Forsskåll, *Clarias gariepinus* (Burchell) and *Labeo victorianus* Boulenger, there has been a rapid increase in total landings from 1980 to a

*Correspondence: Tanzania Fisheries Research Institute, PO Box 475, Mwanza, Tanzania (email: omkumbo@hotmail.com).

peak in the early 1990s. Catches increased five times from 85 914 to 567 660 t and Nile perch contributed over 70% of the total landings (Bwathondi 1990; Ogari & Asila 1990). Catches for other fish species decreased even further and species groups like the haplochromines almost disappeared (Ligtvoet & Mkumbo 1991). With increased fishing pressure, predation and competition among species, the multispecies fishery of Lake Victoria changed to only three species: Nile perch, the pelagic cyprinid – dagaa, *Rastrineobola argentea* (Pellegrin), and the introduced tilapiine, *Oreochromis niloticus* (L.).

In the mid-1990s Nile perch, the dominant species in the fishery, showed signs of decline. Changes in the efficiency of fishing gears, motorisation of canoes and increase in total fishing effort to maintain yield were observed. Extension of fishing grounds was also evident, but all against a continued decrease in catch per unit effort and mean size of fish caught (Mkumbo & Cowx 1999).

With this change in the fishery resources, a number of management measures were effected, including a ban on beach seines and undersized mesh nets (less than 127 mm stretched mesh) in 1994, and a ban on trawlers in 1996. The continuous decline in CPUE and other signs of overfishing have become the focus of a lake-wide stock assessment project under the Lake Victoria Fisheries Research Project (LVFRP) funded by the European Union. Objectives of the project include the derivation of stock biomass, exploitation patterns, and the fish distribution patterns. This chapter examines the trends in the status of the fishery based on experimental trawling and commercial catch data since November 1998, and compares the situation with similar data from the HEST/TAFIRI RV Kiboko 1985–1989 surveys.

8.2 Materials and methods

The Tanzanian part of Lake Victoria was divided into three areas. Area A stretches from Kome and Buhiru islands in the south-west, north-eastwards to the south-west of Ukerewe island, including Speke Gulf and Mwanza Gulf. Area B is from the Tanzania–Kenya border southwards to the north-west of Ukerewe islands. Area C covers the Kagera waters from the Tanzania–Uganda border southwards to Kome island, including Emin Pasha Gulf. The sampling stations were allocated within each area using gridlines of five nautical mile-squares based on degrees and minutes of latitude on hydrographical charts. In each square, one sampling station was allocated, except in very deep waters, where squares of 10–15 nautical miles were sampled. A total of 133 stations covering the entire depth range were established. All stations in each consecutive area were sampled on a 3-month rotational basis because of resourcing constraints, thus the seasonal changes in fish population structure could be determined. The research vessel, RV Victoria Explorer, length 17 m and a 250-HP engine, with a trawl net of 22.6-m head rope and a codend of 25-mm mesh, was used for the present surveys. Towing speed was 3–3.6 knots and trawling duration was 30 min.

Treatment of the catch in the present survey depended on its size. If it was small, the entire catch was sorted into species, weighed, and every fish measured (total length, TL, cm) and weighed (g). If the catch was too large to handle in the time before the following sample was landed, fish above about 35 cm TL were sorted from the catch and measured individually. The remaining small fish, less than 35 cm TL, were sub-sampled by taking three shovelfuls of fish from a thoroughly-mixed heap on deck. Depending on the average size of fish in the sample, the number of shovels could be increased to have a minimum of 200 fish. The weights of the sub-sample and complete sample from which the sub-sample was taken were used to obtain a raising factor. Specimens of Nile perch from one or two hauls in a day (depending on size of the catch) were gutted for sex/maturity and dietary analysis. Maturity was based on the classification of Hopson (1972), where fish in Classes I–III were immature and those in Classes IV–VI were mature.

Quarterly catch rates (kg h^{-1}) from the three sampled areas, A, B and C, were treated separately to assess the temporal distribution of the fish stocks. Mean catch rates from different depth ranges were used to determine the batho-spatial distribution of the stocks. These were then compared to the historical data (Kudhongania & Cordone 1974; HEST/TAFIRI unpublished data). Changes in species composition was determined using the mean catch rates of the different species separately for the three areas, and this was again compared with historical data from RV Ibis and RV Kiboko, to assess the change in fish diversity in the lake. Changes in fish stock abundance were assessed using the mean catch rates of the different research vessels. It is recognised that the capture efficiency and selectivity of the vessel used may vary but it was assumed that the output would illustrate trends in stock abundance.

8.3 Results

8.3.1 *Trends in landing statistics*

Landing records from the Fisheries Department show the abrupt increase of Nile perch catches from 1987 to 1990, and thereafter a declining trend (Fig. 8.1). It has to be noted that fishing effort has been increasing over the years coupled with a continuous decline in total landings, while the total value of the landings has been increasing tremendously (Fig. 8.2). This is explained in the high demand in the fishing industry and the danger of continuous increase in effort if not controlled. Fishing effort especially around the islands seems to be very high and number of canoes as a measure of fishing effort may not be appropriate at present due to the tendency of increasing the number of gillnets per canoe. It was noted that beach seines are still operational gears harvesting considerable amounts of juvenile fish.

Data from beach surveys (Fig. 8.3), also conducted under the LVFRP have indicated that mean length at capture for gillnets greater than 5″ (127 mm) is 47–55 cm TL.

8.3.2 *Stock abundance over time*

Mean catch rates have decreased by almost 50% between 1969/1970 (RV Ibis) and 1999/2000 (RV Victoria Explorer). Underlying this trend, there was a progressive decrease to less than 20% in 1989, when the haplochromines almost disappeared from the catches, followed by an increase during the Nile perch boom to 1996 and then a slight decrease (Fig. 8.4).

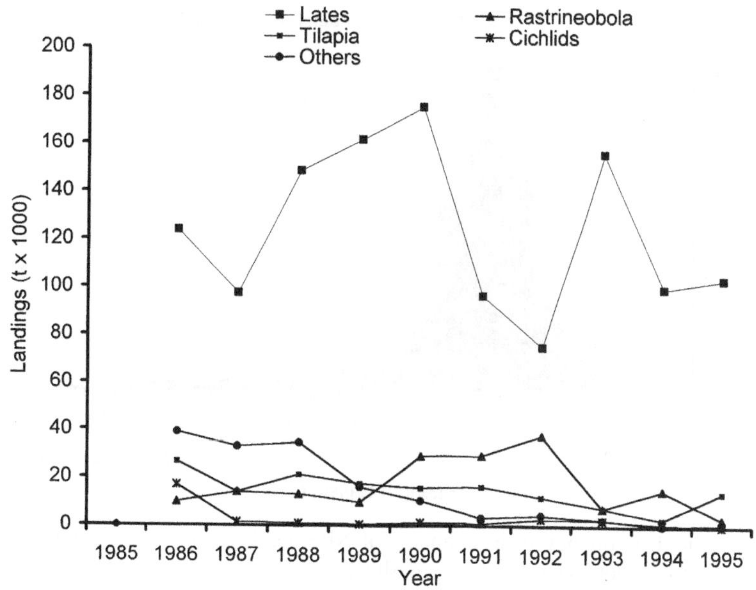

Figure 8.1 Landing by species in Tanzanian waters (t × 1000)

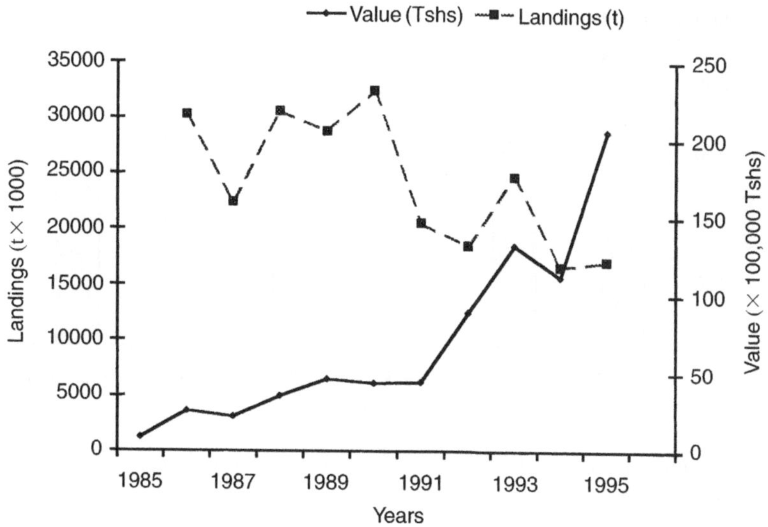

Figure 8.2 Changes in total landings compared with total value in Tanzanian waters between 1985 and 1995

8.3.3 *Temporal distribution of fish stocks*

Catch rates from bottom trawl surveys for 2 months in 1997, 1 month in 1998 and monthly from March 1999 to April 2000 (Fig. 8.5) indicated temporal variation between areas. Area A had more stable catches over the different months sampled

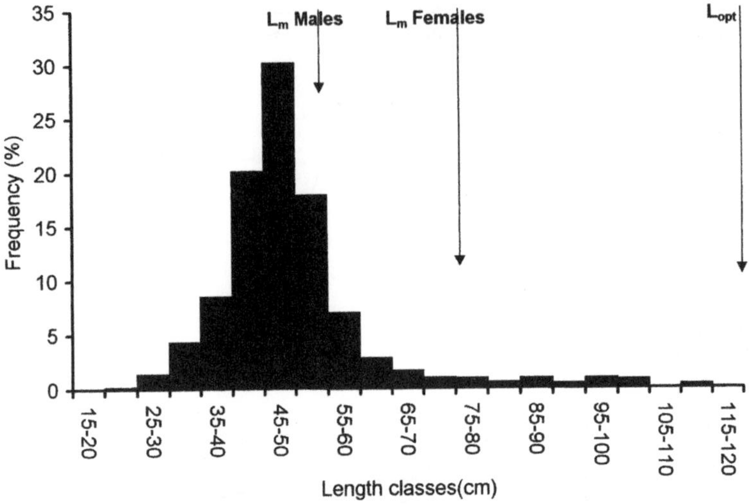

Figure 8.3 Length frequency from commercial Nile perch catches with length at maturity (L_m) and L_{opt} fitted to show the effect of overfishing

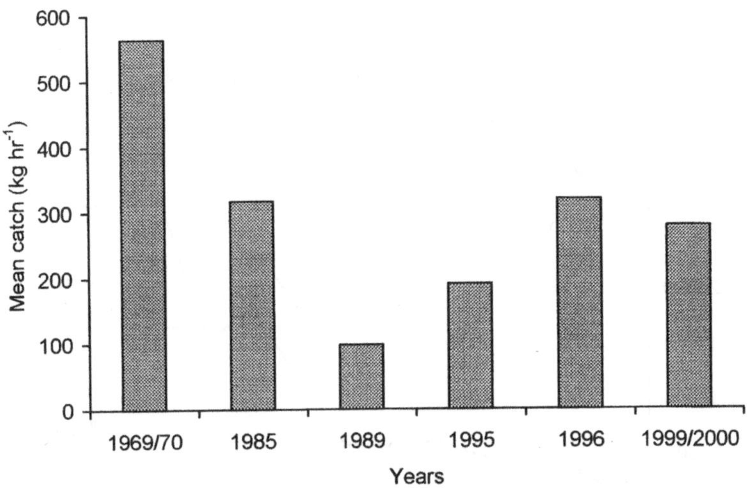

Figure 8.4 Change in mean catch rates of the different research vessels used at different periods in Lake Victoria

compared with Areas B and C. The mean catch varied from 197.2 kg h^{-1} in March 1999 to a high of 273.0 kg h^{-1} in October 1999. Area B catch rates decreased over the study period, although within the same 3-monthly period, catches in the other areas were relatively high. Mean catches decreased from 287.7 kg h^{-1} in December 1997 to 80.0 kg h^{-1} in March 2000. Catches in Area C increased from 74.9 kg h^{-1} in January 1998 to a peak of 415.9 kg h^{-1} in September 1999 and then decreased to 230.7 kg h^{-1} in April 2000.

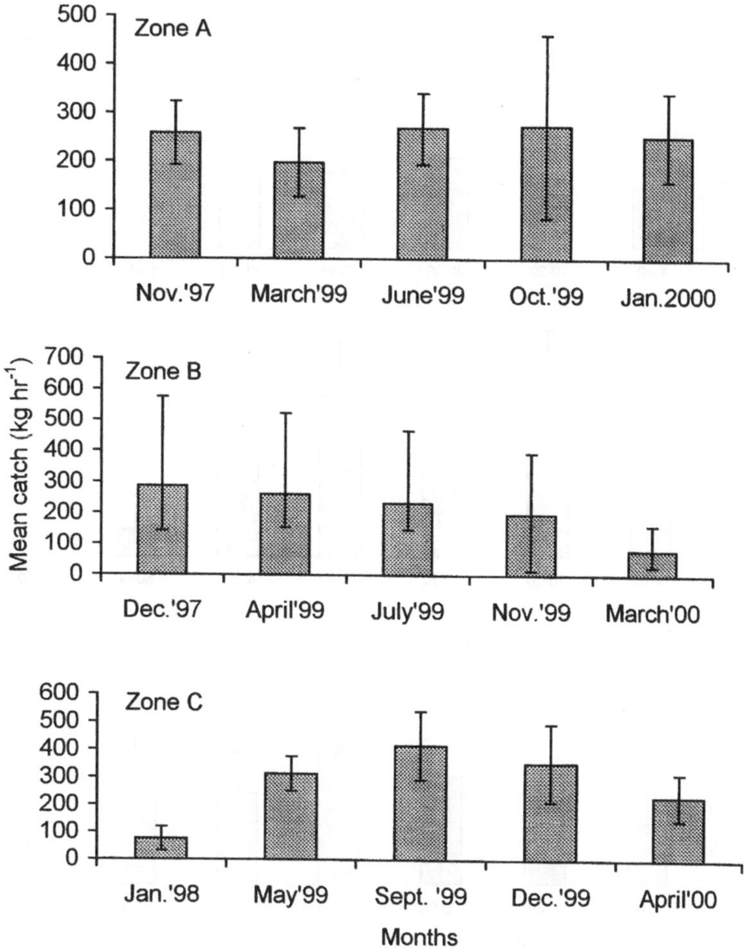

Figure 8.5 Monthly mean catch rates from the three areas (A, B and C) sampled quarterly from November 1997 to April 2000

8.3.4 *Batho-spatial distribution of fish stocks*

Catch rates exhibited showed a marginal, although not significant ($P > 0.05$), increase with depth up to 40 m, although there was a fall off below 50 m in Zones B and C (Fig. 8.6(a)–(c)). A similar trend was observed in the 1969/1970 survey (Fig. 8.6(d)). Mean catch rates in 1969/1970 survey were more than double those of the present survey in waters up to 20 m deep: 298.7 and 282.3 kg h^{-1} in 0–10 and 10–20 m, respectively for the 1999/2000 surveys compared to 644.0 and 935.3 kg h^{-1} for the 1969/1970 survey. At a depth of 60–70 m 1969/1970 catches were comparable to the present catches in the depth ranges 41–50 m.

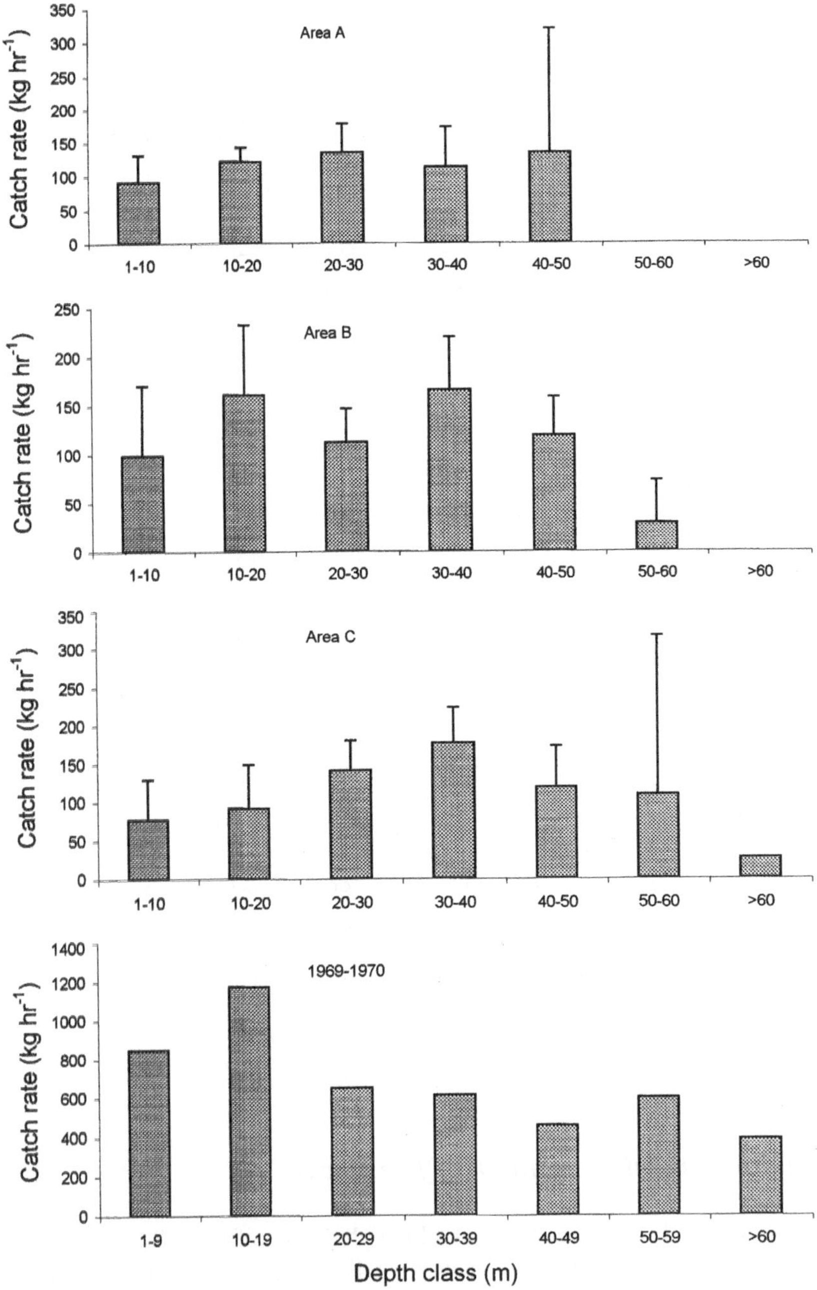

Figure 8.6 Mean catch rate against depth from bottom trawl surveys in Tanzanian sector of Lake Victoria, 1997–2000 surveys in Zones, A, B and C compared to 1969/1970 (Kudhongania & Cordone 1974)

8.3.5 *Population structure*

The size structure of the Nile perch population appears to be similar in the different areas, but with more juveniles in Areas A and B and more adults in Area C (Fig. 8.7). The very high fishing pressure is being compensated by high recruitment. Juveniles seem to occupy all depth ranges during all the different seasons, thus closed areas or close seasons may not be a very effective conservation measure for Nile perch stock.

A total of 3474 fish were analysed for sex and sex maturity; 2792 (80%) were immature, 483 (14%) were mature males and 199 (6%) were mature females. There was a predominance of juveniles in the population with few greater than 50 cm TL. The size

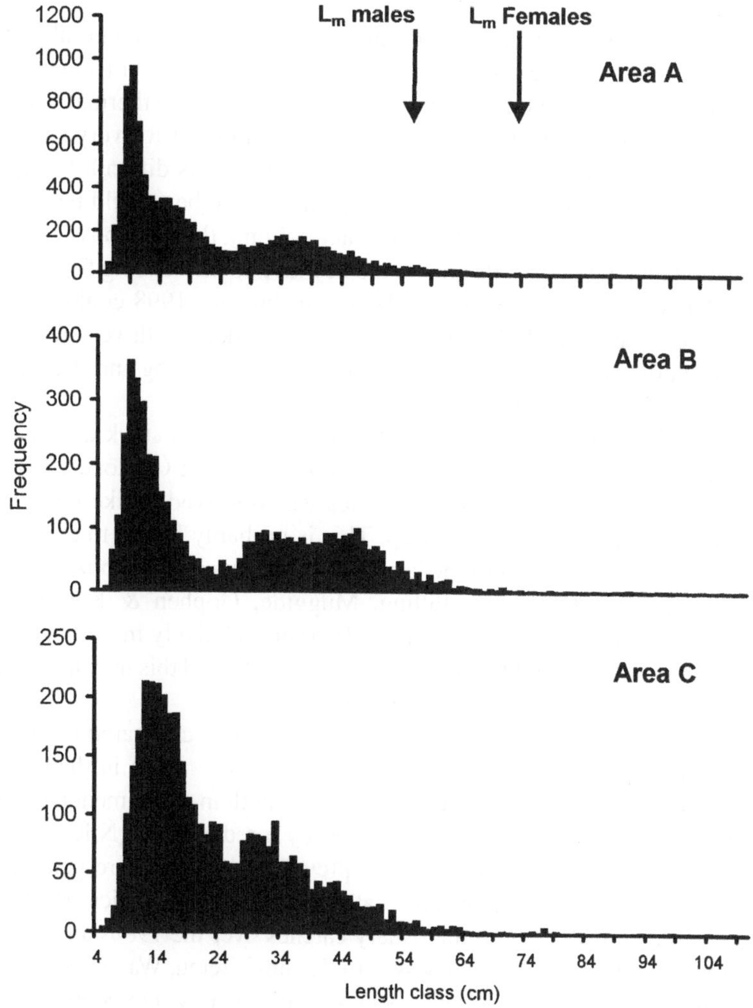

Figure 8.7 Length frequency of *Lates niloticus* caught by trawling in different zones of Tanzanian waters

at first maturity showed a decrease in size over time in both sexes. The size at first maturity was similar in the three zones in Tanzania. Females matured at a slightly greater size (75 cm TL) than males (53 cm TL). The marginal variation in Lm_{50} between the zones was attributable to differences in the time the samples were taken, i.e. each zone was sampled on consecutive months, and the small number of mature fish sampled, especially for females.

8.4 Discussion

8.4.1 *Distribution and abundance of demersal fish stocks of Lake Victoria*

Distribution patterns of fish stocks in Lake Victoria varied temporally between regions. Area A, with relatively shallow and with many vegetation fringed bays, has the most stable catches over the different seasons. This is because thermal stratification and deoxygenation of the lower parts of the water column last for very short periods when the wind velocity is low (Hecky & Bugenyi 1992). It is difficult to explain the trend in Area B without limnological data. July and November would be expected to have high catches due to total mixing of the water column, thus it is possible that thermal stratification and oxygen levels are not the key factors influencing fish distribution in this area. The extreme low catches in Area C in January 1998 could probably be explained by the effect of El'Niño, and being relatively deep with very few bays, the pattern of distribution seems to comply with the periods of mixing and stratification of the water column (Hecky & Bugenyi 1992).

The batho-spatial distribution pattern exhibited a decline in stock abundance with depth, similar to that observed in 1969/1970 (Kudhongania & Cordone 1974). Below 50 m depth, an abrupt drop of mean catch rates was observed unlike the 1969/1970 surveys when the drop occurred 70 m deep. This is probably linked to changes in the physical/chemical characteristics of the lake and the presence of a hypoxic layer below 50 m (Hecky, Bugenyi, Ochumba, Talling, Muggide, Gophen & Kaufman 1994). Haplochromines, the dominant stock in 1969/1970 are relatively more tolerant to low oxygen levels than Nile perch (Ochumba & Kibaara 1989) and this may also contribute to the difference in depth distribution for the two surveys.

Catch rates for the recent experimental trawling were low compared to the pre-Nile perch period, despite a five-fold increase in catches of Nile perch in the early 1990s (Bwathondi 1990). The present decline in catch rates, both in experimental surveys and in total commercial landings, is probably caused by a reduction in Nile perch stocks, although compensation by other species was prevalent, but these are pelagic species and not fully represented in bottom trawl catches or fishery statistics. *Rastrineobola argentea* increased considerably in the fishery catches over the 1980s and early 1990s (Ligtvoet, Mous, Mkumbo, Budeba, Goudswaard, Katunzi, Temu, Wanink & Witte 1995), and is now the second most important species in the fishery. *Oreochromis niloticus* is equally under-represented due to its habitat preference for shallow, non-trawlable areas, but it is third in importance in the fishery (Ogutu-Ohwayo 1995).

8.4.2 *Indicators of fishery overexploitation*

There are several key indicators of intense exploitation in the Nile perch fishery in addition to the declining CPUE. Firstly, there has been a progressive decline in modal length of the catch population over the years. In 1988 the modal length was 70–80 cm TL (Ligtvoet & Mkumbo 1991), while it decreased to 50–60 cm TL in 1992 and even further to 40–50 cm TL in 1994 and it remained around the same level to the end of the 1990s (Nsinda, Mkumbo, Ezekiel 2000). Much of this decline in size at first capture has been linked to a reduction in mesh size, which seems to have stabilised around 12.5 cm.

Change in size at first maturity is a sign of heavy fishing pressure in the stocks. In Tanzanian waters this has decreased from 60 cm TL in males and 95–100 cm TL in females in 1988 (Ligtvoet & Mkumbo 1991) to 50–55 cm TL and 70–80 cm TL in males and females, respectively, in 1999.

The paucity of adult fish, greater than 50 cm TL, in the Nile perch populations in each zone is another sign of overfishing. This suggests good recruitment to the population, probably to compensate for the overfishing, but the relatively small numbers of large, mature fish is of concern as there is potentially a lack of spawning stock. Although Nile perch is a very fecund species and it has lowered its size at first maturity to compensate, its ability to sustain the stocks in the long term under such intense pressure is questionable.

It is well known that there is a very high demand for Nile perch from the filleting factories, but worse is the preference for fillets from juveniles of 0.5–1 kg for some export markets. If the export markets dictate the size at first capture this will inevitably lead to a collapse in stocks because too few fish will reach spawning size. The domestic market at present is equally open for undersized fishes and thus strict measures have to be considered.

A serious threat to a sustainable Nile perch fishery is the use of illegal gears like beach seines. The amount of juveniles being harvested by this gear is alarming. It is to the benefit of all the stakeholders to adhere to the management measures so as to sustain the fishery.

8.4.3 *The future of the fishery*

Since its establishment in Lake Victoria, the Nile perch has exhibited a highly dynamic fishery following its explosion in the 1970s through the boom in the 1980s and early 1990s to the apparent decline presently being experienced. The latter is also linked to signs of overfishing (Ogutu-Ohwayo 1995; Mkumbo & Cowx 1999; present study). Management measures have been introduced to regulate the fishing effort, including control of minimum mesh size, ban of trawling and beach seines, as well as closed fishing areas and close seasons, to protect the spawning stock and ensure recruitment to the fishery. Unfortunately, many of these measures have yet to be enforced, and there is little effort to regulate the fishery *per se*, especially with respect to the use of illegal gears such as beach seines, which exploit undersized fish.

However, direct technical interventions may be an oversimplification of the measures that need to be adopted in the lake to ensure sustainable development of the fisheries, since it takes no account of prey–predator and broader ecosystem interactions, which are important regulatory mechanisms in any natural system. In the lake, there has been a tendency to fish down the trophic levels and as a result instability in the lake environment is prevalent.

Nile perch, currently the most important fishery stock, is the top predator, feeding on a number of trophic levels, mainly primary (juveniles feeding on plankton, Cladocera and cyclopoids, then insect larvae) and secondary levels (*Caridina* and haplochromines and *Rastrineobola*), although the bigger fish are tertiary consumers. *Caridina*, occupying the niche left by haplochromines, has built up in population numbers and is now a major component of the diet of Nile perch. This has reduced predation pressure on haplochromines and is possibly the reason for their recent resurgence. Other fish species are also exploiting the decline in the Nile perch stock abundance, e.g. endemic predators like *Clarias* and *Bagrus*, which probably compete with Nile perch for food. However, fishing has adapted to exploit the newly-emerging haplochromines, preventing any stability in stock structure and influence on ecosystem integrity. Furthermore there is now considerable exploitation of the *Caridina* for animal feeds which may further disturb the trophic interactions. Fishing pressure is also heavily directed at *Rastrineobola* and *Oreochromis*, the other two important stocks in the fishery, but measures must be taken to ensure they are not overexploited to levels which further disrupt the ecosystem integrity.

The intense fishing pressure in Lake Victoria, thus has profound effects on the ecosystem as a whole and the long-term sustainability of the fishery potential. If this is to be maintained an ecosystem-based fisheries management dimension must be integrated into any overall lake management, especially that targeting the fishery resources. Adequate resources should also be made available to support any initiative.

Acknowledgement

The study was funded by the European Union Lake Victoria Fisheries Research Project (Ref.: ACP-RPR 227).

References

Bwathondi P.O.J. (1990) The state of Lake Victoria fisheries, Tanzanian sector. CIFA Report of the Fifth Session of the Sub-Committee for the Development and Management of the Fisheries in Lake Victoria, 12–14 September 1989, Mwanza, Tanzania. Rome: FAO Fisheries Report No. 430, pp. 24–34.

Greboval D. (1990) Socio-economic issues for planning in support of fisheries management. CIFA Report of the Fifth Session of the Sub-Committee for the Development and Management of the Fisheries in Lake Victoria, 12–14 September 1989, Mwanza, Tanzania. Rome: FAO Fisheries Report No. 430, pp. 75–97.

Hecky R. & Bugenyi F.W.B. (1992) Hydrology and chemistry of the African Great Lakes and water quality issues: problems and solutions. *Verhandlungen der Internationale Vereinigung für theoretische und angewandte Limnologie* **23**, 45–54.

Hecky R.E., Bugenyi F.W.B., Ochumba P., Talling J.F., Muggide R., Gophen M. & Kaufman L. (1994) Deoxygenation of the deep water of Lake Victoria, East Africa. *Limnology and Oceanography* **39**, 1476–1481.

Hopson A.J. (1972) A Study of the Nile Perch (*Lates niloticus* (L.), Pisces: Centropomidae) in Lake Chad. London: Overseas Research Publication, 1993 pp.

Kaufman L. (1992) Catastrophic change in species-rich freshwater ecosystems: the lessons of Lake Victoria. *BioScience* **42**, 846–858.

Kudhongania A.W. & Cordone A.J. (1974) Batho-spatial distribution pattern and biomass estimate of the major demersal fishes in Lake Victoria. *African Journal of Tropical Hydrobiology and Fisheries* **3**, 15–31.

Ligtvoet W. & Mkumbo O.C. (1991) A pilot sampling survey for monitoring the artisanal Nile perch (*Lates niloticus*) fishery in southern Lake Victoria (East Africa). In I.G. Cowx (ed.) *Catch Effort Sampling Strategies. Their Application in Freshwater Fisheries Management.* Oxford: Fishing News Book. Blackwell Scientific Publications Ltd, pp. 349–360.

Ligtvoet W., Mous P.J., Mkumbo O.C., Budeba Y.L., Goudswaard P.C., Katunzi E.F.B., Temu M.M., Wanink J.H. & Witte F. (1995) The Lake Victoria fish stocks and fisheries. In F. Witte & W.L.T. van Densen (eds) *Fish Stocks and Fisheries of Lake Victoria. A Handbook for Field Observations.* Cardigan: Samara Publications Limited, pp. 11–53.

Mkumbo O.C. & Cowx I.G. (1999) Catch trends from Lake Victoria–Tanzanian waters. LVFRP/TECH/99/07. Jinja, Uganda: LVFRP, 99–102 pp.

Nsinda P.E., Mkumbo O. & Ezekiel C.N. (1999) Stock assessment of *Lates niloticus* (L.), *Oreochromis niloticus* and *Rastrineobola argentea* (Pellegrin) using fisheries-dependent data from Tanzanian waters of Lake Victoria. LVFRP/TECH/99/07. Jinja, Uganda: LVFRP, 108–117 pp.

Ochumba P.B.O. & Kibaara D.I. (1989) Observations on blue-green algal blooms in the open waters of Lake Victoria, Kenya. *African Journal of Ecology* **27**, 23–34.

Ogari J. & Asila A.A. (1990) The state of Lake Victoria, Kenya waters. CIFA Report of the Fifth Session of the Sub-Committee for the Development and Management of the Fisheries in Lake Victoria, 12–14 September 1989, Mwanza, Tanzania. Rome: FAO Fisheries Report No. 430, pp. 18–23.

Ogutu-Ohwayo R. (1990) The decline of the native fishes of Lake Victoria and Kyoga (East Africa) and the impact of introduced species, especially the Nile perch, *Lates niloticus* and the Nile tilapia, *Oreochromis niloticus. Environmental Biology of Fishes* **27**, 81–96.

Ogutu-Ohwayo R. (1995) Diversity and stability of fish stocks in Lake Victoria, Kyoga and Nabugabo after establishment of introduced species. In T.J. Pitcher & P.J.B. Hart (eds) *The Impact of Species Changes in African Lakes*. London: Chapman & Hall, pp. 59–81.

Ssentongo G.W. (1972) Yield isopleths of *O. esculenta* (Graham) 1928 in Lake Victoria and *O. niloticus* (Linnaeus) 1757 in Lake Albert. *African Journal of Tropical Hydrobiology and Fisheries* **2**, 121–128.

Witte F., Goldschmidt T., Wanink J.H., Oijen M.J.P., van, Goudswaard P.C., Witte-Maas E.L.M. & Bouton N. (1992) The destruction of an endemic species flock: quantitative data on the decline of the haplochromine cichlids of Lake Victoria. *Environmental Biology of Fishes* **34**, 1–28.

Worthington S. & Worthington E.B. (1933) *Inland Waters of Africa*. London: Macmillan & Co. 259 pp.

Chapter 9
Recent trends in the Lake Victoria fisheries assessed by ECOPATH

M.C. VILLANUEVA and J. MOREAU*
Department of Inland Fisheries – INP/ENSAT, Auzeville, Castanet Tolosan, France

Abstract

The ECOPATH IV software with ECOSIM, was used to describe the exploited fish community of the Winam Gulf in the Kenyan sector of Lake Victoria, Africa. Developments in the ecosystem structure that can provide possible explanations for recent trends of evolution of the lake following preliminary assessments made on trophic relationships in the fish community during the mid-1980s using ECOPATH II were demonstrated. Further studies on system dynamics are made using the ECOSIM simulation program that can describe the effects of some fisheries management schemes during the past 10 years and for the future.

Keywords: ecosystem dynamics, fisheries management, *Lates niloticus*, *Oreochromis niloticus*, *Rastrineobola argentea*.

9.1 Introduction

Lake Victoria, East Africa (Fig. 9.1) has undergone dramatic changes in ecosystem structure after the introduction of Nile perch, *Lates niloticus* (L.), more than 40 years ago (Wanink & Goudswaard 1994; Wandera & Wanink 1995). The population explosion of *L. niloticus*, as well as the intensive riparian fishing pressures in combination with increasing eutrophication due to agricultural and waste runoffs (Bundy & Pitcher 1995; Wilson, Medard, Harris & Wiley 1996) all contributed to the decline of the multispecies stock of the lake into three commercially-important species: two totally demersal alien species, the Nile perch and Nile tilapia, *Oreochromis niloticus* (L.), and the native sardinelike, zooplanktivorous, pelagic cyprinid dagaa, *Rastrineobola argentea* (Pellegrin) (Bundy & Pitcher 1995; Wilson *et al.* 1996). In the late 1980s, Nile perch contributed 80–90% by number of total landings in the artisanal fishery and almost 100% in trawl catches (Mkumbo & Ligtvoet 1992; Moreau, Ligtvoet & Palomares 1993). The vast expansion of fish production in the lake was because of the explosion of the Nile perch stock at the end of the 1970s at the expense of endemic species, especially many haplochromine species (Wanink & Goudswaard 1994; Reynolds & Greboval

*Correspondence: Department of Inland Fisheries – INP/ENSAT, BP 107 Auzeville, 31 326 Castanet Tolosan, France (email: moreau@flora.ensat.fr).

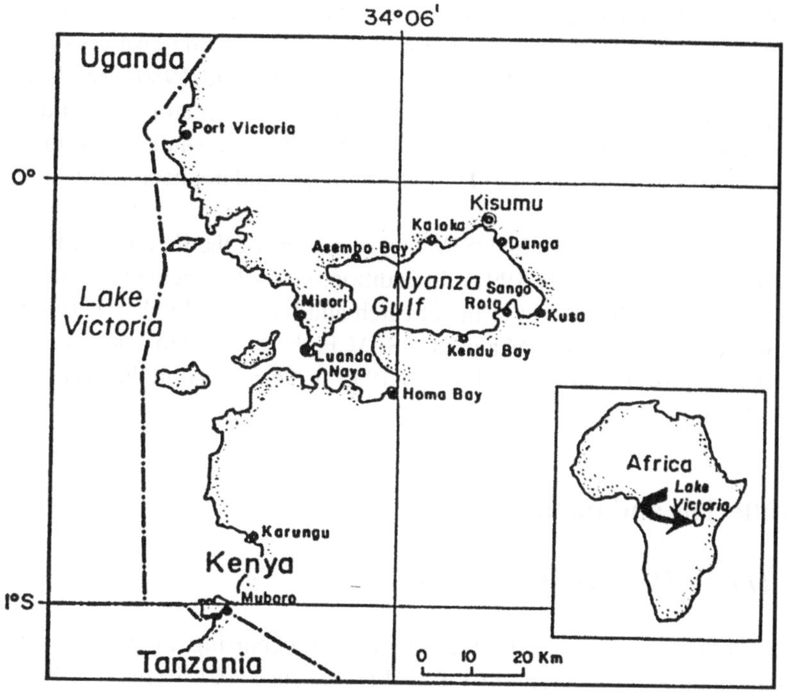

Figure 9.1 General map of the Kenyan sector of Lake Victoria

1995; Wilson *et al.* 1996). The small cyprinid *R. argentea* remained the only indigenous species of commercial importance and became the major prey species of the Nile perch (Ogari & Dadzie 1988; Mkumbo & Ligtvoet 1992). The loss of more than 80% of the demersal cichlids, comprising about 250 species, has promoted shifts in the diet of *L. niloticus* (Ogari & Dadzie 1988). Adult Nile perch, >40 cm total length (TL), now feed on the shrimp, *Caridina nilotica* (L.), *R. argentea*, and exhibit cannibalism, preying on their own juveniles (Ligtvoet 1989; Mkumbo & Ligtvoet 1992).

Analysis of species changes (Bundy & Pitcher 1995) and studies on haplochromine dynamics by Witte, Goldschmidt & Wanink (1995), however, disputed the previous belief that predation by Nile perch is the only factor contributing to the decrease of these demersal fishes. Declines in haplochromine stocks had already been observed before the explosion of the Nile perch due to intense riparian fishing activities and increasing level of eutrophication (Witte, Goldschmidt, Wanink, van Oijen, Goudswaard, Witte-Maas & Bouton 1992; Kudhongania & Chitamwebwa 1995). Goudswaard & Wanink (1993), Wanink & Goudswaard (1994) and Wanink, Goudswaard & Berger (1999) showed that fish-eating birds, such as the pied kingfisher, *Ceryle rudis*, also contributed to the decline of the haplochromines and has subsequently shifted its diet to feed on *R. argentea*, thus competing with the commercial fisheries.

A previous contribution providing comparative descriptions of the ecosystem during the pre-*Lates* and post-*Lates* periods presented quantitative box models for these two different periods in the history of the lake (Moreau *et al.* 1993). The first model defined the role of haplochromine species from the late 1960s to the early 1970s and described the complex food web and trophic relationships existing in the lake ecosystem. The second model was a synthesis of observations from the mid-1980s, emphasising on the ecological importance of Nile perch as the dominant predator. It should be noted that records utilised in these simulations described only the shallow and intensively-exploited Kenyan sector of the lake. Thus, these models did not necessarily apply to the lake as a whole.

This present exercise adopts ECOSIM, a new routine of the ECOPATH IV software (Walters, Christensen & Pauly 1997). In addition, possible mechanisms affecting the changes occurring in the lake in terms of catch and biomass (Pitcher & Bundy 1995) were identified. In using ECOSIM and ECOSPACE, the aim was to assess recent system changes that occurred from 1985–1986 to 1995–1996 and to predict changes in the fishery in the future.

9.2 Materials and methods

9.2.1 *The ECOPATH IV Software*

To construct the structure of trophic interactions occurring in the ecosystem, the steady-state simulation program, ECOPATH IV, was used (Christensen & Pauly 1992; Walters *et al.* 1997). In structuring the model, the various organisms inhabiting the ecosystem have to be grouped into boxes according to their common physical habitat, similar food preferences and life history characteristics. By defining these categories, the model estimates, on an annual basis, parameters such as mean biomass (B), biomass production (P) and biomass consumption (Q) or ecotrophic efficiency (EE) and flow to the detritus pool of every box considered in the ecosystem. It is assumed for the period considered that input to each group is equal to the output from it to deal with equilibrium conditions. It is also required to standardise the data input by applying the same units in each parameter considered (in this case $\mathrm{t\,km^{-2}}$). By establishing an equilibrium condition in the ecosystem, a series of biomass budget equations are determined for each considered group as

Production − all predation on each grouped species − non-predatory mortality − all exports = 0

ECOPATH expresses each term in the budget equation as a linear function of the mean biomass. The resulting budget equations become a system of simultaneous equations following the formula:

$$B_i(P/B)_iEE_i - Y_i - \sigma(B_jQ/B_jDC_{ji}) = 0 \qquad (9.1)$$

where B_i is the biomass of the group i; P/B_i its production/biomass ratio, usually assumed equal to the total mortality (Z); EE_i its ecotrophic efficiency (i.e. the proportion of the ecological production which is consumed by predators and/or exported); Y_i the yield (equal to fishery catch), usually obtained through fisheries statistics; B_j the biomass of predator j; Q/B_j the food consumption per unit biomass of j, a parameter expressing food consumption on an age structured population of fish relative to its biomass, considering that juveniles are numerous compared to adults and consume much more food (compared to their weight; Palomares & Pauly 1998); DC_{ji} the fraction of *i* in the diet of *j*, expressed as percentage of weight or volume.

ECOPATH IV differs from previous versions in that it has additional routines namely, ECOSIM and ECOSPACE. These new routines are simulations that provide descriptions on evolution occurring in the ecosystem considering various fishery management strategies and describe the trophic dynamics of the ecosystem as distributed over space, respectively. In structuring stable models, it is important to define groups belonging to the lower trophic level (i.e. plankton) into sub-groups on the basis of the spatial distribution of fish and related feeding habits.

9.2.2 *Designing the present ECOPATH model of Winam Gulf, Kenya (Lake Victoria)*

In designing the model, input was based on Moreau *et al.* (1993) for the mid-1980s, except where specified below. Actual catch in the Kenyan sector was provided by Pitcher & Bundy (1995) and Mkumbo, Ezekiel, Budeba & Cowx (Chapter 8 of this book). For *Bagrus/Clarias*, the *Q/B* value of 5.5, previously used by Moreau *et al.* (1993), was increased to 6.5 after recomputation using recent data (Froese & Pauly 1999). *Q/B* of zooplankton, *Caridina* and zoobenthos which belong to the second trophic level, were identified as too low leading to high *GE* (food conversion efficiency) values and were increased accordingly. For *R. argentea* a higher *P/B* ratio than used in 1985 (Moreau *et al.* 1993) was adopted because new values of growth and mortality parameters (Manyala, Van den Berghe & Dadzie 1995; Wandera & Wanink 1995) were available.

When considering the data from Wanink & Goudswaard (1994), it appeared relevant to add a box with fish-eating birds. Two cormorant species, *Phalacrocorax carbo* (offshore diver) and *P. africanus* (inshore diver), the African fish eagle, *Haliaeetus vocifer*, and two offshore plungers the white-winged black tern, *Chlidonias leucogaster*, and the pied kingfisher, *Ceryle rudis* were considered. Most of these birds formerly fed on haplochromines but the main food now is *R. argentea* (Goudswaard & Wanink 1993). Information on quantitative food consumption of these birds was taken from Hustler (1997). On a shore line basis, the biomass of these birds are, most likely, within the range of the biomass observed in other African water bodies, such as Lake Kariba. To compute an estimate of the biomass, the density of birds along the shorelines was expressed in km^{-2}. *P/B* and *Q/B* were average values also obtained from Hustler (1997). It should be noted that the biomass value might not be accurate but incorporation of this box into the model should provide an estimate on the impact of this group on the fish community compared to fisheries exploitation.

Nile perch was separated between the adult and the juvenile life stages in relation to *P/B*, *Q/B* and feeding habits. The maximum size of juveniles was assumed to be 40 cm TL because this was the minimum size caught by large mesh size gill nets and long lines and they start to escape predation by adults (Ogari & Dadzie 1988). Juvenile *Lates* are mostly observed in the littoral areas of the lake (Hughes 1986). *P/B* was computed using the FiSAT software (Gayanilo, Sparre & Pauly 1996) on length frequency distributions for relevant sizes using data from Asila & Ogari (1987), which gave an average value of 3.5. For *Q/B*, two computations were performed separately for adult and juveniles using MAXIMS (Jarre, Palomares, Soriano, Sambilay, Christensen & Pauly 1990).

Growth parameters followed Moreau *et al.* (1993) for the period of 1985–1986, i.e. $W_\infty = 76\,000$ g; $k = 0.36$; $\beta = 0.10$; $Z = 0.90$; $W_r = 2000$ g; $W_{max} = 72\,000$ g; $Q/B = 5.03$; and $EE = 0.16$ for adults. The same parameters were used for juveniles except: $Z = 3.5$; $W_r = 25$ g; $W_{max} = 2000$ g; $Q/B = 11.8$ and $GE = 0.29$. (Note that *GE* is higher in juveniles compared to adults which is in agreement with the basis of the method of computation of *Q/B* implemented in MAXIMS publications (Pauly & Palomares 1987).)

EE was assumed to be 0.98 for juveniles since they are caught or cannibalised by adults. Feeding and diet composition of these two groups were established according to Ogari & Dadzie (1988) noting the spatial distribution of the fish. The age of transition from juvenile to adult is about 3 years. The average ratio of transition from juvenile to adult is five, which is achieved by dividing the maximum weight of the juvenile (2 kg) by the average weight of adults (10 kg), as required for a proper use of ECOSIM (Walters *et al.*, 1997).

It is necessary to split the fish yield according to the fishing gears used (Pitcher, Bundy & Neill 1996; Wanink *et al.* 1999; Table 9.1). Gill nets are mainly used for large *Lates* and *O. niloticus*. Long lines are used for adult *Lates* in open waters and bottom

Table 9.1 Landings of the Kenyan sector of Lake Victoria, Africa (derived from Moreau *et al.* 1993 and Ochumba 1995). Note that they have been segregated among various current and potential fishing gears for a proper utilisation of ECOSIM

Groups	*Lates* long line	Lift nets	Gill nets	Beach seines	Hand lines	Trawls	Total
Adult *Lates*	4.00		4.00	2.00			10.00
Juvenile *Lates*			1.00	1.50		0.50	3.00
Bagrus + *Clarias*					0.15		0.15
Protopterus					0.03		0.03
Mormyrids + *Synodontis*					0.03		0.03
Predatory haplochromines				0.01			0.01
Planktivorous haplochromines				0.01			0.01
Benthophagous haplochromines				0.02			0.02
R. argentea		6.40					6.40
O. niloticus			0.80	0.60		0.20	1.60
Other tilapias				0.35			0.35
Total catch	4.00	6.40	5.80	4.49	0.21	0.70	21.60
Trophic level	3.58	2.83	3.33	3.17	3.18	2.96	3.18

zoobenthophagous species of minimum importance in littoral areas. Beach seines are used to catch juvenile *Lates*, Tilapiine fish, and haplochromines. Mosquito nets and scoop nets are mainly used to catch *Rastrineobola*. Trawls were also in use but were banned in the 1990s.

9.3 Results

Table 9.2 shows the current estimates of biomass and gross conversion efficiency and other key features of the present ECOPATH IV model for Lake Victoria (Fig. 9.2), and Table 9.3 summarises the diet composition of the groups considered. Note that biomass values were not changed from Moreau *et al.* (1993) except for *R. argentea.* The increase in biomass estimate can be attributed to the consumption by *L. niloticus* being better accounted for. However, the resulting biomass of *R. argentea* is still within the range suggested by Pitcher *et al.* (1996). The addition of fish-eating birds in the model does not seem to have much impact on the ecosystem structure when compared to Moreau *et al.* (1993).

Variations in biomass and catch with time on a per gear basis used on a specific target group and on a per group basis were analysed with ECOSIM. The trends were the same for any group caught using a single specific gear. Fishing effort was multiplied by two during the first 5 years (as suggested by Pitcher & Bundy 1995) and an additional increase following Mkumbo *et al.* (Chapter 8) for the next 5 years, e.g. until 1995–1996. As a whole, from the mid-1990s to the late 1990s, fishing effort increased by a factor of 2.5–3 according to the fishing gear and/or target species (Fig. 9.3). For the main zoobenthophagous species, *Bagrus/Clarias*, *Protopterus* and *Mormyrops/Synodontis* species, an increase in catch was predicted for the period up to the late 1990s and is expected to remain at that level (Fig. 9.3). Biomass would remain constant except for *Bagrus/Clarias* (a slight decrease by 15%). No records exist regarding the increase in catch of these groups except that between 1994 and 1997 landings increased drastically (Mkumbo *et al.*, Chapter 8) probably due to the high commercial value of these groups when smoked. Njiru, Othina & Cowx (Chapter 21) indicated that infestation by water hyacinth, *Eichhornia crassipes* (Mart), provided sufficient shelter for breeding of these groups that may have permitted the increase in biomass and resulting catch.

The haplochromine stocks seem to be showing signs of recovery since the catches have increased 2.5 times in the past 10 years, but the biomass remained quite constant at a very low level. Further, it appeared that variations in catch or biomass of this group, especially the haplochromine predators, were indirectly affected when running other scenarios (i.e. considering other gears on other target groups) suggesting the sensitivity of this group to variations of the fishing activity with other gears and target fishes. The current increase of abundance of haplochromines is now documented (Mkumbo *et al.*, Chapter 8; Njiru *et al.*, Chapter 21).

Considering dagaa (*R. argentea*), the biomass was expected to have halved after 10 years with the catch reaching a maximum after 6 years (twice that recorded in 1985–1986) and then decreasing slowly by 30% in an apparent pattern of overexploitation, as

Table 9.2 Key features of the ECOPATH model for the Kenyan sector of Lake Victoria. The trophic levels, food consumption flow to detritus and the biomass for all groups (except aquatic birds) have been computed by the model. The inputs, *P/B*, *Q/B* and *EE* are documented in Moreau *et al.* (1993)

Groups	Trophic level	Biomass $t\,km^{-1}$	Production/ biomass $year^{-1}$	Consumption/ biomass $year^{-1}$	Ecotrophic efficiency	Production/ consumption	Flow to detritus $t\,km^{-2}$ $year^{-1}$	Production efficiency $t\,km^{-2}$ $year^{-1}$	Net efficiency
Fish-eating birds	3.88	0.005	0.30	60.0	0.00	0.005	0.062	0.000	0.006
Adult *Lates*	3.58	12.019	0.90	5.0	0.98	0.180	12.236	10.601	0.225
Juvenile *Lates*	3.32	5.055	3.50	11.8	0.98	0.297	12.284	17.340	0.371
Bagrus + *Clarias*	3.21	1.909	0.85	6.5	0.95	0.131	2.563	1.541	0.163
Protopterus	3.13	1.606	0.30	4.3	0.95	0.070	1.406	0.458	0.087
Mormyrids/Synodontis	3.08	0.535	0.90	11.5	0.95	0.078	1.256	0.458	0.098
Predatory haplochromines	3.71	0.195	2.50	8.5	0.95	0.294	0.356	0.463	0.368
Planktivorous haplochromines	2.05	0.196	3.00	41.0	0.95	0.073	1.634	0.558	0.091
Benthophagous haplochromines	3.02	0.423	3.00	21.0	0.95	0.143	1.842	1.207	0.179
R. argentea	2.83	10.301	2.80	25.0	0.95	0.112	52.946	27.400	0.140
O. niloticus	2.06	12.721	0.90	24.6	0.95	0.037	63.160	10.876	0.046
Other tilapias	2.05	0.680	1.20	32.0	0.95	0.038	4.395	0.776	0.047
Zooplankton	2.02	10.718	33.50	140.0	0.95	0.239	335.997	323.136	0.299
Caridina nilotica	2.26	5.964	16.00	64.0	0.95	0.250	81.115	90.658	0.313
Insects/mollucs	2.11	25.128	5.00	25.0	0.80	0.200	150.768	100.512	0.250
Phytoplankton	1.00	5.132	365.00	0.0	0.95	–	93.652	1779.393	–
Benthic production	1.00	12.857	25.00	0.0	0.95	–	16.071	305.357	–
Detritus	1.00	10.000	–	–	0.74	–	0.000	–	–

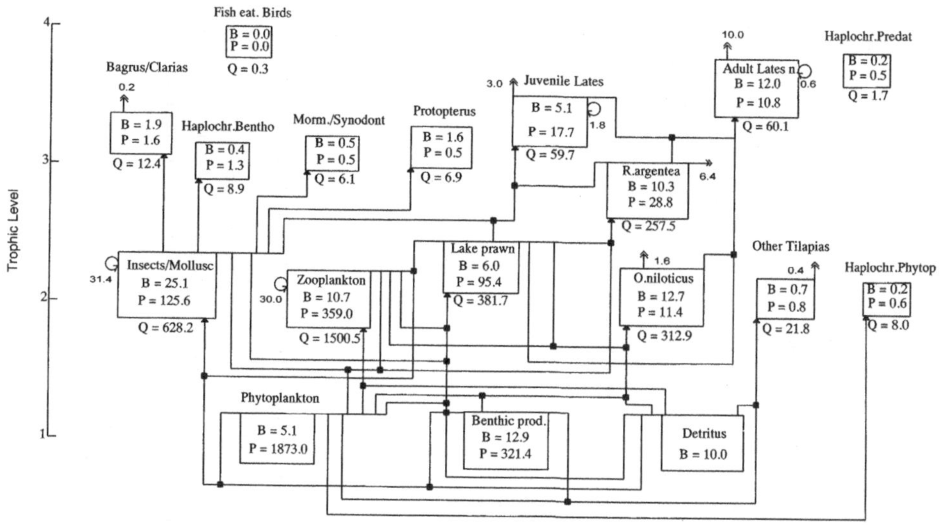

Figure 9.2 The ECOPATH model of Lake Victoria for 1985–1986, in which adult and juvenile *Lates niloticus* have been separated and indicating the biomass of each group and the major flows connecting them. For clarity, less important flows are omitted as are backflows to the detritus box. The horizontal axis of symmetry of each box is aligned with the functional trophic level of this box (see Christensen & Pauly 1992 for details). Actual catch and cannibalism are displayed

predicted by Pitcher *et al.* (1996). Mkumbo *et al.* (Chapter 8) reported, however, an increase of *R. argentea* catch by about three-fold up to the mid-1990s, which was not predicted in the present exercise.

The catch of native tilapias was expected to increase by 20% after 2 years and then decrease progressively to about 33% of the original value. The biomass was predicted to continue to decline to on 12% of the original value. No clear explanation for this possible collapse of native tilapias, which are still regarded as of minor importance in the fish community of the lake, even as a whole, was found.

Fish groups (*O. niloticus* and adults and juvenile *Lates*) caught using multiple gears were analysed on a per species basis as well as by gear utilised. The total catch of *O. niloticus* was predicted to increase by 1.75 after 2 or 3 years and then remain constant, a trend which is in agreement with Mkumbo *et al.* (Chapter 8), and the biomass would decrease regularly by a total of 20% within 10 years. The trends in biomass and catch of *L. niloticus* were difficult to identify, but alternating oscillations in abundance of adults and juveniles were predicted (Fig. 9.4). This exhibits the prey-predatory relationship between the two. Similar patterns are observed for catch with various gears with a slow increase in catch of one of the two groups coupled with a decrease in the catch of the other.

On average, the biomass of the whole Nile Perch population would decrease by 20% during the first 5 years (as suggested by Pitcher & Bundy 1995). An additional limited decrease would take place thereafter leading to a total decrease of 25% of the biomass by the end of the 1990s.

Table 9.3 Diet composition (proportion) of the various groups for ECOPATH in the Kenyan sector of Lake Victoria, Africa (from Ogari & Dadzie 1988, for the two groups of *Lates niloticus*). For other groups see Moreau *et al.* (1993)

	Predator (column number refers to category given in column 1)														
Prey item	1	2	3	4	5	6	7	8	9	10	11	12	13	14	15
1 Fish-eating birds															
2 Adult *Lates*		0.010													
3 Juvenile *Lates*		0.200	0.030	0.040	0.050										
4 *Bagrus* + *Clarias*		0.020		0.010	0.050	0.050									
5 *Protopterus*		0.005		0.005	0.050	0.050									
6 *Mormyrids/Synodontis*		0.005		0.005	0.050	0.050									
7 Predatory haplochromines	0.150	0.003	0.002	0.001	0.001	0.001	0.050								
8 Planktivorous haplochromines	0.150	0.002	0.003	0.002	0.001	0.001	0.100								
9 Benthophagous haplochromines	0.200	0.005	0.010	0.002	0.003	0.003	0.100								
10 *R. argentea*	0.500	0.150	0.150	0.100	0.050	0.050	0.600								
11 *O. niloticus*		0.100	0.040	0.040	0.020	0.040									
12 Other tilapias			0.005	0.005	0.005	0.005									
13 Zooplankton			0.050	0.030	0.050	0.050	0.050	0.020	0.050	0.700	0.020	0.020	0.020	0.200	0.040
14 *Caridina nilotica*		0.450	0.500	0.200	0.150	0.050	0.050		0.150	0.050	0.030				0.010
15 Insects/mollucs		0.050	0.200	0.500	0.650	0.700	0.050	0.030	0.700	0.050		0.030	0.950	0.050	0.050
16 Phytoplankton								0.500		0.200	0.700	0.450		0.100	0.050
17 Benthic production				0.010		0.025		0.250	0.050		0.100	0.300	0.030	0.200	0.300
18 Detritus			0.010	0.050	0.050	0.050		0.200	0.050		0.150	0.200		0.450	0.550
20 Import						0.010									

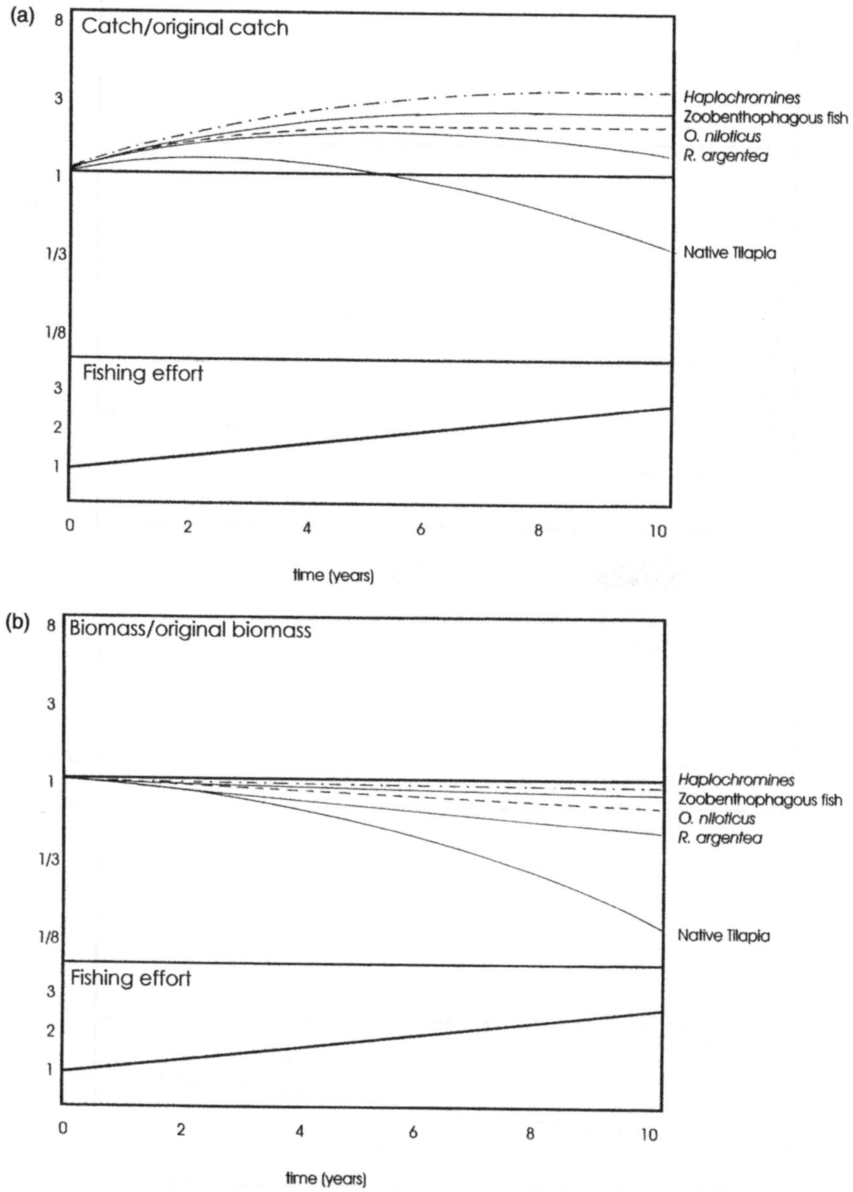

Figure 9.3 The variations of catch and biomass of the fish groups other than *Lates niloticus* over 10 years as described with ECOSIM under various scenarios. Shown here are (a) trends in catch related to the variations of fishing effort using separated fishing gears and their main target species, (b) simultaneous changes in biomass

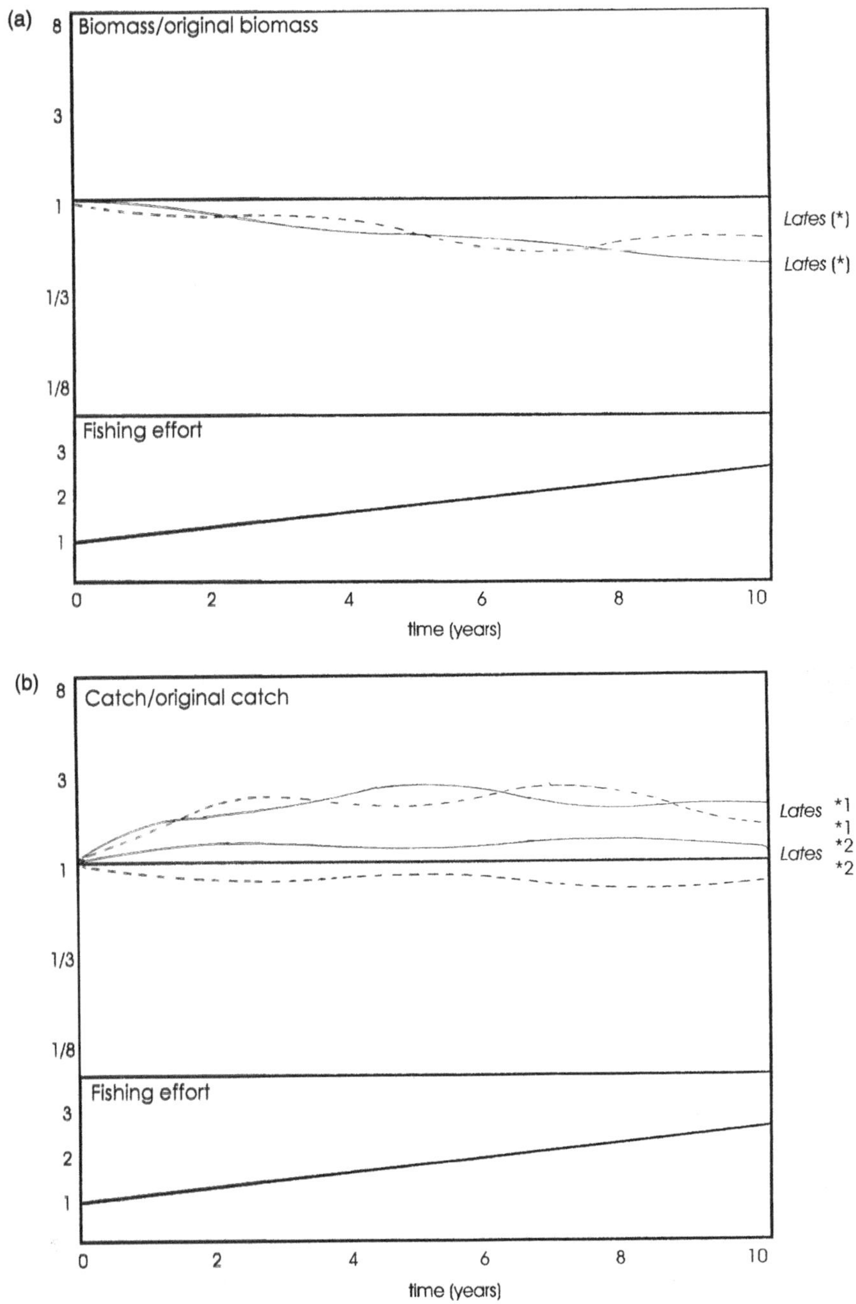

Figure 9.4 Variation in catch and biomass of *Lates niloticus*, juveniles and adults, under various scenarios. (a) Biomasses of adult and juveniles showing opposing trends that are generally decreasing. The presence of (*) indicates that both can assume either one of the trends depending on the group being considered. (b) Catch: two scenarios were considered. The first scenario (*1) was based on the target species which resulted in the same opposing oscillations but generally increasing by two after 5 years and starts to slightly decline after 10 years. The second scenario (*2) was based on target gears. The same oscillating patterns remain, however, at constant level throughout the 10 years considered

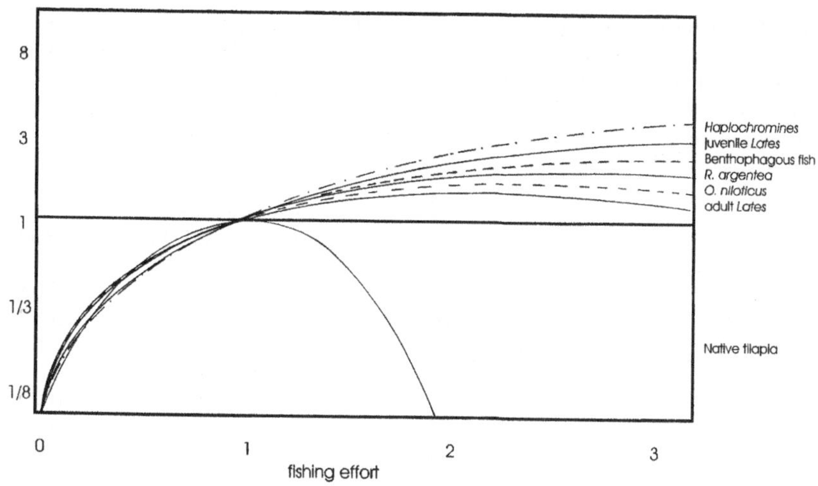

Figure 9.5 Variation of catch of the main fish groups in relationship to the fishing effort. Note the declining trend of the local tilapias and the overfishing trend of *R. argentea*, Nile tilapias and adult *Lates*

When considered as the target group of the simulation exercise, adult *Lates* displayed an increase in catch by a maximum of two-fold after 5 years and a slow and limited decrease thereafter (about 15%), whereas juvenile *Lates* displayed a continuous simultaneous decrease of their catch by 25%. Conversely, when juvenile Nile perch were the target group, their catch appeared to be multiplied by 1.75 after some years but did not exhibit a decreasing pattern at the end of the 1990s. In that case, adult *Lates* showed a continuous decrease in catch by about a total of 25% after 10 years. Considering the information available from Lake Victoria regarding the relative importance of the various fishing gears during the last 10 years (Njiru *et al.*, Chapter 21), this last scenario seems to be the most appropriate one.

To account for possible interactions between the various scenarios related to specific fishing gears, a scenario in which the fishing activity, as a whole with all combined gears, in the Winam Gulf would increase by 2.5 during the 10 years was tested. The results confirmed most of the trends observed with scenarios involving specific gears both in terms of biomass and catch and quoted above. The equilibrium curves (Fig. 9.5) providing the relationship between catch and relative fishing effort showed a collapse of local tilapias, and possible overfishing on *R. argentea*, Nile tilapia and adult *L. niloticus*. Other groups could support a significant relative increase of the current low fishing pressure without risk of collapse.

The catch estimates, when associated with updated estimates of *P/B* for the key groups for the mid-1990s and recomputed values of *Q/B* to keep the gross efficiency constant, led to an updated equilibrium model for the mid-1990s (Table 9.4). It should be noted that the feeding matrix has not been changed. To retain the biomass of zooplankton, *Caridina* and benthos constant, their *EE* was reduced to 0.8, indicating that these groups are no longer fully utilised in the current hypothetical model (Fig. 9.6). In addition, the fish biomasses computed were close to those obtained from ECOSIM (Table 9.4).

Table 9.4 Estimates of catch ($t\,km^{-2}\,year^{-1}$) and biomass ($t\,km^{-2}$) of fish in Lake Victoria (Kenyan sector) in 1995–1996 as derived from ECOSIM simulation from the situation in 1985–1986. The biomasses can be compared with data for 1995–1996 using the equilibrium model. Figures between brackets are from Mkumbo *et al.* (Chapter 8) in which the figure for *Lates niloticus* is for adults and juveniles together

		Biomass ($t\,km^{-2}$)	
Groups	Catch ($t\,km^{-2}\,year^{-1}$)	From ECOSIM	From ECOPATH
Adult *Lates*	7.50	9.00	7.65
Juvenile *Lates*	5.00 (13)	3.75	3.83
Bagrus/Clarias	0.35	1.60	1.62
Protopterus	0.08	1.50	1.37
Mormyrids/Synodontis	0.08	0.48	0.45
Predatory haplochromines	0.03	0.18	0.19
Planktivorous haplochromines	0.03	0.18	0.21
Benthophagous haplochromines	0.05	0.40	0.43
R. argentea	9.00	(16.60)	6.00
O. niloticus	2.80 (2.2)	10.10	10.10
Other tilapias	0.12	0.15	0.43

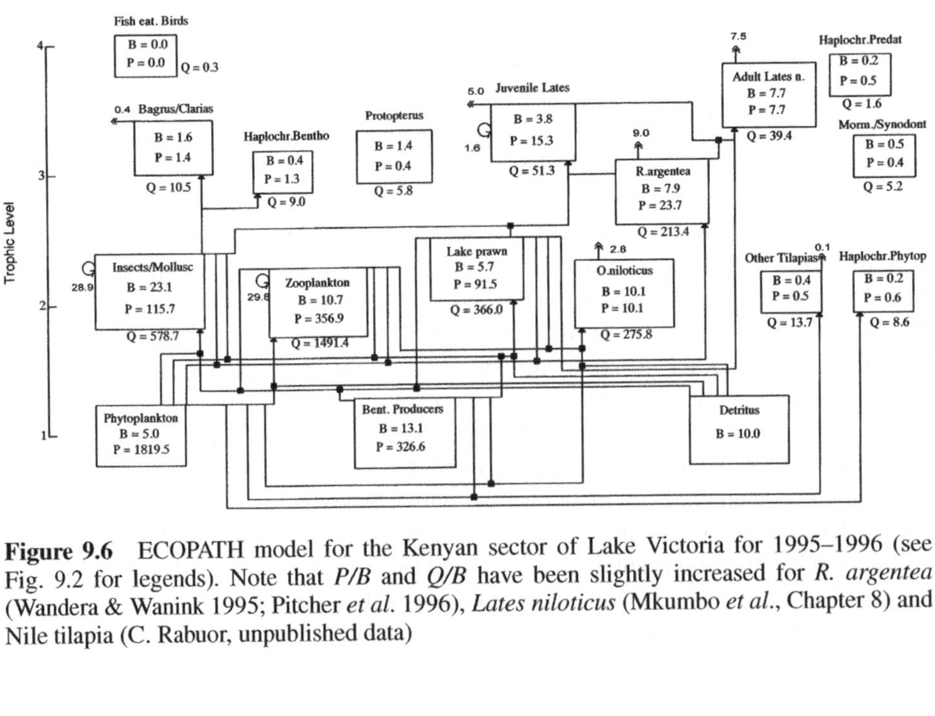

Figure 9.6 ECOPATH model for the Kenyan sector of Lake Victoria for 1995–1996 (see Fig. 9.2 for legends). Note that *P/B* and *Q/B* have been slightly increased for *R. argentea* (Wandera & Wanink 1995; Pitcher *et al.* 1996), *Lates niloticus* (Mkumbo *et al.*, Chapter 8) and Nile tilapia (C. Rabuor, unpublished data)

9.4 Discussion

A group of fish-eating birds were included in the present exercise to show their increasing exploitation on the lake's fish resources (specifically haplochromines and, at present, *R. argentea*) (Wanink *et al.* 1999). There is a need to study the influence of this group due to possible intensive competition it imposes against top predators, for food, at least in the littoral zones. To some extent, it could also contribute to the declining landings of *R. argentea* being observed currently in some parts of the lake (Mkumbo *et al.*, Chapter 8; Njiru *et al.*, Chapter 21).

Possible reasons for the oscillating and yet opposite behaviour in trend of biomass and catch values between the adult and the juvenile *Lates* need investigation. It might be due to a particular stock recruitment relationship and because, although cannibalism takes place, the adult also preys on other competitors of the juveniles, giving them more food resources to exploit and allowing them to grow to the adult stage.

According to Walters *et al.* (1997), an increased density of adult *Lates* would first lead to a decrease of juvenile biomass simply by predation, whereas, at a further stage, it would favour the decrease in density of other predators and competitors. This would result in improving feeding conditions and a simultaneous decrease of the predation pressure for these juveniles.

Njiru *et al.* (Chapter 21) referred to high catches of both Nile perch and *R. argentea* in the Kenyan sector which were not simulated in the present exercise. These figures might be overestimates of the actual catch coming from the Kenyan waters of the lake because Tanzanian and Ugandan fishermen prefer to land their catches on Kenyan landing beaches where they get higher prices for their fishes (R. Welcomme, personal communication).

Although it provides simulation of catch and biomass trends over the study period, ECOSIM has some limits in its predictive power which come from the lack of simulation of the changes in fishing effort with several fishing gears and the resulting influence on the biomass of each fish species. In addition, it would be necessary to incorporate in the scenarios the possible shifts in the water quality, mostly in terms of its effect on primary production.

A new steady-state model for the Kenyan sector could be made incorporating the estimates of actual catch appearing from the present study for the end of the 1990s which are close to the catch statistics available for recent years. The construction of a model for the whole lake has now to be considered, accounting for spatial distribution of the fish groups in the lake and their specific habitats.

References

Asila A. & Ogari J. (1987) Growth parameters and mortality rates of Nile perch (*Lates niloticus*), estimated from length–frequency data in the Nyanza Gulf (Lake Victoria). *FAO Fisheries Report* **389**, 1272–1287.

Bundy A. & Pitcher T.J. (1995) An analysis of species changes in Lake Victoria: did the Nile perch act alone? In T.J. Pitcher & P.J.B. Hart (eds) *The Impact of Species Changes in African Lakes*. London: Chapman & Hall, pp. 111–135.

Christensen V. & Pauly D. (1992) ECOPATH II a software for balancing steady state ecosystem models and calculating network characteristics. *Ecological Modelling* **61**, 169–185.

Froese R. & Pauly D. (eds) (1999) Fishbase 99 Concepts, structure, and sources of the data. ICLARM Publication No. 1555. Manila, Philippines: ICLARM, 324 pp.

Gayanilo Jr F.C., Sparre P. & Pauly D. (1996) The FAO-ICLARM Stock Assessment Tools User's manual. Rome: *FAO Computer Information Series (Fish.)* **8**, 126 pp.

Goudswaard K.P.C. & Wanink J. (1993) Anthropogenic perturbation in Lake Victoria: Effects of fish introductions and fisheries of fish eating birds. *Proceedings of the VIII Pan-African Ornithological Congress*, pp. 312–318.

Hughes N.F. (1986) Changes in the feeding habits of Nile perch, *Lates niloticus* (L.), in the Lake Victoria, East Africa since its introduction in 1960 and its impact on the native fish community of the Nyanza Gulf. *Journal of Fish Biology* **29**, 541–548.

Hustler K. (1997) The Ecology of fish-eating birds and their impact on the inshore fisheries of Lake Kariba. In J. Moreau (ed.) *Advances in the Ecology of Lake Kariba*. Harare: University of Zimbabwe Publications, Zimbabwe, pp. 196–218.

Jarre A., Palomares M.L.D., Soriano M.L., Sambilay M.L., Christensen V. & Pauly D. (1990) A user's manual for MAXIMS. A computer program for estimating the food consumption of fishes from diel stomach contents data and population parameters (ICLARM Software 4), Manila, Philippines: ICLARM, 27 pp.

Kudhongania A.W. & Chitamwebwa D.B.R. (1995) Impact of environmental changes, species introductions and ecological modifications on the fish stocks of Lake Victoria. In T.J. Pitcher & P.J.B. Hart (eds) *The Impact of Species Changes in African Lakes*. London: Chapman & Hall, pp. 19–32.

Ligtvoet W. (1989) The Nile perch in Lake Victoria: A blessing or a disaster? *Annals Museum Royale Afrique Centre, Science et Zoologique* **257**, 151–156.

Manyala J.O., Vanden Berghe E. & Dadzie S. (1995) Mortality, exploitation rate and relative yield of *Rastrineobola argentea* (Pellegrin 1904) in the Winam Gulf of Lake Victoria (Kenya). *Science and Marine* **59**(3–4), 555–563.

Mkumbo O.C. & Ligtvoet W. (1992) Changes in the diet of Nile perch, *Lates niloticus* (L.), in the Mwanza Gulf, Lake Victoria. *Hydrobiologia* **232**, 79–83.

Moreau J., Ligtvoet W. & Palomares M.L.D. (1993) Trophic relationship in the fish community of Lake Victoria, Kenya, with emphasis on the impact of Nile perch (*Lates niloticus*). In V. Christensen & D. Pauly (eds) *Trophic models of aquatic ecosystems*. Manila: ICLARM Conference Proceedings, **26**, pp. 144–152.

Ogari J. & Dadzie S. (1988) The food of the Nile perch, *Lates niloticus* (L.), after the disappearance of the haplochromine cichlids in the Nyanza Gulf of Lake Victoria (Kenya). *Journal of Fish Biology* **32**, 571–577.

Palomares M.L. & Pauly D. (1998) Predicting food consumption of fish populations as functions of mortality, food type, morphometrics, temperature and salinity. *Marine and Freshwater Research* **49**, 447–453.

Pauly D. & Palomares M.L. (1987) Shrimp consumption by fish in Kuwait waters. A methodology, preliminary results and their implication for management and research. *Kuwait Bulletin of Marine Sciences* **9**, 101–125.

Pitcher T.J. & Bundy A. (1995) Assessment of the Nile perch fishery in Lake Victoria. In T.J. Pitcher & P.J.B. Hart (eds) *The Impact of Species Changes in African Lakes*. London: Chapman & Hall, pp. 163–180.

Pitcher T.J., Bundy A. & Neill W.E. (1996) The fishery for *Rastrineobola argentea* in Lake Victoria: estimation of potential yields using a new approximation model based on primary production. *Fisheries Research* **28**, 133–149.

Reynolds J.E. & Greboval D.F. (1995) Socio-economic effects of the evolution of Nile perch fisheries in Lake Victoria: a review. FAO: *CIFA Technical Paper* **17**, 148 pp.

Walters C., Christensen V. & Pauly D. (1997) Structuring dynamic models of exploited ecosystems from trophic mass balanced assessments. *Reviews in Fish Biology and Fisheries* **7**, 139–172.

Wandera S.B. & Wanink J.H. (1995) Growth and Mortality of Dagaa (*Rastrineobola argentea*, Fam. Cyprinidae) in Lake Victoria. *NAGA, The ICLARM Quarterly* **18**(1), 42–45.

Wanink J.H. & Goudswaard K.P.C. (1994) Effects of Nile perch (*Lates niloticus*) introduction into Lake Victoria, East Africa, on the diet of Pied Kingfishers (*Ceryle rudis*). *Hydrobiologia* **279/280**, 367–376.

Wanink J.H., Goudswaard K.P.C. & Berger M.R. (1999) *Rastrineobola argentea*, a major resource in the ecosystem of lake Victoria, pp. 295–309. In W.L.T. van Densen & M.J. Morris (eds) *Fish and Fisheries of Lakes and Reservoirs in Southeast Asia and Africa.* Otley (Yorshire, UK): Westbury Academic & Scientific Publishing, 432 pp.

Wilson D.C., Medard M., Harris C.K. & Wiley D.S. (1996) Potentials for comanagement of the Nile Perch Fishery-Lake Victoria, Tanzania. *Proceedings of the Voices from the Common Conference of the International Association for the Study of Common Property*, Berkeley, CA. 5–8 June 1996.

Witte F., Goldschmidt T., Wanink J.H., van Oijen M., Goudswaard K., Witte-Maas E. & Bouton N. (1992) The destruction of an endemic species flock: quantitative data on the decline of the haplochromine cichlids of Lake Victoria. *Environmental Biology of Fishes* **34**, 1–28.

Witte F., Goldschmidt T. & Wanink J.H. (1995) Dynamics of the haplochromine cichlid fauna and other ecological changes in the Mwanza Gulf of Lake Victoria. In T.J. Pitcher & P.J.B. Hart (eds) *The Impact of Species Changes in African Lakes.* London: Chapman & Hall, pp. 83–110.

Chapter 10
Fish population structure and its relation to fisheries yield in small reservoirs in Côte d'Ivoire

L. TITO de MORAIS*
Centre IRD de Bouaké, Bouaké, Côte d'Ivoire

Abstract

Lack of reliable data on the fisheries of northern Côte d'Ivore reservoirs is a major limitation for management of the fish stocks. As individual reservoirs cannot be studied due to the high costs and lack of trained investigators, empirical and production models have to be used to evaluate exploitation rates and fisheries potential. Daily yield (DY, in kg fishing $day^{-1} year^{-1}$) and total catch (TC, in kg $year^{-1}$) appear to be better descriptors than annual yield (in kg $ha^{-1} year^{-1}$). TC is related to variables linked to nutrient availability (watershed area, conductivity and Morphoedaphic Index). DY is related to fish species richness and to reservoir area, indicating that fish population structure could be a key to understanding the patterns observed. Correlation between fish population structure and yield was significant. It was hypothesised that overall catch, TC, is related to nutrient availability, but that catch per unit effort is related to the presence of given species in the reservoirs, and particularly to the presence of bottom feeding species.

Keywords: fish population structure, fisheries, reservoirs, yield models.

10.1 Introduction

At first glance, small-scale reservoir fisheries appear to be of marginal importance in the national fisheries statistics of southern countries. Côte d'Ivoire, for example, declares almost no yield from its small-scale reservoirs to FAO statistics. This is mostly related to a lack of knowledge of its importance rather than a true image of the fisheries situation in such water bodies. A preliminary study (Da Costa, Traore & Tito de Morais 1998) indicated that small-sized (<10 ha) and medium-sized (10–100 ha) reservoirs may contribute between 16% and 46% of the 13 447 t $year^{-1}$ of total freshwater fisheries production in Côte d'Ivoire. The large variation in these estimates depends upon the variation in yield estimates in the reservoirs. Several models and relationships between annual fish yield and morphometric and biological data have been proposed for tropical reservoirs (see Amarasinghe 1996 for review). But to be operational, such models need to be adapted to local conditions, and plausible biological criteria for such

*Correspondence: Centre IRD, BP 1386, Dakar, Senegal (email: tito@ird.sn).

relationships have to be tested (Amarasinghe 1996). The present chapter is the first attempt to adapt empirical and production models to Côte d'Ivoire small-reservoir fisheries.

10.2 Study area

The aim of this study was to test for different relationships between fisheries yield and lake and watershed morphology and fish community-related variables. Such variables are more easily obtained in the field than are catch and effort data. The collection of the latter requires more than 1 year of observation of daily catch in each reservoir. Conversely, morphometric data can be obtained by remote sensing, map observation, or by means of a single field trip. Fish assemblage data already exist or can be obtained with few (three–four) sampling sessions in each reservoir.

The major applications of such relationships are in ecological monitoring and in fisheries management. Southern countries lack funding and skilled investigators to efficiently monitor small-scale fisheries, which are numerous and scattered. It is essential to distinguish between myths and reality when it comes to discussing the fishing importance of small-scale reservoirs. Arguments related to the importance of fish as a by-product of water supply reservoirs are widespread. Yet, no one can really give accurate figures of fisheries yield for small- and medium-size reservoirs in the north of Côte d'Ivoire. The only regional study appears to be that of Baijot, Moreau & Bouda (1994) in Burkina Faso, where the reservoirs are larger (15–1500 ha) than in Côte d'Ivoire and the fisheries organisation is totally different due to a lack of marine fisheries.

10.3 Materials and methods

Over 300 reservoirs have been built in Côte d'Ivoire, mostly for water supply for cattle and irrigation (Table 10.1). Reservoirs for watering cattle are generally smaller than those devoted to other uses. Two hundred and sixty-nine of the 297 small reservoirs built for cattle are in the northern region. Nearby inhabitants have been using the reservoirs for other purposes such as fishing, hygiene and recreation. As people living in the area are mostly from the Senoufo group, they consider themselves as cultivators and are reluctant to fish. Foreigners such as the Bozos from Mali do the fishing. The Bozos pay a levy in fish, money or other form, to exploit the reservoirs. Da Costa *et al.* (1998) showed that fishing is an important (with regard to local standards) source of income for local people. Fish is also important in nutritional terms in the northern region of the country, as other sources of protein are expensive or unavailable.

The traditional fishing gears used in the small reservoirs in the north of the Côte d'Ivoire are mainly (by order of importance) gillnets, traps, hooked lines (baited or unbaited), and cast nets. Beach seines are seldom used. In most cases, at any given period, one single fisherman practises in each given reservoir. When a fisherman leaves a reservoir, another one rapidly replaces him. Fishermen may change from one reservoir to another, or travel back to Mali for a while. All fishermen use similar fishing

Table 10.1 Number of reservoirs in Côte d'Ivoire according to river basin and main use

	Main use			
River basin	Irrigation	Drinking water	Cattle watering	Total
Agnébi	3	4	5	12
Bandama	33	6	115	154
Bia	0	0	2	2
Comoé	3	5	57	65
Marahoué	3	1	8	12
N'zi	29	9	22	60
Niger	2	4	60	66
San Pedro	1	5	2	8
Sassandra	5	0	3	8
Black Volta	0	0	23	23
Total	79	34	297	410

Only small-size (area <10 ha) and medium-size (area 10–100 km^2) reservoirs are listed.
Sources: Anonymous 1992 and Gourdin 1999.

gears and similar fishing strategies, allowing use of fishing day as a measure of fishing effort in the reservoirs.

Due to staffing limitations, rather than collecting superficial data from many reservoirs, fishing activity at seven reservoirs was studied in detail. These reservoirs (Table 10.2) are in the same area and belong to two of the largest river basins in Côte d'Ivoire: the Bandama (five dams) and the Comoé (two dams). Three are medium-sized reservoirs, and four small-sized reservoirs. Overall species richness (SR) is similar in the two rivers (88 and 91 species) and fish SR varies from 10 to 24 in the reservoirs.

10.3.1 *Fish catches*

The study lasted 27 months from April 1996 to June 1998. One observer was appointed to each reservoir to record daily fish landings, registering: fisherman's name, number of fish by species, total weight by species and by sub-sample, gear used (when available), duration of fishing activity. Other information such as weather conditions, fish selling prices, activity of fishermen when not fishing, were also recorded but are not presented here. The data were enumerated as total catch per year (TC, kg year^{-1}), annual yield (AY, kg ha^{-1} year^{-1}), unit effort (fishing days) and daily catch per unit effort (CPUE), i.e. daily yield (DY, kg day^{-1} ha^{-1}). Unit effort was chosen to compare yield values between reservoirs. The only effort variable available in this context, fishing day, was used. This means that if nobody fishes in a given reservoir, the value for that day is 0. If only one person fishes the value is 1 for that day. If two persons fish the value is 2 and so on. So in the larger reservoirs (e.g. Katiali and Tiné) where there are several fishermen, it is possible to have more than 365 fishing days in 1 year. Even though this variable is not very precise, it has been used widely in the fisheries studies in Africa.

Table 10.2 Characteristics of river and reservoir used in this study

River or reservoir name	Area (km^2)[b]	Maximum *D* (m)[b]	Type	Main river basin	SR	WA (km^2)
Bandama	–	–	River	Bandama	88[a]	97 000[a]
Comoé	–	–	River	Comoé	91[a]	78 000[a]
Tiné	0.45	10.0	Medium dam	Bandama	18[c]	9.4[b]
Katiali	0.24	2.5	Medium dam	Bandama	33[c]	16.8[b]
Sambakaha	0.15	4.2	Medium dam	Bandama	31[c]	21.2[b]
Gboyo	0.07	2.9	Small dam	Bandama	29[c]	10.2[b]
Korokara T	0.07	3.8	Small dam	Bandama	22[c]	1.1[b]
Nambingué	0.10	2.9	Small dam	Comoé	23[c]	11[b]
Tiaplé	0.07	3.3	Small dam	Comoé	20[c]	6[b]

After: [a]Lévêque 1997. [b]Gourdin 1999. [c]This study.

Fishermen generally combine their catch when they land. It is seldom possible to separate the catch according to the fishing gears used, although fishermen generally use the same fishing gears and strategies every day.

10.3.2 *Fish population structure*

SR and fish population structure were investigated using bimonthly beach seine and gillnet samples between January 1998 and September 1999. Gillnets were set overnight using two, 13-net gangs. Gillnets were monofilament nets, 25 m in length, 2 m in height. Mesh sizes were 10, 12.5, 15, 17.5, 20, 22.5, 25, 30, 40, 50, 60, 70, 80 mm, knot-to-knot. Beach seining (14-mm multifilament mesh size) was performed, in the morning around 07.00 h.

10.3.3 *Statistical analysis and tests*

Significance of morphometric and production models was tested by correlation. Several paired combinations of both were tested: SR on reservoir area (RA, ha), SR on watershed area (WA, km^2), SR on the ratio WA/RA, catch on SR, catch on RA, catch on WA, catch on the ratio WA/RA, catch on depth (*D*, m), catch on conductivity (*C*, μS), catch on fishing intensity (FI, fishing days km^{-2}), and catch on morpho-edaphic index (MEI) ($D \times C$). Catch was tested as TC per year (kg $year^{-1}$), AY (kg ha^{-1} $year^{-1}$) and as DY (kg day^{-1} ha^{-1}).

Correspondence analysis of the matrix of fish abundance per reservoir was used to investigate the relationship between yield and species composition. Reservoirs were plotted in the factorial space of the correspondence analysis on the matrix of species abundance per reservoir. The clustering of reservoirs according to their scores in the factorial space was compared to the grouping of reservoirs according to yield. The clustering distance used was the Euclidian distance. The clustering method used was average linkage for yield and Ward's method (Ward 1963) for scores. Mantel's test (Mantel 1967) was used to test for the similarity between the two distance matrices (yield and scores). Mantel's test is a permutation test (1000 random permutations) on the Z value (Manly 1994):

$$Z = \sum_{i=2}^{n} \sum_{j=2}^{i-1} m_{ij} e_{ij},$$

where m and e are the elements of two $n \times n$ distance matrices with i as the row index and j the column index; m is the canonical distance matrix between reservoirs calculated on the DY; e is the canonical distance matrix between reservoirs calculated on the scores of the reservoirs in the factorial space. Factorial space is that of the correspondence analysis on the matrix of fish species abundance per reservoir.

To estimate the capability of fish to feed upon bottom living food items, mouth orientation was noted for each species. The percent of fish species with a downward oriented mouth was calculated in each reservoir.

The correlation tests were performed using the 'R' software and the multivariate and Mantel tests using the 'ADE' software (Thioulouse, Chessel, Dolédec & Olivier 1997).

10.4 Results

The most abundant fish species captured by the small-scale fisheries in the reservoirs were (by order of abundance in the artisanal catches): *Oreochromis niloticus* L., *Sarotherodon galilaeus* (L.), *Clarias* spp. (mostly *C. anguillaris* L.), *Chrysichthys* spp. (mostly *C. nigromarginatus* Lacepède), *Tilapia zillii* (Gervais), *Heterotis niloticus* Cuvier, and *Marcusenius* sp. (mostly *M. senegalensis* Steindachner) (Table 10.3).

A 27-month survey of the artisanal fisheries was performed, but in all reservoirs except Katiali, fishing activity stopped for several weeks during the survey. Only two continuous periods were available for analysis in all reservoirs: one lasting 1 year and

Table 10.3 TCs (all fishing gears combined) (during 27 months) of the eight main fish groups in the small-scale fisheries in seven reservoirs from northern Côte d'Ivoire

Fish taxa	TC (seven reservoirs) (kg)
Oreochromis niloticus	11 207
Sarotherodon galilaeus	8983
Clarias sp. (*anguillaris*)	6530
Chrysichtys sp. (*nigromarginatus*)	7215
Tilapia zillii	4746
Heterotis niloticus	2463
Other tilapines	1236
Marcusenius sp. (*senegalensis*)	779

one lasting only 6 months. Only those continuous surveys are taken into account hereafter (Table 10.4). Only the 1-year survey was available in the Tiaplé reservoir whereas two 1-year periods were available in the Katiali reservoir.

It was apparent that fishing days is a reliable measure of the fishing effort (Table 10.4). Whenever two independent values of DY are available they are very similar (except in Korokara reservoir where fishing activity was very irregular).

SR exhibited no correlation with RA ($r^2 = 0.05$, $P = 0.641$), but it was correlated to WA ($r^2 = 0.64$, $P = 0.030$) and to the ratio WA/RA ($r^2 = 0.568$, $P = 0.050$).

Morphometric and production linear models showed a complex picture (Table 10.5). AY (kg ha^{-1} year^{-1}) showed little or no correlation with the variables studied.

Table 10.4 Artisanal fisheries data from the seven studied reservoirs

Reservoir	Area (ha)	Duration of study	Total yield (kg)	AY (kg ha^{-1} year^{-1})	Effort (fishing days)	DY (kg day^{-1} ha^{-1})
Gboyo	7	1	4442	635	213	3.0
		0.5	1926	550	103	2.7
Korokara T	7	1	384	55	81	0.7
		0.5	1925	550	196	1.4
Tiaplé	7	1	1069	153	93	1.6
		0.5	No data	No data	No data	No data
Nambingué	10	1	1825	183	238	0.8
		0.5	1480	296	219	0.7
Sambakaha	15	1	7572	505	195	2.6
		0.5	3555	474	94	2.5
Katiali	24	1	4082	170	472	0.4
		1	2434	101	250	0.4
Tiné	45	1	5811	129	538	0.2
		0.5	1726	77	243	0.2

Table 10.5 Correlation (r^2) and ANOVA results (probability of F) in linear models between different pairs of variables

	AY (kg ha^{-1} year^{-1})		DY (kg day^{-1} ha^{-1})		TC (kg year^{-1})	
	$P > F$	r^2	$P > F$	r^2	$P > F$	r^2
SR	0.157	0.173	**0.049**	**0.308**	0.120	0.205
RA (ha)	0.079	0.254	**0.036**	**0.340**	0.358	0.077
WA (km^2)	0.577	0.029	0.429	0.058	**0.014**	**0.433**
WA/RA	**0.023**	**0.389**	**0.002**	**0.586**	0.205	0.141
D (m)	0.267	0.110	0.198	0.145	0.371	0.073
C (μS)	0.237	0.124	0.131	0.194	**0.006**	**0.508**
FI	0.107	0.218	0.500	0.042	0.665	0.018
MEI	0.116	0.350	**0.030**	**0.505**	**0.028**	**0.509**

Significant values are marked in bold.

Table 10.6 Dominant fish species and trophic status in reservoirs grouped by yield level and SR level

Reservoirs	Dominant species	Trophic status	Yield level	SR level
Gboyo and Sambakaha	*Chrysichthys nigrodigitatus*	Bottom feeder		
	Chrysichthys maurus	Bottom feeder		
	Hepsetus odoe	Pelagic feeder	High	High
	Marcusenius senegalensis	Bottom feeder		
	Mormyrus hasselquisti	Bottom feeder		
	Synodontis schall	Bottom feeder		
	Synodontis bastiani	Bottom feeder		
Tiaplé	*Barbus sublineatus*	Pelagic feeder	Intermediate	Intermediate
	Auchenoglanis occidentalis	Pelagic feeder		
Katiali	*Chromidotilapia guentheri*	Pelagic feeder	Low	High
	Polypterus endlicheri	Pelagic feeder		
Nambingué	*Labeo coubie*	Pelagic feeder	Low	Low
Korokara	*Hemichromis bimaculatus*	Pelagic feeder		
Tiné	*Lates niloticus*	Pelagic feeder	Low	Low

DY, which is a measure of CPUE, was correlated with SR, RA, WA/RA and MEI. The last was also correlated to TC, as were WA and *C*.

The clustering of reservoirs according to their scores in the factorial space of the correspondence analysis on the matrix of species abundance per reservoir, showed good agreement with the grouping obtained from DY values in each reservoir (Fig. 10.1). The two groupings were obtained independently (fishing inquiries for yield, and survey net fishing for species abundance) and the Mantel test indicated a significant correlation ($P = 0.015$) between the two distance matrices. This indicates a relationship between fish population structure (presence and abundance) and fisheries yield.

Characteristic species of each reservoir were noted (Table 10.6). Species are considered characteristic of a given group of reservoirs when more than 90% of the TC of the species came from that group of reservoirs. Gboyo and Sambakaha were the reservoirs with the highest per cent of bottom feeding fishes, 43% and 48% of the total species present respectively. Apart from Tiné (40% bottom feeding species), reservoirs having a low per cent of benthic feeding species (Katiali – 39%; Korokara – 30%; Nambingué – 29%) were also the reservoirs showing low yields (Tables 10.4 and 10.6).

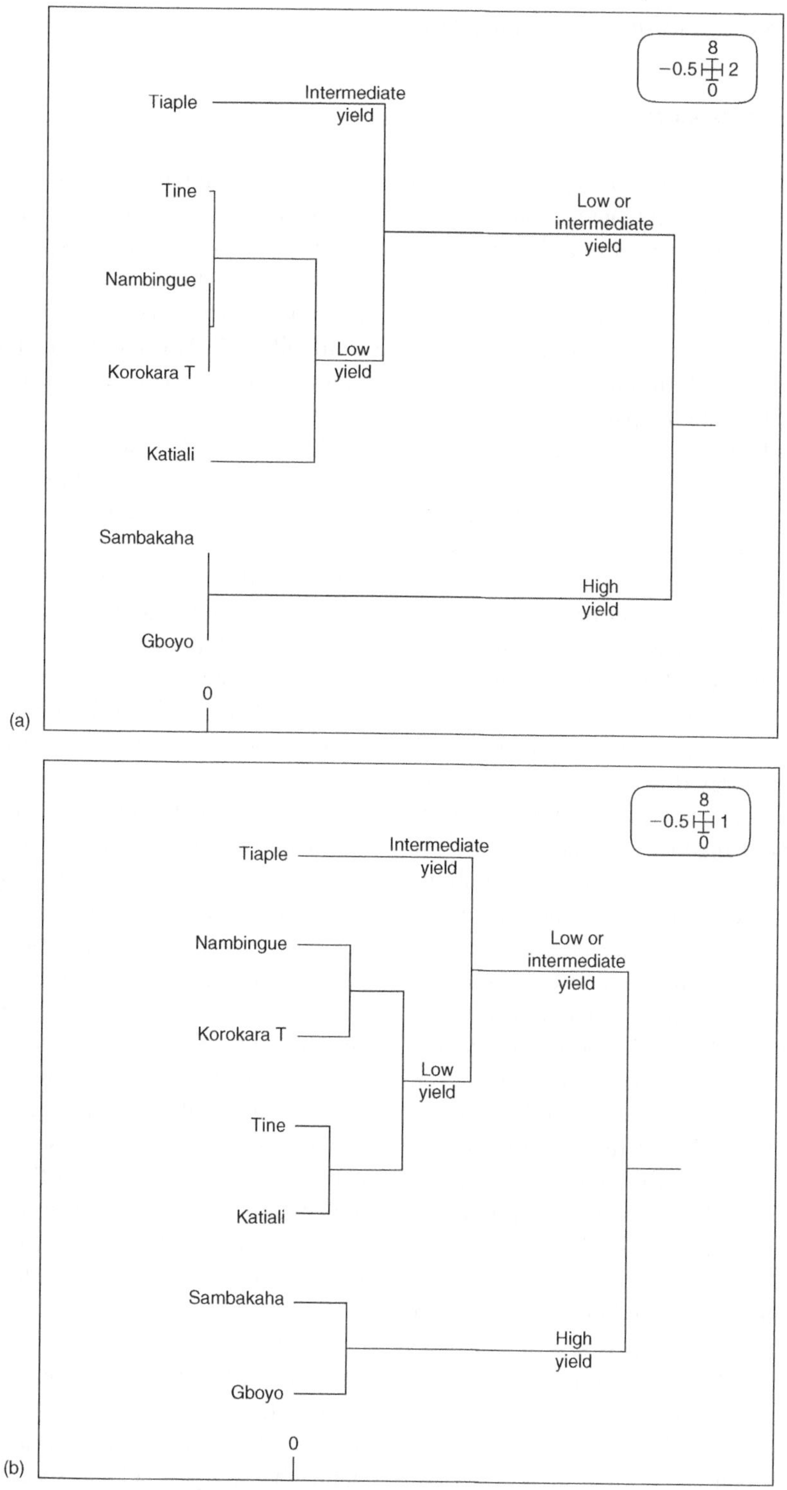

Figure 10.1 (a) Clustering of reservoirs according to their scores in the factorial space of the correspondence analysis on the matrix of species abundance per reservoir; (b) clustering of reservoirs according to yield

10.5 Discussion

Annual fish yield does not appear to be a good descriptor for small-scale reservoir fisheries in the northern Côte d'Ivoire. The pattern of yield observed among reservoirs is complex and the AY (kg ha^{-1} $year^{-1}$) does reflect the variability of the system. DY, which is a measure of fishing effort, and TC (kg $year^{-1}$), which is a measure of the overall productivity of the system, appear to be better descriptors. Not surprisingly, TC was related to variables linked to nutrient availability (WA, *C* and MEI). DY was related to fish SR and to RA, indicating that fish community structure could be a key to understanding the patterns observed.

Several authors (see Lévêque 1997 for review) have raised the question of the relationship between SR and productivity for which there appears to be no simple relationship. Some simple systems, both natural and artificial, appear to be more productive than heterogeneous ones, but similar levels of production can be observed in species-rich ecosystems and in ecosystems with a low level of diversity (Lévêque 1997). An alternative hypothesis could be that SR is not the relevant variable. Ecological processes may function perfectly well with very few fish species. Lake Nakuru (Vareschi & Jacobs 1984, cited by Lévêque 1997) is an extreme example of very high production (over 2000 kg ha^{-1} $year^{-1}$) in a mono-specific lake. What may be of importance are the species biomass (relative abundance of species) and their capacity to ensure an efficient energy flow in any given environment.

Indeed, in this study, correlation between fish population structure (presence and abundance) and yield (kg day^{-1} $year^{-1}$) was significant. This hypothesised that overall catch, TC, is related to nutrient availability (approximated here by water conductivity), but that CPUE is related to the presence of given species in the reservoirs, and particularly to the presence of bottom feeding species. The most efficient fisheries in the small- and medium-sized reservoirs in the Côte d'Ivoire are those where bottom feeding fish species dominate. The dominance of bottom feeding species is an indicator of good physical conditions in the deeper zones of the reservoirs. Such reservoirs certainly experience a better use of the trophic resources than reservoirs where only a few bottom feeding fish species live. Benthic resources thus become available at any time to the upper levels through fish consumption. In reservoirs where bottom layers cannot be used by fish (due to low levels of oxygen, for instance), the utilisation of the energy locked on the bottom depends upon winds strong enough to induce an upward transfer of nutrients. The possibility of use of bottom resources by fish is thus expected to increase productivity in the reservoirs. The dominance of bottom feeding fish species is thus an indicator of high fisheries yield.

Turner (1996) discussed a similar pattern but relating the efficiencies of fisheries to primary productivity. He concluded that the presence of the predator *Lates niloticus* (L.) affected transfer efficiencies, leading to less efficient fisheries in lakes where *Lates* is present than in lakes where this predator is absent. Turner investigated very large African lakes where fisheries yield is mainly based upon pelagic species (like *Limnothrissa*) and primary productivity is expected to play a major role. In this study, in small reservoirs, the pelagic zone is reduced and the availability of bottom energy appears to account for the major differences in fisheries yield. Primary productivity (estimated by its surrogate-conductivity) of course plays a role in fixing the limits to the overall catch, but the fisheries efficiency and yield are determined by the availability of bottom energy.

In this study, the presence of *Lates niloticus* can also be suspected of being responsible for the low efficiency of the fishery in Tiné reservoir. Tiné was the largest reservoir selected, but it had the lowest SR and the lowest yield. Any conclusion about this reservoir has thus to be tempered. Tiné reservoir is virtually emptied every 3 years by the rice cultivators who manage the reservoir. It is the only reservoir selected to be managed that way. The low numbers of fish species observed in Tiné is therefore likely to be the result of events occurring when the reservoir almost dries up. The SR of that fish community is drastically reduced and this situation may continue when the lake has been filled again. Its fish population can be considered to be in a permanent pioneering state and unable to support an efficient and stable fishery. This reservoir deserves a specific study, which is being undertaken.

Acknowledgements

Field data were obtained, thanks to several inquirers and to the technical and fishing teams of Idessa/CNRA in Bouaké (A. Daha, E. Koffi Koffi, J. N'Guessan and L. Kadja). Jean Raffrray, A. Tito de Morais and Y. Kerroux were of invaluable help in data processing. Kouassi S. Da Costa helped through several discussions and provided comments. Both 'R' (http://lib.stat.cmu.edu/R/CRAN) and 'ADE' software (http://www.plib/univ-lyon1/ADE-4) are freely available on the web. I wish to express gratitude to their numerous developers for such impressive work in the benefit of the scientific community.

References

Anonymous (1992) Direction et Contrôle des Grands Travaux. Inventaire des barrages hydro-agricoles et à autres vocations existant en Côte d'Ivoire. Report, DCGTx, Abidjan, Côte d'Ivoire, multigr, 57 pp.

Amarasinghe U.S. (1996) Stock assessment in Sri Lankan reservoirs: a review. In I.G. Cowx (ed.) *Stock Assessment in Inland Fisheries.* Oxford: Fishing News Books, Blackwell Science, pp. 345–356.

Baijot E., Moreau J. & Bouda S. (1994) Aspects hydrobiologiques et piscicoles des retenues d'eau en zone soudano-sahélienne. Centre Technique de Coopération Agricole et Rurale (CTA) ACP/CEE. Commission des Communautés Européennes (CCE), DG VIII D5, Bruxelles, 250 pp.

Da Costa K.S., Traoré K. & Tito de Morais L. (1998) Effort de pêche et production exploitée dans les petites retenues du Nord de la Côte d'Ivoire. *Bulletin Français de la Pêche et de la Pisciculture* **71**, 65–78.

Gourdin F. (1999) Premiers éléments pour une caractérisation hydrologique à diverses échelles du Nord de la Côte d'Ivoire: du local au régional. Report IRD Bouaké, Côte d'Ivoire. Mult. pag. (in French).

Lévêque C. (1997) *Biodiversity, Dynamics and Conservation. The Freshwater Fish of Tropical Africa.* Cambridge: Cambridge University Press, 438 pp.

Manly B.F. (1994) *Multivariate Statistical Methods. A Primer*, second edition. London: Chapman & Hall, 215 pp.

Mantel N. (1967) The detection of disease clustering and a generalized regression approach. *Cancer Research* **27**, 209–220.

Thioulouse J., Chessel D., Dolédec S. & Olivier J.M. (1997) ADE-4: a multivariate analysis and graphical display software. *Statistics and Computing* **7**, 75–83.

Turner G.F. (1996) Maximization of yields from African lakes. In I.G. Cowx (ed.) *Stock Assessment in Inland Fisheries*. Oxford: Fishing News Books, Blackwell Science, pp. 465–481.

Vareschi E. & Jacobs J. (1984) The ecology of lake Nakuru (Kenia). V. Production and consumption of consumer organisms. *Oecologia (Berlin)* **61**, 83–98.

Ward J.H. (1963) Hierarchical grouping to optimize an objective function. *Journal of the American Statistical Association* **58**, 238–244.

Chapter 11
Review of the fisheries in the Brazilian portion of the Paraná/Pantanal basin

M. PETRERE Jr*
UNESP – Departamento de Ecologia, Rio Claro (SP), Brazil

Â.A. AGOSTINHO, E.K. OKADA and H.F. JÚLIO Jr
UEM/NUPELIA, 87020-900 Maringá (PR), Brazil

Abstract

The status of the commercial fisheries in the Brazilian part of the Paraná river basin, including the Pantanal is reviewed. The Paraná is nearly completely regulated by a cascade of dams built for hydroelectric power generation. Only 230 km of main river remain as running waters. The basin has the highest population density in the country, and a high concentration of industry and intensive agriculture. Despite the variable quality of data, fish stocks have been severely degraded by flow regulation and pollution. Large migratory fishes are rare or even absent in some stretches of the main river and in some important tributaries. Fisheries are now dominated by small, low commercial value specimens preadapted to lacustrine habitats. Species introductions are also a major concern.

Keywords: Brazil, commercial fisheries, inland fisheries, Paraná river.

11.1 Introduction

The continent of South America, and Brazil in particular, has a large number of rivers which have been subjected to intense damming in the last 30 years. The first hydroelectric power plant built in Brazil was on the Paraibuna river, in the city of Juiz de Fora, in the State of Minas Gerais. This reservoir was inaugurated in 1889, generating 252 kW. By 1980, 154 large dams had been built generating 33 140 MW. By 2000 nominal generating capacity reached 78 139 MW and the eventual target is an estimated 106 450 MW. Presently, 90% of energy consumed in the country is of hydroelectric origin and 70% of the reservoirs are concentrated in the southeast region mostly in the Paraná basin (Figs 11.1 and 11.2) (Paiva 1982; ELETROBRÁS 1991).

Among the main river basins in South America, the Paraná is the most intensively dammed. It is expected that by the end of the decade, 69 hydroelectric impoundments with areas greater than 200 ha will be created in the Brazilian portion of the basin alone. The 45 existing reservoirs in the basin have transformed the main Paraná river and its tributaries (Grande, Paranaíba, Tietê, Paranapanema, Iguaçu) into a cascade of

*Correspondence: UNESP – Departamento de Ecologia, CP 199, 13506-900 Rio Claro (SP), Brazil (email: mpetrere@life.ibrc.unesp.br).

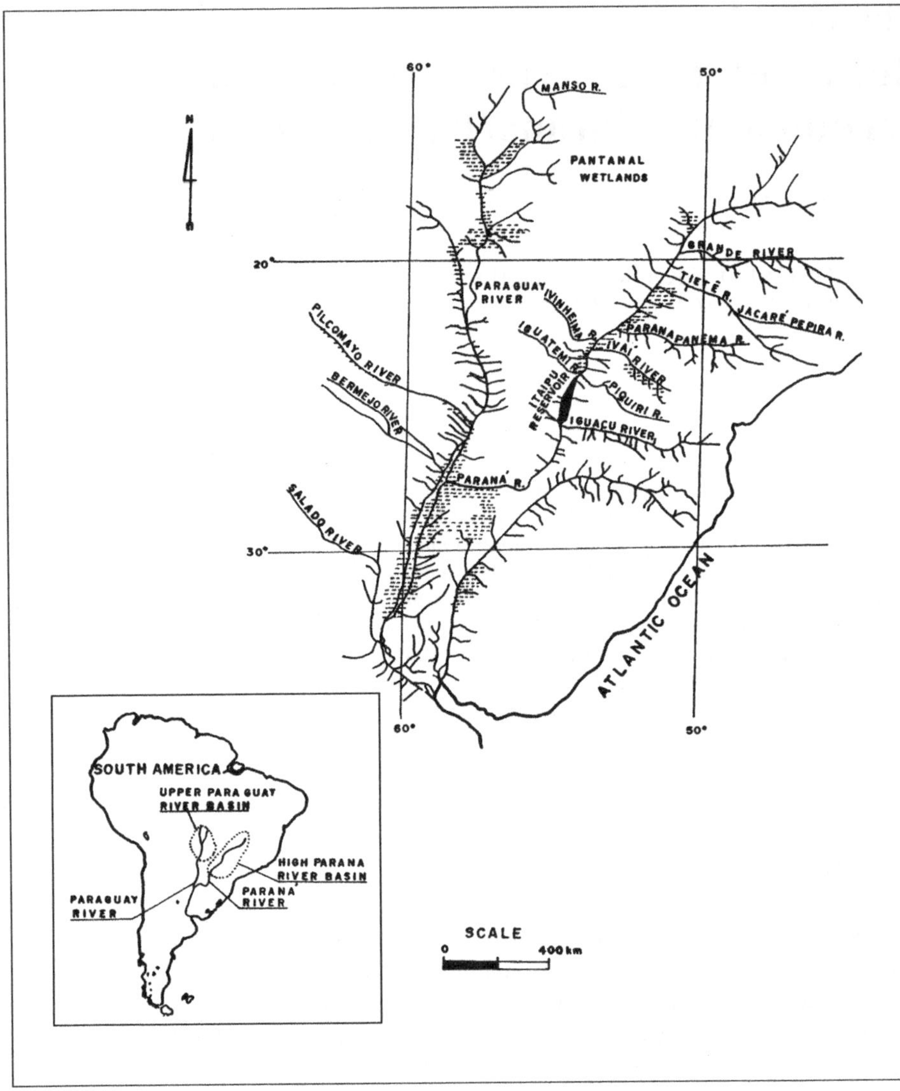

Figure 11.1 Map of the Paraná basin showing location of reservoirs

lakes (Table 11.1). Of the 809 km of the Paraná flowing in the Brazilian territory, 250 km is running water. The remaining Brazilian 30 km below the Itaipu reservoir (Fig. 11.2) will also be dammed by the Argentinean–Paraguayan reservoir of Corpus.

Data on fish stocks are sparse and it is only since the building of the Itaipu reservoir, and with the Convention Universidade Estadual de Maringá (UEM/ITAIPU BINACIONAL), that information on fish landings in the reservoir area have been made available. The present paper examines information on Itaipu, the Tietê reservoirs and the Pantanal, which is part of the Great Paraná basin.

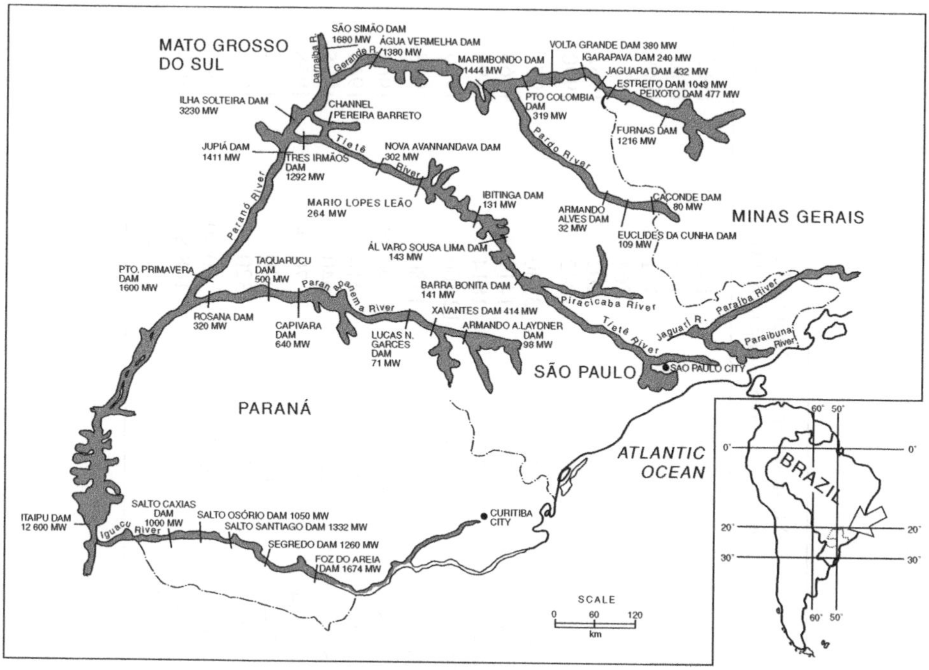

Figure 11.2 Location of main reservoirs on the Paraná Tietê rivers

11.2 Paraná basin

The Paraná basin has a catchment of 2.8 million km^2 draining all south-central South America, from the borders of the Andes to the Serra do Mar adjacent to the Atlantic Coast. Its main river, the Paraná, is 4695 km long, being the 10th longest river in the world, and the second largest in South America, after the Amazon. Its major tributary is the Paraguay river (2550 km). The climate may be classified as sub-tropical to humid-tropical with 1 or 2 dry months, with an average monthly temperature >20°C. Precipitation is >1500 mm $year^{-1}$ (IBGE 1990). River discharge in its upper course varies from 8400 to 13 000 $m^3 s^{-1}$ (minimum of 2550 $m^3 s^{-1}$ and maximum 33 740 $m^3 s^{-1}$). It is formed by the rivers Paranaíba and Grande (Figs 11.1 and 11.2), which rise in the Brazilian shield but then cut into the sedimentary strata. Most of the tributaries of the Upper Paraná present high deposits of sediments in two main areas along the Grande–Paraná channel in the region of Cachoeira (Falls) de Marimbondo and in the stretch of the Três Lagoas and Guaíra in the Paraná river (Souza Filho 1993). Sediment yields reach 3 million t $year^{-1}$ (Stevaux 1994). The Paraná floodplain is 20 km wide in places along its right bank from the city of Três Lagoas to the mouth of the Ivaí river. It has many lagoons with a wide range of habitats. Its main left bank tributaries are the Tietê, Paranapanema, Ivaí, Piquiri and Iguaçu, and in the right bank the rivers Verde, Pardo, Ivinheima, Amambaí and Iguatemi (Fig. 11.1). In this stretch, the river exhibits two distinct morphological patterns. The first comprises the upper course with an

Table 11.1 Some characteristics of selected reservoirs of the Paraná river basin

River	Reservoirs	Inundated area (ha)	Perimeter (km)	Discharge ($m^3 s^{-1}$)	Volume ($10^6 m^3$)	Mean water residence time	Nominal generating capacity (MW)	Year filled
Tietê	Barr bonita	33 430	788	402	2566	73.9	141	1962
	Bariri	5461	203	443	60	1.5	143	1965
	Ibitinga	12 216	375	525	56	1.2	131	1969
	Promissao	53 000	1423	640	2128	38.5	264	1974
	N. Avanhandava	21 700	462	688	380	60.4	302	1982
	Três Irmaos	81 700		733	3600	56.8	1292	1990
Paraná	Ilha Solteira	123 100	1513	5121	12 328	29.0	3230	1973
	Jupiá	35 200	482	6158	1230	2.3	1411	1968
	Porto Primavera	260 000		6931	4643	7.7	1814	1998
	Itaipú	135 000	3060	9670	29 000	35.0	12 600	1983
Grande	Água Vermelha	64 400	1190	2056	5169	29.1	1380	1978
Paranapanema	Jurumirim	54 648	1286	200	3165	183.2	97.75	1961
	Xavantes	42 760	1085	307	3041	114.6	414	1969
	Capivara	64 405	1550	1024	5724	64.7	640	1975
	Salto Grande	1587	81	413	29.4	0.8	70.38	1958
	Taquaruçu	64 405	1550	1024	5724	64.7	640	1992
	Rosana	27 614		1195	407	3.9	320	1996
Pardo	Cacondew	3737	269	56.8	504	102.7	80.40	1965
	Euclides da Cunha	114	16	88.9	4.68	0.6	108.80	1979
	Limoeiro	269	21	90.3	16.36	2.1	32.20	1979

approximate extension of 619 km between the confluence of rivers Paranaíba and Grande and the now inundated Guaíra Falls, which originally was a geographical barrier to fish dispersal. In this sector the river has an irregular course with variable width, numerous islands and bars and an extensive floodplain, mainly on its right bank. In the second sector, now nearly completely inundated, the river runs through a basaltic canyon down to the city of Posadas in Argentina where, it turns west, receives the Paraguay river, then runs south through a very extensive floodplain receiving the Uruguay river near the Atlantic Ocean in its estuary known as La Plata river. The waters of the Paraná, as many other large tropical rivers, have low salinity, low calcium and high silica concentration. The upper river exhibits high turbidity during floods, albeit suspended solids remain relatively low (5–100 $mg\,L^{-1}$), reducing light penetration. Secchi disc readings during low water are around 1.0 m. Average conductivity is below 50 $\mu S\,cm^{-1}$, varying with water levels and inside reservoirs; pH is around neutrality; dissolved oxygen concentrations towards supersaturation. Dominant ions are $HCO_3 > Cl^- > SO_4$ and $Ca^- > Mg^{2+} > Na^+ > K^+$ (Bonetto 1986). Although pollution problems are not intense in the main river, mercury concentration, at least in the region of Porto Rico, exceed the maximum concentrations permissible for protection of aquatic communities (Rodrigues, Lenzi, Luchese & Rauber 1992).

The Pantanal is a humid area of extensive alluvial plains with an average altitude below 200 m. It originates from subsidence associated with the rising of the Andes. Its estimated area is around 140 000 km^2 in the States of Mato Grosso and Mato Grosso do Sul (Fig. 11.1), and is fed by 12 main rivers of which the Paraguay and Cuiabá are the most important. The dominant climate is tropical with relative humidity varying from 60% to 80%, and an average annual temperature of around 25°C. Due to the influence of polar air masses, temperature may drop as low as 0°C for short periods. Average precipitation is about 1200 $mm\,year^{-1}$ and is highest in January. The dry season extends from May to September and the rainy season from October to March. The main soils are classified as hydromorphic podzols (Petrere 1992). In the Pantanal the only reservoir to be constructed is the Manso dam (38 000 ha, 210 MW, 0.55 $MW\,km^{-2}$) on the river with the same name, which joins the Cuiabazinho river to form the Cuiabá river.

The aquatic landscape of the Pantanal exhibits floodplain lakes (baías) of varying shape and dimension seasonally connected to the main river. Sandy strips (cordilheiras) are located between the baías with average elevation of 2 m above high water level. Depressions between the cordilheiras are called vazantes, which drain the water at the beginning of the dry season. The corixos are small intermittent creeks. The salinas are inundated depressions which store rain water, are brackish, and are not influenced by the main river – and do not contain fish (Petrere 1992).

11.3 The fish fauna

The Paraná–Paraguay basin has about 600 species (Bonetto 1986). Ferraz de Lima (1981) estimated that there are 400 fish species in the Pantanal, predominantly Characiformes and Siluriformes, the main families probably being of pre-Gondwania origin (Garavello 1986).

Experimental fishing between 1983 and 1996, using gillnets, long lines, cast nets and trawl seines in the different biotopes in the Paraná river between the mouths of the rivers Paranapanema and Iguaçu, including the whole area of the 170-km long Itaipu reservoir, collected 261 fish species. Of these, 114 species (43.7%) were Characiformes and 110 (42.2%) Siluriformes, although the latter contributed with more than half the fish caught by number. Other orders which also appeared in the catches were: Perciformes (22 species), Cyprinodontiformes (6 species), Rajiformes (3 species), Clupeiformes (2 species), Pleuronectiformes (1 species), Synbranchiformes (1 species), Cypriniformes (1 species, introduced) and Atheriniformes (1 species, introduced). The main families were Characidae (68 species), Pimelodidae (39 species) and Loricariidae (26 species). The fish fauna is more diverse in creeks (153 species), and channels and lagoons (125 species), followed by the main river channel and Ivinheima river where 117 and 100 species were caught, respectively. There was a decreasing richness from its upper third (80 species) of River Iguaçu ichthyofauna toward the dam (53 species). Twenty eight small-sized adult species were restricted to the small creeks in the region of the reservoir and above it. From the species caught below the reservoir (90 species), 16 are exclusive, i.e. they did not occur above it. From the total of 172 species, 20 were responsible for 94% of biomass of the catches (Agostinho, Júlio & Petrere 1994). Terminal desiccating lagoons also exhibit a high species diversity, but numbers reduce by some 50% as pools dry up (Verissimo 1994).

Detritivorous and piscivorous are responsible for more than 75% of the catches in the floodplain, while in the reservoir the fish communities are dominated by piscivores, one of planktophagous species and insectivores. The fish fauna in lagoons is dominated by *Loricariichthys*, *Hoplosternum*, *Leporinus lacustris* Campos, young of *Prochilodus lineatus* (Val.) and others rheophilic species. In the main channel of the Paraná river the typical species are *Paulicea luetkeni* (Steindachner), *Loricaria*, and adults of large migratory fishes (*Pseudoplatystoma corruscans* (Spix & Agassiz), *Salminus maxillosus* Val.). In large tributaries, the characteristic species associated with meandering biotopes belong to Doradidae, Ageneiosidae, *Schizodon*, *Hoplias*, *Rhaphiodon*, Auchenipteridae *Pimelodus*, *Roeboides*; in rapid waters, *Leporinus ambliirhynchus* Garavello & Britski, *Schizodon nasutus* Kner, *Galeocharax knerii* Steindachner, *Apareiodon*, *Myloplus*; and in small streams *Steindachneridion*, small specimens of Cheirodontidae, Tetragonopterinae, small Pimelodidae, Loricariidae and Trichomycteridae are typically found. In the Itaipu reservoir, with a large floodplain upstream, the main species are *Plagioscion squamosissimus* (Heckel) (introduced) and *Hypophthalmus edentatus* Spix & Agassiz in the middle sector; *Pterodoras granulosus* (Val.), *Rhinelepsis aspera* Spix & Agassiz, *Prochilodus lineatus* in the upper section and *Iheringichthys* and *Auchenipterus* throughout the reservoir (Agostinho & Julio Jr 1999).

Barrella (1989) studying the fish fauna of the Jacaré Pepira river, the only unpolluted tributary of the Tietê river (Fig. 11.1), during the dry season, employed a combination of different fishing gears and collected 52 species. Beaumord (1991) sampled the fish community of the River Manso, which together with River Cuiabazinho forms the River Cuiabá (Fig. 11.1), from September 1987 to July 1989, employing gillnets of different mesh sizes set from 18 : 00 to 6 : 00 h and found 80 fish species, of which 72 were identified.

Catella & Petrere (1996) studied the Baía da Onça, a floodplain lake of the River Aquidauana in the Pantanal of the State of the Mato Grosso do Sul. From July to December 1988, when the lake was separated from the main river, they collected 75 fish species and verified the presence of many small-sized adult species. The most abundant families in terms of number of individuals were Characidae (79.5%), Curimatidae (13.1%) and Pimelodidae (3.6%). In terms of biomass, the most abundant were Curimatidae (38.3%), Characidae (28.1%), Loricariidae (10.4%), Pimelodidae (10.1%) and Erythrinidae (4.5%). Catches made with trawl seines showed that the highest abundance occurs at sunset and sunrise. The lowest abundance was detected in May ($2.2\,g\,m^{-2}$), the highest in July ($43.6\,g\,m^{-2}$) with intermediary rates in December ($16.7\,g\,m^{-2}$). The authors interpreted these figures through the direction of the migration lake–river–lake, depending on the water regime of the River Aquidauana. The fish communities which occupied the lake in the periods of July 1988 and June 1989 were distinct in relation to species composition with a proportional similarity PS = 39%, showing the complex and poorly studied dynamics of habitat colonisation in tropical rivers following annual floods.

11.4 The fisheries

Given the available data, the fisheries in the Paraná basin may be divided into six sectors, in:

(1) the Grande river, the former of Paraná river;
(2) the Tietê reservoirs;
(3) the lotic waters of the River Paraná;
(4) Jupiá reservoir;
(5) Itaipu reservoir; and
(6) the Pantanal.

11.4.1 *The fisheries in the River Grande*

The artisanal fishermen in the River Grande below the small Marimbondo reservoir (Fig. 11.2) utilise four main fishing gears (Castro 1992). Cast nets are used mostly in the high water period from November to March for catching corimba, the main fish species in the region, which schools below the dam during its reproductive migration. As a result, the catch per fishing trip is highest at this time of the year. Long lines are used in the transition periods (April and October), when the catch per trip decreases sharply. This fishery is markedly territorial, and catches mainly barbado *Pinirampus pirinampu* (Spix & Agassiz), which sells at twice the price of corimba, compensating for the low catches. Gillnets, and hooks and line are used in the dry season (May to September), when the catch per trip is low and most of the fish caught have low market value, obliging many fishermen to give up fishing for alternative occupations. The most important fish species in order of abundance are: corimba (42%), barbado (20%) and mandi-guaçu *Pimelodus maculatus* Lacepède (17%), all migratory species. There is also an intense sport fishery, which does not compete with the artisanal fishery because it tends to concentrate upon different fishing sites and fish species. In the high water season there is a conflict between the artisanal and sport fisheries, as the power boats used by the tourists interfere with the long lines used by the commercial fishermen.

The artisanal fisheries of Água Vermelha reservoir (Table 11.1, Fig. 11.2) are based on gillnets (Corrêa dos Santos, Ferreira & Torloni 1993). Total annual catches in 1990 and 1991 were 119 and 259 t, respectively. The average catch for 34 fish species was 678 kg fisherman^{-1} month^{-1}, the most abundant of which were mandi-guaçu (34%), curvina (28%); acarás: acará-geo *Geophagus brasiliensis* (Quoy & Gaimard) and acará-geo-bengala *Geophagus* sp. (9%); Nile tilapia *Oreochromis niloticus* (L.) (9%) and traíra *Hoplias malabaricus* (Bloch) (7%).

11.4.2 *The fisheries in the River Tietê*

The River Tietê (Fig. 11.2) is the main affluent of the left margin of the River Paraná and has its head streams in the town of Salesópolis. It is 1050 km long and it is very polluted where it crosses the city of São Paulo, receiving industrial and domestic sewage from an urban concentration of 12 million people. The River Tietê has been under intense cultural stress since the 1920s linked to industrialisation of the city and State of São Paulo. At the end of the 1940s it was no longer possible to swim in its waters. Presently it receives 4500 t year^{-1} of sewage from the Great São Paula area.

There are around 24 million inhabitants in the Tietê river basin (72 000 km^2), 96% of whom live in urban areas. It is the most densely populated region of South America, with 210 towns and cities, and where the majority of the industries of the country are located (São Paulo 1990).

In addition to pollution, the Tietê river and its major tributaries are impaired by:

(1) bank erosion, due to felling of marginal vegetation;
(2) sedimentation, mainly due to sugar cane plantations, where the soil is ploughed every year and left bare until the cane starts to grow;
(3) dams, at present, the Tietê river has six reservoirs along its length (Table 11.1, Fig. 11.2), whose total capacity is 2273 MW. Tundisi, Matsumura-Tundisi, Henry, Rocha & Hino (1988), in comparing the trophic state of the Tietê reservoirs, concluded that all are eutrophic, although the condition of the reservoirs slowly improves from eutrophic to mesotrophic as the water approaches the confluence with the River Paraná.

Below the Nova Avanhandava reservoir, the water is visually clean, compared with Barra Bonita reservoir. The cascade of reservoirs plays the role of settling lagoons, retaining particulate material, N and P. This situation may continue to degrade as the cities of the basin continue to grow without sewage treatment or land use control. There are no commercial fisheries in the River Tietê above Barra Bonita reservoir. There is a marginal sport fishery which becomes more intense during high floods, when better oxygenation of the water permits upstream movement.

After the main period of dam construction during the 1960s, CESP (Companhia Energética de São Paulo) translocated several Brazilian fish species exotic to the basin into these reservoirs. Those included the apaiari *Astronotus ocellatus* (Agassiz), curvina *Plagioscion squamosissimus*, tucunaré *Cichla* sp. and sardinha *Triportheus angulatus* (Spix & Agassiz) from the Amazon basin and the shrimp camarão sossego *Macrobrachium jelskii* (Miers) from north-east Brazilian reservoirs. Alien carp, *Cyprinus carpio* L., and tilapias (*Oreochromis niloticus*, *Oreochromis hornorum* (Trewavas), *Tilapia rendalli* (Boulenger)) were also introduced. Of these only the curvina, and on a lesser scale tucunaré, appear to have been successful (Torloni, dos Santos, Carvalho Jr & Corrêa 1993a; C.E.C. Torloni, personal communication).

In Barra Bonita reservoir, fish are mostly caught by gillnets with different mesh sizes but are strongly predated upon by the pirambeba *Serrasalmus spilopleura* Kner. The total annual catches were: 1989 – 122 t; 1990 – 254 t; 1991 – 129 t, with an average production of 809 kg fisherman^{-1} month^{-1} (Carvalho, dos Santos, Gonçalves & Torloni 1993a). Here 39 fish species were caught, the most important being curvina (24.7%); corimba (22.7%); traíra (11.9%); the piavas: piava-catinguda *Leporinus friderici* (Bloch), piava-da-asa-amarela *L. cf. paranensis* and piava-três-pintas *Schizodon borelli* (Boulenger) (10.2%); the mandis mandi-guaçu, mandi-chorão *Pimelodella* sp., mandi-boca-de-velha *Iheringichthys labrosus* (Lütken), mandi-serrote *Rhinodoras dorbignyi* (Krøyer) (8.4%); the saguirus: saguiru-branco *Steindachnerina insculpta* (Fernándes-Yépes); saguiru-curto *Cyphocharax modesta* (Fernándes-Yépes) and saguiru-comprido *C. nageli* (Steindachner) (7.3%).

In Ibitinga reservoir, fishermen only employ gillnets with different mesh sizes (Corrêa *et al.* 1993). The total annual catches are: 1989 – 22 t; 1990 – 64 t; 1991 – 24 t. The average production is 327 kg fisherman^{-1} month^{-1}. Here 41 fish species were caught, the most important of which were: curvina (22.5%); mandis (15.9%); lambaris: lambari-prata *Astyanax schubarti* (Britski), lambari-tambiú *A. bimaculatus* (L.) and lambari-corintiano *Moenkhausia intermedia* Eigenmann (16%); corimba (11.7%); traíra (11%) and piavas (8.5%).

In Promissão reservoir, fishermen usually employ gillnets and cast nets (Cruz, Moreira, Verani, Girardi & Torloni 1990; Torloni, dos Santos, Moreira & Giradi 1991). Gillnets have variable lengths from 30 to 40 m, an average height of 1.5 m and meshes from 7 to 14 cm bar. Each fisherman utilises an average of 2000 m of gillnets, which are set at sunset and lifted at sunrise. They fish 240 days per year on average, with a major concentration of effort from October to March. Forty-two fish species were caught in this reservoir, six of which are responsible for 85% of total landings. The total annual catches were: 1986/1987 – 267 t, 1987/1988 – 266 t, 1989/1990 – 255 t. The most important fish species in 1988/1989 were mandi-guaçu, corimba, curvina and lambaris. Characiformes predominated in every year making up 75% of the catches, followed by Siluriformes (15%) and Perciformes (10%). Torloni, Carvalho Jr, Corrêa, Santos & Cruz (1993b) gave slightly different figures for these annual catches. Average production was 888 kg fisherman^{-1} month^{-1}.

Moreira, Santos, Silva & Torloni (1993) described the fisheries in Nova Avanhandava reservoir, where fishermen only use gillnets of different mesh sizes to catch 42 different fish species. The total annual catches were: 1988 – 76 t; 1989 – 53 t; 1990 – 41 t;

1991 – 44 t. The average production was 457 kg fisherman^{-1} month^{-1}. The main species landed were: curvina (29%); mandis (26%); corimba (15%); traíra (7%), pirambeba (4%) and lambaris (3%).

11.4.3 *The fisheries in the River Paraná*

Carvalho, Santos, Deus & Torloni (1993b) described the fisheries in Jupiá reservoir (Table 11.1, Fig. 11.2) where the professional fishermen exclusively employ gillnets set between 16 : 00 and 6 : 00 h. The total catches were 1989 – 162 t; 1990 – 182 t and 1991 – 152 t. The average production was 737 kg fisherman^{-1} month^{-1}. Thirty-four fish species appear in the catches, the most important of which were corimba (37%); mandi-guaçu (12%); curvina (11%); acarás (9%); piavas (5%); cascudos: cascudo-chinelão *Rhinelepis aspera* Spix & Agassiz, cascudo-chita *Hypostomus regani* (Ihering), cascudo-voador *Loricaria vetula* (Thering) and cascudo-caborja *Callichthys callichthys* (L.) (5%); pirambeba (4%) and barbado (4%).

Information about the lotic fisheries of the Paraná river are sparse. Preliminary surveys carried out by the Universidade Estadual de Maringá, revealed three kind of fisheries in the region:

(1) artisanal fisheries by fishermen from Port Rico and Guaíra, small towns at the river bank;
(2) sport fisheries performed by citizens, from larger urban centres around the region;
(3) subsistence fisheries performed by small farmers and part-time workers who cut sugar cane in the harvest time and live on the innumerable islands of the river cultivating cereals or living in small settlements at the river margin.

The professional fisheries are different in character from those practised in reservoirs in the basin. They target preferentially the large Pimelodidae, such as the pintado *Pseudoplatystoma corruscans* and barbado, Characidae (dourado *Salminus maxillosus*), Anostomidae (piaparas *Leporinus elongatus* (Val.), *L. obtusidens* (Val.)), Prochilodontidae (corimba) and Erythrinidae (traíra). More recently Doradidae (armado) have appeared in catches in the middle reaches of the river through its dispersion up river after the building of the Itaipu reservoir. It seems that the pintado is still reasonably abundant. Marques (1993) studied its biology by sampling the commercial fisheries in the town of Porto Rico (Fig. 11.1) and measured 4800 individuals totalling 24 t in 1987–1988. The large Pimelodidae and dourado are caught with hooks baited with live fishes. Bait fish are drawn from species which are more resistant to manipulation and harsh conditions while being held in the boats before being utilised. These typically consist of species such as the morenita *Gymnotus carapo* L., caboja *Hoplosternum littorale* (Hancock) or some abundant floodplain fishes such as corró *Leporinus lacustris*, muçum *Synbranchus marmoratus* Bloch and young corimba. Fisheries for pintado, which is the most sought after species, use a strategy called 'anzóis de galho', where hooks are set in the early hours of the night to avoid pirambeba attacks on the baits. The movements of pintado schools in this stretch of the river are sometimes followed by fishermen for more than 100 km (Buck 1988). Although long lines employed for catching dourado, jaú and barbado, the fisheries are more usually performed during the day or at sunset with hook and line.

Corimbas and armados are essentially caught with gillnets, whilst armados are also caught with long lines. Some variations in these fisheries are observed during the piracema (reproductive up river migration) for species like corimba and cascudo-preto, when they are caught by beach seining on sandy banks. During the winter, when there is a decrease in catch per unit effort in the main river channel, fishermen fish in the floodplain lakes and channels. At this time of the year smaller specimens of large fish are caught together with traíra. Professional fishermen also catch morenita in lagoons covered by macrophytes for sale to sport fishermen. The professional fisheries are forbidden during the months of November–February and during the whole year on the right margin where the floodplain is wider (20 km). Nets with meshes below 7.0 cm stretched mesh are not allowed, although mesh control is always difficult. The absence of data does not allow inferences about fishing yields.

Sport fisheries operate mostly at weekends throughout the year. They mainly target dourado, piracanjuba *Brycon orbignyanus* (Val.), pacu *Piaractus mesopotamicus* (Holmberg), piaparas, pintados and jaú. Sport fishing is restricted to the main river channel and its main tributaries, to hook and line or pole and line, baited with live baits for catching dourado and jaú, or fruits of the season for the remaining species. Annual fishing contests are crowded events in some river-side towns.

The subsistence fisheries are practised by virtually all islanders and by a considerable portion of the river-side population, fish being the main protein source for these people. The islanders employ gillnets and to a lesser degree hook and line or pole and line, to catch medium sized specimens, while in smaller settlements fishing is carried out by women and children using pole and line for catching small Pimelodidae, such as mandis and *Characidium* as lambarís.

11.4.4 *Fisheries in Itaipu reservoir*

Itaipu reservoir is the largest in the Paraná basin and up to now the largest in South America in nominal power generating capacity (Figs 11.1 and 11.2, Table 11.1). Its gates were closed in 1982. Professional fisheries in the reservoir only started in February 1984. Prior to that date, fishing was forbidden with the intention of avoiding intense catches of juvenile dourado, piracanjuba, etc. The Universidade Estadual de Maringá established a system for collecting professional catch and effort data in the reservoir from 1987.

The number of professional fishermen working in the reservoir oscillates around 1000, of which 35% are permanent. In 1997 there were 941 and in 1998, 619. The fall was due to IBAMA effort restrictions, mainly in the fluvial zone of the reservoir (Agostinho, Okada & Ambrosio 1999a). Most (73%) work in the upper half of the reservoir, due to the increased abundance of fish. The average profit of the fishing activity is only US\$25 fisherman^{-1}month^{-1}. In 1998, 91.3% were affiliated in seven professional fishermen associations. (FUEM/ITAIPU BINACIONAL 1989; Agostinho *et al.* 1994; 1999a). Fishing effort is controlled mainly by mesh regulations, where meshes smaller than 7 cm stretched mesh are forbidden. Professional and sport fisheries are forbidden in all the reservoir's tributaries. In the reservoir, professional fisheries are allowed all year round and sport fisheries are negligible. The gillnets (meshes ranging from 7 to 24 cm, between knot to knot, 50 m length in average, depth 2.0 m), are the most important gear employed and in 1998 about 711 247 m^2 (mostly 8 cm mesh) were set in the reservoir by 80% of the fishermen. They are mainly set for sardela *Hypophthalmus edentatus*, corimba, corvina, and more recently, *Pterodoras granulosus* (Val.). Long lines are set only in the transitional lotic/lentic zone by 43% (in 1998) of the fishermen, baited with seasonal fruits for catching the armado. In 1998, 72 800 hooks were used, nearly double that used in 1993 (49 055 hooks) (Agostinho, Okada & Ambrosio 1999b). The average number of hooks per long line is 75 (150 m length of nylon filament), with different sizes. The cast net, which is 3 m high, is employed only in the reservoir entrance for catching cascudo-preto and cascudo abacaxi *Megalancistrus aculeatus* (Perugia).

Total catches were fairly constant from 1987 to 1998: 1563, 1500, 1727, 1427, 1589, 1573, 1542, 1297, 1373, 1411 and 1192 t respectively. The CPUE (kg fisherman^{-1} day^{-1}) was 23.2, 20.2, 19.3, 15.9, 15.9, 12.8, 11.6, 12.0, 12.1 and 11.2, respectively (Okada, Agostinho & Petrere Jr 1996; Agostinho *et al.* 1999b). The average fish yield for the period was 11.2 kg ha^{-1} year^{-1} and the average price per kilo was US$ 0.60. Nearly 50 fish species were caught in different years, of which nine were responsible for 90% of the landings. The most important were sardela, armado, corvina, corimba and mandi, which were responsible for 78.7% of the catches (Table 11.2); (Agostinho *et al.* 1994, 1999b; Okada *et al.* 1996).

11.4.5 *Fisheries in the Pantanal*

There are three main cycles of fish migration in the Pantanal, which, on the whole, determine fishing strategies:

(1) the Iufada which designates lateral fish migration coming from the floodplain lakes towards the main river at the end of the dry season. For a short period they concentrate in large schools in the lake mouths, thus becoming very vulnerable to fishing at that time. In the River Cuiabá the phenomenon occurs in April–June, generally during the full moon;
(2) the 'piracema' which designates fish migration up-river, occurring in Cuiabá river in October–November; and
(3) the 'rodada' which describes the behaviour exhibited by fish at reproduction. Spawning for the vast majority of the species happens from December–February (Aguirre 1945; Ferraz de Lima 1984; 1986; Petrere 1989; 1992; EMBRAPA/CPAP 1991).

Professional and sport fisheries are traditional in the whole region. The main gears employed are gillnets, trawl seines, long lines, cast nets, igaratéia (a special type of multiple hook) and harpoons. There are legal restrictions for the employment of nets during the reproductive season (Aguirre 1945; Ferraz de Lima 1981; Ferraz de Lima & Chabalin

Table 11.2 Catch (t) and catch per unit effort (CPUE = t $fisherman^{-1}$ $year^{-1}$) for the main species in the professional fisheries on the Itaipu reservoir

	Period																					
	1987		1988		1989		1990		1991		1992		1993		1995		1996		1997		1998	
Species	Catch	CPUE	Catch	CPUE	Catch	CPUE	Catch	CPUE	Catch	CPUE	Catch	CPUE	Catch	CPUE	Catch	CPUE	Catch	CPUE	Catch	CPUE	Catch	CPUE
P. lineatus	494.4	8.60	227.2	4.17	253.8	4.22	197.7	3.10	215.3	3.20	257.5	4.17	190.9	3.18	83.8	1.68	102.4	1.83	97.9	1.75	101.8	1.78
H. edentatus	241.9	5.76	463.5	8.52	443.9	6.87	392.6	5.98	515.3	7.25	411.8	6.43	364.8	4.81	232.4	3.23	215.6	3.16	180.9	2.82	154.2	2.55
P. squamosissimus	232.4	4.21	225.9	3.66	257.2	3.66	251.5	3.46	304.2	3.81	317.1	4.19	323.7	3.71	240.6	2.76	256.6	2.89	236.7	2.98	177.8	2.46
P. granulosus	176.5	3.17	168.8	2.79	288.8	3.98	225.6	3.07	233.6	3.03	289.1	3.46	334.0	3.40	398.6	4.47	455.2	4.92	544.8	5.44	466.8	4.83
H. malabaricus	23.2	0.74	19.8	0.58	23.3	0.61	28.0	0.67	21.6	0.53	23.8	0.68	32.9	0.71	48.3	0.97	46.6	0.96	57.0	1.08	42.8	0.92
R. aspera	62.7	2.20	46.2	1.95	81.1	3.22	75.3	2.79	63.0	2.58	30.0	1.40	26.6	1.10	40.0	0.91	21.8	0.68	14.4	0.46	11.2	0.42
P. luetkeni	59.5	1.39	56.8	1.59	78.0	2.16	31.4	1.15	24.1	0.93	17.1	0.83	9.7	0.96	3.1	0.51	4.3	0.82	2.3	0.44	2.6	0.32
P. pirinampu	49.8	1.13	41.2	0.97	23.8	0.72	13.6	0.47	25.0	0.67	26.4	0.76	20.1	0.56	33.8	1.05	46.2	1.36	48.7	1.51	52.8	1.56
P. corruscans	4.9	0.10	2.9	0.07	6.4	0.16	6.5	0.20	5.6	0.20	7.0	0.29	4.4	0.24	1.7	0.26	1.5	0.21	1.5	0.14	5.1	0.38
P. maculatus	42.1	1.12	66.1	1.35	75.4	1.34	46.4	0.87	45.9	0.81	54.3	0.96	68.5	0.98	52.4	0.76	42.6	0.62	45.0	0.70	33.2	0.57
Total	1563		1500		1727		1430		1589		1663		1542		1297		1373		1411		1192	

Source: Agostinho *et al.* (1994; 1999); Okada *et al.* (1996).

1984; Silva 1986; Resende 1988). In the State of Mato Grosso do Sul, where traditionally gear restrictions are more severe due to the pressure of sport fishermen, the cast net is allowed only for catching corimba. Other fish species are only caught by hook and line.

Collection of statistics of fish landings in the region has been discontinuous. In 1983 the official yield was 7505 t. Of these 2069 t (28%) were landed in the State of Mato Grosso do Sul, where there is a marked preference for catching large migratory catfishes represented by the pintado and cachara, *Pseudoplatystoma fasciatum*; 5436 t (72%) were landed in the State of Mato Grosso, where the corimba was the predom inant species. This means that professional fisheries were more important in this State as corimba is not traditionally a sport fish, being an illiophage it is not caught by hooks. Silva (1986) estimated that clandestine fisheries in the State of Mato Grosso do Sul may reach 50% of the official landings. Half of the production is exported, mainly to the State of São Paulo. Sport fishermen are much more numerous in the State of Mato Grosso do Sul and are allowed to catch a TAC of 30 kg plus one fish of any size. They generally come from the States of São Paulo, Paraná and Rio de Janeiro.

Catella, Peixer & Palmeira (1996) showed that from May 1994 to April 1995, 1434 t of fish were recorded for the whole State of Mato Grosso do Sul, of which 72% was caught by sport fishermen and the rest by professionals. The main species caught were pacu *Piaractus mesopotamicus* (656 t), pintado *Pseudoplatystoma coruscans* (254 t), cachara *Pseudoplatystoma fasciatum* (104 t), piranha (77 t), piavuçu *Leporinus macrocephalus* (69 t), barbado *Pinirampus pirinampu* (Spix & Agassiz) (69 t), dourado *Salminus maxillosus* (63 t), jau *Paulicea luetkeni* (42 t) and corimba *Prochilodus lineatus* (21 t). The following rivers were most heavily exploited: Paraguay (43%), Miranda (26%), Aquidauana (8.5%). Around 46 000 sport fishermen visited the State, mainly from July to October, spending about 4–6 days per trip and caught between 20 and 27 kg $trip^{-1}$. Professional fishermen spend 4–7 days per week fishing and catch between 45 and 82 kg $fisherman^{-1} trip^{-1}$.

11.5 Discussion

Yields from seven Paraná basin reservoirs are shown in Table 11.3. Their production are low (average = 4.51 kg $ha^{-1} year^{-1}$) when compared with international reservoirs (58.4 kg $ha^{-1} year^{-1}$ for African lakes (Bayley 1988) 99.5 kg $ha^{-1} year^{-1}$ for African reservoirs (Marshall 1984) and 151.8 kg $ha^{-1} year^{-1}$ for 17 north-east Brazilian reservoirs with inundated areas >1000 ha (Paiva, Petrere Jr, Petenate, Nepomuceno & de Vasconcelos 1994)). Possible reasons for these low productions include the following.

(1) Comparatively low fishing intensity (Table 11.3), 0.2 ± 0.2 fishermen $km^{-2} year^{-1}$ [CV (coefficient of variation) = 113%, $n = 7$]. This contrasts with the north-east reservoirs with 3.2 ± 2.6 fishermen $km^{-2} year^{-1}$ [CV = 80%, $n = 17$] and with African lakes 1.5 ± -1.3 fishermen $km^{-2} year^{-1}$ [CV = 85%, $n = 31$] (Henderson & Welcomme 1974). Welcomme (1990) in a review of the status of fisheries in South American rivers, considered that fishermen densities less than

Table 11.3 Comparison among the yield characteristics of seven different reservoirs in the Paraná river basin

Reservoirs	Jupiá Grande river[1]	A. Vermelha Grande river[2]	Barra Bonita Tietê river[3]	Ibitinga Tietê river[4]	Promissão Tietê river[5]	N. Avanhandava Tietê river[6]	Itaipú Paraná river[7]
Yield ($t\,year^{-1}$)	165	189	202	37	222	54	1800
Inundated area (ha)	35 200	64 400	33 430	11 400	53 000	21 700	135 000
Production ($kg\,ha^{-1}\,year^{-1}$)	4.7	2.9	6.0	3.2	3.7	2.5	13.3
Number of active fishermen	49	66	79	26	80	39	1000
Number of fishermen km^{-2}	0.14	0.10	0.23	0.23	0.15	0.18	0.74
CPUE kg $fisherman^{-1}\,month^{-1}$	737	678	809	327	888	457	187
CPUA $kg^{-1}\,fisherman^{-1}\,year^{-1}\,ha^{-1}$	0.25	0.13	0.03	0.34	0.20	0.26	0.02

[1]Carvalho Jr *et al.* (1993a), Torloni (1993).
[2]Corrêa *et al.* (1993).
[3]Carvalho Jr. *et al.* (1993b).
[4]Corrêa *et al.* (1993b).
[5]Torloni *et al.* (1993b).
[6]Moreira *et al.* (1993).
[7]Relatório Annual de Projeto (1990).

0.5 fishermen km^{-2} would mean an underfished floodplain. Density was as high as 29.8 fishemen km^{-2} for the Oueme in Africa, the average for a set of 19 floodplains being 4.36 ± -8.06, (CV of 185%). Welcomme (1990) also considered 1.0 fisherman km^{-2} appropriate for African reservoirs. Comparing the higher CV = 185% of floodplains with the above CVs for lakes and reservoirs which are nearly 50% lower, perhaps indicates that for fishermen, these biotopes are more predictable habitats than floodplains.

(2) Systematic stocking – the lack of systematic fish stocking in the Paraná basin reservoirs compared with the Brazilian NE reservoirs where stocking is a common practice. According to fishermen from Pereira de Miranda reservoir in the State of Ceará, stocking is efficient for the purpose of raising commercial landings, mainly of the migratory fish species. The local fishing community is highly organised and always pressing DNOCS (Departamento Nacional de Obras Contra as Secas) to continue stocking. Stocking to maintain fisheries in Cuban reservoirs is only effective where natural reproduction of the stocked species is low (Quiros & Mari 1999) and used to overcome this bottleneck.

(3) Possible low densities of lacustrine-adapted species such as tilapia which could occupy a probably empty plankton-grazing niche in the open waters of the reservoirs (see discussion by Fernando 1991 and Fernando & Holcik 1991). All the Tietê, Grande and Paraná river reservoirs have tilapias of which *O. niloticus* is by far the most important. In the Tietê river reservoirs, millions of tilapia fry were introduced to take advantage of cultural eutrophication. The only reservoirs supporting considerable commercial production of tilapia are Água Vermelha and Marimbondo, on the River Grande. In these impoundments a fishery for tilapia and corvina is allowed (Decree number 33/IBAMA, from 31/7/1992) using a particular net called rede louca (crazy net) where the nets are hand made with sliding (11 mm stretched mesh) knots using a fine filament. This net is set in a sinuous curve while fishermen beat the water forcing the fish to become entangled. According to professional fishermen from different river systems in Brazil, tilapias are not caught by passive nets because they retreat as soon as they come into contact with the gear (Câmara, dos Santos, Campos, Barbosa 1988). In 1988, CESP officers took fishermen skilled in crazy net fishing to Promissão reservoir where they did not catch any tilapia after a whole days fishing. Tilapia nests are not seen in any of the Tietê reservoirs and are only observed in marginal lagoons. They are probably predated on by lambaris and pirambeba, as soon as parental care ceases. Predation pressure by these species, which are absent in the north-east reservoirs, would be maximal in the Tietê reservoirs which do not have a well developed littoral zone and where macrophytes are not abundant (C.E.C. Torloni, personal communication). Hahn (1991) explained the rarity of pirambebas in Itaipu, where tilapias are also never caught, by a combination of low littoral development and low occurrence of macrophyte cover.

The most successfully introduced species in the Paraná basin is the corvina, which is now widely distributed in South America (Goulding & Ferreira 1984). It belongs to the family Sciaenidae, which in Brazil alone is represented by 37 marine and 10 freshwater species belonging to the genera *Pachyurus*, *Pachypops* and *Plagioscion* (Nomura 1984; Hahn 1991). The main reasons for the success of corvina are as follows.

(1) Reproduction – being predominantly a species inhabiting lentic habitats, which is able to reproduce in still waters, it was pre-adapted to live in reservoirs.
(2) Feeding – Hahn (1991) studying the feeding habits of corvina in Itaipu reservoir considered the species as a generalist piscivore, preying preferentially upon the sardela, the dentudo *Roeboides paranensis* (Pignalberi) and *Astyanax bimaculatus* and to a lesser extent insects (mostly Odonata nymphs), crustacean Decapoda, Arachnidae and vegetative remains. As the fish become larger they become piscivorous, so occupying the majority of niches in the reservoir. Being a generalist, with a wide geographical range, corvina feeding habits change widely to eat whatever is available, although Goulding & Gerreira (1984) characterised it as a specialised shrimp eater, consuming other items only when shrimps are less abundant. Lowe-McConnell (1987) suggested that despite the numerous feeding specialisation presented by many piscivores, in tropical regions they are very flexible in exploiting alternative resources.
(3) Habitat – despite its external morphology, corvina live in meso-pelagic habitats, Hahn (1991) argued that there is some evidence that corvina is preferentially a pelagic species in Itaipu reservoir and the present author has noticed its increased occurrence in impoundments with well developed pelagic zones. Where commercial catches occur preferentially in these zones, corvina has a high proportion of sardela in its diet. So corvina feeds from mid-waters to the surface, where resources are probably most abundant in Itaipu.

Studies of fish communities are only now starting in the Pantanal (Catella & Petrere 1996; Súarez 1998). Information about fisheries in this important region is urgently needed. A recent problem was the accidental introduction, perhaps less than 10 years ago, of tucunaré (*Cichla* sp.) in the region. This fish is able to reproduce and is probably establishing a sustainable population for commercial fisheries. It is still not known what the impact of this voracious predator upon the fish fauna in the Pantanal will be (Ferraz de Lima 1993).

In the last 30 years the fisheries in the Paraná river have undergone tremendous changes in their strategy due to the disappearance of the large migratory species and these being replaced by small sized sedentary ones with low commercial value. Even in the lotic stretches of the river these large migrators are now rare and this tendency coincided with river regulation, pollution and introduction of exotic species. Pollution, including siltation, has degraded breeding sites in small tributaries, which are vital in the life cycle of several migratory species.

The impoundments, which occupy more than 70% of the main river reaches and to the same extent in most of its tributaries, are effective barriers for migratory species. The combined effect of the cascade of reservoirs has eliminated from the upper reaches of the river the large Pimelodidae (pintado *Pseudoplatystoma corruscans* and jaú *Paulicea luetkeni*) and Characidae (dourado *Salminus maxillosus*, pacu *Piaractus mesopotamicus* and piracanjuba *Brycon orbignyanus*). With the exception of the pacu, which is still abundant in the Pantanal, these species are still important in the commercial catches from the remnant 230 km of running waters. In this stretch the spawning areas are located in the upper portion of the non-regulated larger tributaries (Ivinheima, Piquiri, Ivaí, Iguatemi) and the nursery areas are located in the adjacent floodplain which extends for 20 km from the margins of the main river (Agostinho, Vazzoler, Gomes & Okada 1993; Nakatani, Baumgartner & Cavicchioli 1997). The maintenance of these stocks, the last in the Brazilian stretch of the River Paraná, will only be possible with effective measures to protect the spawning and nursery areas through reservoir operation to assure adequate flows to connect the main river with the floodplain water bodies (Agostinho & Zalewski 1996). A major concern is the recent intensification of agriculture in the upper Ivinheima, now the most important tributary for the spawning of the large migrators, with the consequent loss of the riparian vegetation and of large floods in some years (1986–1987, 1995–1996) due to increased water retention in the reservoirs induced by man (Nakatani *et al.* 1997).

Six exotic species have been reported in the area, most of them introduced by accident due to fish-culture, i.e. blackbass, *Micropterus salmoides* (Lacepède), carp, *Cyprinus carpio* (L.) tilapias, the African catfish, *Clarias gariepinus* (Burchell), and channel catfish, *Ictalurus punctatus* (Rafinesque), besides the corvina already cited. With the raising of the backwater level above Itaipu, 16 fish species which were originally unable to jump Sete Quedas cascades have dispersed upstream. Although at the moment there are no studies in the area, the parasite copepod *Lernaea cyprinacea* is a major threat to natural fish populations since it was introduced by infected carps.

The dams, combined with intense human occupation of the Paraná basin, have contributed to the reduction of fish catches and the disappearance of large migratory fish species, mainly in the upper courses of the River Paraná.

The lack of historical data on the fish fauna hampers a better understanding of impacts over the last 30 years. Catch and effort data from the Tietê reservoir suggest that the yield was lower in those reservoirs with exotic species, but corvina is more abundant (CESP 1996). In Itaipu reservoir, catches of curvina are negatively correlated with catches of sardela *Hypophthalmus edentatus* (Agostinho & Júlio Jr 1996).

References

Agostinho A.A. & Júlio Jr H.F. (1996) Ameaça ecologica: peixes de outras águas. *Ciência Hoje* **21**, 36–44.

Agostinho A.A. & Júlio Jr H.F. (1999) Peixes da bacia do alto rio Paraná. In R.H. Lowe-McConnell (ed.) *Estudos ecológicos de comunidades de peixes tropicais.* São Paula: EDUSP, pp. 374–400.

Agostinho A.A. & Zalewski M. (1996) *Upper Paraná River Floodplain: importance and Preservation.* EDUEM/NUPELIA, Maringá, Brazil, pp. 374–400.

Agostinho A.A., de M. Vazzoler A.E.A., Gomes L.C. & Okada E.K. (1993) Estratificación espacial y comportamento de *Prochilodus scrofa en distintas* fases del ciclo de vida, en la planicie de inundación del alto Rio Paraná y embalse de Itaipu, Paraná, Brasil. *Revue d'Hydrobiologie Tropicale* **26**, 79–90.

Agostinho A.A., Júlio Jr H.F. & Petrere M. (1994) Itaipu reservoir (Brazil): impacts of the impoundment on the fish fauna and fisheries. In I.G. Cowx (ed.) *Rehabilitation of Freshwater Fisheries.* Oxford: Fishing News Books, Blackwell Science, pp. 171–184.

Agostinho A.A., Okada E.K. & Ambrosio A.M. (1999a) A pesca no reservatório de Itaipu. In A.A. Agostinho, E.K. Okada & A.M. Ambrosio (eds) *Reservatório de Itaipu: Aspectos Biológicos e Socioeconômicos da Pesca.* Maringá-PR, Brasil: UEM-Nupélia, pp. 1–73.

Agostinho A.A., Okada E.K. & Ambrosio A.M. (1999b) Rendimento pesqueiro no reservatório de Itaipu. In A.A. Agostinho, E.K. Okada & A.M. Ambrosio (eds) *Reservatório de Itaipu: Aspectos Biológicos e Socioeconômicos da Pesca.* Maringá-PR, Brasil: UEM-Nupélia, pp. 74–134.

Aguirre A. (1945) A caça e a pesca no Pantanal de Mato Grosso. *Div. Caça e Pesca, Min. Agric.*, **1**, 1–46.

Barrella W. (1989) *Estrutura da Comunidade de Peixes da Bacia do Rio Jacaré Pepira (SP) em Diferentes Biótopos.* MSc Thesis, UNICAMP, Campinas (SP), Brasil, 171 pp.

Bayley P.B. (1988) Accounting for effort when comparing tropical fisheries in lakes, river-floodplains and lagoons. *Limnology and Oceanography* **39**, 963–972.

Beaumord A.C. (1991) *As Comunidades de Peixes do Rio Manso, Chapada dos Guimarães, MT: Uma Abordagem Ecologica Numérica.* MSc Thesis, UFRJ, Rio de Janeiro (RJ), Brasil, 182 pp.

Bonetto A.A. (1986) The Paraná river system. In B.R Davies & K.F. Walker (eds) *The Ecology of River Systems.* Rotterdam: Dr Junk Publications, pp. 541–556.

Buck N. (1988) Relatório Final de Andamento do Projecto: "Autoecologia do pintado *Pseudoplatystoma corruscans*". FINEP, Rio de Janeiro, 72 pp.

Câmara J.J.C, dos Santos R.A., Campos E.C. & Barbosa J.M. (1988) "Pesca de Batida": um método eficiente para captura de Tilápias Preta e do Nilo, utilizado na Reprêsa de Marimbondo, Rio Grande, limite Centro-Norte do Estado de São Paulo. Boletim Técnico 11, Instituto de Pesca, São Paulo, 12 pp.

Carvalho Jr A.A., dos Santos J.J., Gonçalves J.L. & Torloni C.E.C. (1993a) Produção pesqueira e composição das capturas no reservatório da UHE Barra Bonita, CESP, São Paulo. *Resumos do X Encontro Brasileiro de Ictiologia*, USP, São Paulo, p. 105.

Carvalho Jr A.A., dos Santos J.J., de Deus E.F. & Torloni C.E.C. (1993b) Produção pesqueira e composição das capturas no reservatório da UHE Souza Dias (Jupiá), CESP, São Paulo. *Resumos do X Encontro Brasileiro de Ictiologia*, USP, São Paulo, p. 105.

Castro F. (1992) *Aspectos Ecológicos da Pesca Artesanal no Rio Grande à Jusante da Usina Hidrelétrica de Marimbondo.* MSc Thesis, UNICAMP (SP), Brasil, 175 pp.

Catella A.C. & Petrere M. (1996) Feeding patterns in a fish community of Baia da Onça, a floodplain lake of the Aquidauana River, Pantanal, Brazil. *Fisheries Management and Ecology* **3**, 229–237.

Catella A.C., Peixer J. & Palmeira S.S. (1996) Sistema de Controle da Pesca de Mato Grosso do Sul. SCPESCA/MS – I Maio/1994 a Abril/1995. EMBRAPA – CPAP/SEMADES – MS, Documentos 16, Corumbá (MS), 49 pp.

CESP (1996) Aspectos Limnológicos, Ictiológicos e Pesqueiros de Reservatórios da CESP no perido de 1986 a 1994. CESP, São Paulo, Brazil, 79 pp.

Corrêa A.R.A., dos Santos J.J., Ferreira A.S. & Torloni C.E.C. (1993) Produção pesqueira e composição das capturas no reservatório da UHE José Ermirio de Moraes (Água Vermelha), CESP, São Paulo. *Resumos do X Encontro Brasileiro do Ictiologia*, USP, São Paulo, p. 109.

Cruz J.A., Moreira J.A., Verani J.R., Girardi L. & Torloni C.E.C. (1990) Levantamento da ictiofauna e aspectos da dinâmica de populações de algumas espécies do reservatório de Promissão – SP (1 etapa). *CESP/UFSCar, Report*, 78 pp.

ELETROBRÁS (1991) Plano Diretor de Meio Ambiente do Setor Elétrico 1991/1993. Centrais Elétricas Brasileiras, Rio de Janeiro, 2v. 284 pp.

EMBRAPA/CPAP (1991) Avaliação da Contaminação Ambiental da Bacia Hidrográfica do Rio Miranda. Relatório Final, Corumbá, MS, 174 pp.

Fernando C.H. (1991) Impacts of fish introduction in tropical Asia and America. *Canadian Journal of Fisheries and Aquatic Sciences* **48**(Suppl. 1), 24–32.

Fernando C.H. & Holcik J. (1991) Fish in reservoirs. *Int. Revue ges. Hydrobiol.* **76**, 149–167.

Ferraz de Lima J.A. (1981) A pesca no Pantanal de Mato Grosso (Rio Cuiabá: Ecologia e biologia pesqueira). *Anais do II Congresso Brasileiro de Engenharia de Pesca*, Recife, pp. 503–516.

Ferraz de Lima J.A. (1984) Piracema e reprodução no Pantanal. *Jornal da Pesca*, SUDEPE, **3**, 6.

Ferraz de Lima J.A. (1986) A pesca no Pantanal de matogrosso (Rio Cuiabá: movimento cíclico dos peixes). *Resumos do XIII Congresso Brasileiro de Zoologia*, Cuiabá, p. 393.

Ferraz de Lima J.A. (1993) Recursos pesqueiros em ambientes inundáveis (Rio Cuiabá: Pantanal de Mato Grosso). *Resumos do X Encontro Brasileiro de Ictiologia*, USP, São Paulo, pp. 302–310.

Ferraz de Lima J.A. & Chabalin E. (1984) O Mercado de Peixes de Cuiabá. Prefeitura Municipal de Cuiabá, Secretaria Municipal de Serviços Públicos, 96 pp.

FUEM/ITAIPU BINACIONAL (1989) Ecologia de Populações de Peixes no Reservatório de Itaipu, nos Primeiros Anos de sua Formação – 5°. Relatório. Maringá Paraná, Fundação Universidade Estadual de Maringá, 3v. 480 pp.

Garavello J.C. (1986) Fauna terrestre e aquática. *Anais do I Simpósio sobre Recursos Naturais e Sócio-Econômicos do Pantanal*, Corumbá, EMBRAPA-CPAP, 221 pp.

Goulding M. & Ferreira E.J.G. (1984) Shrimp-eating fishes and a case of prey-switching Amazon rivers. *Revista Brasilia Zoologia* **2**, 85–97.

Hahn N.S. (1991) *Alimentação e Dinâmica da Nutrição da Curvina* Plagioscion squamosissimus *(Heckel 1840) e Aspectos da Estrutura Trófica da Ictiofauna Acompanhante no Rio Paraná.* PhD Thesis, UNESP, Rio Claro (SP), Brazil, 287 pp.

Henderson H.F. & Welcomme R.L. (1974) The relationship of yield to Morpho-edaphic Index and numbers of fishermen in African inland fisheries. *CIFA Occasional Paper* **1**, 19 pp.

IBGE (1990) Geografia do Brasil. Região Sul. Rio de Janeiro (RJ) Brasil, 186 pp.

Lowe-McConnell R.H. (1987) *Ecological Studies in Tropical Fish Communities.* Cambridge University Press, UK, 382 pp.

Marques E.E. (1993) *Alimentação, Dinâmica da Nutrição e Reprodução do Pintado* Pseudoplatystoma corruscans *(Siluriformes, Pimelodidae) na Planície de Inudação do Alto Paraná.* MSc Thesis, UFPr, Curitiba (PR), Brasil, 104 pp.

Marshall B.E. (1984) Predicting ecology and fish yields in African reservoirs from preimpoundment physico-chemical data. *CIFA Technical Paper* **12**, 72 pp.

Moreira J.A., dos Santos J.J., da Silva D.C. & Torloni C.E.C. (1993) Produção pesqueira e composição das capturas no reservatório da UHE Nova Avanhandava, CESP, São Paulo. *Resumos do X Encontro Brasileiro de Ictiologia*, USP, São Paulo, p. 108.

Nakatani K, Baumgartner G. & Cavicchioli M. (1997) Ecologia do ovos e larvas de peixes. In A.E.A.M. Vazzoler, A.A. Agostinho & N.S. Hahn (eds) *A Planície de Inundação do Alto Rio Paraná: Aspectos Limnológicos e Socioeconômicos.* Maringá-PR, Brazil: EDUEM, pp. 121–134.

Nomura H. (1984) *Dicionário de Peixes do Brasil.* Brasília, Brazil: Editerra, 482 pp.

Okada E.K., Agostinho A.A. & Petrere Jr M. (1996) Catch and effort data and the management of the commercial fisheries of Itaipu reservoir in the upper Paraná river, Brazil. In I.G. Cowx (ed.) *Stock Assessment in Inland Fisheries.* Oxford: Fishing News Books, Blackwell Science, pp. 154–161.

Paiva M.P. (1982) *Grandes Represas do Brasil.* Brasília: Editerra, 292 pp.

Paiva M.P., Petrere Jr M., Petenate A.J., Nepomuceno F.H. & de Vasconcelos E.A. (1994) Number of predatory fish species and the fish yield of large North-eastern Brazilian reservoirs. In I.G. Cowx (ed.) *Rehabilitation of Freshwater Fisheries.* Oxford: Fishing News Books, Blackwell Science, pp. 120–129.

Petrere M. (1989) River fisheries in Brazil: a review. *Regulated Rivers: Research and Management* **4**, 1–16.

Petrere M. (1992) A pesca no Pantanal. In F.D. Rilla (ed.) *Memorias. II. Taller Regional de Humedales. III Reunion de Miembros Sudamericanos de la IUCN.* Brasil: Paraty, pp. 79–82.

Quiros R. & Mari A. (1999) Factors contributing to the outcome of stocking programmes in Cuban Reservoirs. *Fisheries Management and Ecology* **5**, 241–254.

Rsende E.K. (1988) Recursos pesqueiros: Diggnóstico e propostas de pesquisas. EMBRA PA, Corumbá, MS, 51 pp.

Rodrigues B.A., Lenzi E., Luchese E.B. & Rauber T. (1992) Níveis de concentração de mercúrio total nas águas dos rios Paraná e Baía – região de Porta Rico. *Acta Limnologia, Brasil* **4**, 255–260.

São Paulo (1990) Plano Estadul de Recursos Hídricos: Primeiro Plano do Estado de São Paulo. Síntese. Conselho Estadul de Recursos Hídricos, DAEEE, São Paulo, 97 pp.

Silva M.V. (1986) Mitos e Verdades sobre a Pesca no Pantanal Sul-Matogrossense. FIPLAN-MS, Campo Grande (MS), 146 pp.

Souza Filho E.E. (1993) Aspectos da Geologia e Estratigrafia dos Depósitos Sedimentares do Rio Paraná entre Port Primavera (MS) e Guajra (PR). Tese de Doutorado, USP (SP), Brasil, 214 pp.

Stevaux J.C. (1994) The Upper Paraná River (Brazil): geomorphology, sedimentology and paleo climatology. *Quaternary International Journal* 21, 143–161.

Torloni C.E.C., dos Santos J.J., Moreira J.A. & Giradi L. (1991) Pesca Artesanal e Produção Pesqueira no Reservatório da UHE Mário Lopes Leão, Promissão – SP. CESP – São Paulo, *Série Pesquisa e Desenvolvimento* **62**, 17 pp.

Torloni C.E.C., dos Santos J.J., Carvalho Jr A.A. & Corrêa A.R.A. (1993a) A pescada-do-Piauí *Plagioscion squamosissimus* (Heckel 1840) nos reservatórios da CESP – Companhia Energética de São Paulo. *Resumos do X Encontro Brasileiro de Ictiologia*, USP, São Paulo, p. 112.

Torloni C.E.C., Carvalho Jr A.A., Corrêa A.R.A., dos Santos J.J. & Cruz J.A. (1993b) Produção pesqueira e composição das capturas no reservatório da UHE Mário Lopes leão (Promissão), CESP, São Paulo. *Resumos do X Encontro Brasileiro de Ictiologia*, USP, São Paulo, p. 107.

Tundisi J.G., Matsumura-Tundisi T., Henry R., Rocha O. & Hino K. (1988) Comparações do estado trófico de 23 reservatórios do Estado de São Paulo: eutrofização e manejo. In J.T. Tundisi (ed.) *Limnologia e Manejo de Represas*, ACIESP/FAPESP, CNPq, São Paulo, Brasil, pp. 165–204.

Welcomme R.L. (1990) Status of fisheries in South American rivers. *Interciencia* **15**, 337–345.

Chapter 12
Fish and fisheries of Kaptai Reservoir, Bangladesh

G.C. HALDAR, K.K. AHMED, M. ALAMGIR, J.N. AKHTER and M.K. RAHMAN*
Bangladesh Fisheries Research Institute, Riverine Station, Chandpur – 3602, Bangladesh

Abstract

Fish species diversity, historical catch statistics, fishing methods, efficiency of gears, significance of major carps stocking, fishing community structure, marketing system and management tools pertaining to Kaptai Reservoir are described. A total of 73 species of fish, two species of Crustacea and one species of dolphin were identified. The fishery is multi-species and multi-gear in nature. Fish production ranged between 55 and 113 kg ha^{-1} $year^{-1}$. Total fish landings from the reservoir increased steadily linked to an increased contribution of pelagic clupeids, while the importance of major carps declined, despite an intensive stocking programme. The gears now operating in Kaptai Reservoir are mainly traditional and are similar to the gears used in other inland fisheries. No mechanised fishing crafts have been introduced to the reservoir except carrier boats. Average catch per unit effort for all types of gears varied from 2.91 to 32.16 kg $unit^{-1}$ day^{-1}. Fishermen are resource poor, and are dependent on middlemen for cash and other resources. The marketing system of Kaptai Reservoir is complicated compared to other fishery and agricultural marketing systems. The management strategies for Kaptai Reservoir basically pertain to a close season during the breeding period, licensing and stocking.

Keywords: catch statistics, fishing methods, management, marketing, recruitment.

12.1 Introduction

Kaptai Reservoir (22°22′–23°18′N; 92°00′–92°26′E) was impounded in 1961, damming the River Karnaphuli at Kaptai in the Chittagong Hill Tracts (Fig. 12.1). The primary purpose of the reservoir was hydropower generation, while fisheries, navigation, flood control and irrigation are secondary activities. It is one of the largest man-made lakes in south-east Asia (Fernando 1980), covering an area of approximately 58 300 ha (68 800 ha at full surface level). The maximum and mean depths of the reservoir are 35 and 9 m, respectively, mean annual water level fluctuation is 8.14 m and the water reserve is $524.7 \times 10^6 m^3$ (Aquatic Research Group 1986).

The shoreline of the reservoir is rocky and strewn with remnants of submerged wooden logs. The bottom of the reservoir is mostly uneven with little clay, except in the inundated river bed. The reservoir area was once part of undulating valleys and the

*Correspondence: Bangladesh Fisheries Research Institute, Riverine Station, Chandpur – 3602, Bangladesh (email: frirs@bdmail. net).

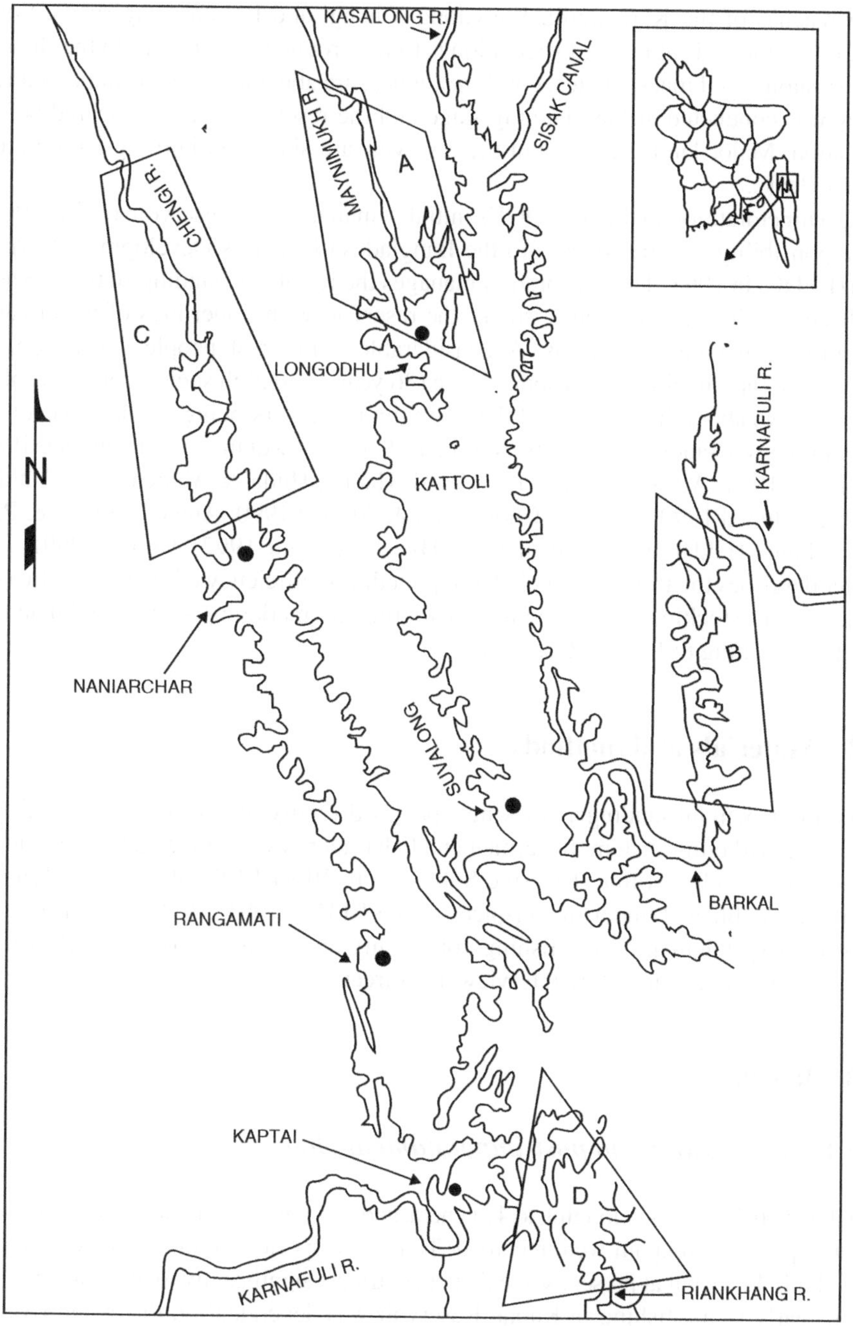

Figure 12.1 Map of Kaptai Reservoir showing major in-flowing rivers, landing centres and major carps spawning grounds

lower reaches of the Karnaphuli, Kassalong, Chengi and Reinkhiyang rivers. Some aquatic floating and submerged vegetation is found around the reservoir. In the shallow regions, submerged macrophytes, e.g. *Najas najas*, are abundant. No pre-impoundment survey was conducted at the planning stage and the reservoir was not cleared before inundation. Major land uses around the reservoir are slash and burn agriculture and forestry (teak).

The Directorate of Fisheries (DOF) initially managed the reservoir (1961–1963), but responsibility was transferred to the Bangladesh Fisheries Development Corporation (BFDC) in 1964. The corporation manages the ice plants, landing stations, where it collects royalties from catches, stocks the reservoir with fingerlings of major carps and enforces fishing regulations. A considerable number of people depend on the reservoir for their livelihood. Despite its age (36 years) and high socio-economic value, little is known about the fisheries of the reservoir, most work being on the fish ecology and limnology (Sandercock 1966; Rahman 1980; Chowdhury & Mazumder 1981; Hye 1983; 1988; ARG 1986; Haldar, Mazid, Haque, Huda & Ahmed 1989; Haldar, Mazid & Ahmed 1990; Alamgir, Chowdhury & Ahmed 1990; Ahmed, Haldar & Paul 1991; Ahmed Haldar, Saha & Paul 1994; Hye & Alamgir 1992; Ahmed 1999). This chapter describes the trends in catch, fishing methods, efficiency of gears, significance of major carps stocking, fishing community structure, marketing system and management tools relating to Kaptai Reservoir.

12.2 Materials and methods

Data on commercial landings, stocking, species diversity, fish production, catch per unit effort, marketing system, fishermen and fish traders were collected by Bangladesh Fisheries Research Institute, Riverine Sub-Station (BFRI-RSS 1993; 1994). Reports and articles published under the auspices of the BFDC (1991; 1996) were also used. The role of middlemen on the fishing community and status of commercial exploitation of fish were evaluated by interview and direct observation.

12.3 Results

12.3.1 *Fish fauna and post-impoundment changes*

A total of 73 fish species belonging to 47 genera, and 25 families, plus two species of prawn and one species of dolphin were recorded. Thirty-one species are of commercial importance (Table 12.1), but at present some do not contribute significantly to the total landings.

In its early years, fishing in Kaptai Reservoir was dependent on the extent riverine stocks, but since the early 1980s Chinese carps, *Hypophthalmichthys molitrix* (Val.), *Ctenopharyngodon idella* (Val.), common carp, *Cyprinus carpio* L., and silver barb, *Puntius gonionotus*, have been introduced. In 1985, tilapia, *Oreochromis niloticus* (L.) escaped from cage culture experiments and African catfish, *Clarias gariepinus* (Burchell), was also accidentally introduced.

Table 12.1 Changes in fish populations in the Kaptai Reservoir, Bangladesh

Introduced	Disappeared	Dwindled	Commercially important	Now dominant
Cyprinus carpio	*Silonia silondia*	*Tor tor*	*Labeo rohita*	*Corica soborna*
Hypophthalmichthys molitrix	*Clupisoma garua*	*Johnius coitor*	*Catla catla*	*Gudusia chapra*
Ctenopharyngodon idella	*Bagarius bagarius*	*Setipinna phasa*	*Cirrhinus mrigala*	*Gonialosa manminna*
Puntius gonionotus	*Puntius sarana*	*Ompok bimaculatus*	*Labeo calbasu*	*Mystus aor*
Oreochromis niloticus	*Labeo angra*	*Amblypharyngodon mola*	*Tor tor*	*Mystus cavasius*
Clarias gariepinus	*Channa orientalis*	*Labeo bata*	*Labeo gonius*	*Labeo gonius*
	Pangasius pangasius		*Cirrhinus reba*	*Cirrhinus reba*
	Platanista gangetica		*Notopterus notopterus*	*Labeo calbasu*
			Notopterus chitala	*Notopterus chitala*
			Channa striatus	*Notopterus notopterus*
			Channa marulius	*Oreochromis niloticus*
			Rohtee cotio	*Macrobrachium lamarri*
			Xenentodon cancila	
			Wallago attu	
			Ompok bimaculatus	
			Ailia coila	
			Eutropiichthys vacha	
			Mystus aor	
			Mystus cavasius	
			Mystus bleekeri	
			Mystus vittatus	
			Gudusia chapra	
			Corica soborna	
			Mastacembelus armatus	
			Gonialosa manminna	
			Heteropmeustes fossilis	
			Clarias batrachus	
			Setipinna phasa	
			Johnius coitor	
			Amblypharyngodon mola	
			Ambasis nama	

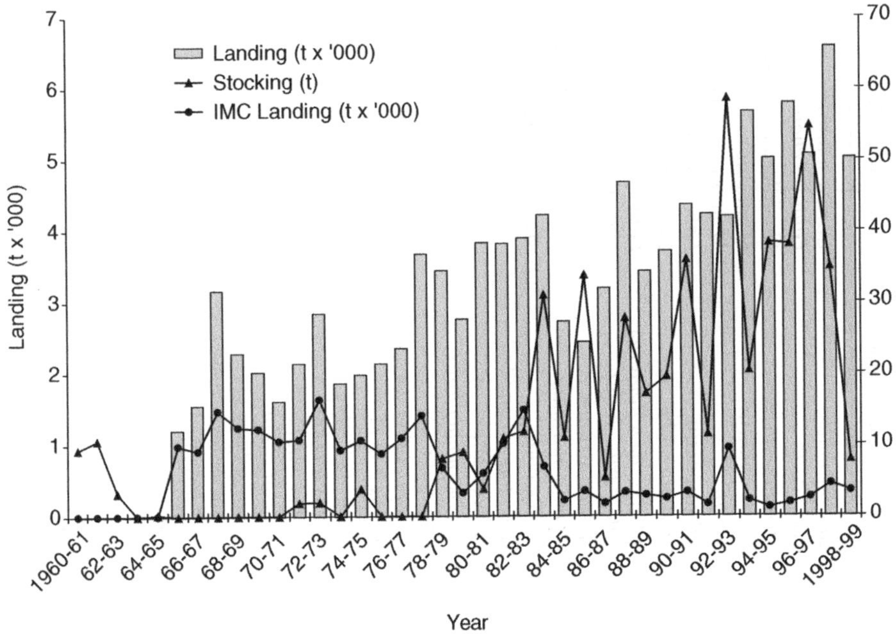

Figure 12.2 Total and major carps landing and stocking of major carps in Kaptai Lake during 1965–1966 to 1997–1998

12.3.2 *Historical catch statistics*

During the first 3 years of impoundment (1962–1965) there was no commercial fishing. To build up the fishery about 2.3 million major carp (*Labeo rohita* Hamilton, *Catla catla* (Hamilton), *Cirrhinus mrigala* Buch and *Labeo calbasu* (Hamilton)) fry were stocked and fishing commenced in 1965–1966. Since inception of the fishery, fluctuations in the annual landings have been observed peaking every 4–5 years (see Fig. 12.2). Initially, fish yield was low but thereafter it increased steadily to more than 5000 t in the 1990s.

Predicted yield based on a linear relationship of the annual catch trend ($Y = 1483.02 + 114.67X$; $r = 0.87$) underestimated catches in recent years, mainly as a result of improved management of the lake, although catches of major carps are considered unsatisfactory.

12.3.3 *Fish yield*

Fish yield of the Kaptai Reservoir ranged between 55 and 113 kg ha^{-1}, mean of 80.0 ± 17.3 kg ha^{-1}, for the period 1986/1987 to 1998/1999) (Table 12.2). Hye (1983) reported a mean annual landing of 57 kg ha^{-1} but, similar to the Aquatic Research Group (ARG 1986), estimated that about 30% of the total catch was not recorded due to local

Table 12.2 Production (kg ha^{-1}) and per cent composition of fish of Kaptai Reservoir

Year	Major carps	Kechki	Chapila	Catfish	Featherback	Tilapia	Dried fish	Miscellaneous	Production
1986–1987[a]	4.3	–	23.2	3.4	7.6	0.02	45.8[c]	15.8	54.6
1987–1988	8.2	–	16.6	4.7	6.1	0.01	46.1	18.3	69.8
1988–1989[b]	8.3	9.4	11.3	14.2	8.4	0.03	40.7	07.7	60.0
1989–1990	6.5	16.4	12.2	11.9	8.7	0.08	34.2	10.0	63.7
1990–1991	7.5	25.2	15.0	5.9	6.6	0.57	27.0	12.2	75.3
1991–1992	3.7	17.9	19.8	5.7	5.0	0.87	41.3	07.8	72.2
1992–1993	2.3	32.3	19.6	5.0	3.0	1.44	29.4	07.0	71.1
1993–1994	3.7	34.6	21.1	4.6	2.5	0.89	24.9	07.6	99.5
1994–1995	2.2	32.8	18.5	3.4	1.8	1.05	32.2	08.2	89.3
1995–1996	3.0	24.5	15.7	4.0	1.7	0.69	30.1	20.2	99.4
1996–1997	5.1	38.7	17.5	4.9	2.5	0.95	16.9	13.6	86.0
1997–1998	5.9	28.0	27.6	3.1	2.2	1.16	20.5	11.7	113.0
1998–1999	6.4	40.0	25.0	3.1	1.4	1.27	13.8	9.1	86.1
Average	5.2	27.2	18.7	5.7	4.4	0.69	31.0	11.5	80.0

[a] First landing of tilapia.
[b] First landing of kechki.
[c] Dried fish are being expressed as raw fish converting four times of the original value.

consumption and bypassing the landing sites. Therefore, if the annual yield is adjusted by adding 30% unrecorded fish, mean production in the reservoir was 104.0 kg ha^{-1}.

12.3.4 *Species composition*

In the early years, major carps (*Labeo rohita, Calta catla, Cirrhinus mrigala* and *Labeo calbasu*) contributed greatly (50–80%) to the total catch, but their contribution declined when catches of two pelagic clupeids, kechki, *Corica soborna* Hamilton, and chapila, *Gudusia chapra* (Hamilton), increased substantially to account for about 65% of the total catch in 1998/1999 (Table 12.2). Major carps now contribute only about 5% to the total catch. A drop in landings of catfishes *Wallago attu* (Bloch) and feather-backs, *Notopterus* spp., was also noticed in the recent years (Table 12.2).

During the last 13 years, 31% of total landings was used to produce dried fish (Table 12.2). Dried fish is a heterogeneous assemblage of short-lived, easily-caught, small- or moderate-sized, low-priced fish of all categories. *Gudusia chapra, Gonialosa manmina* (Hamilton), and *Corica soborna* constitute the bulk (approximately 80% of the total weight) of dried fish. Fingerlings of major carps are also found occasionally within this category. Other species that are dried include: *Rohtee cotio* (Hamilton), *Mystus aor* (Hamilton), *Channa striatus* (Bloch) and *Channa marulius* (Hamilton). Kaptai Reservoir is one of the two major habitats for *Tor tor* (Hamilton) (the other is Kangsha River in north-east Bangladesh), although it is rarely caught.

12.3.5 *Significance of major carps and stocking*

Major carps are the most favoured, commercially-important fish in Bangladesh because of their price, taste and demand. Major carps are also the main target species in Kaptai Reservoir. To replenish the stock, and also to forestall possible breeding failures, fingerlings of major carps have been stocked since the beginning of the fishery. In the first 10–12 years major carps contributed 50–80% of the total catch, but this has decreased steadily. An intensive stocking programme has been undertaken since the 1980s but the catch of major carps has not increased (Fig. 12.2).

A simple linear relationship between the fingerlings stocked and yield of major carps proved negative ($r = -0.68$; $P < 0.05$) suggesting stocking is ineffective. The reasons for the failure are not clear but may be due to high mortality immediate after stocking, predation, illegal fishing or overfishing.

To colonise empty, vacant or new ecological niches and to utilise aquatic weeds, about 1.6 million fingerlings of alien carps, viz. common carp (685 550 pieces), silver carp (914 142) and grass carp (7216) were stocked in the Kaptai Reservoir between 1981 and 1985. The commercial catch of these three species between 1983 and 1988 was only 3.17 t (common carp, 0.02 t; silver carp, 2.01 t; and grass carp, 1.14 t) again demonstrating failure in the stocking exercise.

The introduction of tilapia, *Oreochromis niloticus*, in 1985 originated from an abortive cage culture trial. After the incident, the catch of tilapia increased rapidly from its initial year of harvesting in 1986–1987 (0.02%) to 1992–1993 (1.4%) after which it time stabilised (Table 12.2).

African catfish has also been accidentally introduced into the reservoir from surrounding culture ponds. These catfishes are also caught commercially, but are of concern because of their predatory feeding habits and competition for food resources.

12.3.6 *Natural breeding*

Natural breeding is the main source of fish recruitment in Kaptai Reservoir. Of the 73 species, most breed naturally in the reservoir. The major carps do not breed in confined waters, although they reach maturity and ripen. However, during the monsoon, when there is heavy rainfall in the surrounding hills, torrential currents prevail in the feeder rivers and the upper reaches become inundated providing an opportunity for natural breeding of major carps. Several attempts were made to identify the breeding grounds of major carps of the feeder rivers of Kaptai Reservoir (Hye & Alamgir 1992; Azadi 1985) but only four sites have been confirmed in different feeder rivers (ARG 1986; Fig. 12.1A–D), viz:

(1) Kassalong range, Myanimukh and upwards;
(2) Barkal range, Foremukh to Jagahnathchari and upwards;
(3) Chengi range, Nannearchar and upwards; and
(4) Reankhyong range, Belaichari to Chakrachari and upwards.

However, considering the poor recruitment in the 1990s it is assumed that natural spawning no longer takes place, probably due to siltation, water level fluctuation, lack of heavy rainfall at the proper time, low current velocity during the breeding season and fishing pressure. Detail information on breeding of major carps was provided by the BFRI during the post-spawning season through *in situ* observation of fishermen's catch (BFRI-RSS 1994). From the survey only the fingerlings of *Labeo calbasu* were observed. Fingerlings of *Tor tor* and *Labeo gonius* (Hamilton) were also observed in the upper reaches but no fingerlings of *Labeo rohita*, *Calta catla* and *Cirrhinus mrigala* were found. *Labeo calbasu* and *Labeo gonius* have never been stocked in the reservoir, but large quantities of this fish are being harvested every year presumably based on natural recruitment. It thus appears that there is a bottleneck to recruitment of all major carps except *Labeo calbasu.*

12.3.7 *Fishing crafts and gears*

In Kaptai Reservoir, basically two types of fishing crafts, the large country boat (baro nauka) (10–12 m length; 1–2 m breadth) and the small country boat (dinghi nauka) (6–8 m length; 0.7–1.2 m breadth) are found. Except for their size, these two types of craft are more or less similar in form. Baro nauka are generally used for seine netting and are operated by between five and 10 fishermen. Dinghi nauka are used for gill nets, lift nets, push nets, hooks and lines, and are operated by one to two fishermen. Another type of boat occasionally found, which the tribal people popularly call ekgaischha nauka (boat from one timber), made by grooving a large timber log, is used for small-scale netting operations. All of the above crafts are made of indigenous timbers and are manually operated.

Different fishing methods are employed in different seasons. Mode of operation of gears is dependent on water level, rainfall, abundance of fish species and lunar cycle. In general, fishermen select the gear type and mesh size to capture the desired species and size of fish. Type of nets, their lengths, depths and mesh sizes vary depending on choice and capital of the persons involved, as well as on the abundance of fish at harvest. Four major types of gears, the nets, hook and line, fish traps, and wounding gear, are found in Kaptai Reservoir. These gears are not equally used for commercial fishing. Three types of seine nets, the large-meshed seine net (*tengra jal*), mosquito seine net (*kechki jal*) and small-meshed seine net (*deka jal*) and two types of gill net, the small-meshed gill net (*chapila jal*) and large-meshed gill net (*vasa jal*) are found in Kaptai Reservoir. One type of push net (*thela jal*), which consists of a triangular net forming a bag shape, and one type of lift net (*dharma jal*) are also used. Four types of hook and line, cluster hooks (*jhoomka borshi*), long line (*chara borshi*), hand line (*chip borshi*) and reel line (*wheel borshi*) are employed.

A shift in gear operation has occurred in recent years. With the appearance of kechki (*Corica soborna*) since 1980, a large number of very small-meshed mosquito nets (beach seine) are being operated and their number is increasing gradually, spreading throughout the reservoir. Introduction of four-boat lift nets is also a new addition in this context. This gear consists of four-mesh size nets from the top to the bottom so different sizes of fishes can be caught by same gear at any one time. An unusual fishing method locally known as fud (brush shelter), has also been introduced in the reservoir since the early 1990s. At present about 1000 brush shelters are in operation in the reservoir (Ahmed 1999). BFRI-RSS (1993) estimated 679 gill nets, 305 seine nets, 93 lift nets, 18 push nets and 212 hook and line rigs of different categories operate on the reservoir.

12.3.8 *Catch per unit effort and catch composition of different gears*

The average catch per unit effort of different gears varied from 2.9 to 32.2 kg unit^{-1} day^{-1}, with an average for all the gears of 11.4 kg unit^{-1} day^{-1} (Table 12.3). Among the different gears, the mosquito seine net (beach seine) was the most efficient gear (32.2 kg unit^{-1} day^{-1}) followed by the small-meshed seine net and lift net (30.9 and 24.1 kg unit^{-1} day^{-1}). The range of fish caught by the lift net, small-meshed seine net and push net is diverse, whereas the mosquito seine net and hook and line are very selective. The main species targeted, the major carps, were mainly caught by reel line (94%), hand line (89%), cluster hooks (92%) small-meshed gill nets (70%), large-meshed gill nets (77%) and large-meshed seine nets (67%). The clupeids were caught mainly by the mosquito seine nets (96%) and small-meshed seine nets (56%).

Table 12.3 Catch per unit effort (kg unit^{-1} day^{-1}) and catch composition (%) of different gears of Kaptai Reservoir

Gear	Catch per unit effort	Catch composition								
		Major carps	Catfish	Clupeid	Schilbeid	Featherback	Tilapia	Shrimp	Snakehead	Others
Lift net	24.1	19.7	39.6	0.0	14.9	10.1	2.5	0.0	0.0	13.2
Large-meshed seine net	8.7	66.7	5.7	0.0	0.0	27.0	0.0	0.0	0.0	0.6
Small-meshed seine net	30.9	6.4	13.2	56.4	0.0	0.6	3.7	0.0	0.0	19.1
Mosquito seine net	32.2	0.0	0.0	96.1	0.0	0.0	0.0	0.0	0.0	3.9
Push net	3.6	0.5	31.3	0.0	0.0	8.3	0.0	49.8	0.0	10.3
Large-meshed gill net	3.5	77.4	13.5	0.0	0.0	5.0	4.0	0.0	0.0	0.2
Small-meshed gill net	8.4	70.5	18.2	0.0	0.0	0.0	0.0	0.0	0.0	11.2
Cluster hooks	3.7	92.0	0.0	0.0	0.0	0.0	1.5	0.0	0.0	6.5
Long line	4.6	0.0	23.5	0.0	0.0	19.6	0.0	0.0	45.6	11.3
Hand line	3.4	89.0	0.0	0.0	0.0	0.0	3.5	0.0	0.0	7.5
Reel line	2.9	94.2	0.0	0.0	0.0	0.0	0.0	0.0	0.0	5.8
Average	11.4	47.0	13.2	13.9	1.4	6.4	1.0	4.5	4.2	8.1

12.3.9 *The fishermen*

Three types of fishermen: owner-1, owner-2 and crew or labour are engaged in fishing in Kaptai Reservoir. Owner-1, fishermen are financially solvent, and own their boats and gear. Owner-2 fishermen have purchased their boats and gears by taking loans from fish traders. The labour fishermen are hired on a daily or monthly basis by the owner-1 or owner-2 fishermen. They do not share the catch. Both local and migratory fishermen are found in Kaptai Reservoir. About 38% of the owner fishermen come from different districts of Bangladesh to the reservoir during the fishing season. Crew or labour fishermen also come from outside regions. Overall the immigrant fishermen play a vital role in the exploitation of the reservoir.

Both subsistence and commercial fishing is carried out on Kaptai Reservoir. Local fishermen are resource poor and usually engage in subsistence fishing. Subsistence fishermen usually sell their catch in the local markets, but the commercial fishermen have to sell their catch to the commission agents or to the fish traders. The fishermen of Kaptai Reservoir are mainly Muslim (69%), followed by Hindu (22%) and tribal (9%). About 5560 fishermen exploit Kaptai Reservoir (Ahmed 1999).

12.3.10 *Fish traders*

In the Kaptai Reservoir fishery, the middlemen have a group of regular employees or hired labour engaged in purchasing, carrying, selling fish and organising loans to the fishermen. The middlemen groups are usually called fish traders (locally called saodagar). About 185 fish traders regulate trading on Kaptai Reservoir. Of them, 30 traders handle the major share of the commercial landings (Ahmed 1999), and are known as master traders. The master traders are influential, invest big capital by disbursing dadon (advanced credit disbursement in cash) to the owner fishermen at the start of each fishing season. In the early years the fish business was controlled by only seven fish traders (ARG 1986), but at present their number has increased resulting in a more competitive business environment.

12.3.11 *Marketing and royalty collection system*

Kaptai Reservoir is surrounded by high hills and there were only two (Chittagong–Rangamati and the Chittagong–Kaptai) access points to the reservoir from the plains. Hence, BFDC established two fish-landing centres, one at Rangamati and other at Kaptai for royalty collection and to ease management of the lake.

The fishermen operating in different fishing areas usually sell their catch to the commission agents or to the fish traders and never come to the landing centres. At the landing centres, fish traders first settle the royalty with the BFDC and again sell or transport the fish to different city markets. The small fish traders collect fish by manually propelled boats locally called sampan. They sell all the fish to the master traders since their quantity is small and they lack facilities to transport the fish to the city markets.

Some opportunistic commission agents are also engaged at different fishing sites. These agents purchase fish from the fishermen and sell it to the agents of master traders. But fishermen who receive dadon have to sell their catch to the selected agents of the fish trader. The fish traders always maintain links with fishermen providing them ice on behalf of master fish traders. These agents also disburse ice, drums for holding fish, and other fishing-related information to the fishermen. The master traders carry fish to the city markets and a small quantity is also exported. The BFDC's portion is sent to preservation plants and then sold to the wholesalers and retailers. As such, catches are handled by at least two to three intermediaries before they reach the consumer (Fig. 12.3).

There is no fixed price for fish; it varies within local and wholesale markets. Around the reservoir, the fish traders or commission agents fix the price of fish at as low a rate as possible.

The BFDC collects revenue from fish trader's giving licences at the start of the fishing season in each year. In addition, BFDC collects royalties on total fish landings from the fish traders. The royalties are collected in kind (kg of fish) for the high-priced, large fishes and in cash (Bangladesh Taka kg^{-1}) for low-priced, small-sized fishes. The average share of the BFDC ranges from 33.3% to 40% depending on species and total landing. Usually they collect a higher percentage of royalty from the high-priced and larger fishes. This system seems to be highly impractical and exploitative of the fishermen and fish traders.

12.3.12 *Management measures*

Kaptai Reservoir fisheries management basically pertains to implementing close seasons, issuing licences, fish act implementation and stocking. To protect natural recruitment, fishing is prohibited from mid-May to mid-August every year because June–July is the peak breeding season. However, in distant places it is almost impossible to ban fishing completely during the breeding season. Subsistence fishermen and tribal people fish at that time for their own consumption and local marketing. In Kaptai Reservoir, capture of fish is regulated by a minimum size, viz. *Calta catla* 2.0 kg, *Labeo rohita* 1.0 kg, *Cirrhinus mrigala* 0.75 kg and *Labeo calbasu* 0.5 kg. This fishing regulation is not enforced. Although continuous effort is given to increase major carp fry and fingerlings, yield has decreased. The recapture rate from lakes and reservoirs is assumed to be 20–25% of the total stocking, but for Kaptai Reservoir this target has never been achieved, and is less than 1%. The reasons behind this may be due to undersized fingerlings being stocked, carrying fish fry and fingerlings from distant places, high rate of predation and under reporting. The BFRI suggested stocking the reservoir with major carp fingerlings of greater than 23 cm, but this was not followed up because of scarcity of appropriate-sized fingerlings.

12.4 Discussion

Fish yield from Kaptai Reservoir, when compared with the production of lakes in Thailand (Bhukaswan & Chookajorn 1988), Java (Indonesia) (Baluyut 1984) and India

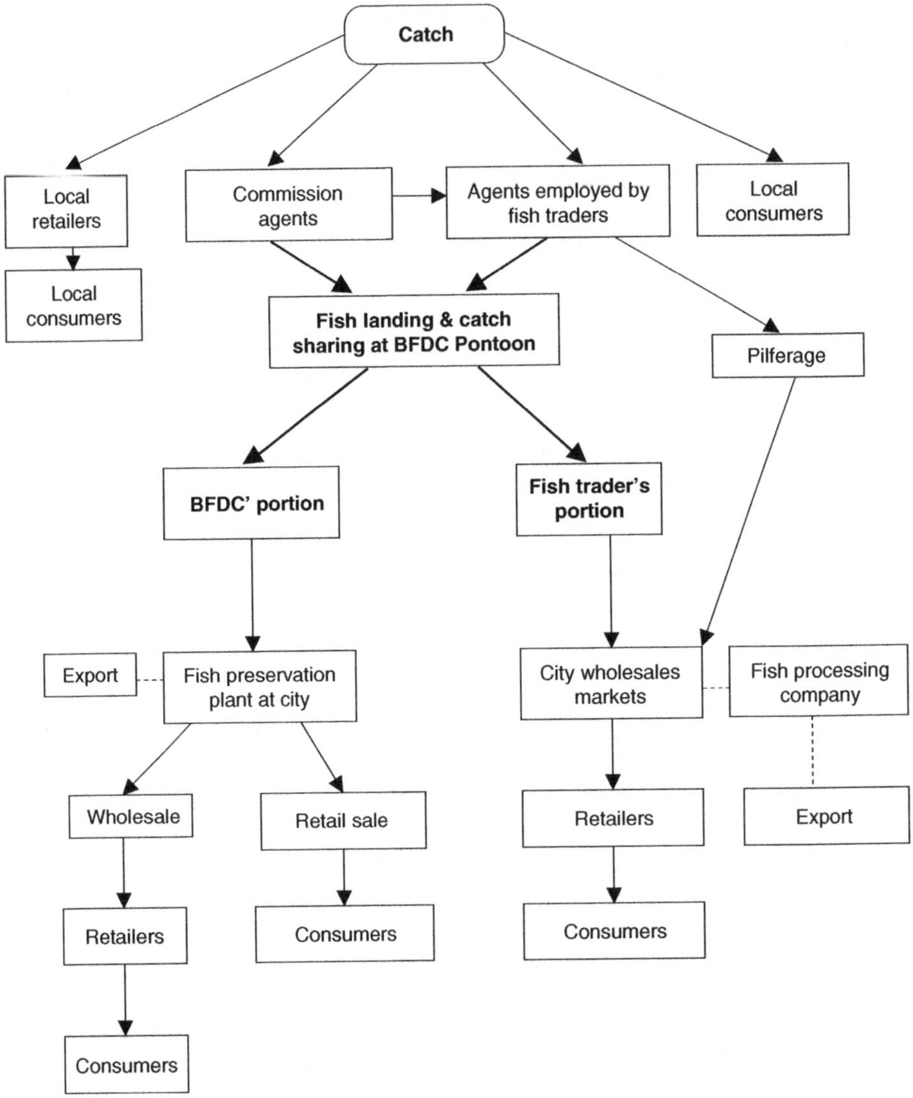

Figure 12.3 Fish marketing channel of Kaptai Reservoir

(Sharma 1983) suggests the reservoir is moderately productive. To increase and sustain its fishery potential, a rational management policy is necessary for Kaptai Reservoir. The effective implementation of open-water fisheries management requires a thorough knowledge of the fish stocks and their dynamics to regulate fishing effort and/or fish catch at a desirable level. Effective fisheries management can only be conducted if the production potential is at least approximately known. These assessments require specific know-how and skills but the BFDC does not have adequate technical manpower. Data collection on size and catch composition are difficult due to the distant, scattered and isolated location of fishing areas. Moreover, fishermen are reluctant to give their actual catch data, for fear of paying more fees for licences, royalties, etc.

Control of detrimental fishing through enforcement of fishing rules and regulations, proper stocking with locally-raised, optimum-sized fingerlings, declaration of fish reserves for the protection of natural spawning stocks, replenishment of natural spawning grounds, regulation of harmful gears are other possible measures to increase yield. However, logistic support is required to implement these actions. Under these circumstances, agencies including fishermen, the fish trading community, local leaders, public representatives, government departments particularly concerned with research, non-government organisations and law enforcing agencies should be brought together to coordinate their efforts and views for sound management of the fishery. A detail research programme is necessary to assess the stock, stocking impacts, replenishment of natural spawning and nursery grounds and on other relevant aspects of the fishery.

References

Ahmed K.K., Haldar G.C. & Paul S.K. (1991) Studies on the bottom macro fauna of Kaptai Lake. *Bangladesh Journal of Zoology* **6**, 49–56.

Ahmed K.K., Haldar G.C., Saha S.B. & Paul S.K. (1994) Studies on the primary production of Kaptai Lake. *Bangladesh Journal of Zoology* **22**, 69–77.

Ahmed K.K. (1999) Options for the Management of Major Carp Fishery in the Kaptai Reservoir, Bangladesh. Unpublished PhD Thesis, School of Environment, Resources and Development, Asian Institute of Technology, Bangkok, Thailand, 298 pp.

Alamgir M., Chowdhury S.H. & Ahmed A.S. (1990) New record of ichthyofauna of Kaptai Lake. *Chittagong University Studies, Part II: Science* **14**, 22–28.

Aquatic Research Group (ARG) (1986) Hydrobiology of the Kaptai Reservoir. *FAO/UNDP Final Report No. DP/BGD/79/015/FI*, 192 pp.

Azadi M.A. (1985) Hydrological conditions influencing the spawning of major carps in the Halda River, Chittagong, Bangladesh. *Bangladesh Journal of Zoology* **13**, 163–172.

Baluyut E.A. (1984) A review of inland water capture fisheries in Southeast Asia with special reference to fish stocking. *FAO Fisheries Report No. 288*, 57 pp.

BFDC (Bangladesh Fisheries Development Corporation) (1991) Karnafuli Reservoir: Fisheries Development and Marketing Project. A Report on Total Landing of Fishes from Kaptai Lake (1965–1990), 48 pp.

BFDC (Bangladesh Fisheries Development Corporation) (1996) Annual Progress Report (in Bangla), 6 pp.

BFRI-RSS (Bangladesh Fisheries Research Institute, Riverine Sub-Station) (1993) Annual Report Riverine Sub-Station, Rangmati, Bangladesh. 8 pp.

BFRI-RSS (Bangladesh Fisheries Research Institute, Riverine Sub-Station) (1994) Annual Report Riverine Sub-Station, Rangmati, Bangladesh. 20 pp.

Bhukaswan T. & Chookajorn T. (1988) Reservoir fisheries management in Thailand. In S.S. De Silva (ed.) *Fishery Management and Development in Asia.* IDRC 264E, pp. 214–223.

Chowdhury S.C. & Mazumder A. (1981) Limnology of Kaptai Lake-I: physico-chemical features. *Bangladesh Journal of Zoology* **9**, 52–72.

Fernando C.H. (1980) Fisheries potential of man made lakes in Southeast Asia and some strategies for its optimization. In BIOTROP. Anniversary Publication, Bogor, BITROP, pp. 23–28.

Haldar G.C., Mazid M.A., Haque M.K.I., Huda M.S. & Ahmed K.K. (1989) A review on the fisheries fauna of the Kaptai lake. *Bangladesh Journal of Zoology* **14**, 127–135.

Haldar G.C., Mazid M.A. & Ahmed K.K. (1990) In S.S. De Silva (ed.) *Reservoir Fisheries of Asia. Proceedings of the 2nd Asian Reservoir Fisheries Workshop held in Hangzhou, People's Republic of China, 15–19 October 1990.* IDRC, Ottawa, Canada. IDRC-291E, pp. 2–11.

Hye M.A. (1983) Fishery potentials of Kaptai lake. *Agricultural Development Agencies of Bangladesh News* **10**(6), 2–8.

Hye M.A. (1988). Socio-economic condition of Kaptai Lake fishermen. Bangladesh. *Journal of Extension Education* **3**, 23–33.

Hye M.A. & Alamgir M. (1992) Investigation on the natural spawning of carps in Lake Kaptai. *Bangladesh Journal of Zoology* **14**, 27–33.

Rahman M.A. (1980) Need for reservoir fisheries management in Bangladesh. In: S.S. De Silva (ed.) *Reservoir Fisheries Management and Development in Asia*: Proceedings of a Workshop held in Kathmandu, Nepal, 23–28 November 1987. IDRC, Ottawa, on Canada. IDRC-264E. pp. 164–168.

Sandercock F.K. (1966) Chittagong hill-tracts soil and land use survey 1964–65. Vol. 4, Fisheries. Canadian Colombo Plan Project No. F 475, East Pakistan Agricultural Development Corporation, 67 pp.

Sharma K.P. (1983) Multipurpose use of water resource in relation to the inland fisheries of India. *FAO Fisheries Report No. 228*, pp. 150–166.

Section II
Anthropogenic activities/rehabilitation and mitigation

Chapter 13
Coexistence of introduced *Perca fluviatilis* and *Coregonus lavaretus* in Santa Croce Lake (Italy), and propositions for fishery management

R. LORO*
Bioprogramm S.c.r.l., via Aldo Moro, Fontanelle, Italy

A. MAMCARZ
Department of Lake and River Fisheries, University of Varmia and Mazury, Olsztyn-Kortowo, Poland

Abstract

Populations of the introduced *Perca fluviatilis* L. and *Coregonus lavaretus* (L.) in Santa Croce Lake, Italy, were studied between September 1996 and August 1997. In the recent past anglers have recorded a decline in perch catches at the expense of whitefish. Perch grow rapidly during the first year, based on a zooplanktonic diet, while in later years food availability appears inadequate. The whitefish, introduced at the beginning of the 20th century, also has a zooplanktonic diet and grows to a large size. It was concluded that there was no competition between the two species, but the poor growth of perch was due to the low abundance of benthic foods and small-sized prey fishes. To overcome this deficiency, it was suggested to increase the cyprinid larval production by the creation of artificial spawning sites and stocking with bleak, *Alburnus alburnus alborella* De Filippi.

Keywords: competition, perch, stocking, whitefish.

13.1 Introduction

A large proportion of the Alpine reservoirs in Northern Italy contain fish species that were introduced in the 20th century. Apart from salmonids, whitefish (*Coregonus* spp.) and perch, *Perca fluviatilis* L., represent the major interests of the recreational fishery sector. In one of these lakes, Santa Croce, perch was introduced in 1939 (Fossa 1988) and it is now the most important target species for local anglers. Before 1970 the species was also exploited by 10 commercial fishermen.

Whitefish was introduced in 1901–1902 (Zanetti, Loro, Siligardi & Turin 1994), when *Coregonus schinzi helveticus* L. was stocked from the Brescia fishery station. Later, around 1945, a second introduction, *Coregonus wartmanni coeruleus*, was made (Fossa 1988). This species is not commercially exploited but in the last 5 years local

*Correspondence: Bioprogramm S.c.r.l., via Aldo Moro, 12 31043 Fontanelle (TV), Italy (email: r.loro@oderzo.nettuno.it).

anglers have seen its numbers increase rapidly during the reproductive period, when males and females spawn along the shoreline.

During the same period a steady decrease in perch catches was observed and the local angling association was stimulated to study the status and feeding requirements of perch and whitefish in an attempt to understand why the perch catches have declined. This chapter examines the results with particular reference to potential feeding interactions between perch and whitefish.

13.2 Materials and methods

Santa Croce Lake (Fig. 13.1) is located in the Dolomites (altitude 386 m), north-east Italy. It was formed by the natural damming of a valley with stone material from recent mountain glacial activity. Between 1926 and 1928, the lake was enlarged by damming the outflow for hydropower production. It has a surface area of 7.8 km^2 and volume of 150 million m^3. Its watershed is small (154 km^2), but it receives a large amount of water from the River Piave through an artificial channel, but the turnover ratio (% water exchange) remains low at 0.117 year. The maximum depth is 41 m, with an average depth of only 19 m.

The fish community consists of whitefish, perch, pike, *Esox lucius* L., brown trout, *Salmo trutta* L., marbled trout, *Salmo marmoratus* Cuv., bleak, *Alburnus alburnus*

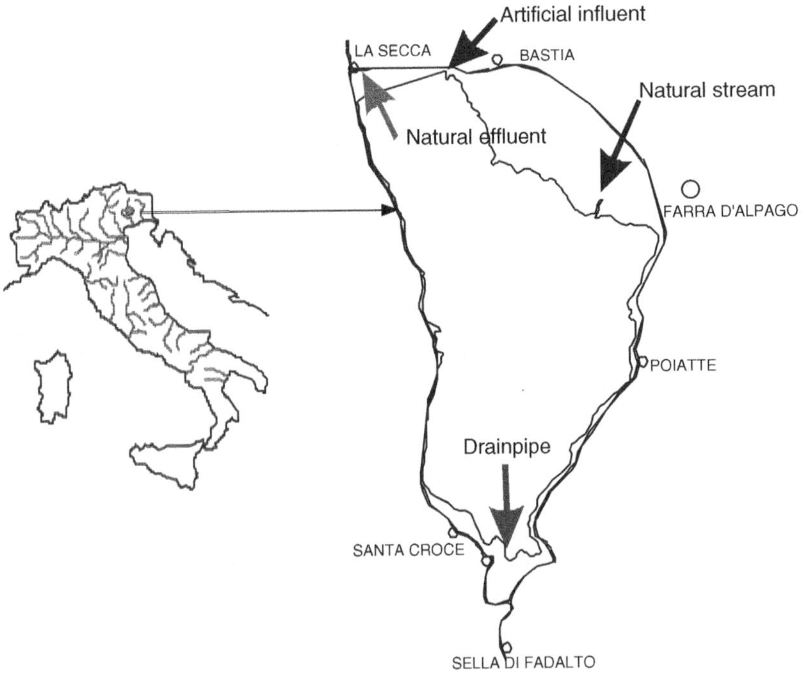

Figure 13.1 Map showing location of Lake Santa Croce

Table 13.1 Size categorisation of perch and whitefish in samples

Age group	Perch total length (mm)	Whitefish total length (mm)	Specimens studied
0+	<100	<200	30
1+	100–150	200–300	30
2+	150–200	300–400	30
3+ and older	>200	>400	30

alborella De Filippi, chub, *Leuciscus cephalus* (L.), rudd, *Scardinius erythrophthalmus* (L.), and *Rutilus pigus* Lacepède.

Surveys were made monthly from December 1996 to September 1997 using fleets of gill nets comprising 13 nets (two nets 30 × 2 m, two nets 30 × 4 m, four nets 50 × 4 m, two nets 20 × 5 m, and three nets 30 × 5, 50 × 5 and 50 × 6 m, respectively) fishing on the bottom and in the water column. The mesh size ranged from 10 to 60 mm (knot-to-knot). Gill nets were set at night and fish were removed in the early morning.

Specimens of perch and whitefish caught were ranked into four length classes for further analysis (Table 13.1). In total, 937 whitefish and 835 perch were collected and analysed. Total length (mm) and weight (nearest g) were measured. Fish age was determined from scales in whitefish, and from opercula in perch; sex and maturity were determined for each fish. The stomachs of fish were fixed in 4% formalin for later dietary analysis. The number of prey of each species or animal group in each of the stomachs was counted in the laboratory.

Benthic invertebrates and zooplankton were sampled in the same area as the fish were caught. Benthos was collected with a Petersen grab, sieved using a 0.2-mm mesh net and fixed in 4% formalin.

Zooplankton were sampled using a standard plankton net of 1 m length and 20 cm mouth diameter with a 50 μm mesh net. Vertical hauls were taken on each sample date from a water depth of 5–11.5 m (2.5 × Secchi disk depth). All samples were preserved in 4% formalin and examined under a microscope.

13.3 Results

13.3.1 *Invertebrate fauna*

The zooplankton fauna exhibited typical seasonal fluctuations. Rotifers, mainly represented by *Kellikottia*, *Asplanchna* and *Synchaeta* spp., were more common in summer and autumn (Table 13.2). Cladocera, mainly *Daphnia longispina*, were an important food organism for fish and peaked in abundance over the summer (June–September). Copepoda were dominated by juvenile life stages of cyclopids (nauplii and copepodites) and adults of *Cyclops strenuus*, *Mesocyclops leuckarti* are abundant from early summer to late autumn (7.7–36.0 ind. L^{-1}). In general zooplankton resources in the lake were not abundant. Numbers of organisms ranged from 5 L^{-1} in December–January to

Table 13.2 Seasonal changes in the zooplankton composition of Lake Santa Croce (ind. L^{-1})

Species	Oct.'96	Nov.'96	Jan.'97	Mar.'97	Apr.'97	May'97	Jun.'97	Jul.'97	Aug.'97	Sep.'97
Kellikottia longispina	4.08	2.8				2.43	13.47	6.88	20.03	0.94
Keratella cochlearis							0.17			
Asplanchna priodonta	5.18	36.04	2.56			2.19	22.72	5.28		7.44
Synchaeta sp.	9.79	154.97		1.98			3.06	3.2	0.46	0.51
Daphnia longispina	6.24	8.01	1.11	1.06	3.68	12.72	28.91	7.52	4.43	17.34
Bosmina longirostris	2.57									
Leptodora kindti									0.38	
Cyclops strenuus	0.94				2.57	2.43	3.3	2.08	1.98	3.61
Eudiaptomus sp.	0.12	1.2						0.48	0.2	0.22
Mesocyclops leuckarti	3.67	4						2.08	0.86	3.4
Thermocyclops oith.		13.21					0.17			
Nauplius	1.51	8.01	1.11	1.06	2.57	0.81	2.72	1.12	0.46	7.95
Copepoditi	22.32	21.22	0.33	1.06	2.57	6.72	8.5	15.52	4.63	36.63
Rotifera	19.05	193.81	2.56	1.98	0	4.62	39.42	15.36	20.49	8.89
Cladocera	8.81	8.01	1.11	1.06	3.68	12.72	28.91	7.52	4.81	17.34
Copepoda	28.56	47.64	1.44	2.12	7.71	9.96	14.69	21.28	8.13	51.81
Total	56.42	249.46	5.11	5.16	11.39	27.3	83.02	44.16	33.43	78.04

250 ind. L^{-1} in late autumn. During winter and early spring (to April) abundance of organisms was low (10 ind. L^{-1}).

The benthic invertebrate community, dominated by Tubificidae, Lumbriculidae and Naididae (Oligochaeta), was more stable throughout the year, but still exhibited seasonal changes with maximum density of benthic organisms reaching over 30 000 ind. m^{-2} in the summer (Table 13.3). The density of Chironomidae and *Pisidium* increased to 5.9% and 32.1%, respectively, in the summer.

13.3.2 *Perch*

Most perch captured were less than 5 years of age. The number of older fish was very limited and the oldest fish caught was 8+ (Table 13.4). The growth rate was very high, especially in the first year of life.

The percentage of empty perch stomachs was very high in all size groups, except the youngest fish 0+ and 1 age groups between May and September (Fig. 13.2). Generally, the number of empty stomachs increased with perch length and was particularly high in winter (October–March).

The diet of young fish was dominated by *Cyclops* sp. and to a lesser extent *Daphnia* sp. More than 100 specimens per stomach were noted between May and October, with a peak of 350 in July. The main food component during winter (from November to April) was detritus. The diet of the second length class was similar to smaller perch, except for *Daphnia* sp. which was the most dominant zooplankton item. In one case there was two larval whitefish in a stomach and in another case a perch larva was found.

Stomachs in the third length class only contained food items between April and September. Other than detritus, six food components were found, i.e. perch larvae, bleak (*Alburnus alborella*), unidentified fish, Chironomidae, insects and *Asellus* sp. The main diet of larger perch was fish, mainly perch larvae and juveniles, and bleak (*Alburnus alborella*). Other food items included Chironomidae, insects and occasionally Trichoptera or *Lumbricus* sp.

The maximum body length of fish eaten by adult perch was 8.4 cm (*Alburnus alborella*), but generally prey size did not exceed 6 cm in length. Bigger perch preferred bigger preys but this relationship was not significant (Fig. 13.3). The maximum body length of perch consumed was about 6 cm, which suggests that young perch were too big to be potential food for bigger perch in the autumn of their first year of life.

Perch in the lake matured very early. Males were ready to spawn in age 1+, females in age 2+, and GSI was high at about 30% or more before spawning (March–April) in females and 25% in males in October, suggesting considerable investment in reproduction. The GSI of males fell markedly overwinter, possibly reflecting poor food resources during this period and reabsorption of nutrients from the gonads.

13.3.3 *Whitefish*

Whitefish ranged in age from 0+ to 7+, but most were 2+ and 3+ (Table 13.4). Growth was considered slow, reaching only 230 mm by the end of the second year of

Table 13.3 Seasonal changes in the benthos composition of Lake Santa Croce (ind. m^{-2})

Species/group	Oct.'96	Nov.'96	Jan.'97	Mar.'97	Apr.'97	May'97	Jun.'97	Jul.'97	Aug.'97	Sep.'97
Pisidium	972	1736	1667	139	1111	1181	556	10417	5625	486
Musculium	69	0	0	0	0	0	0	0	0	0
Chironomidae	0	69	69	0	0	69	208	972	1389	69
Ceratopogonidae	0	69	0	0	0	0	0	139	208	69
Simuliidae	69	0	0	0	0	0	0	0	0	0
Tubificidae	4861	4792	2917	1111	4306	1875	1944	15 903	8125	486
Lumbriculidae	0	2222	1667	486	0	556	972	0	3889	1042
Naididae	208	833	1389	0	347	486	347	4444	3472	278
Helobdella	0	0	0	0	0	0	69	208	694	0
Valvata	0	0	0	0	0	0	0	139	69	0
Nudibranchia	69	0	0	0	0	0	0	0	0	0
Mermenthidae	0	0	0	0	278	0	0	139	0	0
Pselaphochernes litoralis	0	0	0	0	0	0	0	69	0	0
Total	6250	9722	7708	1736	6042	4167	4097	32 431	23 472	2431

Table 13.4 Growth of perch and whitefish in Lake Santa Croce

	Age							
	0+	1+	2+	3+	4+	5+	6+	7+
Perch								
Sample size		196	251	245	151			
Mean total length (±SD)(mm)		100 (16)	143 (14)	177 (20)	200 (23)			
Mean total weight (±SD)(g)		6 (1)	29 (7)	83 (12)	173 (21)			
Whitefish								
Sample size	25	149	281	249	159	65	7	2
Mean total length (±SD)(mm)	192 (13)	230 (41)	314 (38)	389 (44)	430 (28)	439 (19)	452 (31)	500 (35)
Mean total weight (±SD)(g)	51 (10)	105 (65)	268 (107)	502 (193)	630 (128)	659 (105)	760 (210)	1105 (120)

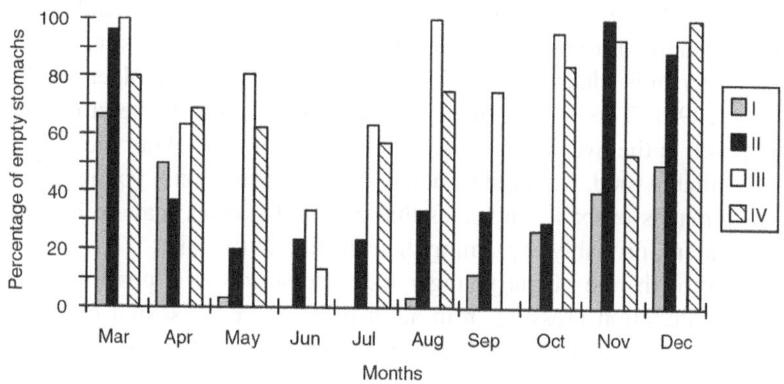

Figure 13.2 Percentage of empty guts in perch in four size groups

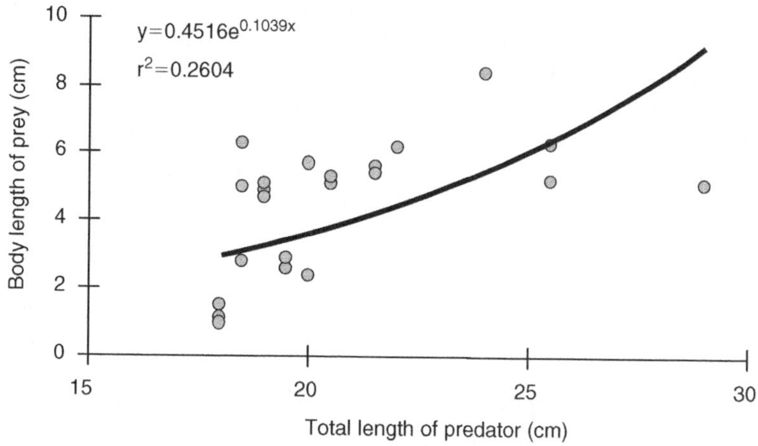

Figure 13.3 Relationship between body length of prey and total length of perch

life. The main food components of all age classes of whitefish were *Daphnia*, *Cyclops* and *Leptodora*. Additional food components were occasionally *Alona*, Chironomidae larvae and Insecta. *Pisidium* was present in the stomachs in late autumn and winter for the young age classes, but early spring for older fish. Fish eggs were found in the stomachs of second and third size classes of whitefish during the winter period. However, most size classes, but especially for larger individuals, had empty guts during early spring (March), suggesting a shortage of food.

13.4 Discussion

The perch population exhibited very fast growth in its first year of life, especially compared to other European populations (Thorpe 1977). However, the slowing down of growth was not typical of other populations (e.g. Terlecki 1986). This possibly stems from the perch's switch in diet from plankton feeding to piscivory as they get larger (Thorpe 1977). The great abundance of zooplankton (*Cyclops* sp. and *Daphnia* sp.) is a rich food resource for juvenile perch, but the relative absence of small fishes as a food base for larger individuals is manifest in the poor growth of older fish (Persson & Greenberg 1990). This is supported by the lack of food in the stomachs of many younger perch in the overwinter period and older fish throughout the year, suggesting a lack of available food. It should be noted, the presence of only young perch and bleak in perch stomachs probably indicates that cyprinid species may be having problems with recruitment, probably explaining their decline in the lake.

The fast growth rate of the younger fish also serves as a possible mechanism to either avoid predation pressure from adult perch or reach sexual maturity faster, or both. This is a typical r-related life history strategy in stressed populations (Cowx 1988), an interpretation supported by the great reproductive effort exhibited by the perch early in their life (Persson 1990).

Although whitefish is a zooplantivore, it probably does not exert direct competition on the perch population because dietary overlap is prevalent only at the younger life stages. There appears to be an adequate supply of small zooplankton for these juvenile fishes hence the good growth, but as the fish get larger strong interspecific competition for the larger zooplankton species restricts growth. This is possibly why whitefish have empty stomachs in winter and early spring or feed on fish eggs and benthic organisms (*Pisidium*, Nematoda), which are not typical of its zooplanktivorous diet. It is possible that commercial exploitation of the whitefish might remove some of this competition for food and improve the growth rate.

Some of the problems with the lack of food and poor recruitment of cyprinids can be linked to the rapid water level fluctuations experienced in the lake. These are typically found in hydroelectric reservoirs because of their mode of operation (Tundisi 1993). In Lake Santa Croce, large changes in water level prevent the growth of macrophytes in the littoral zone and thus the lake has restricted production of fish food, and limited spawning habitat and refugia for fry and young fish. Consequently only opportunistic or plastic species, such as bleak, are able to maintain their presence. The water fluctuations in the lake are less critical for perch and whitefish because the main fluctuations occur outside the egg incubation period for these species.

As a result, it is proposed that angling associations enhance the bleak stocks, through stocking, to increase the food resources for adult perch. It is also proposed to restore the shallow parts of lake where aquatic plants and macrophytes were abundant, as a place for fish spawning and also as a habitat for many food species, such as Chironomidae, Trichoptera, and littoral zooplankton. In addition, spawning sites for several lake cyprinid fishes should be enhanced by construction of floating artificial spawning habitats located around the lake.

References

Cowx I.G. (1990) The reproductive tactics of roach, *Rutilus rutilus* (L.) and dace, *Leuciscus leuciscus* (L.) populations in the rivers Exe and Culm, England. *Polskie Archiwum Hydrobiologii* **37**, 193–208.

Fossa I. (1988) Pesci, anfibi, rettili, uccelli, mammiferi del bellunese. – Tipografia Piave, Belluno (in Italian).

Persson L. (1990) A field experiment on the effects of interspecific competition from roach, *Rutilus rutilus* (L.), on age at maturity and gonad size in perch, *Perca fluviatilis* L. *Journal of Fish Biology* **37**, 899–906.

Persson L. & Greenberg L.A. (1990) Interspecific and intraspecific size class competition affecting resource use and growth of perch, *Perca fluviatilis*. *Oikos* **59,** 97–106.

Terlecki J. (1986) Perch (*Perca fluviatilis* L.). In M. Brylinska (ed.) *Freshwater Fishes of Poland.* Warsaw: PWN, pp. 371–378 (in Polish).

Thorpe J.E. (1977) *Synopsis of Biological Data on the Perch* (Perca fluviatilis *L. and* Perca flavescens *Mitchill*). Rome: FAO Fisheries Synopsis, 138 pp.

Tundisi J.G. (1993) Theoretical basis for reservoir management. *Verhandlungen der Internationale Vereinigung Limnologie* **25**, 1153–1156.

Zanetti M., Loro R., Siligardi M. & Turin P. (1994) Il lago di Santa Croce: studi limnologici – 1993. Amm. Prov. Di Belluno; A.P.S. Bacino di pesca no 7 'Alpago', 102 pp. (in Italian).

Chapter 14
Compensatory responses of fish populations in a shallow eutrophic lake to heavy depredation pressure by cormorants and the implications for management

J.R. BRITTON*, J.P. HARVEY, I.G. COWX
University of Hull International Fisheries Institute, Cottingham Road, Hull, UK

T. HOLDEN, M.J. FELTHAM, B.R. WILSON and J.M. DAVIES
Liverpool John Moores University, Byrom Street, Liverpool, UK

Abstract

Through their protection under EU law, cormorant numbers have increased markedly on inland waters in the UK and the rest of Europe since the 1980s, although numbers have stabilised in recent years. It is believed that cormorant depredation has resulted in a deleterious impact on the freshwater fish populations of some inland fisheries, although few data have been able to prove this. This chapter describes the impact of intense cormorant depredation on the fish populations of Holme Pierrepont Rowing Course, Nottingham, between 1995 and 1998. Cormorant depredation allegedly reduced the fish stocks in the winter of 1993/1994, resulting in poor angling catches in the 1994 World Angling Championship. Despite subsequent, heavy depredation by over-wintering cormorants recorded in this study, the fish populations compensated for the losses incurred by cormorants by accelerating their growth rate. This minimised the size window when they were most vulnerable to depredation, lowered their age of sexual maturity and increased their fecundity for age. This ensured long-term sustainability in the fish populations in the face of the depredation and was due to a decrease in inter- and intra-specific competition in the fish populations. The implications of this on the management of fish populations subject to cormorant depredation are discussed.

Keywords: roach, common bream, cormorants, depredation, growth rates, compensatory mechanisms.

14.1 Introduction

Two sub-species of cormorant, *Phalacrocorax carbo carbo* (L.) and *Phalacrocorax carbo sinensis* (Blumenbach) inhabit UK waters. In the UK, cormorants are traditionally associated with coastal and estuarine waters, but in recent years an increasing number has been found over-wintering inland (Veldkamp 1996). There was an estimated increase

*Correspondence: Environment Agency, National Fisheries Laboratory, Huntingdon, UK (email: rob.britton@environment-agency.gov.uk).

of 5–10% per annum between 1970 and 1987 in the UK but this accelerated to approximately 25% between 1987/1988 and 1990/1991 (Sellers 1979; Feare 1988; Kirby, Gilburn & Sellars 1995). This increase is in part due to protection of the species under the EU Birds Directive (EEC/79/409) enacted in the UK through the Wildlife and Countryside Act 1981. Licensed shooting of cormorants is allowed in the UK, 'for the purposes of preventing serious damage to … fisheries' (Wildlife and Countryside Act 1981).

Inland roosting cormorant colonies require an abundant food source to meet the birds' daily nutritional requirements (340–520 g day^{-1} for *P. sinenis*; Kirby, Holmes & Sellers 1996 and 400–800 g day^{-1} for *P. carbo*; Feltham, Davies, Wilson, Holden, Cowx, Harvey & Britton 1999). This is generally derived from water bodies in the vicinity of the colony (30 km radius). As a consequence, over-wintering foraging on inland waters has caused considerable concern amongst fishery managers and anglers, who claim the depredation has an adverse effect on their catch rates through the reduction in the numbers of angler-exploitable fish or wounding of larger fish (Britton 1999).

Although a great deal of effort has been put into estimating the 'impact' of cormorants at freshwater fisheries (Kirby *et al.* 1996; Russell, Dare, Eaton & Armstrong 1996; Feltham *et al.* 1999), few previous studies have moved beyond estimating the number or mass of fish removed from a particular fishery by cormorants. Little effort has been given to the effect of that depredation on the fish population dynamics. This study was carried out to estimate the impact of cormorants on the fish populations at a large, inland water body in the UK.

14.2 Materials and methods

14.2.1 *Study area*

The study carried out between September 1995 and August 1998 took place on the Rowing Course situated at the National Water Sports Centre, Nottingham, adjacent to the River Trent (Fig. 14.1). The Rowing Course is 2215 m long and 135 m wide (total water area = 28.68 ha), with a maximum depth of 2 m. Recreational, catch and release angling is practised on the Rowing Course. However, no fishery management occurs on the site as it is used primarily as a centre of excellence for water-sports.

The Rowing Course is within the locality of an over-wintering, inland cormorant roost sited at Attenborough Gravel Pits, Nottingham (Fig. 14.1). During their over-wintering period (October–March), cormorants from the roost visited the Rowing Course daily for foraging purposes (Feltham *et al.* 1999). Anecdotal evidence suggested that cormorants first foraged regularly at the site in winter 1993/1994 (M. Thompson, personal communication).

14.2.2 *Assessment of cormorant feeding ecology*

Cormorant abundance/occupation at the Rowing Course was monitored by counting the number of birds from dawn until dusk, 1 day each month. Cormorant feeding

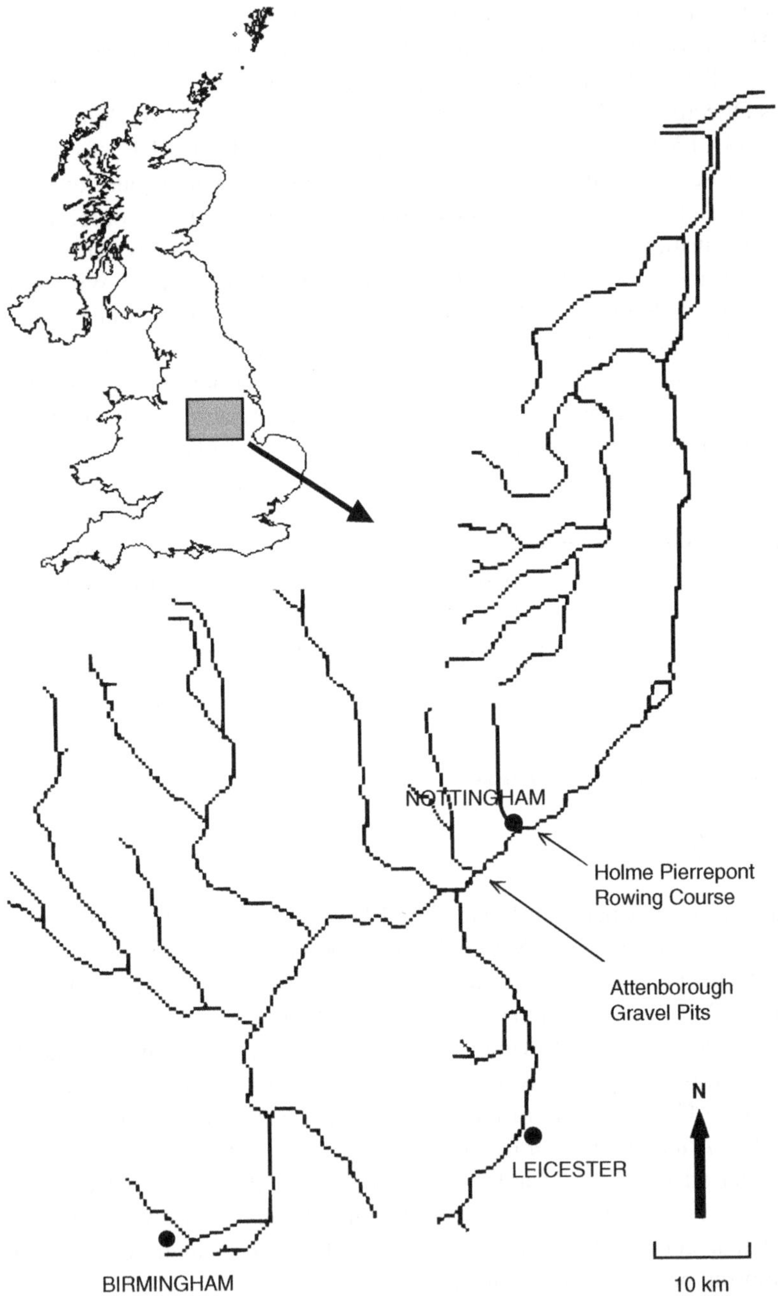

Figure 14.1 The River Trent catchment showing location of Holme Pierrepont Rowing Course and the cormorant roost at Attenborough gravel pits

observations were also carried out at the Rowing Course, on average 1–2 times per week. These commenced at dawn and continued whilst the birds were active or until weather conditions made further foraging bouts unlikely, for example, low light intensity and high wind conditions (Feltham *et al.* 1999). During feeding observations, prey species ingested were identified along with approximate size in relation to cormorant bill-length, allowing classification into 50-mm size classes. This enabled site specific length–weight relationships to be used to estimate the mass of fish removed during each cormorant foraging bout. These data were utilised in a Monte-Carlo Simulation (MCS) model to calculate the biomass of each species removed by the cormorants during each over-wintering period (Feltham *et al.* 1999). The MCS model also incorporated data on the cormorant foraging dynamics and abundance at the site (Feltham *et al.* 1999).

The proportion of the fish removed by cormorants from the fish standing crop of the Rowing Course was determined by comparison of fish standing crop, following a winter of cormorant depredation, with the actual standing crop that would have been present had cormorants not foraged (Britton 1999; Feltham *et al.* 1999).

14.2.3 *Assessment of fish population dynamics*

Data on fish populations of the Rowing Course were collected using boom boat electric fishing (Harvey & Cowx 1995). Surveys were carried out on a quarterly basis over the 3-year period, in September, January, March and June of each year. Species composition, fork length (mm) and weight (g) were recorded, with a scale sample removed for age and growth analysis. Data were also collected from anglers' catches (Britton 1999).

To ascertain how the fish populations responded to the effects of cormorant depredation, the following indices were calculated:

- back-calculated growth rates (Bagenal & Tesch 1978);
- relative growth index (Kempe 1962);
- mortality rates (Ricker 1975);
- year-class strengths (Cowx & Frear, in press).

14.3 Results

14.3.1 *Cormorant seasonal occupancy at Holme Pierrepont*

Cormorant numbers at the Rowing Course peaked each year from December to February, with numbers decreasing dramatically in April (Fig. 14.2), similar to the trend in numbers at the roosting site at Attenborough (Feltham *et al.* 1999). No cormorants were observed at the Rowing Course between May and September of each study year. Peak winter counts at the Rowing Course decreased during the study period, reflecting a decline observed at the Attenborough roost (Feltham *et al.* 1999).

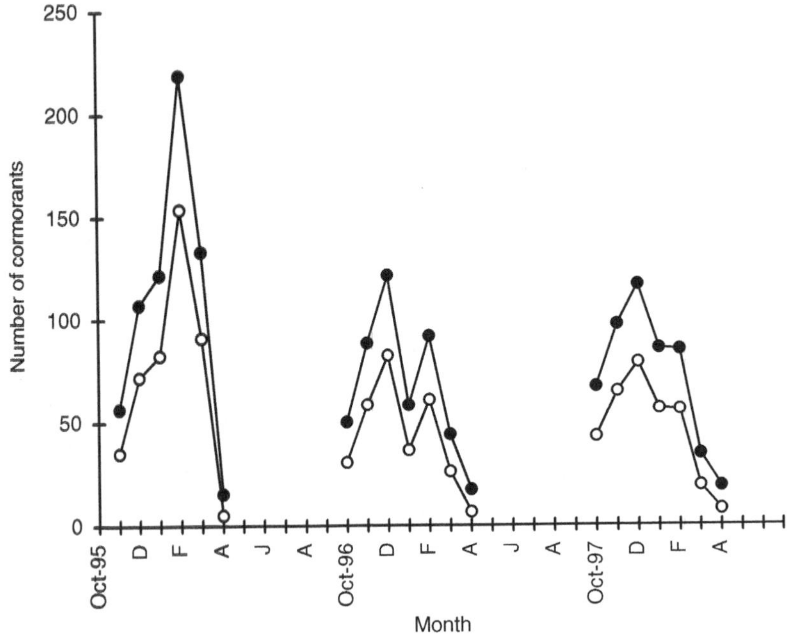

Figure 14.2 Peak early morning (–O–) and estimated cumulative daily counts (–●–) of cormorants feeding at Holme Pierrepont Rowing Course, 1995–1998

14.3.2 *Fish species ingested by cormorants*

Cormorants feeding at the Rowing Course mainly ingested roach, *Rutilus rutilus* (L.), common bream, *Abramis brama* (L.) and perch, *Perca fluviatilis* L., although the proportion of each species eaten varied between study years (Fig. 14.3). The majority of fish of all species ingested were <100 mm, although fish up to 250 mm were taken (Fig. 14.3).

The abundance of roach in the Rowing Course was reduced by 62% in number and 72% in biomass after three winters of cormorant depredation (Table 14.1). Common bream abundance was reduced by 51% in number and 67% in biomass after the three winters of cormorant depredation and perch by 65% in number and 75% in biomass (Table 14.1).

14.3.3 *Population parameters of fish*

The growth rates of roach and common bream were fast compared to standard growth (Hickley & Dexter 1979) (Table 14.2). The growth rate of roach from years prior to the first appearance of cormorants on the Rowing Course was much slower than when cormorants were foraging on the lake (Table 14.2). A shift in growth rate in common bream was difficult to highlight because of the longevity of the species (15+ years)

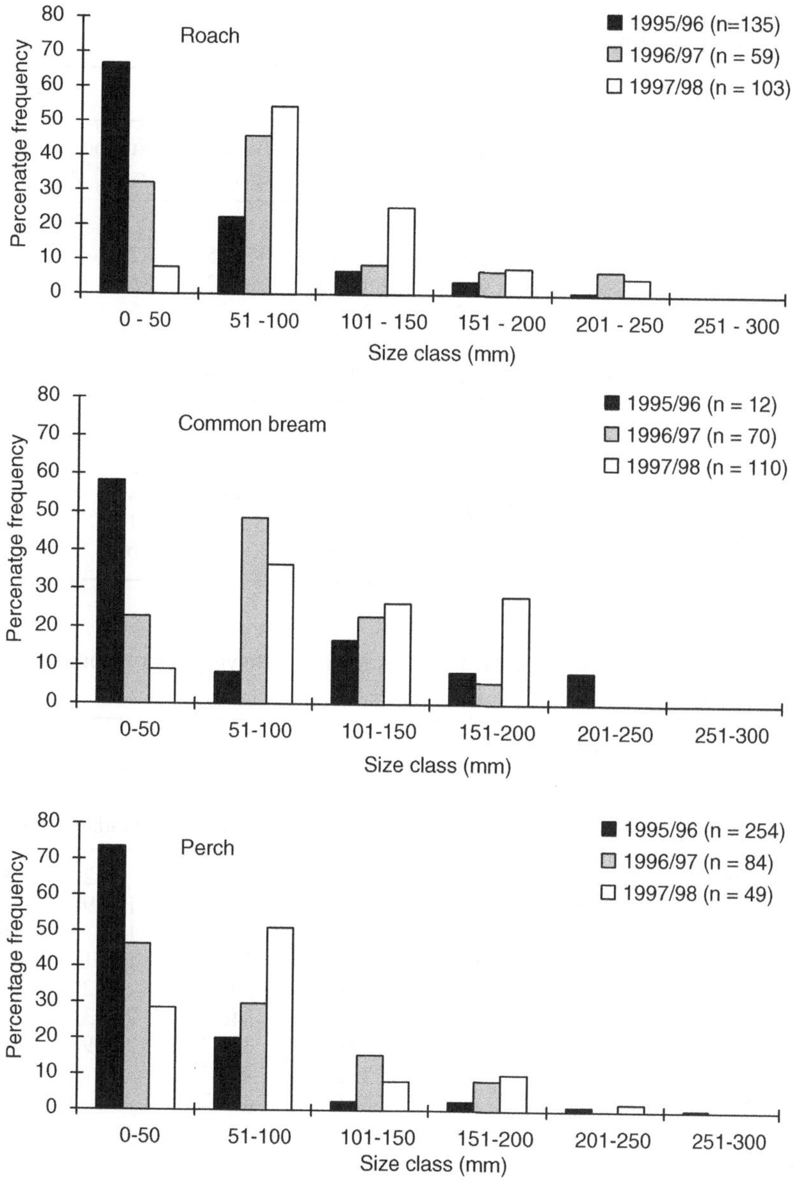

Figure 14.3 Diet composition of cormorants at Holme Pierrepont Rowing Course, 1995–1998

and required calculation of the relative annual growth rate. The growth rate of perch at the Rowing Course was also fast when compared with standard growth curves (Cowx 1999) (Table 14.2). However, there were insufficient numbers of fish from the year-classes produced prior to cormorants foraging on the lake to allow comparative growth rates and the relative growth rate to be calculated.

Table 14.1 Standing crop reduction attributable to cormorant depredation at Holme Pierrepont Rowing Course, 1995–1998

	1996		1997		1998	
	Number (n ha^{-1})	Biomass (g ha^{-1})	Number (n ha^{-1})	Biomass (g ha^{-1})	Number (n ha^{-1})	Biomass (g ha^{-1})
Roach						
With cormorants	225.1	8246.9	1173.6	47 065.3	322.2	17 762.7
No cormorants	416.1	21 962.3	1560.5	70 723.2	856.6	62 882.6
% Reduction	45.9	63.4	24.8	33.5	62.4	71.8
Common bream						
With cormorants	419.7	48 300.1	209.8	24 150.1	419.7	48 300.1
No cormorants	456.9	55 253.3	481.0	59 501.1	861.7	147 305.7
% Reduction	8.1	12.6	56.4	59.4	51.3	67.2
Perch						
With cormorants	230.8	14 967.4	253.6	14 809.5	249.6	13 200.2
No cormorants	1299.1	48 834.4	928.2	57 588.4	715.6	53 146.3
% Reduction	82.2	69.3	72.3	74.3	65.1	75.2

Table 14.2 Growth (mm) of roach, bream and perch in the Rowing Course compared to standard growth estimates

	Roach			Bream		Perch	
Age (years)	Standard[a]	Pre-cormorants	Post-cormorants	Standard[a]	Post-cormorants	Standard[b]	Post-cormorants
1	50.0	65.4	89.9	50.0	75.3	71.2	99.5
2	91.9	109.2	132.1	97.2	148.3	109.9	150.9
3	127.0	142.8	174.3	142.0	210.4	138	192.5
4	156.4	171.2	205.9	184.0	263.2	166.1	226.2
5	181.1	194.6	229.6	224.3	308.1	184.2	253.5
6	201.2	214.8	247.3	262.2	346.2	202.3	275.6
7	219.0	231.8	260.7	298.1	378.7	216.9	
8	233.5	246.1		332.0	406.3		
9	245.6	258.2		364.1	429.7		
10	255.7	268.4		390.6			

[a]Hickley & Dexter 1979. [b]Cowx 1999.

A marked change in the growth pattern in the roach and common bream populations was apparent in 1993/1994 (Figs 14.4 and 14.5). The relative annual growth of roach was poor prior to cormorants foraging on the lake compared to the period of cormorant depredation (Fig. 14.4). Furthermore, between 1992 and 1995, roach relative annual growth rate increased by 90% (Fig. 14.4). The historical growth rate of common bream was also poor when compared to the cormorant depredation period, with a marked increase in 1994 (Fig. 14.4).

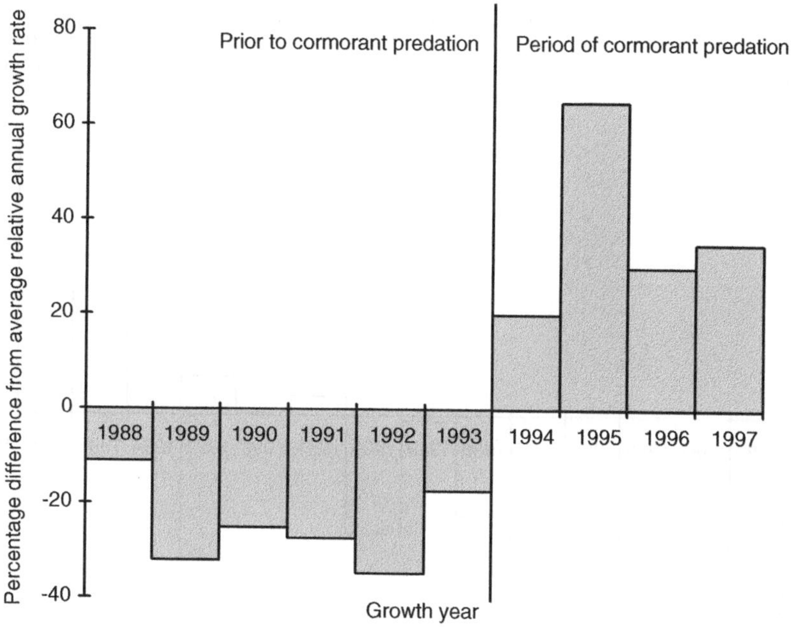

Figure 14.4 Relative annual growth of roach in Holme Pierrepont Rowing Course, 1988–1997

The mortality rates for roach, common bream and perch at the Rowing Course were high, indicating low annual survival of fish. Mortality rates for roach and common bream were, however, within the range found in other UK fisheries (Table 14.3).

Strong year-class strengths of roach occurred in 1988, 1992, 1995 and 1996, while poor year-class strengths occurred in 1989, 1990, 1991, 1993 and 1994 (Fig. 14.6). The year-class strengths (data not presented) for common bream were dominated by the 1984–1988 and 1994–1997 year classes only, resulting in skewed data (Britton 1999; Feltham *et al.* 1999).

14.4 Discussion

14.4.1 *Compensatory responses*

Fish populations are highly plastic and exert compensatory responses to stressful events, enabling the maintenance of long-term sustainability in the population (Wootton 1990). An example of compensation in a fish population experiencing catastrophic fish losses is an increased growth rate in the surviving individual fish, due to decreased competition, for example, in the commercially-exploited vendace, *Coregonus albula* L., populations in Lake Pyhajarvi, Finland (Sarvala, Helminen & Hirvonen 1994). Following an exploitation rate of 90%, losses were compensated by increased growth rates, increased fecundity for age and a reduced age at maturity. The shift in the

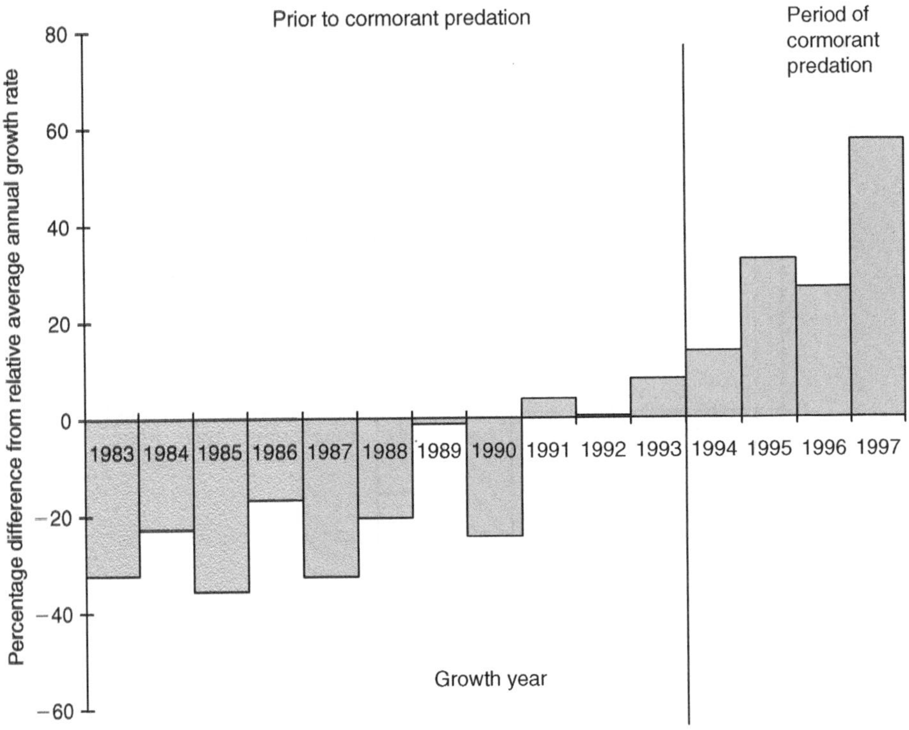

Figure 14.5 Relative annual growth of common bream in Holme Pierrepont Rowing Course, 1983–1997

Table 14.3 Mortality and survival rates of roach and common bream in Holme Pierrepont Rowing Course, 1995–1998, compared with other UK fisheries (Cowx 1999)

Species	Venue	Mortality (Z)	Survival (S)
Roach	Rowing Course	0.86	0.42
	7 UK fisheries	0.20–1.40	
Common bream	Rowing Course	0.92	0.40
	7 UK fisheries	0.33–1.23	
Perch	Rowing Course	0.96	0.39
	2 UK fisheries	0.46–0.53	

population dynamics of the roach and common bream in the Rowing Course in response to the cormorant depredation showed similar compensatory mechanisms.

Growth rates of roach, common bream and perch, the species heavily predated upon by cormorants, were extremely fast compared with standard growth curves (Hickley & Dexter 1979; Cowx 1999). Furthermore, relative growth rates of roach and common bream have increased markedly since cormorants first foraged on the lake in 1994

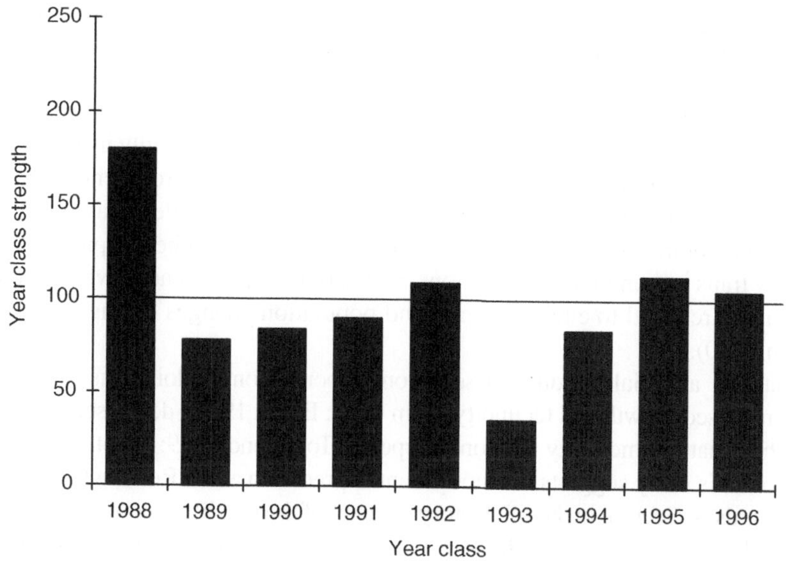

Figure 14.6 Year-class strength of roach in Holme Pierrepont Rowing Course

(Figs 14.4 and 14.5). The increased growth rates of the surviving fish were likely to have occurred as a direct result of reduced fish density due to cormorant depredation (Table 14.1). It is probable that the reduction in population density resulted in decreased levels of inter- and intra-specific competition, allowing the fish to compensate for the losses attributable to the cormorant depredation by increasing somatic growth.

Roach ingested by cormorants were primarily <100 mm (Fig. 14.3). As roach attained lengths >100 mm in less than 2 years (Table 14.2), these fish were mainly vulnerable to cormorant depredation only during their first winter of life. Common bream <250 mm were ingested by cormorants (Fig. 14.3). However, cormorant depredation concentrated on fish below 100 mm in winter 1995/1996 (66%) and 1996/1997 (70%). The growth rate of common bream was very rapid, achieving 200 mm by the end of the third year of life (Table 14.2). Thus, bream were mainly vulnerable to cormorant depredation during their first and second winters of life.

Therefore, the compensatory mechanism of increased growth rates in the exploited fish populations minimised losses to cormorants by reducing the time of vulnerability to depredation. This was in contrast to roach in the adjacent River Trent, where they were vulnerable to cormorant depredation throughout their life span due to much slower growth rates (Britton 1999).

This compensatory response would be advantageous in ensuring the sustainability of the populations in the presence of the heavy over-wintering cormorant depredation due to:

- decreased age at maturity;
- increased fecundity for age;
- reduced time period over which fish were vulnerable to cormorant depredation.

Fish life history strategies maximise offspring survival and are controlled by fluctuations between birth and death rates, and the availability of resources. These give rise to the two distinct life history strategies of r and K. The r-strategists rely on their ability to colonise new habitats, make use of short-lived resources and maximise fitness by increasing growth rates and improving their ability to reproduce rapidly in an uncrowded environment (Wootton 1990). K-strategists live in stable environments where it is important for organisms to persist and out-compete rivals by subtle behavioural means, and the main controlling factors are biological. Considerable phenotypic plasticity in life history traits is seen in fish populations, which is an important adaptive trait allowing individuals to respond to environmental and population changes during their lifetime (Wootton 1990).

Exploitation artificially induces r-selection responses on exploited fish populations, causing increased growth and fecundity (Alm 1959; Barret 1971), decreased age at maturity and high natural mortality (Begon, Harper & Townsend 1989; Wootton 1990). The output from this study suggests that cormorant depredation at the Rowing Course caused a shift towards r-selection in the life history strategy of the roach and common bream populations. This compensated for the fish losses due to cormorant depredation by rapid development and early reproduction, with a short life span. This was revealed by fast growth rates (Table 14.2), which accelerated markedly after 1993/1994, when cormorant depredation was first observed on the fishery (Figs 14.4 and 14.5, Table 14.2).

The compensatory processes observed at the Rowing Course were only thought to be prevalent in the roach and common bream population when strong year classes of fish were produced, for example, 1995 and 1996. If cormorants were to depredate on a series of weak year classes, then the compensatory process may not occur due to the removal of a large proportion of the weak year class, leaving too few surviving fish to exploit the low competitive environment. This is an important consideration for the management of the fishery, as angler catches were shown to be dependent on a small number of year classes in the population. For example, angler catches were dominated by roach of the 1996 year class during August 1998 (Britton 1999).

14.4.2 *Management*

There are three main management strategies that can be considered for the Rowing Course in the light of the impact of cormorant depredation on the recreational fishery:

- do nothing;
- management of the cormorant population;
- management of the fish population.

The compensatory mechanism in the fish populations exposed to the heavy cormorant depredation pressure suggests the do-nothing approach may be appropriate especially since angler catches also improved in the post-1994 period, associated with the improved growth rates allowing fish to surpass the critical size of exploitation rapidly. This was believed to be due to the compensatory mechanism that ensured sufficient survivors, from the 1995 and 1996 year classes of roach, were able to exploit the uncrowded environment and subsequently provide an adequate angler-exploitable fish stock.

A series of strong year classes in the fishery was considered to be required for the compensatory mechanism to be effective. The physical conditions in the summer of 1998, with low seasonal temperatures, may have resulted in poor recruitment and weak year classes of the exploited species. As a result, the years when these fish would be expected to contribute positively to angler catches, 2000 and 2001, assuming similar fast growth rates, may result in poor angler catches due to:

- low number of recruits in spring 1999;
- subsequent intense cormorant depredation;
- strong relationship between angler catches and year-class strength of exploited species.

As a result, the do-nothing approach may result in future angling catches declining from their 1998 level.

Large numbers of cormorants foraged daily at the Rowing Course, hence management of their population by scaring or licensed shooting may reduce their presence and depredation pressure. However, due to the current management policy of the lake, its funding and its location, there are few viable methods that could be successfully implemented (Britton 1999). Shooting is not permissible as the site has 24-h public access with no areas where firearms could be used legally. Furthermore, scaring structures that are mounted terrestrially are likely to be vandalised if permanent.

Due to large losses of fish attributable to cormorant depredation between 1995 and 1998 (Table 14.1), management of the fish population may be the only way of successfully minimising cormorant depredation. This could be carried out in two ways:

- provision of 'cormorant-proof' fish refuges to minimise losses;
- improve spawning and nursery habitat in the lake to maximise recruitment success.

During the over-wintering periods, large numbers of fish utilised the area immediately underneath, and around, boat pontoons for refuge (personal observation). Although these provided overhead cover, cormorants were able to dive underneath the pontoons to feed successfully, with 69% of foraging bouts successful (minimum one fish ingested) between 1995 and 1998 (Feltham *et al.* 1999). Therefore, strategic placement of fencing and/or netting may allow fish to retreat under the pontoons for refuge during a cormorant foraging bout whilst ensuring the cormorants cannot gain access. Before such action is taken, observations on cormorant foraging success about different designs of refuges should be undertaken to determine the best design.

Artificial fish refuges could be distributed at a number of locations throughout the lake. These would aim to encourage winter shoaling in alternative areas of the lake, preferably in areas deeper than those observed under the boat pontoons, where depths did not exceed 1.5 m. Brushwood or tyre reefs could be utilised, as these are also likely to be used as spawning media by fish, and can be easily lifted for maintenace.

In theory, a strategy to minimise cormorant depredation impact on the Rowing Course could be developed. However, in reality, due to the management policy of the lake and its location, there are few viable methods that could be successfully implemented. Therefore, management of the fish population, via provision of refuges and suitable spawning habitats, presents the only viable and realistic option.

Acknowledgements

The authors wish to thank the English Sports Council for permission to use Holme Pierrepont Rowing Course as a study site, and the staff of the National Water Sports Centre for their valuable assistance. Thanks are also extended to the Environment Agency for granting permission to undertake electric fishing surveys. The work was funded by MAFF, project contract number VC106.

References

Alm G. (1959) Connection between maturity, size and age in fishes. *Report to the Institute of Freshwater Research, Drottningholm* **40**, 5–145.

Bagenal T.B. & Tesch F.W. (1978) Age and Growth. In T.B. Bagenal (ed.) *Methods for the assessment of Fish Production in Fresh Waters*. IBP Handbook No. 3, third edition. Oxford: Blackwell Scientific Publications, pp. 101–136.

Barret (1971) The control of fish predators in the Welsh Water Authority Area. Report for the WWA Directorate of Scientific Services (1980), 16 pp.

Begon M., Harper J.L. & Townsend C.R. (1989) *Ecology – Individuals, Populations and Communities*. Oxford: Blackwell Scientific Publications.

Britton J.R. (1999) *The impact of cormorants* (Phalacrocorax carbo carbo *(L.) and* Phalacrocorax carbo sinensis *(Blumenbach)) on inland fisheries in the UK*. PhD Thesis, University of Hull.

Cowx I.G. (1999) Factors influencing coarse fish populations in lowland rivers – literature review. *Environment Agency R&D Report*, Bristol, 185 pp.

Cowx I.G. & Frear P. Evaluating year class strength in unexploited inland fish populations (in press).

Feare C.J. (1988) Cormorants as predators at freshwater fisheries. In *Proceedings of the Institute of Fisheries Management Annual Study Course West Bridgford* **18**, 18–42.

Feltham M.J., Davies J.M., Wilson B.R., Holden T., Cowx I.G., Harvey J.P. & Britton J.R. (1999) Case studies of the impact of fish-eating birds on inland fisheries in England and Wales. Report to the Ministry of Agriculture, Food and Fisheries, contract number VC0106. London: MAFF, 207 pp.

Harvey J.P. & Cowx I.G. (1995) Electric fishing in large deep waters: User manual. *NRA R&D Report 0034/7/ST*. Bristol: NRA, 46 pp.

Hickley P. & Dexter K.F. (1979) A comparative index for quantifying growth in length of fish. *Fisheries Management* **10**, 147–151.

Kempe O. (1962) The growth of roach (*Leuciscus rutilus* L.) in some Swedish lakes. *Report of the Institute for Freshwater Research, Drottningholm* **44**, 42–104.

Kirby J.S., Gilburn A.S. & Sellers R.M. (1995) Status, distribution and habitat use by Cormorants *Phalacrocorax carbo* wintering in Great Britain. *Ardea* **83**, 93–102.

Kirby J.S., Holmes J.S. & Sellers R.M. (1996) Cormorants *Phalacrocorax carbo* as fish predators: an appraisal of their conservation and management in Great Britain. *Biological Conservation* **75**, 191–199.

Ricker W.E. (1975) Computation and interpretation of biological statistics of fish populations. *Bulletin of the Fisheries Research Board of Canada*, **191**, 382 pp.

Russell I.C., Dare P.J., Eaton D.R. & Armstrong J.D. (1996) Assessment of the problem of fish-eating birds in inland fisheries in England and Wales. *Report to the Ministry of Agriculatre, Fisheries and Food.* Project number VC0104, 130 pp.

Sarvala J., Helminen H. & Hirvonen A. (1994). The effect of intensive fishing on fish populations in Lake Pyhäjärvi, south-west Finland. In I.G. Cowx (ed.) *Rehabilitation of Freshwater Fisheries.* Oxford: Fishing News Books, Blackwell Scientific Publications, pp. 77–89.

Sellers R.M. (1979) The changing status of the cormorant in Gloucestershire. *Gloucestershire Bird Report* **1979**, 56–61.

Veldkamp R. (1996) Cormorants, *Phalacrocorax carbo*, in Europe: a first step towards a European management plan. Report for the National Forest and Nature Agency, Denmark and the National Reference Centre for Nature Management, The Netherlands. 105 pp.

Wootton R.J. (1990) *Ecology of Teleost Fishes*. London: Chapman and Hall, 404 pp.

Chapter 15
Restoration of Ormesby Broad through biomanipulation: ecological, technical and sociological issues

M.L. TOMLINSON, M.R. PERROW*
ECON, Biological Sciences, University of East Anglia, Norwich, UK

D. HOARE, J.-A. PITT
Environment Agency Biology Laboratory, Haddiscoe, Great Yarmouth, UK

S. JOHNSON
Environment Agency, 79 Thorpe Road, Norwich, UK

C. WILSON
Broads Authority, 18 Colegate, Norwich, UK

D. ALBOROUGH
Essex & Suffolk Water, Chelmsford, Essex, UK

Abstract

The biomanipulation of zooplanktivorous and benthivorous fish in Ormesby Broad, a moderately large (55 ha), shallow (<2.5 m) lake, began in 1995. This was to improve water quality for commercial supply and to develop a luxurious submerged plant community for conservation purposes. To involve the local community, a consultation group, voluntary warden scheme and a fish conservation group were instigated. During biomanipulation, a number of novel techniques were used to remove 73% of the stock in the first year, increasing to 99% after the second. This was supported by periodic reactive winter removals and annual attempts to prevent undesirable fish from spawning successfully. Biomanipulation led to an increased representation of larger, more efficient zooplankton grazers but with little impact upon the chlorophyll *a* concentration and an unpredicted increase in phosphorus concentrations. Zooplankton grazing and nitrogen-limitation appears to have triggered a shift to blue-green, *Aphanizomenon* spp., rather than green algae, with a positive effect on water clarity. The concurrent marked increase in macrophyte cover may have been more closely linked with the removal of benthivorous fish, which are known to uproot seedlings whilst foraging, rather than light climate. Despite an improvement in suitable habitat, the recruitment of macrophyte-associated fish species has been sporadic and desirable piscivorous species such as pike, *Esox lucius* L., and perch, *Perca fluviatilis* L., have declined. The manipulation is now entering a critical phase as the development of the fish community must proceed in step with the attainment of a self-sustaining macrophyte-dominated state. Further control of bream, *Abramis brama* (L.), stocks, perhaps supplemented by stocking of desirable species, is planned.

Keywords: benthivorous fish, fish removal, lake restoration, zooplanktivorous fish.

---

*Correspondence: ECON, School of Biological Sciences, University of East Anglia, Norwich NR4 7TJ, UK (email: m.perrow@uea.ac.uk).

15.1 Introduction

Anthropogenic eutrophication (an increase in the rate of supply of the nutrients, nitrogen, N, and phosphorus, P) has compromised the ecological, recreational and commercial value of many shallow lakes; with replacement of clear water dominated by submerged macrophytes by turbid, algal dominated water. Restoration of the former has thus become a priority for water managers, particularly in North America (e.g. National Research Council, US 1992; Hanson & Butler 1994) and north west Europe (e.g. Jeppesen, Jensen, Kristensen, Søndergaard, Mortensen, Sortkjær & Olrik 1990; Meijer, Jeppesen, van Donk, Moss, Scheffer, Lammens, van Nes, van Berkum, de Jong, Faafeng & Jensen 1994; Moss, Stansfield, Irvine, Perrow & Phillips 1996).

Experience has shown that simply reducing external and/or internal nutrient loading is unlikely to be successful without further intervention. This is partly because the nutrient level at which submerged macrophytes typically dominate, $<50\,\mu g\,L^{-1}$ total phosphorus (TP), is rarely attainable. At intermediate nutrient concentrations (100–$150\,\mu g\,L^{-1}$ TP), which are more readily achieved through nutrient reduction, macrophytes and planktonic algae may exist as alternative stable states (Scheffer, Hosper, Meijer, Moss & Jeppesen 1993). Within this range, buffering mechanisms within the ecological community tend to preserve the algal-dominated state, even in the face of nutrient reduction. The principal causative agents of this effect are fish. Both zooplanktivorous (e.g. roach, *Rutilus rutilus* (L.)) and benthivorous (common bream, *Abramis brama* (L.), and carp, *Cyprinus carpio* L.) species may be responsible: zooplanktivorous fish controlling the grazing potential of cladoceran zooplankton through selective predation (Brooks & Dodson 1965) which would otherwise control planktonic algae (Jeppesen *et al.* 1990; Turner & Mittelbach 1990; Jeppesen, Søndergaard, Kanstrup, Petersen, Eriksen, Hammershøj, Mortensen, Jensen & Have 1994); and foraging benthivorous fish causing significant re-suspension of fine sediments and nutrient release into the water column (Tatrai & Istvanovics 1986; Breukelaar, Lammens, Klein Breteler & Tatrai 1994).

Biomanipulation is thus typically required as an extreme perturbation to push one stable equilibrium to another. This most often comprises drastic (>75% – Hosper & Meijer 1993; Perrow, Meijer, Dawidowicz & Coops 1997) reduction of both zooplanktivorous and/or benthivorous taxa, partly as it is often unclear which is primarily responsible for turbid conditions, and also because many species are capable of functioning as zooplanktivores and benthivores at different life stages or through adoption of different foraging strategies.

The objective of biomanipulation is to provide a clear water period of sufficient length to allow macrophytes to establish sufficient cover or biomass to become self-sustaining (Perrow *et al.* 1997a). The establishment of an appropriate fish community is then seen as the cornerstone of restoration (Meijer *et al.* 1994). A high proportion of piscivorous species, including pike, *Esox lucius* L., and/or perch, *Perca fluviatilis* L., in Europe (Meijer *et al.* 1994; Meijer, Lammens, Raat, Klein Breteler & Grimm 1995; Perrow *et al.* 1997a), is desirable. Control of zooplanktivorous species through both direct predation (Spencer & King 1984; Grimm & Backx 1990) and indirect effects on distribution and behaviour (Werner, Gilliam, Hall & Mittelbach 1983; Jacobsen &

Perrow 1998) may then occur. This may ultimately enhance the stability of the clear-water, macrophyte-dominated state.

The aim of this chapter is to review the experiences of biomanipulating Ormesby Broad, at 55 ha one of the larger lakes in the Norfolk Broads. The planning requirements and the methods used are outlined, the expectations in relation to the experiences for the fish community reviewed and the prognosis for long-term recovery discussed.

15.2 Study site

The Norfolk Broads comprises around 50 shallow (1–2 m) lakes (broads), which ancestrally, were dominated by submerged macrophytes, including low-growing communities dominated by *Chara* species. Eutrophication initially favoured dense growths of taller-growing macrophytic forms (e.g. *Potamogeton* spp.) but these were suddenly replaced by planktonic algae some 30–40 years ago. Only three lakes now retain their original macrophyte populations (Broads Authority 1997) and many of the remainder are hypereutrophic. The loss of macrophytes led to considerable changes in lake structure and function (Moss 1983) with severe impact on macroinvertebrate, bird and fish communities (de Nie 1987). In particular, the famous Broads recreational fishery went into decline, with a virtual loss of species such as rudd, *Scardinius erythrophthalmus* (L.), and perch. Anglers' catches declined dramatically (Moss, Leah & Clough 1979). A few remaining large pike (>13.6 kg) are now the focus of angling attention.

Ormesby Broad forms the uppermost of a chain of three similarly sized basins called the Trinity Broads, encompassing 71 ha with a catchment of approximately 35 km^2. As a result of isolation from the main river system, all the basins have maintained better water quality than many other broads, although an increasing nutrient gradient exists across the basins, with Ormesby being of lower status. The lake of highest status, Filby, has a history of persistent blue-green algal blooms. In Ormesby, there is some, largely anecdotal, evidence of alternative clear and turbid periods. Such a phenomenon has been described for a number of lakes (Blindlow, Andersson, Hargeby & Johansson 1993; Scheffer *et al.* 1993). Changing patterns of fish recruitment and/or summer or winter fish-kill have been directly implicated in changes in lake status by some authors (Brönmark & Weisner 1992). In theory, biomanipulation offered a good chance of allowing the persistent, but low, stock of macrophytes to dominate, and the long-term future of Ormesby as a conservation resource to be secured. Previous attempts at restoration in the Norfolk Broads were often compromised by the tardy re-colonisation of macrophytes as a result of insufficient inocula (Moss *et al.* 1996; Perrow, Schutten, Howes, Holzer, Madgwick & Jowitt 1997). In addition, likely improvements in water quality were eminently desirable to the lake managers, Essex & Suffolk Water, which used the lake waters as a potable supply. The biomanipulation of Ormesby Broad was undertaken as a demonstration project under the EU LIFE programme, through collaboration by the Broads Authority, Environment Agency and Essex and Suffolk Water. The current project represents the first commercial application of biomanipulation in the UK.

Prior to biomanipulation the isolation of Ormesby from the other lake basins had to be undertaken. This was achieved by a permeable barrier constructed of pebbles supported in an angle-iron frame with a wire grill (Holzer, Perrow, Madgwick & Dunsford 1997). A warden was employed to manage the site, be involved in all aspects of the project and to disseminate findings to the local community, particularly anglers.

15.3 Materials and methods

15.3.1 *Planning and sociological issues*

Removal of fish is an emotive issue in the UK as a result of strong angling interests. Anglers must be convinced that the temporary loss of (generally) small zooplanktivores and a few large benthivores is of benefit to their pastime in the longer term, as a wider range of more desirable species, including large predators, become established. As a result of angling and conservation interests, and animal welfare considerations, only non-destructive methods of fish capture were acceptable. Fish were therefore captured and transported to other sites. This requires detailed consideration of logistical issues and consent from the Environment Agency.

The full co-operation of land and water owners and users is essential to the success of any project. Full and frank public consultation is an important step in this process. Unfortunately, several factors conspired to make this less than ideal in the initial stages of the project. First, Essex and Suffolk Water were in the process of acquiring the bulk of the land (and water) in the Trinity system at the inception of the project. As a result of sensitive negotiations, not all information was in the public domain. Second, financial support from the EU became available before a full planning process had been implemented. Third, the requirement to initiate biomanipulation in the winter months when fish are conveniently aggregated, led to the development of a trial barrier and some fish manipulation prior to full public consultation, although consultation with the likely affected user groups had taken place.

As a result, there was a considerable degree of misinformation and a feeling of mistrust from the local community towards the project and the partnership (Haddon 1995). The major bone of contention proved to be the fish barrier. Even though a boat pass had been installed over the barrier, the principal that local navigation rights (associated with the deeds of many properties) had been infringed was established, irrespective that these were rarely used. As compensation, an existing jetty within Ormesby was re-opened to boat hire and most importantly, it was agreed that the barrier would be removed after 5 years. This effectively gave the broad a short period to respond and stabilise.

To further strengthen the relationship between the community and the partnership it was agreed that a consultation group of 18 people representing each local parish, organisations (e.g. the Inland Drainage Board), all user groups (anglers, sailors, wildfowlers) and landowners (principally farmers) was set up. This meets 3–4 times a year with dissemination of all results. In addition, to directly involve the local community in the management of the site, 12 voluntary wardens were enrolled, under control of the permanent warden, to undertake wide-ranging duties from patrolling to site maintenance and assistance with data gathering. Finally, to further involve local anglers, representing the major user group on the site, the Trinity Broads Fisheries Conservation Group was initiated. Members recorded details of pike captured by rod and line, which were individually marked with a Panjet inoculator. Members were supplied with basic materials (scales, tape measure and tweezers with which to take scales). The group aimed to have one group day per month during the main pike season of October–March when an Environment Agency fisheries officer was present to mark the fish captured.

15.3.2 *Fish monitoring, removal strategy and techniques*

Point abundance sampling (PAS) by electric fishing was used to monitor fish populations immediately prior to biomanipulation, and annually thereafter during the winter months (Table 15.1). The principal advantage of this technique over conventional techniques is that it allows a wide range of habitats, from open water to littoral margins to small drainage channels in which the fish congregate in winter, to be sampled (Perrow, Jowitt & Zambrano González 1996). At each point, a large (40 cm) anode was immersed rapidly and all stunned fishes collected with a long handled (2 m) hand net. The sampling area was quantified as 1.3 m^2 using a volt meter to determine at what distance from the anode the voltage gradient was reduced to 0.12 V, which is the minimum effective voltage at which inhibited swimming occurs (Copp & Peňáz 1988). High-frequency

Table 15.1 Number and biomass (kg) of fish removed from Ormesby Broad, from 1995 to 1998

		Winter/ spring 1995	Summer 1995	Winter/ spring 1996	Winter/ spring 1997	Winter/ spring 1998
Roach	Number	217 969		28 378	1	10 867
	Biomass	6351		652	<1	123
Bream	Number	25 147	1 447 000	302	72	207 843
	Biomass	2246	798	549	207	333
Rudd	Number	319		309		10
	Biomass	10		10		1
Hybrid	Number	1820			1	
	Biomass	119			1	
Carp	Number			49		
	Biomass			1		
Gudgeon	Number	9057		263	1	
	Biomass	53		3	<1	
Perch	Number	23 859		1489	147	
	Biomass	141		41	1	
Ruffe	Number	24 087		3225	3	41
	Biomass	168		18	<1	<1
Total	Number	302 258	1 447 000	34 015	225	218 761
	Biomass	9088	798	1274	210	457

(600 Hz) pulsed DC (rectangular wave at 300 V with a variable duty cycle of 0–50%) electric fishing equipment (Electracatch WFC11) was used.

The timing of the annual survey corresponded to the period in which biomanipulation was best undertaken (Perrow *et al.* 1997a). This is a result of:

(1) the aggregation of the small fish in sheltered inflows, dykes, littoral margins and boatyards (Jordan & Wortley 1985), lending itself to efficient removal of a large proportion of the stock;
(2) reduced handling stress and thus mortality of small fish, especially in cold water;
(3) the opportunity to use the entire growing season ahead to generate a beneficial effect.

The initial survey indicated that the majority of the fish in Ormesby were aggregated in one long (5 km) drainage system. This became the focus of the initial removal. The technique adopted here was to create sections of 50–200 m in length between stop-nets, starting from the inflow and working away from the broad. Electric fishing was then used to drive fish into pre-set seine nets, first one way and then the other, until no further fish were caught. As access for the transportation equipment (two, 5000-L tanks equipped with aeration on a 4 m trailer) was limited, fish were driven for 500 m or more on occasion to allow direct loading from the seine net to the waiting tank.

Attention then shifted to the large bream known to exist at the site. To tackle these a systematic approach using a 'scare line' technique was developed. The scare line itself consisted of floating polypropylene rope (1 cm diameter) with 2 m long, weighted (with stones) yellow fluorescent kite tails secured at 2-m intervals along its 150-m length. This was pulled down the length of a section of broad, demarcated by stop-nets, between three small rowing boats. Another boat, with an outboard motor, drove back and forwards just behind the scare line to add to the disturbance and to pass instructions between boats. Prior to the drive a net trap had been constructed to receive the fish. This was achieved by setting a 100-m long net from each bank at an angle of about 45° into the broad. The ends of the nets were thus about 3 m from each other. The resulting funnel led into a net corral, which had a 100-m seine net (fitted with a cod end) set inside it. Once the drive was complete, the entrance to the corral was closed by pulling the free end of one of the stop-nets forming the funnel across it. This net was then secured with ropes and the free end of the 100-m seine net was pulled across the entrance of the trap to form a second seal. After preparations had been made to receive fish, the seine was pulled by four or five people, from a large (5 m) floating platform which had been anchored alongside the net corral. This technique accounted for the majority (380) of the 429 large bream (total biomass of 1121 kg) removed in the first year of biomanipulation, with one drive alone accounting for 206 individuals (522 kg total biomass). Other species captured included roach, tench, *Tinca tinca* (L.), pike and perch.

Attempts were also made on an annual basis to prevent the remaining adult fish, particularly bream, from spawning successfully. This involved putting in nets at suitable locations to act as artificial spawning medium. Observations suggested this was often preferred to natural substrata. Once eggs had been laid, the net was removed and dried. Adult fish were also captured passively using large Dutch-style fyke nets, checked every few hours.

In the summer of 1995, following the initial biomanipulation, it became clear that large numbers of small bream had recruited successfully. These became naturally concentrated at the barrier. Over a period of several days these were simply captured with large hand-nets and placed over the barrier. Processing of sub-samples indicated that 1 447 000 individuals (798 kg) were transferred in such a manner.

Winter surveys allowed an assessment of whether the residual stock recruited successfully or not, and allowed the need for further removal to be determined. Additional removal from the dyke systems was undertaken in the winters of 1995/1996 and 1997/1998.

Over time it became clear that the potential for bream to recruit successfully was not diminished by the removal of a total of 690 adults over the three years. Consequently, an attempt to estimate the numbers remaining in the broad, as the basis of their future management, was made, by mark-recapture techniques in 1998. This utilised catches made using the scare line and large fykes, particularly as fish came to the margin to spawn. All captured fish were individually marked using a Panjet at a variety of locations on the ventral surface. A population estimate was calculated by the Petersen method: $N = ((n_1 + 1)(n_2 + 1)/(m_2 + 1)) - 1$.

15.3.3 *Water quality monitoring*

Water quality and zooplankton (see below) samples have been routinely undertaken in all three basins of the Trinity system since 1978. This provided an effective means of evaluating the impact of biomanipulation in Ormesby as a before and after comparison, and in relation to the un-manipulated Rollesby and Filby Broads.

Prior to 1997, samples were taken every two weeks over the summer period (water temperature >10°C), and monthly over winter. After 1997, samples were taken every 2 weeks throughout the year. Water required for analysis of soluble components was filtered through GF/C filters on site.

The water quality parameters analysed included: TP ($mg\,L^{-1}$) and soluble reactive phosphorus (SRP) ($mg\,L^{-1}$) by the molybdenum blue reaction method; total oxidised nitrogen (TON) ($mg\,L^{-1}$); and chlorophyll *a* ($\mu g\,L^{-1}$) by cold acetone extraction and spectrophotometric pigment analysis. Full details of the sampling methodology are given by Moss, Balls, Irvine & Stansfield (1986).

15.3.4 *Zooplankton*

Zooplankton was sampled at the same frequency as water quality parameters. Three locations were sampled in Ormesby. At each location, five samples were taken with a 90-cm long, perspex tube (7 L) inserted vertically into the water column, at approximately 20-m intervals. All water was filtered through a 60 μm mesh net. Zooplankters were rinsed into a 300-mL storage pot and immediately preserved in 70% industrial methanol. Secchi disc depth was recorded during collection of zooplankton samples and a mean visible depth calculated.

In the laboratory, sub-samples were counted and identified in a 5-mL counting ring under a 25 × binocular microscope. All *Daphnia* were identified to species where possible with other taxa identified to genera. As for water quality, full details of zooplankton analysis are provided by Moss *et al.* (1986).

15.3.5 *Macrophytes*

For macrophytes and filamentous algae, samples were taken along permanent transects with a grapnel towed behind a boat in late summer (August/September) from 1983. Frequent stops were made along each transect and the abundance of each macrophyte species and filamentous algae estimated on a 1–5 scale from trace to abundant (i.e. <5% to 75–100% cover). The mean score per transect for each group (rooted submerged macrophytes, rooted floating-leaved macrophytes, floating macrophytes and macro-algae was summed, deriving an index of macrophyte cover. Scores >100 were obtained by more diverse communities with a number of co-dominant species. For full details see Kennison, Dunsford & Schutten (1998).

15.4 Results

15.4.1 *Fish*

The high abundance and biomass prior to biomanipulation, 1.77 ind. m^{-2} and 27.44 g m^{-2} respectively, was reduced to 0.49 ind. m^{-2} and 5.84 g m^{-2} by the end of 1995 (Fig. 15.1). The initial removal had the desired effect on the fish community with the recruitment of tench and its consequent numerical dominance of the community, supplemented by perch. Perch and pike dominated by biomass. Continued removal, including small perch, further reduced the abundance and biomass to very low levels (0.01 ind. m^{-2} and 1.04 g m^{-2}) by the end of 1996. There was no requirement for further winter removal until 1997/1998, when the annual survey indicated 0+ bream had successfully recruited, with a density of 0.40 ind. m^{-2} present. The presence of two cohorts of under-yearling bream, the majority with a fork length of 40–45 mm with a few others up to 80 mm, suggested the spawning operation that year had been successful, but that a second spawning went uncontrolled. Over 200 000 under-yearling bream were then removed (Table 15.1). The need to derive a management plan for the remaining adult bream was identified. As a first step in this process, the mark-recapture exercise in 1998 estimated the adult bream population at 881 (95% confidence limits, 488–1176), greater than the number previously removed. Notable results of the 1998 survey were the failure of bream to recruit effectively and the successful recruitment of rudd for the first time since surveys began (Fig. 15.2). However, despite continuing efforts to prevent fish recruitment, bream recruited again in 1999, increasing overall abundance from 0.12 to 0.37 ind. m^{-2}. As a result of the small size of these fish the biomass remained virtually unchanged (2.78 and 2.88 g m^{-2}, respectively). By 1999, although still at relatively low density and low biomass, the community began to show

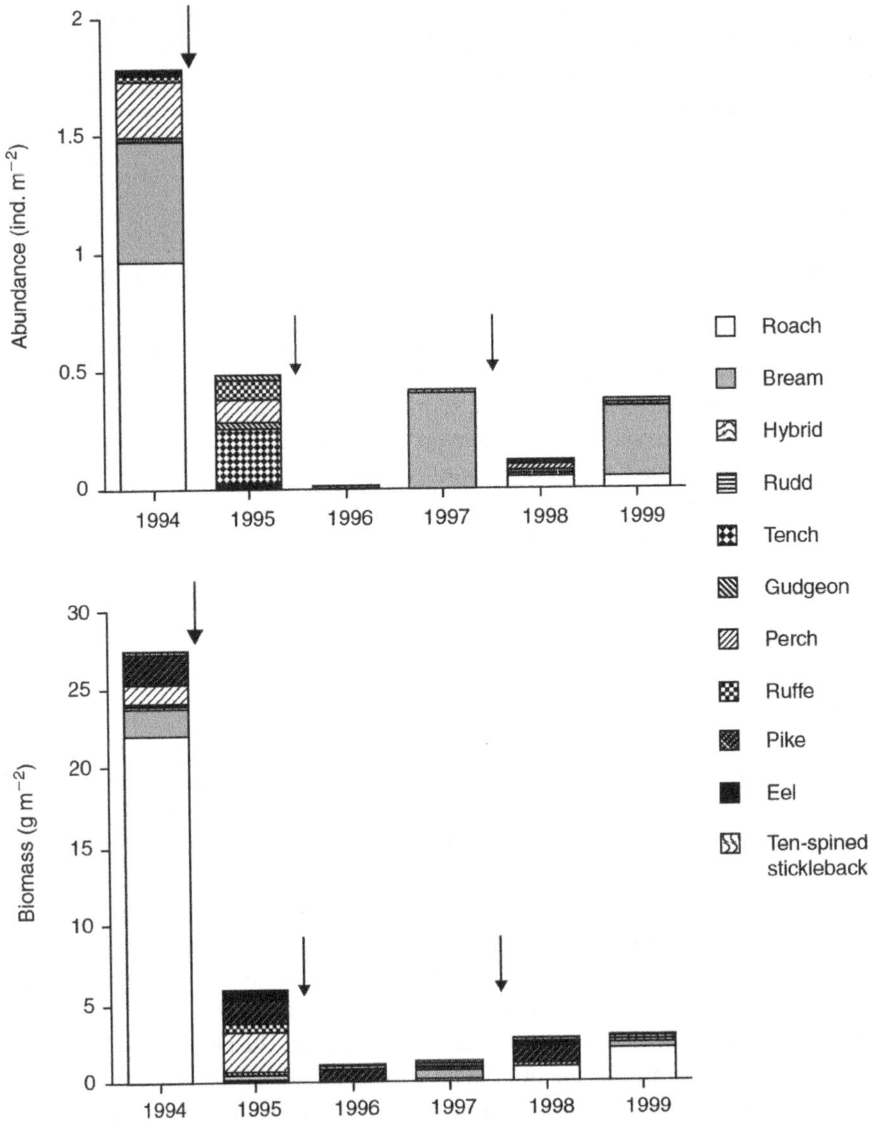

Figure 15.1 Abundance (ind. m^{-2}) and biomass (g m^{-2}) of all fish species in Ormesby Broad before (1994) and after (1995–1999) biomanipulation. The initial removal is indicated by a bold arrow, with subsequent interventions marked with smaller arrows

signs of shifting back to its original structure with roach dominant. The steady decline of the pike population since biomanipulation began (Fig. 15.2) was undoubtedly a factor in the dominance of the biomass by the relatively small-bodied roach in 1999. Populations of perch also failed to recover sufficiently to have any bearing on the dominance by cyprinids.

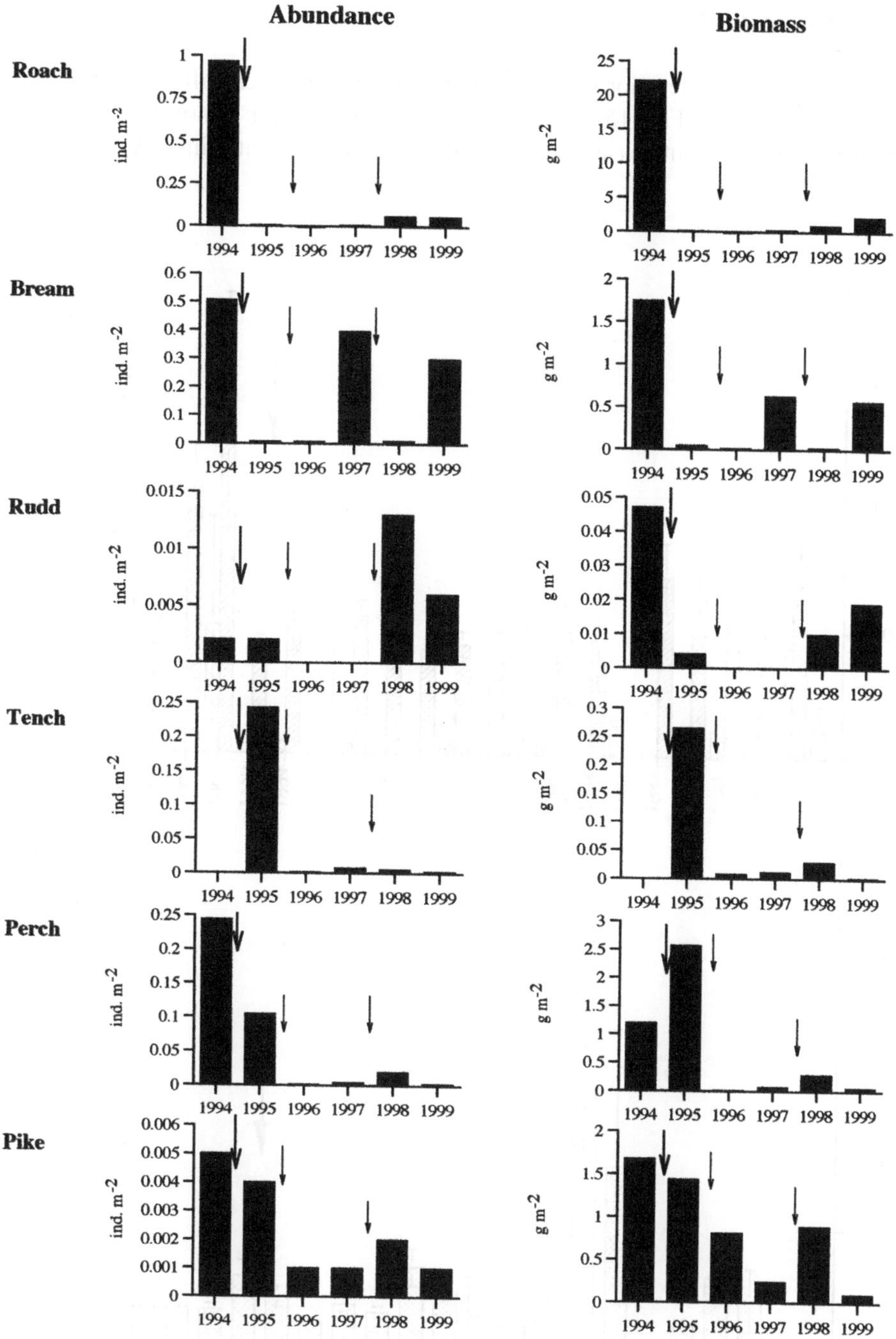

Figure 15.2 Mean abundance (ind. m^{-2}) and biomass ($g\ m^{-2}$) of individual fish species before (1994) and after (1995–1999) biomanipulation in Ormesby Broad. The initial removal is indicated by a bold arrow, with subsequent interventions marked with smaller arrows

15.4.2 Water quality

TP levels fluctuated little prior to biomanipulation and were generally low (between 0.04 and 0.07 mg L^{-1}; Fig. 15.3). Following the initial major removal of fish, the summer mean for 1996 increased sharply to 0.136 mg L^{-1}, albeit primarily caused by

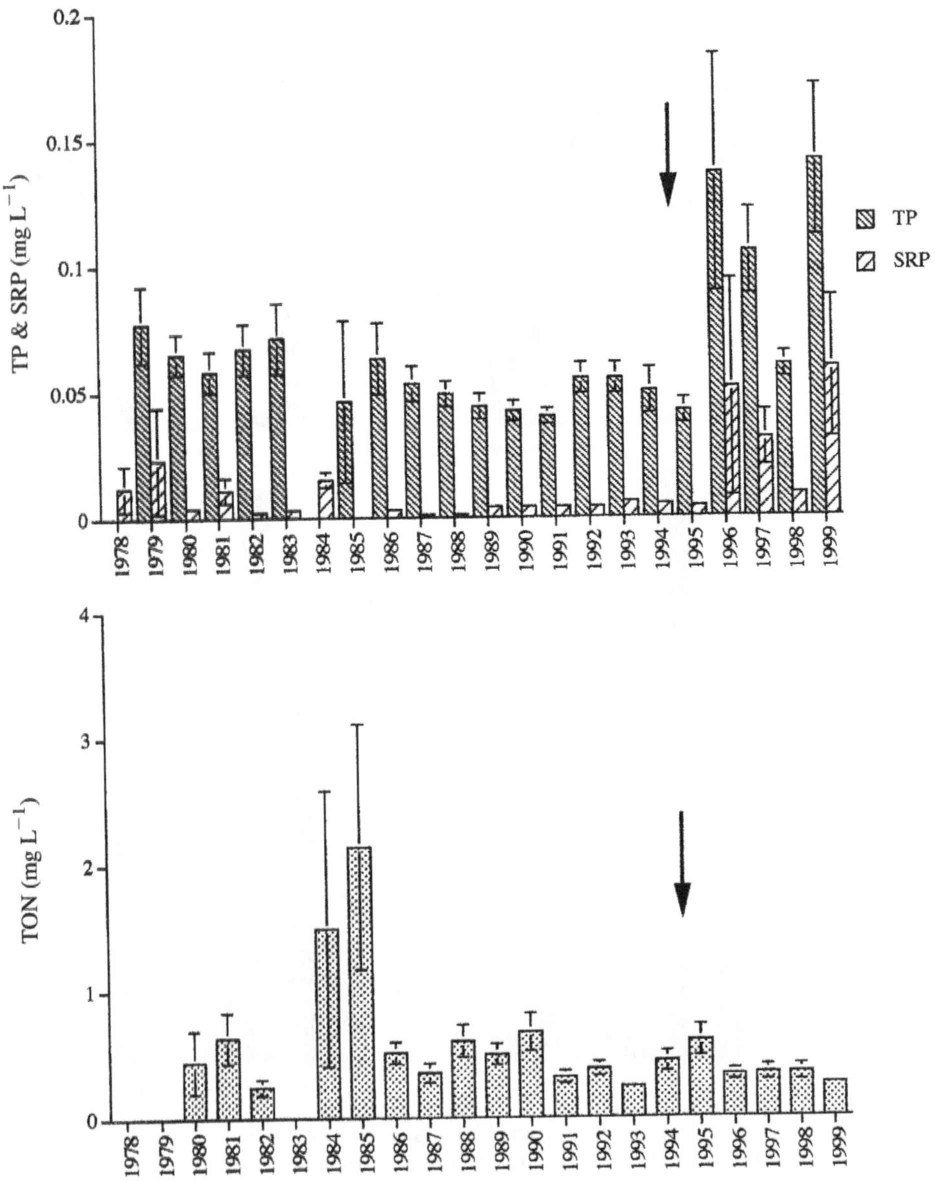

Figure 15.3 Mean (±1 SE) summer TP and SRP (mg L^{-1}) and TON (mg L^{-1}) in Ormesby Broad from 1978 to 1999. The onset of biomanipulation is indicated by an arrow

a large single event. Subsequently, the summer means have fluctuated (0.06 mg L^{-1} in 1998 and 0.141 mg L^{-1} in 1999) at a higher level than prior to manipulation. The pattern for SRP mirrors that of TP, being at very low levels prior to manipulation and increasing dramatically following manipulation, but showing considerable differences between years and within the growing season.

In a similar manner to phosphorus, levels of nitrogen (TON) were generally low and fluctuated little prior to biomanipulation, with the exception of 1984/1985, when summer means increased to 1.5 and 2.1 mg L^{-1}, respectively (Fig. 15.3). Following manipulation the levels decreased and levelled off at between 0.2 and 0.3 mg L^{-1}.

In keeping with generally low nutrient levels, chlorophyll *a* concentrations indicative of algal biomass, remained low, and more or less constant at around 40 $\mu g\,L^{-1}$ from 1978 to the late 1980s (Fig. 15.4). From 1989 to 1991, chlorophyll *a* fell to below 10 $\mu g\,L^{-1}$, subsequently increasing back to previous levels immediately prior to biomanipulation. Following biomanipulation there was some fluctuation (between 17 and 36 $\mu g\,L^{-1}$), but always at a low level.

From the late 1970s to the early 1980s, the mean visible Secchi depth decreased from 1.1 to 0.5 m (Fig. 15.4), without an apparent change in chlorophyll *a* concentrations. Following a period when no Secchi disc depths were recorded, the mean visible depth in 1989 was 1.5 m, corresponding to very low chlorophyll *a* concentrations. Since that time, Secchi depth has more or less been the inverse of chlorophyll *a* concentrations. Following manipulation, although there was some fluctuation, the visible depth was generally maintained at 1 m or more.

15.4.3 *Zooplankton*

Amongst the grazing Cladocera, relatively small species, such as *Daphnia hyalina/longispina*, dominated the community prior to manipulation (Fig. 15.5). Densities were generally high (apart from in 1991) at around 100 ind. L^{-1}. Following manipulation there was a shift to larger species, notably *Daphnia magna*, although the overall density reduced to between 25 and 50 ind. L^{-1} on average.

15.4.4 *Macrophytes*

The macrophyte index fluctuated slightly from 1983 to 1993, with the greater abundance of filamentous algae responsible for the increases in 1989 and 1991 (Fig. 15.6). Species composition of the macrophyte community remained relatively unchanged after biomanipulation with *Ceratophyllum demersum*, *Elodea canadensis*, *Zannichellia palustris* and *Potamogeton* spp. dominating. However, following manipulation, the cover of all species increased markedly (Fig. 15.5). The cover of filamentous macro-algae also immediately increased following biomanipulation, but fell subsequently.

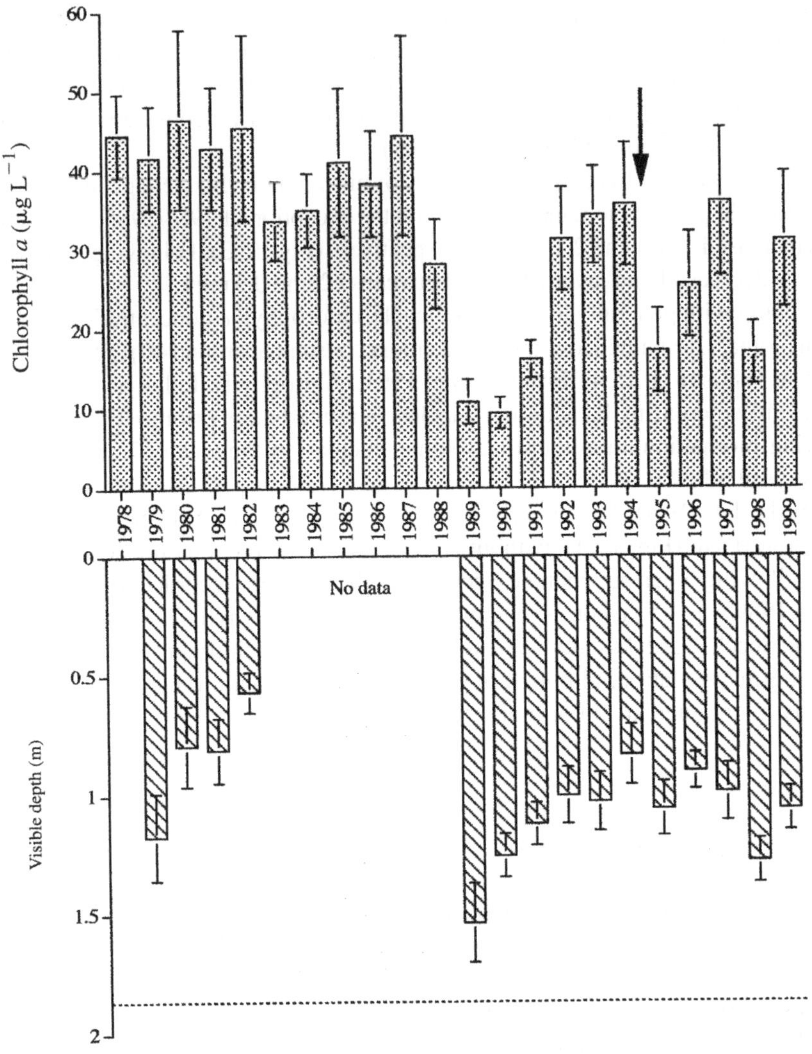

Figure 15.4 Mean (±1 SE) summer chlorophyll *a* concentration (μg L^{-1}) and Secchi depth (m) in Ormesby Broad from 1978 to 1999. The onset of biomanipulation is indicated by an arrow

15.5 Discussion

Prior to manipulation, Ormesby Broad was characterised as a relatively low nutrient system (<0.1 mg L^{-1} TP and around 0.5 mg L^{-1} TON in the summer months). The density of grazing zooplankton, albeit small species, was relatively high and chlorophyll *a* concentrations were generally low (<40 μg L^{-1}). Secchi depth was, however, not high, which may have had an impact on the relatively depauperate macrophyte community in what was one of the deepest of the shallow lakes in the Norfolk Broads. Secchi depth

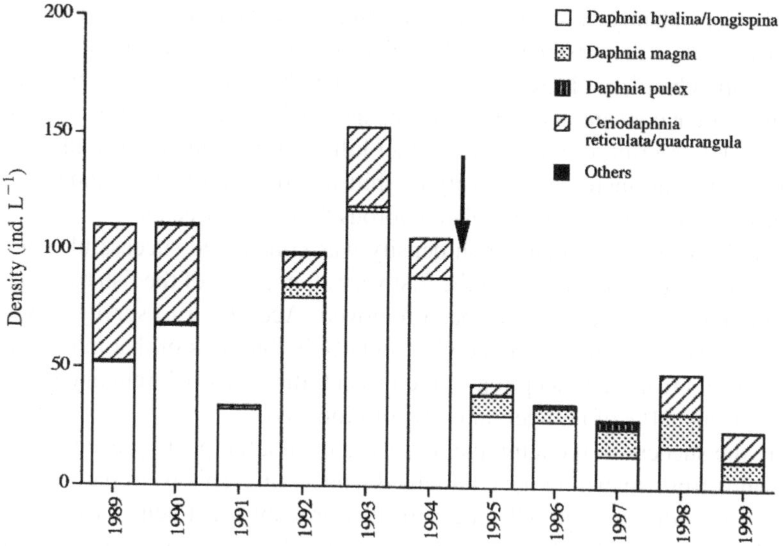

Figure 15.5 Mean summer density (ind. L^{-1}) of grazing cladocerans in Ormesby Broad from 1989 to 1999. The onset of biomanipulation is indicated by an arrow

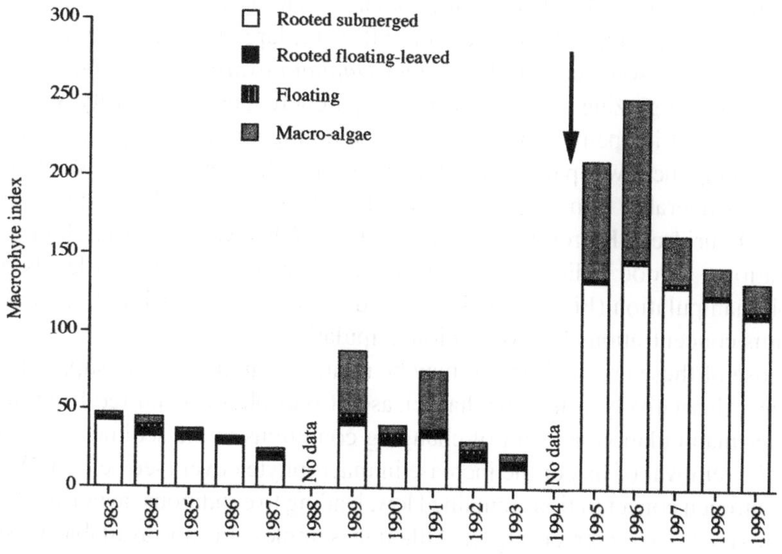

Figure 15.6 An index of macrophyte cover, divided into functional groups, in Ormesby Broad from 1983 to 1999. The onset of biomanipulation is indicated by an arrow

fluctuated independently of chlorophyll, indicating a contribution from suspended solids from either wind or disturbance from benthivorous fish. A reduction in chlorophyll *a* with a corresponding large increase in the Secchi depth in the late 1980s, irrespective of changes in nutrient concentration, points to changes in the fish community, perhaps a failure of 0+ fish to recruit, or winter or summer kill of larger adults. Since this event was not manifested as an increase in density or shift in structure of the grazing cladoceran community, it is suggested non-zooplanktivorous fish were affected.

The initial operation to reduce fish density was considered successful, vindicating the approach adopted. By the end of the first year alone, 73% of the fish by number and 79% by biomass appeared to have been removed. According to surveys, by the end of 1996 the stock was only 1% by number and 4% by biomass of that initially present. Over the 5 years from the inception of the project, the fish stock remained effectively suppressed (only 21% of number and 10% of biomass).

Despite the success in reducing the density, annual attempts to prevent fish (particularly bream) from spawning have remained a necessity. This is, in part, the result of an increase in individual growth rate and thus probably fecundity of the remaining individuals. For example, 2-year-old roach in 1994 prior to biomanipulation were around 100 mm in length, whereas in 1998 fish of that age reached 150 mm. The current bream population appears to exceed that removed and must be controlled to prevent massive recruitment and a return to pre-manipulation densities. However, this must be undertaken sensitively as these large bream provide angling interest for the local community. The high-level involvement of the latter in management of the fishery is testament to the public strategy adopted after a difficult start.

The removal of the bulk of the fish population had an effect on the grazing zooplankton community (Fig. 15.5). The more efficient, large bodied (>3 mm) *Daphnia magna*, replaced, to some extent, the smaller *Daphnia hyalina*. However, it is unclear if this led to greater grazing potential, with an apparent reduction in density. Whether the latter is real or not is open to question, as the sampling strategy adopted was not capable of sampling the lower part of the water column or the edges of the lake effectively, where large vulnerable forms are known to refuge from their predators during the day (Jeppesen, Lauridsen, Kairesalo & Perrow 1998). Whatever the case, chlorophyll *a* concentrations, a good indicator of algal biomass, fluctuated between levels present before biomanipulation (Fig. 15.4). This was despite the counter-intuitive increase in phosphorus concentrations following biomanipulation.

The cause of these are unclear, but may be related to an increase in supply from the catchment. Alternatively, it may be that release of phosphorus occurred from the sediments after manipulation, a result of changing conditions at the sediment surface as a result of the removal of fish or the increase in macrophyte cover (see below). Whatever the case, nitrogen concentrations remained low, leading to conditions favouring nitrogen-fixing blue-greens over green algae. Although samples are not available to support the dominance of blue-greens, observations suggested *Aphanizomenon* spp. dominated the sparse algal community after manipulation. A shift in the 480/663 nm ratio (carotenoids/chlorophyll) to >1.3 since biomanipulation, also supports the case for nitrogen deficient phytoplankton or dominance of blue-greens. This would also be expected with a shift in the grazer community to large-bodied grazers such as *D. magna*.

Patchy, albeit bulky, colonies of blue-greens such as *Aphanizomenon* seemed to have little impact on Secchi depth. Secchi depths >1 m may have favoured the sudden increase in the cover of macrophytes, which was the target of manipulation. However, such values were achieved in the past, with little impact. Consequently, it seems unlikely that an improved light regime was the only explanation for the increase in macrophyte cover. Of the fish-related mechanisms, large benthivores are known to uproot plants whilst foraging (Ten Winkel & Meulemans 1984). This may be particularly important early in the growing season when seedlings are small and is always likely to be a problem in the fine, fluid sediments of the broads. Although only about half of the dominant benthivorous bream were removed, it may be that this triggered the increase in macrophyte cover.

Although macrophytes are establishing, it appears several key elements of the desirable ecological community capable of generating self-sustaining conditions are lacking, notably a high proportion of piscivorous fish. Pike populations decreased after biomanipulation began, despite all captured individuals being returned and suitable habitat in the form of structured littoral margins and submerged macrophytes, from which to ambush prey, being present. Whereas anglers are currently enjoying good sport with large fish, the recruitment of younger pike appears to be poor. With the recruitment of small bream in several years, it seems unlikely that a lack of food is the problem for 0+ fish. However, such small bream are unlikely to be taken easily by large pike and it may be that levels of cannibalism have increased as a result (Grimm 1983). The population of the other potentially piscivorous species, perch, is also currently low in the broad. This appears to result from the removal of small fish in the initial stages of manipulation. Despite a revised policy of retaining perch, in keeping with their potential impact as a piscivore in the Broads (Perrow, Hindes, Leigh & Winfield 1999), populations have yet to recover.

15.6 Concluding remarks

The biomanipulation of Ormesby was successful in reducing fish density and biomass and triggering an increase in macrophyte cover. However, this may not have been through the anticipated top-down trophic cascade response of zooplankton–phytophyton–macrophytes, but primarily through the removal of benthivorous fish and a consequent lack of disturbance of macrophyte seedlings. Although macrophytes are recovering, an increase in species diversity and seasonal succession is desirable to ensure good coverage and the stability of the macrophyte-dominated state within and between seasons. The biomanipulation is now entering a critical phase. Shifts in the fish community must be in step with the development of macrophytes. Further control of the bream population must be instigated. Steps (e.g. stocking) may have to be taken to replace roach and bream with macrophyte-associated species such as rudd and tench, the recruitment of which has been sporadic and populations have remained low. These in turn may help buffer cannibalism within the pike population. Perch too, may have to be stocked should populations be slow to recover. Provided an alternative fish community can become established the prognosis for the long-term recovery of Ormesby remains good.

Acknowledgements

The initial biomanipulation was undertaken as a demonstration project under the EC LIFE programme project (LIFE 92-3/UK/031 & NRA Project 475) on the Restoration of the Norfolk Broads: Development of Biomanipulation Techniques & Control of Phosphorus Release from Sediments (Project Managers Jane Madgwick & Dr Geoff Phillips), in a partnership between the Broads Authority, Environment Agency and Essex & Suffolk Water. The latter have provided financial support for the project subsequently. We thank the staff of the Environment Agency and the Broads Authority for providing data and the staff of ECON, notably Adrian Jowett and Andy Hindes, for assistance with data analysis, field sampling and fish removals. The Agency fisheries team also assisted with removals.

References

Blindlow I., Andersson G., Hargeby A. & Johansson S. (1993) Long-term pattern of alternative stable states in two shallow eutrophic lakes. *Freshwater Biology* **30**, 159–167.

Broads Authority (1997) *The Broads Plan.* Broads Authority Consultation Document, Broads Authority, Norwich, UK, 175 pp.

Brönmark C. & Weisner S.E.B. (1992) Indirect effects of fish community structure on submerged vegetation in shallow, eutrophic lakes: an alternative mechanism. *Hydrobiologia* **243/244**, 293–301.

Breukelaar A.W., Lammens E.H.R.R., Klein Breteler J.G.P. & Tatrai I. (1994) Effects of benthivorous beam (*Abramis brama* L.) and carp (*Cyprinus carpio* L.) on sediment resuspension and concentrations of nutrients and chlorophyll *a. Freshwater Biology* **32**, 113–121.

Brooks J.L. & Dodson S.I. (1965) Predation, body size, and composition of plankton. *Science* **150**, 28–35.

Copp G.H. & Peñáz M. (1988) Ecology of fish spawning and nursery zones in the flood plain, using a new sampling approach. *Hydrobiologia* **169**, 209–224.

de Nie H.W. (1987) The decrease in aquatic vegetation in Europe and its consequences for fish populations. *EIFAC/CECPI Occasional Paper No. 19*, 52 pp.

Grimm M.P. (1983) Regulation of biomasses of small (<41 cm) Northern pike (*Esox lucius* L.), with special reference to the contribution of individuals stocked as fingerlings. *Fisheries Management* **14**, 115–134.

Grimm M.P. & Backx J.J.G.M. (1990) The restoration of shallow eutrophic lakes, and the role of northern pike, aquatic vegetation and nutrient concentration. *Hydrobiologia* **200/201**, 557–566.

Haddon M. (1995) Broadland brouhaha. *Water Bulletin* **645**, 7–8.

Hanson M.A. & Butler M.G. (1994) Responses to food web manipulation in a shallow waterfowl lake. *Hydrobiologia* **279/280**, 457–466.

Hosper S.H. & Meijer M.-L. (1993) Biomanipulation, will it work for your lake? A simple test for the assessment of chances for clear water, following drastic fish-stock reduction in shallow, eutrophic lakes. *Ecological Engineering* **2**, 63–72.

Holzer T., Perrow M.R., Madgwick F.J. & Dunsford D. (1997) Practical aspects of Broads restoration. In F.J. Madgwick & G.L. Phillips (eds) *Restoration of the Norfolk Broads Final Report BARS 14e.* Broads Authority and Environment Agency, Norwich, UK, 41 pp., ISBN 0 948119 35 7.

Jacobsen L. & Perrow M.R. (1998) Predation risk from piscivorous fish influencing the diel use of macrophytes by planktivorous fish in experimental ponds. *Ecology of Freshwater Fish* **7**, 78–86.

Jeppesen E., Jensen J.P., Kristensen P., Søndergaard M., Mortensen E., Sortkjær O. & Olrik K. (1990) Fish manipulation as a lake restoration tool in shallow, eutrophic, temperate lakes 2: threshold levels, long-term stability and conclusions. *Hydrobiologia* **200/201**, 219–227.

Jeppesen E., Søndergaard M., Kanstrup E., Petersen B., Eriksen R.B., Hammershøj M., Mortensen E., Jensen J.P. & Have A. (1994) Does the impact of nutrients on the biological structure and function of brackish and freshwater lakes differ? *Hydrobiologia* **275/276**, 15–30.

Jeppesen E., Lauridsen T.L., Kairesalo T. & Perrow M.R. (1998) Impact of submerged macrophytes on fish-zooplankton interactions in lakes. In E. Jeppesen, Ma. Søndergaard, Mo. Søndergaard, K. Christoffersen (eds) *The Structuring Role of Submerged Macrophytes in Lakes*. Ecological Studies Series, New York: Springer Verlag, pp. 91–114.

Jordan D.R. & Wortley J.S. (1985) Sampling strategy related to fish distribution, with particular reference to the Norfolk Broads. *Journal of Fish Biology* **27**(Suppl. A), 163–173.

Kennison G.C.B., Dunsford D.S. & Schutten J. (1998). Stable or changing lakes? A classification of aquatic macrophyte assemblages from a eutrophic shallow lake system in the United Kingdom. *Aquatic Conservation: Marine and Freshwater Ecosystems* **8**, 669–684.

Meijer M.-L., Jeppesen E., van Donk E., Moss B., Scheffer M., Lammens E.H.R.R., van Nes E., van Berkum J.A., de Jong G.J., Faafeng B.A. & Jensen J.P. (1994) Long-term responses to fish-stock reduction in small shallow lakes: interpretation of five-year results of four biomanipulation cases in The Netherlands and Denmark. *Hydrobiologia* **275/276**, 457–466.

Meijer M.-L., Lammens E.H.R.R., Raat A.J.P., Klein Breteler J.G.P. & Grimm M.P. (1995) Development of fish communities in lakes after biomanipulation. *Netherlands Journal of Aquatic Ecology* **29**, 91–101.

Moss B. (1983) The Norfolk Broadland: Experiments in the restoration of a complex wetland. *Biological Reviews* **58**, 521–561.

Moss B., Leah R.T. & Clough B. (1979) Problems of the Norfolk Broads and their impact on freshwater fish communities. In *Proceedings of the First British Freshwater Fisheries Conference*. University of Liverpool, 67–85 pp.

Moss B., Balls H., Irvine K. & Stansfield J.H. (1986) Restoration of two lowland lakes by isolation from nutrient-rich water sources with and without removal of sediment. *Journal of Applied Ecology* **23**, 391–414.

Moss B., Stansfield J.H., Irvine K., Perrow M.R. & Phillips G.L. (1996) Progressive restoration of a shallow lake – a twelve year experiment in isolation, sediment removal and biomanipulation. *Journal of Applied Ecology* **33**, 71–86.

National Research Council, US (1992) *Restoration of Aquatic Ecosystems-Science, Technology and Public Policy*. Washington, DC: National Academy Press, 552 pp.

Perrow M.R., Jowitt A.J.D. & Zambrano González L. (1996) Sampling fish communities in shallow lowland lakes: point-sample electricfishing versus electrofishing within stop-nets. *Fisheries Management and Ecology* **3**, 303–313.

Perrow M.R., Meijer M.-L., Dawidowicz P. & Coops H. (1997a) Biomanipulation in shallow lakes: state of the art. *Hydrobiologia* **342/343**, 355–365.

Perrow M.R., Schutten J., Howes J.R., Holzer T., Madgwick F.J. & Jowitt, A.J.D. (1997b) Interactions between coot (*Fulica atra*) and submerged macrophytes: the role of birds in the restoration process. *Hydrobiologia* **342/343**, 241–255.

Perrow M.R., Hindes A.M., Leigh S. & Winfield I.J. (1999) Stability of fish populations after biomanipulation. R&D Technical Report W199, Environment Agency, c/o WRc, Swindon, Wilts, UK, 109 pp.

Scheffer M., Hosper S.H., Meijer M.-L., Moss B. & Jeppesen E. (1993) Alternative equilibria in shallow lakes. *Trends in Ecology and Evolution* **8**, 275–279.

Spencer C.N. & King D.L. (1984) Role of fish in regulation of plant and animal communities in eutrophic Ponds. *Canadian Journal of Fisheries and Aquatic Sciences* **41**, 1851–1855.

Tatrai I. & Istvanovics V. (1986) The role of fish in the regulation of nutrient cycling in lake Balaton, Hungary. *Freshwater Biology* **16**, 417–424.

Ten Winkel T.E.H. & Meulemans J.T. (1984) Effects of fish upon submerged vegetation. *Hydrobiological Bulletin* **18**, 157–158.

Turner A.M. & Mittelbach G.G. (1990) Predator avoidance and community structure: interactions between piscivores, planktivores and plankton. *Ecology* **71**, 2241–2254.

Werner E.E., Gilliam J.F., Hall D.J. & Mittelbach G.G. (1983) An experimental test of the effects of predation risk on habitat use in fish. *Ecology* **64**, 1540–1548.

Chapter 16
Impacts and subsequent control of an introduced predator: the case of pike, *Esox lucius*, in Chew Valley Lake

A. IBBOTSON*
Centre for Ecology and Hydrology, Winfrith Newburgh, Dorchester, Dorset, UK

C. KLEE
Bristol Water, Bridgwater Road, Bristol, UK

Abstract

In 1990 pike, *Esox lucius* L., were illegally introduced into Chew Valley Lake, a shallow eutrophic lake which supports a successful trout fishery. The first major spawning appeared to take place in 1991, leading to the establishment of a population of fast growing pike. The diet of pike >560 mm consisted of 90% trout by biomass. A mark–recapture study indicated that the 1996 trout season would begin with 3.3 kg ha^{-1} of pike large enough to consume trout. It was concluded that this would not be catastrophic to the trout fishery but if the population reached levels found in other lakes up to 50% of stocked trout might be consumed. The impacts of the introduction were most realised through a decline in capture efficiency of stocked trout by anglers, which fell by an average of 14% for rainbow trout, *Oncorhynhynchus mykiss* (Walbaum). Brown trout, *Salmo trutta* L., were more susceptible to predation with the efficiency of capture of these falling to 11% in 1996. The number of visits by anglers and the number of trout caught per visit have exhibited an upward trend since 1985. Annual pike removal exercises, introduced from 1996, appear to be improving the fishery performance.

Keywords: angling performance, introduced species, predation, trout.

16.1 Introduction

Chew Valley Lake is a shallow, productive, hard-water lake with a mean depth of 4.3 m and a maximum depth of 11.5 m, in the south west of England. It is primarily a water storage reservoir constructed by Bristol Waterworks Company in the valley of the River Chew in 1953. At top water level it covers an area of 490 ha. However, water levels can fluctuate widely over the year and a relatively small drawdown can result in large areas of the littoral zone becoming exposed. However, there are large areas of emergent vegetation in the shallow sheltered margins of the lake, particularly on the south and east shores.

*Correspondence: Centre for Ecology and Hydrology, Dorset, Winfrith Technology Centre, Winfrith Newburgh, Dorchester, Dorset DT2 8ZD, UK (email: ati@ceh.ac.uk).

Since its construction, there have been few studies of the fish populations in the lake. Wilson, Maxwell, Mance, Sleigh & Milne (1975) stated that as well as stocked brown trout, *Salmo trutta* L., and rainbow trout, *Oncorhynchus mykiss* (Walbaum), there were roach, *Rutilus rutilus* (L.), perch, *Perca fluviatilus* L., eel, *Anguilla anguilla* (L.) and three-spined stickleback, *Gasterosteus aculeatus* L., present in the early 1970s. The roach population has gone through marked variations in abundance (Wilson 1971; Wilson *et al.* 1975) with heavy infestations of the parasite *Ligula intestinalis* (L.) being blamed for the declines. However, in most years recruitment seems to be sufficient to produce very large shoals of roach fry in the late summer and autumn. These form a significant part of the diet of the trout (Wilson *et al.* 1975b) and probably of the other piscivores, perch and eel.

The trout fishery at Chew Valley Lake has a good reputation for being a well managed fly fishery. Rainbow trout and brown trout are stocked at a catchable size each year and each stocking makes the greatest contribution to catches in the same year. In addition, brown trout have been stocked as fingerlings, contributing to catches in future years (Fig. 16.1).

In 1991, a small number of pike, *Esox lucius* L., between 2 and 4 kg were captured in the lake for the first time by anglers, and a subsequent electric fishing survey by the National Rivers Authority, resulted in a few more captures. Since 1991, the numbers of pike captured by anglers increased, prompting a netting survey in October 1994. This revealed large numbers of pike between 23 and 30 cm in fork length, indicating the presence of a spawning population. Concerns about the threat to the trout fishery resulted in a study to determine the potential impacts of the pike population, instigation of control measures and monitoring of the performance of the fishery in subsequent years. An annual removal programme was started in 1996 using gill nets (5 and 6 in) set during February and March in known pike spawning areas. Nets were checked regularly and pike were kept alive and moved to other still waters where they were more appreciated.

This chapter examines the impacts of pike on the fishery and the changes brought about by the control exercise.

16.2 Materials and methods

Samples of pike for ageing, dietary analysis and mark–recapture experiments were obtained by seine netting in the autumn 1994 and 1995, whilst water temperatures were comparatively high and the reservoir level was low, plus gill netting (with 10.2, 12.7 and 15.2 cm knot-to-knot mesh size) and electric fishing in the marginal areas during spring 1995 (see Table 16.1 for sampling effort plus numbers of fish caught). A total of 999 pike captured were sexed, measured (fork length nearest cm), weighed (g) and had scales removed for ageing and back-calculation of lengths.

The stomachs of 318 pike captured either with seine nets or gill nets between November 1994 and September 1995 were examined. All pike examined were between 223 and 930 mm fork length, as smaller pike were not captured by the methods used. Prey items were identified to species and measured to the nearest cm.

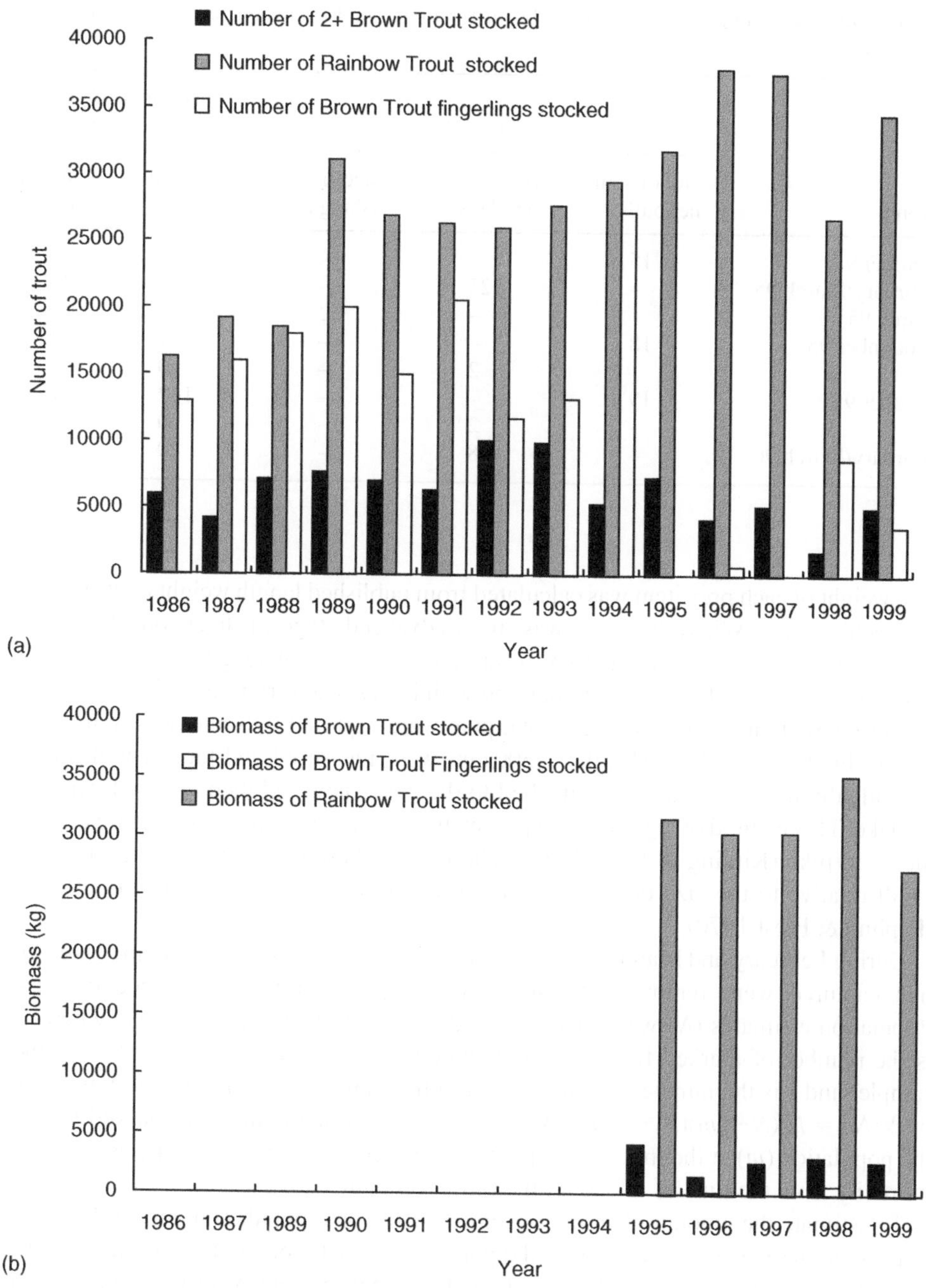

Figure 16.1 The (a) numbers and (b) biomass of brown and rainbow trout stocked into Chew Valley Lake between 1985 and 1999

Table 16.1 Numbers of pike captured by different methods between September 1994 and March 1996

Month	Method by effort			Number of pike captured by size	
	No. of seine net pulls	No. of gill net days	Electric fishing days	<40 cm (FL)	≥40 cm (FL)
October 94	15	–	–	334	7
February/March 95	–	125	–	0	266
March 95	–	–	2	13	14
September 95	18	–	–	157	34
	–	2	–	0	2
October 95	19	–	–	162	17
	–	2	–	0	5
February/March 96	–	58	–	0	220

The weight of each prey item was calculated from published length weight relationships for each species. Where the prey was in an advanced stage of digestion, they were identified from the remaining bony tissue and estimates of lengths were made by comparing the size of the bones remaining with bones taken from fish of known size.

During September and October 1995, pike were captured by seine netting in areas where this was possible. All pike >300 mm were measured and tagged with metal individually numbered jaw tags attached to the upper jaw, and then released live into the lake. This method of tagging has a proven track record with mark–recapture experiments in pike (Kipling & Frost 1970). The loss of these tags is negligible; they are small relative to the size of the fish and probably do not impede them in any way (Kipling & Frost 1970).

During February and March 1996, 12.7 and 15.2 cm gill nets were set in the lake. All fish captured were removed, measured and inspected for the presence of tags. Population estimates (N) were made using the Petersen method, $N = mc/r$; where m is the number of marked fish in the population; c is the total number of fish in the sample; and r is the number of marked fish in the sample. The variance of N is given by $V(N) = N^2(N - m)(N - c)/mc(N - 1)$. To estimate the number of marked fish in the population (m) at the time of recapture, it was necessary to make some estimate of mortality from the time of release to the time of capture.

To estimate the amount of trout consumed by the pike population during the 1996 season, biomass was estimated from the population size based on the mark and recapture study and the mean weight of fish captured. Since there were insufficient data to estimate accurately mortality from this population, it was decided that the most reliable method of estimating the annual food requirements was using the estimated biomass figure and to assume that:

(1) the production to biomass ratio was similar to that found in the River Frome (Mann 1980), i.e. 0.51;

(2) the total food intake requirements for a given production were the same as for those found in the River Frome (Mann 1982) (=2.9 g g^{-1} production).

Annual food requirements (g) are equal to biomass (*B*) (g) × 0.51 × 2.9. The proportion that constituted trout was estimated by multiplying food requirement by the proportion of diet (biomass) consisting of trout, taken from the study of stomach contents.

The Chew Valley Lake fishery maintains records of the numbers of fish stocked and the numbers of anglers that visit. Each angler is required to complete a return of the fish caught. Standard statistics, which measure the performance of the fishery, were calculated from these figures between 1985 and 1999. These statistics included catch per angler visit and efficiency of capture for both rainbow and brown trout.

16.3 Results

16.3.1 *Age, growth and age at maturity*

The majority of fish captured were <1-year old (Fig. 16.2, Table 16.2). Six fish were from the 1989 cohort, two from 1987 and one from 1986. The three oldest fish all showed a marked increase in growth rate during the summer of 1990. Since 1991 was the first year when recruitment was identified, it was deduced that the pike had probably been introduced in the spring of 1990.

With the exception of the 1993 year class, the mean length of pike at the end of their first year was between 30 and 35 cm (FL). By the time they reached 4 years of age, females achieved a mean length of 69 cm and males a mean length of 57 cm. Of those fish captured during the spring of 1995, all males were mature before their first birthday and all females were mature before their second birthday.

16.3.2 *Diet of pike*

The diet of pike in Chew Valley Lake was predominately fish. The most common food item found in the stomachs (size range 223–930 mm) was roach fry; in 28% of the 318 stomachs examined. Fry of other species such as perch and other small fish were also found, but less often. However, as a proportion of the total biomass of fish depredated by the pike population these small fish contributed <2%.

Cannibalism appeared to be negligible with only 1.3% of the stomachs examined containing pike. Cannibalism contributed only 4% to the total biomass of food items consumed. By contrast, although only 3.7% of the stomachs examined contained trout, the larger size of this prey item resulted in trout contributing >50% to the total biomass of food consumed. The relative importance of each of the prey species to the diet becomes clearer when the stomach contents analysis is considered in relation to pike size and season.

Roach fry, the most commonly eaten prey, was the most important prey item to the smaller size classes of pike (size range 223–<400 mm) (Fig. 16.3), but their importance

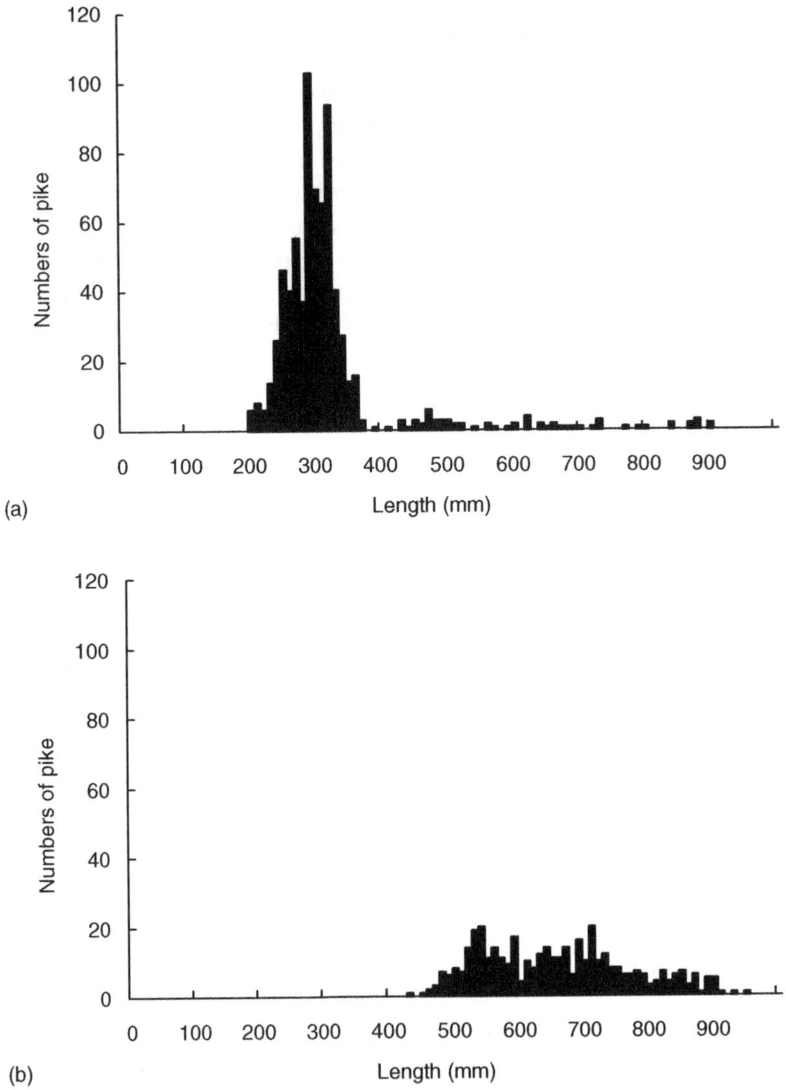

Figure 16.2 Length–frequency distribution of pike captured by (a) seine net and (b) gill nets in Chew Valley Lake between 1994 and 1996

declines as the pike increase in size. In the largest size category (size >600 mm) roach fry only contributed 0.2% to the total biomass of food consumed (Fig. 16.3) and was replaced by trout.

There were differences in the diet of pike captured during the trout fishing season and those captured outside the fishing season (Fig. 16.3). This was most noticeable in the larger size groups of pike, which ate mostly adult roach outside the fishing season but, in biomass terms, 97% trout during the trout fishing season. The diet of the

Table 16.2 Numbers of pike from different year classes captured by seine and gill net between September 1994 and October 1995

	Year class									
Capture date	1995	1994	1993	1992	1991	1990	1989	1988	1987	1986
Winter 1994	–	246	111	60	128	17	6	–	2	1
Summer 1995	318	15	10	15	16	3	–	–	–	–

Data from the gill netting in 1996 is not included because these fish were not aged.

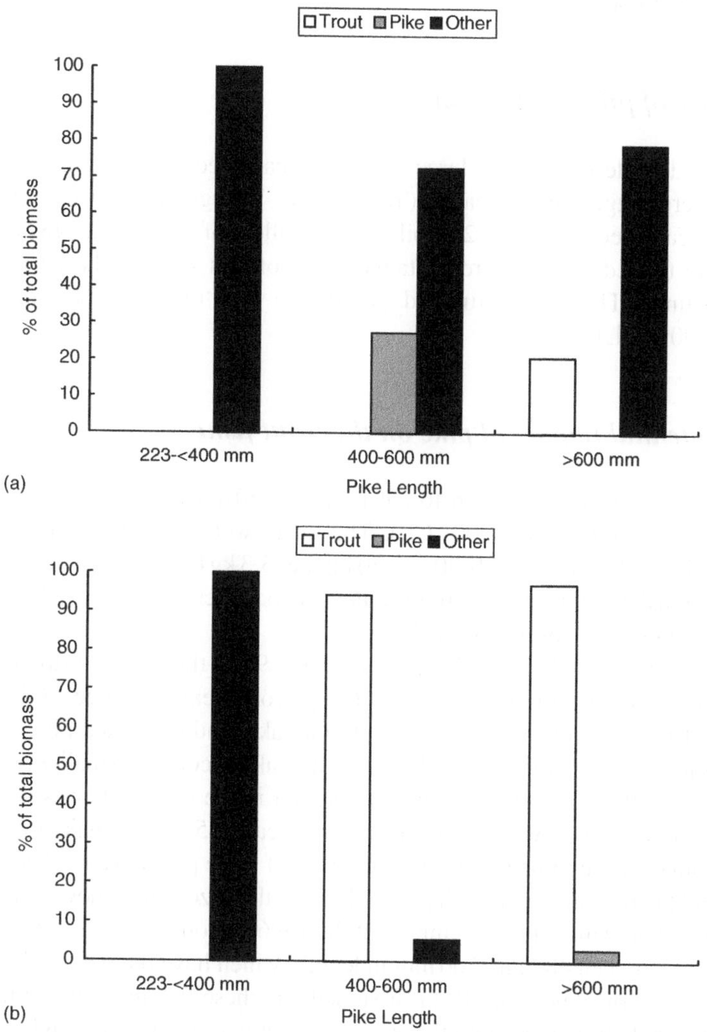

Figure 16.3 The biomass of each food item consumed by pike (a) out of trout season and (b) in trout season in Chew Valley Lake

Table 16.3 Numbers of pike tagged in September and October 1995 and the numbers recaptured in February and March 1996, for different size groups

	Numbers of each pike size group tagged and number recaptured with tags (cm)						
	30–<40	40–<50	50–<60	60–<70	70–<80	80–<90	90<100
Tags	107	21	5	10	2	2	0
Recap	0	0	2	4	1	0	0

smallest size class of pike was much more stable with roach fry forming the bulk of the food eaten in both periods.

16.3.3 *Size of pike population*

Despite considerable effort, few large pike were captured for tagging. Only 40 pike >400 mm were tagged and released (Table 16.3). The gill netting of February and March 1996 captured a total of 220 pike, the smallest of which was 482 mm. Seven of these were tagged. There were 22 tagged pike of this size and larger released the previous autumn. Thus the estimated population size of fish >482 mm (*N*) was 691 (±348; 90% CL).

16.3.4 *Potential impact of pike on the trout fishery*

The number of pike large enough to eat trout (>560 mm) immediately prior to the 1996 season was 467 ± 243 (90% CL). The mean weight of these pike was 3.5 kg giving an estimated biomass of 1630 ± 850 kg, i.e. 3.3 kg ha^{-1}. Using the assumptions above, this translates into a food requirement of between 1150 and 3700 kg, with a mean of 2400 kg $year^{-1}$ or 4.9 kg ha^{-1}.

The proportion of this food made up of trout was 90%, thus it was estimated that the 1996 pike population would consume 2160 kg trout $year^{-1}$ (range 1035–3330 kg). In 1996, 35 000 kg of trout were stocked into the lake, and estimated pike depredation was 6.2% (range 3.0–9.5%) of these. This does not take account of smaller pike growing and reaching a sufficient size to consume trout later in the season. For example, a 470-mm female on average grows to 580 mm within 1 year. A 510-mm male grows on average to 570 mm over the same period. The numbers of these pike have not been estimated precisely, and the time at which each pike will reach this size is not known. However, the number of fish >482 mm at the beginning of the 1996 season was known; 691 fish, or an additional 224 between 482 and 560 mm in length, which have the potential to reach the minimum size to consume trout. If it is assumed that these fish reach the minimum size to consume trout half way through the season, and at that point they weigh 1.9 kg each, then the additional contribution of these fish to trout depredation will be 280 kg, increasing estimated depredation to 2440 kg or 7.0% of stocked fish.

16.3.5 *Control of the pike population*

The 1996 gill netting removed an estimated 37% of the pike large enough to consume trout during 58 gill net days at a rate of 3.8 pike per gill net day (Fig. 16.4). Since then the numbers of pike removed has held steady, but the rate per gill net day has fallen to 2.1 fish per gill net day.

16.3.6 *Performance of the fishery*

Although there is considerable variation in the number of angler visits between years, there has been a generally increasing trend since 1985 (Fig. 16.5). The introduction of pike in 1991 coincided with a drop in angler visits, but despite a fall off in visitors between 1992 and 1995, the numbers have remained higher than prior to the introduction of pike.

Similarly catches per angler visit have also risen over the same period and recently have exceeded three fish per visit, well above the level experienced before pike were

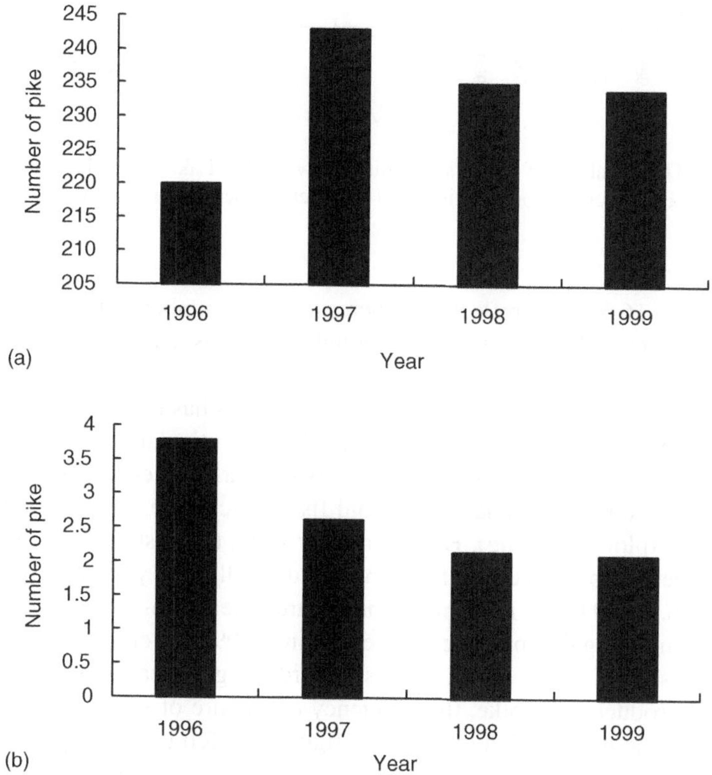

Figure 16.4 (a) The total number of pike captured in control measures between 1996 and 1999 and (b) the catch of pike per gill net day in each year between 1996 and 1999

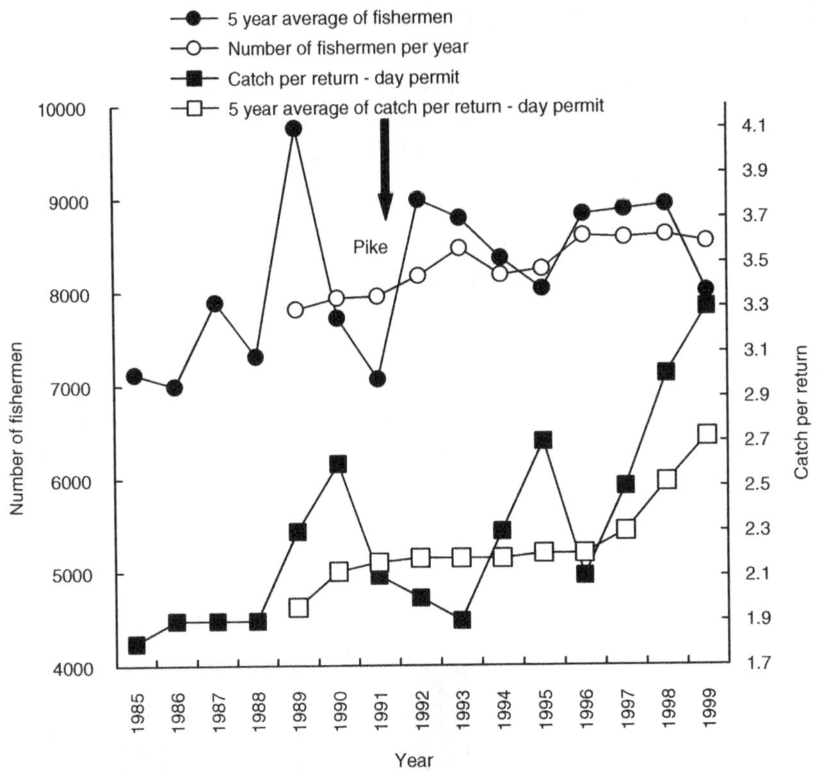

Figure 16.5 The numbers of anglers visiting Chew Valley Lake 1985–1999 and the mean catch per visit. Pike were introduced in 1990 but their presence was not widely known until 1991

introduced (Fig. 16.5). This reflects the increase in the number of fish stocked which has generally increased since 1985 so that roughly twice as many fish are now stocked (Fig. 16.1).

However, the efficiency of capture of the stocked fish has declined markedly since pike were introduced to the lake (Fig. 16.6). This is particularly the case for brown trout. Prior to the introduction of pike, returns regularly exceeded 100%. This was achieved by the addition of small 1-year-old fish, which then grew on to a size that anglers could exploit (ranching). However, after 1991, the first year when pike were caught in large numbers, the efficiency of capture fell each year to a low point of 11.8% in 1996. In 1994 the ranching of brown trout ceased as it was not considered worthwhile continuing that practice. However, since 1996 the efficiency of capture of these fish has started to rise again, the last recorded figure reaching 46%. Similarly, prior to the introduction of pike, the efficiency of capture of rainbow trout averaged 75%, but since the introduction of pike this figure also fell to a low of under 49% in 1997 and averaging 61% between 1991 and 1998. Note, there was an increase in 1999 returning to a capture efficiency of 76%, the same level found prior to the introduction of pike.

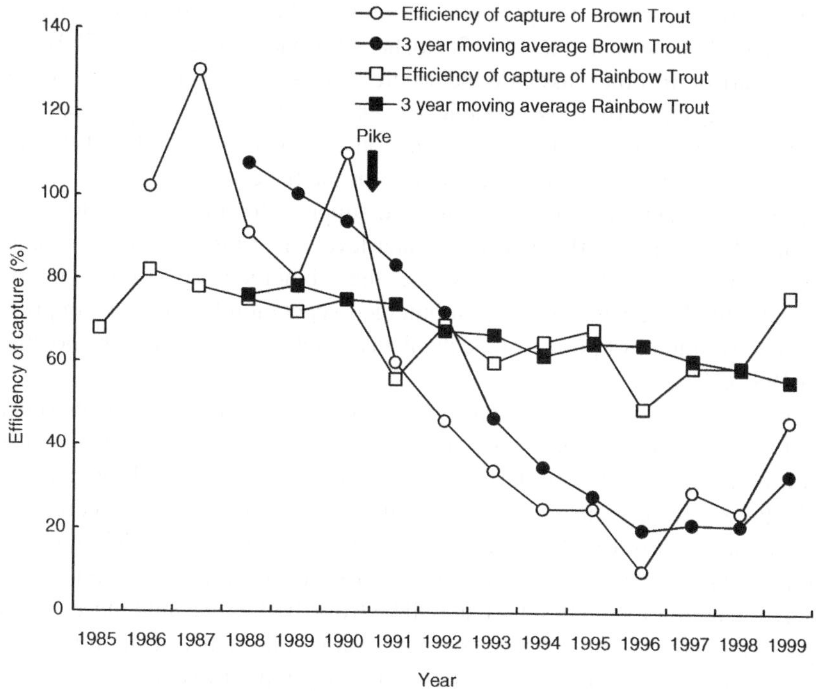

Figure 16.6 The efficiency of capture by anglers of rainbow trout and brown trout in Chew Valley Lake between 1985 and 1999. Trend lines are drawn for each statistic. Pike were introduced in 1990 but their presence was not widely known until 1991

16.4 Discussion

Chew Valley Lake is rich in nutrients and has ample supply of prey fish for pike. At all sizes the pike should find plenty of food. The smaller fish can feed on the abundant roach and perch fry and once they reach a larger size there is a constantly replenished supply of trout. Therefore it is not surprising that the growth rate of these fish is at the upper end of the range found in circumpolar pike populations (Casselman 1996).

The predicted impact of pike on the stocked trout in the reservoir was only about 7% loss through predation in 1996. This figure on its own might not be particularly worrying. However, its estimation is based on a production/biomass estimate taken from the River Frome (Mann 1980) which is a river with an established population. A new and young population such as the one in Chew Valley Reservoir might be expected to have a far higher production/biomass ratio. This would seem reasonable considering the growth rates of the individual pike. The exact figure is not known but production could at least equal biomass with pike doubling their weight each year. This would increase the potential losses of stocked trout to 14%. This figure is close to the differences in the efficiency of capture by anglers of the stocked rainbow trout, which fell from an average of 75% prior to the introduction of pike to 61% afterwards.

It is not known how large the pike population would expand to without any control measures. The 1996 estimated level of 3.3 kg ha^{-1} for pike >3 years old (560 mm) was low in comparison to other lakes (Windermere – 9.5 kg ha^{-1}, Kipling & Frost 1970; Slapton Ley 31 kg ha^{-1}, Bregazzi & Kennedy 1980). Should the pike population reach the depredation level in Slapton Ley it might consume as much as 50% of the stocked trout population, which would be catastrophic for the fishery.

There are some factors which might help to suppress the pike population in Chew Valley Lake. These include fluctuation in water levels. During the summer months the reservoir often drops to a level which exposes the submerged vegetation. The juvenile pike are particularly dependent on the submerged vegetation for cover from predation and this is one of the most powerful mechanisms of self-regulation in pike populations (Grimm 1981).

It has been shown that for some pike populations at least predation from older age classes of fish has a significant impact on recruitment of 0+ fish (Grimm 1994). Thus, there is a potential risk that by targeting the older age classes with annual gill net removals, the reduced predation pressure will feed back into stronger recruitment of the 0+ fish, as occurred with regular netting in Lake Windemere (Kipling & Frost 1972) presenting a greater problem in future years. However, it is argued that most of the population regulation on the 0+ fish will be from the 1+ fish, and these will not be the target of any removal programme, because fish will normally be approaching their third birthday before they are large enough to be caught in gill nets.

The impacts of a predator like pike on the Chew Valley Lake trout fishery could manifest themselves through a number of mechanisms. The number of angler visits may respond negatively either to changes in the quality of the fishery such as falls in the number of trout caught per visit or to a perceived loss of quality merely as a psychological reaction to the presence of pike. However, through the period that pike have been present both the catches per visit and the number of anglers visiting the lake have exhibited an increasing trend (Fig. 16.5). The only exception to this was in 1991 when angler visits appeared to be below trends. This was the first year that it became widely known that pike were present in the lake in significant numbers. Apart from that year numbers of angler visits has continued to increase suggesting no long term impact on either the actual or psychological quality of the fishery.

Where pike appeared to have an impact was in the efficiency of capture of the stocked fish by the anglers. This was most marked for brown trout where capture rates by anglers only reached 11% of stocked fish in 1996. Further the ranching of brown trout appeared to contribute little to future year's catches and this practice was stopped as a direct result of the introduction of the pike. Indeed the number of catchable brown trout stocked was also reduced through the 1990s as a result of the decline in angler catch efficiency of this species. A similar reduction, although much less marked, was also recorded for rainbow trout. Since both angler visits and catches per visit have been rising at the same time as the efficiency of capture has fallen there may have been a compensatory increase in the number of rainbow trout stocked (Fig. 16.1). Although there are some limits to the sources of fish that can be stocked in the lake there is some flexibility through the season on the number of fish stocked. Usually the lake managers stock rainbow trout three times during the season and the numbers put in is partially a response to recent angling performance. Thus it would seem that the lake managers have compensated for the loss of trout to predation by increasing the number of fish that are available to both pike and anglers. This can be achieved as the costs of stocking more fish is a fairly low proportion of the total costs of the fishery. Indeed most of the fish stocked are reared by the fishery and therefore have a lower cost than the normal market price.

Brown trout are stocked at an average size of 0.56 kg whereas the rainbow trout tend to be stocked between 0.75 and 1 kg. This may explain the greater susceptibility of the brown trout to pike predation. Being smaller there will be a greater proportion of the pike population able to catch and handle the brown trout than for the rainbow trout. There may also be behavioural explanations as brown trout and rainbow trout are thought to make use of different habitats in lakes. Rainbow trout tend to use the mid-water more than brown trout that live closer to the bottom (Idyll 1942; Brown, Oldham & Warlow 1980; Albertova 1978, 1982; Warlow & Oldham 1982; Lucas 1993). Brown trout also have a greater tendency to piscivory and these two behaviours may bring them into contact with the pike more frequently than the rainbow trout.

The apparent ability of the control measures to remove >30% of the pike above a size capable of eating stocked trout has encouraged this process to continue annually. This apparently, high capture rate was possible because the pike concentrate at the south west end of the lake during the spawning period immediately before the trout fishing season. It is difficult to be certain that this control programme is having an effect but there is some evidence of this from the numbers of pike captured each year per unit effort, which has declined since 1996. However, probably the best evidence comes from the recovery in the efficiency of capture for both brown trout and rainbow trout since the low point of 1996.

Acknowledgements

The authors thank all those that helped with the fieldwork and sample processing, in particular Kim Lucas, Bill Beaumont, Brian Dear, Todd Lewis and Matthew Byrne. The study was funded by contract from Bristol Water plc.

References

Albertova O. (1978) The food of brown and rainbow trout in the water-supply reservoir Hubenov. *Vertebrata Zpravy* **1978**, 69–74.

Albertova O. (1982) The food of rainbow trout and brown trout in the water-supply dam reservoirs Lucina and Rimov. *Prace Vuhr Vodnany* **11**, 127–134.

Bregazzi A.R. & Kennedy C.R. (1980) The biology of pike *Esox lucius* L., in a southern eutrophic lake. *Journal of Fish Biology* **17**, 91–112.

Brown A.E., Oldham R.S. & Warlow A. (1980) Chironomid larvae and pupae in the diet of brown trout (*Salmo trutta*) and rainbow trout (*Salmo gairdneri*) in Rutland Water, Leicestershire. In D.A. Murray (ed.) *Chironomidae*. Oxford: Pergamon Press, pp. 199–124.

Casselman J.M. (1996) Age, growth and environmental requirements of pike. In J.F. Craig (ed.) *Pike; Biology and Exploitation*. Fish and Fisheries Series 19. London: Chapman & Hall, pp. 69–101.

Grimm M.P. (1981) Intraspecific predation as a principal factor controlling the biomass of northern pike (*Esox lucius* L.). *Fisheries Management* **12**, 77–79.
Grimm M.P. (1994) The influence of aquatic vegetation and population biomass and recruitment of 0+ and 1+ northern pike (*Esox lucius* L.). In I.G. Cowx (ed.) *Rehabilitation of Freshwater Fisheries*. Oxford: Fishing News Books, Blackwell Science, pp. 226–234.
Idyll C. (1942) Food of rainbow, cutthroat, and brown trout in the Cowichan River system, B.C. *Journal of the Fisheries Research Board of Canada* **5**, 448–458.
Kipling C. & Frost W.E. (1970) A study of the mortality, population numbers, year class strengths, production and food consumption of pike *Esox lucius* L., in Windermere from 1944 to 1962. *Journal of Animal Ecology* **39**, 115–157.
Kipling C. & Frost W.E. (1972) Certain effects of an annual removal of pike (*Esox lucius*) from Windermere. *Verhandlungen der Internationale Vereinigung für theoretische und angewandte Limnologie* **18**, 1122–1127.
Lucas M.C. (1993) Food interrelationships between brown trout, *Salmo trutta* L., and rainbow trout, *Oncorhynchus mykiss* (Walbaum), in a small put-and-take stillwater fishery. *Aquaculture and Fisheries Management* **24**, 355–364.
Mann R.H.K. (1980) The numbers and production of pike (*Esox lucius*) in two Dorset rivers. *Journal of Animal Ecology* **49**, 899–915.
Mann R.H.K. (1982) The annual food consumption and prey preferences of pike *Esox lucius* in the River Frome, Dorset. *Journal of Animal Ecology* **51**, 81–95.
Warlow A.D. & Oldham R.S. (1982) Temporal variations in the diet of brown trout (*Salmo trutta* L.) and rainbow trout (*S. gairdneri* R.) in Rutland Water. *Hydrobiologia* **88**, 199–206.
Wilson R.S. (1971) The decline of a roach *Rutilus rutilus* (L.) population in Chew Valley Lake. *Journal of Fish Biology* **3**, 129–137.
Wilson R.S., Maxwell T.R.A., Mance G., Sleigh M.A. & Milne R.A. (1975) Biological aspects of Chew Valley and Blagdon Lakes, England. *Freshwater Biology* **5**, 379–393.

Chapter 17
The potential to control fish community structure using preference for different spawning substrates in a temperate reservoir

M. ZALEWSKI* and P. FRANKIEWICZ
Department of Applied Ecology, University of Lodz, Lodz, Poland

Abstract

The relationship between hydrological and ecological processes helps to understand the reason for lack of macrophytes and its knock on effect on spawning success and recruitment of phytophilous fish species. To overcome the problem of poor recruitment in reservoirs, spawning substrate made up of branches of trees are commonly used but little is known of preferences of fish to the different tree species. To test this, artificial spawning sites made of branches of four different trees (spruce, pine, juniper and birch) were placed in the Sulejow Reservoir Poland. Roach, *Rutilus rutilus* (L.) preferred branches of juniper and spruce against those of pine and birch as spawning substratum. By contrast, perch, *Perca fluviatilis* L., preferred birch against coniferous branches. This suggests that alteration of the availability of spawning substrata in the littoral zone may be an effective way of controlling the density and structure of juvenile fish communities in reservoirs, which in turn may be used to control eutrophication.

Keywords: eutrophication, *Perca fluviatilis*, recruitment, *Rutilus rutilus*, spawning substrate.

17.1 Introduction

The structure of a fish community in a reservoir plays a crucial role from the point of view of controlling algal blooms, which are the main symptoms of water eutrophication (e.g. Hrbacek, Dvorakova, Korinek & Prochazkova 1961; Brooks & Dodson 1965; Shapiro & Wright 1984; Persson, Andersson, Hamrin & Johansson 1988). Changing this structure by predator population enhancement and/or reduction of zooplanktivorous fish populations can help to control some extent algal blooms, via the cascading effect (e.g. Carpenter, Kitchell & Hodgson 1985; McQueen, Post & Mills 1986; Hambright, Drenner, McComas & Hairston Jr 1991). Changes in fish community structure may result not only from direct actions (stocking, fishing), but also from alteration of fish spawning success. Water level manipulations during the spawning period are a productive way of using the latter for management of fish communities in reservoirs (Zalewski, Brewinska-Zaras & Frankiewicz 1990a; Zalewski, Brewinska-Zaras,

*Correspondence: Department of Applied Ecology, University of Lodz, Banacha 12/16, 90-237 Lodz, Poland (email: mzal@boil.uni.lodz.pl).

Frankiewicz & Kalinowski 1990b). However, in many reservoirs the water level is lowered around the spawning period of fish thus isolating the fish from flooded vegetation in the littoral zone which may act as spawning substrate (Ploskey 1985; Zalewski *et al.* 1990a, b). An alternative way of regulating fish spawning success, may be to provide the target fish species with its preferred spawning substratum. Changing fry community structure and density by manipulating spawning substratum should lead to change in the food web and dynamics of the trophic cascade (Frankiewicz, Dabrowski & Zalewski 1996).

This chapter examines the preferences of different artificial spawning substrates, based on branches of different tree species, by roach, *Rutilus rutilus* (L.), and perch, *Perca fluviatilis* L., as a preliminary for addressing whether the spawning success of populations can be manipulated in reservoirs prone to eutrophication.

17.2 Materials and methods

Investigations were conducted on the Sulejow Reservoir, situated in Central Poland on the River Pilica, a major tributary of the River Vistula. Due to large fluctuations in water level, the reservoir's area may vary from 630 to 2380 ha, and as a consequence the reservoir is almost devoid of littoral macrophytes. The average depth is 3.3 m, and the maximum depth, close to the dam, is 11 m. Experiments on spawning substratum preference of roach and perch were conducted in and around Tresta Bay. The location was chosen because of earlier observations of fish spawning activity in this part of the reservoir. Spawning substratum preferences of roach and perch were investigated using rectangular frames, where each wall was made of a different substrate: juniper *Juniperus communis* L.; spruce, *Picea abies* (L.); pine, *Pinus sylvestris* L.; and birch, *Betula pendula* Roth. Such artificially-prepared spawning sites were exposed at three stations, at 1, 2, 3 and 4 m depths, with 2–3 replications for each depth. The structures were placed in the water column a few days before the spawning period. At the end of spawning these structures were removed from the reservoir and five randomly chosen pieces of each substratum were preserved in 4% formalin. In the laboratory eggs were separated from the substratum, the substratum was weighed and the eggs counted. The relative importance of a given spawning substratum was expressed as a number of eggs per 1 g of substratum wet weight. This procedure was employed for roach. In the case of perch, which lays its eggs in ribbon-shaped batches, the number of batches was recorded on given substratum according to station and depth.

17.3 Results

The substratum preference of roach was consistent. Significant differences in egg numbers between substratum type occurred at all three stations (one-way ANOVA, in all three cases $P < 0.001$). Juniper and spruce were the preferred substratum (Fig. 17.1(a)); number of eggs on pine and birch branches were always significantly lower. For any given substratum type roach generally preferred the littoral zone (Fig. 17.1(b)), although in the case of pine differences were not significant and in the case of juniper

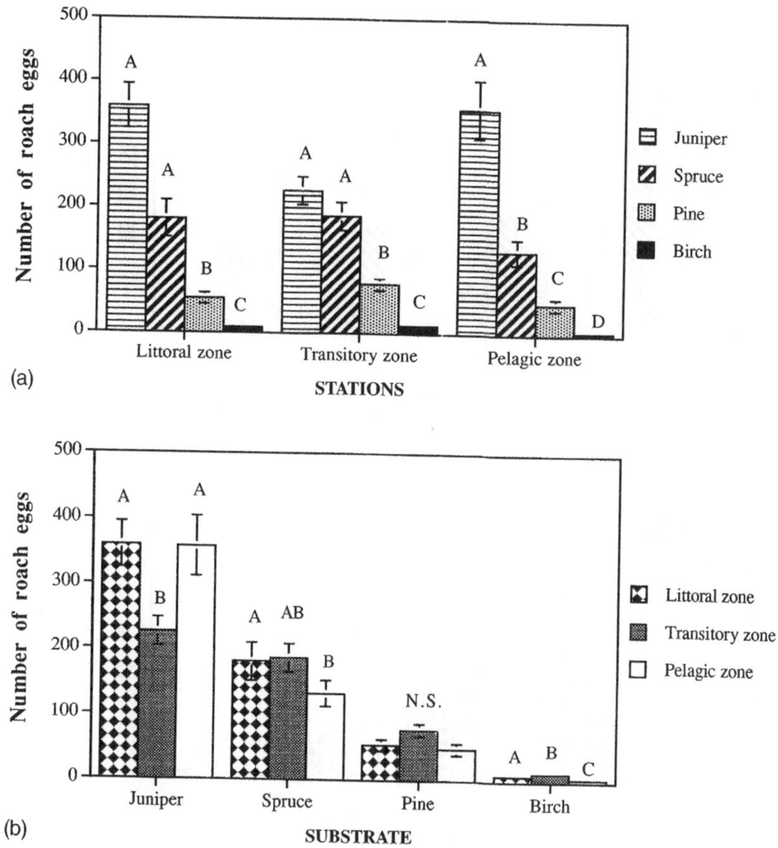

Figure 17.1 (a) Preference for spawning substratum type by roach in different lake zones and (b) change in use of different substratum types in relation to lake zones

there were no differences in egg numbers between the littoral and the pelagical zone. Significant differences were found between substrata at all depths (Fig. 17.2(a)) (one-way ANOVA, at all four depths $P < 0.001$). Juniper and spruce were again preferred to pine and birch. However, in the case of juniper, similar numbers of eggs were observed at all depths, while for the other substratum types egg numbers generally declined with depth (Fig. 17.2(b)).

Substratum preferences of perch were opposite to those of roach. Egg batches of this species were almost exclusively found on birch branches (Fig. 17.3). Spruce and pine were selected sporadically, while juniper was never chosen by perch as a spawning substratum.

17.4 Discussion

Other studies on the utilisation of artificial spawning substrata in lakes have noted that egg density on such substratum was always much higher than on natural spawning areas,

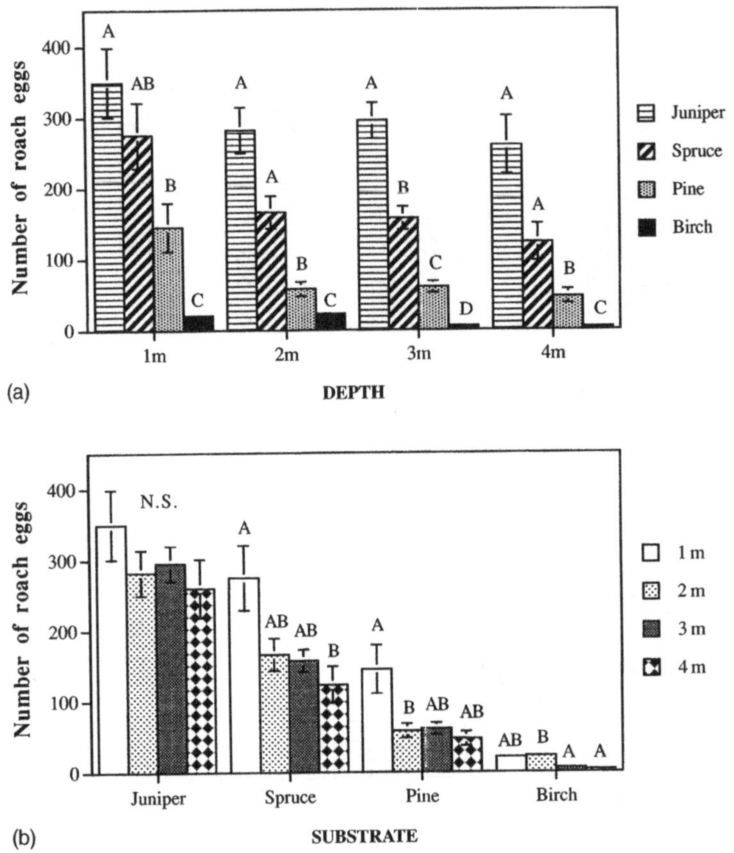

Figure 17.2 (a) Utilisation of spawning substratum type by roach at different depths zones and (b) change in use of different substratum types in relation to lake depth

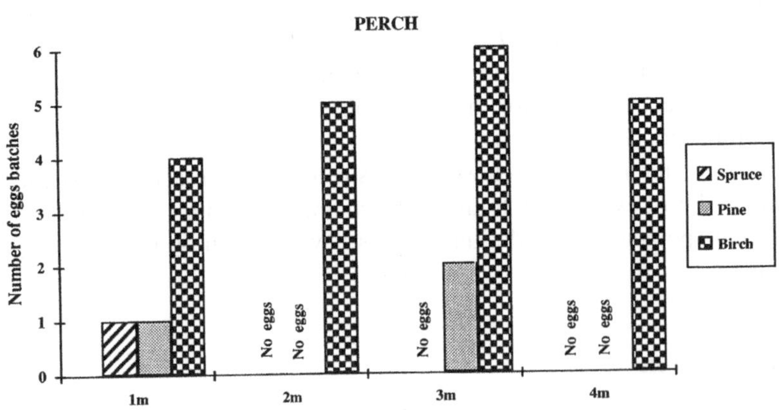

Figure 17.3 Utilisation of spawning substratum type by perch at different depth zones

for both perch (Lang 1987; Zeh, Ritter & Ribi 1989) and roach (Zuromska 1967). Conifer branches were frequently used as an artificial spawning substratum by both roach and perch (Echo 1955; Zeh *et al.* 1989; Delos Reyes, Arzbach & Braum 1992; Gillet & Dubois 1995), but significant differences in egg densities between them have only been reported by Nash, Hendry & Cragg-Hine (1999). Thus the observation that juniper was the most suitable spawning substratum for roach, and was chosen with the same intensity despite water depth and distance from the shore may have practical implications for fishery management. Conversely, the preference of perch for birch branches over conifer as a spawning substratum may have similar value. This preference probably resulted from perch spawning behaviour. Hergenrader (1969) and Treasurer (1981) found that submerged plants are a key spawning substratum, because females release ribbons of eggs whilst swimming rapidly amongst macrophytes. Consequently, dense branches like juniper or spruce could have disturbed this behaviour, but birch probably closely matched perch spawning needs because the branch structure allows easy penetration. Thus, by using suitable artificial spawning substrata it should be possible to influence spawning success of perch and roach, which often suffer from a scarcity of appropriate vegetated grounds in the littoral zone of most reservoirs. Moreover, providing fish species with a preferred substratum should make it possible to isolate spawning grounds within a given area and to achieve almost unmixed eggs of a target fish species. This suggests that alteration of the availability of spawning substrata in the littoral zone may be an effective way of controlling the density and structure of juvenile fish communities in reservoirs, which in turn may be used to control eutrophication.

References

Brooks J.L. & Dodson S.I. (1965) Predation, body size, and composition of plankton. *Science* **150**, 28–35.

Carpenter S.R., Kitchell J.F. & Hodgson J.R. (1985) Cascading trophic interactions and lake productivity. *Bioscience* **35**, 634–639.

Delos Reyes M.R., Arzbach H.H. & Braum E. (1992) In situ development of perch eggs, *Perca fluviatilis* L. (Pisces, Percidae) in a small eutrophic lake, Lake Plussee, Holstein, Germany. *International Revue gesten Hydrobiologie* **77**, 467–481.

Echo J.B. (1955) Some ecological relationships between yellow perch and trout in Thompson Lake. *Transactions of the American Fisheries Society* **84**, 239–248.

Frankiewicz P., Dabrowski K. & Zalewski M. (1996) Mechanism of establishing bimodality in a size distribution of age-0 pikeperch, *Stizostedion lucioperca* (L.) in the Sulejow Reservoir, Central Poland. *Annales Zoologici Fennici* **33**, 321–327.

Gillet C. & Dubois J.P. (1995) A survey of the spawning of perch (*Perca fluviatilis*), pike (*Esox lucius*), and roach (*Rutilus rutilus*), using artificial spawning substrates in lakes. *Hydrobiologia* **300/301**, 409–415.

Hambright K.D., Drenner W.C., McComas S.R. & Hairston Jr. N.G. (1991) Gape-limited piscivores, planktivore size refuges, and the trophic cascade hypothesis. *Archiv für Hydrobiolgie* **121**, 389–404.

Hergenrader G.L. (1969) Spawning behaviour of *Perca flavescens* in aquaria. *Copeia* **1969**, 839–841.

Hrbacek J., Dvorakova M., Korinek V. & Prochazkova L. (1961) Demonstration of the effect of the fish stock on the species composition of zooplankton and the intensity of metabolism of the whole plankton association. *Verhandlungen der Internationale Vereinigung für theoretische und angewandte Limnologie* **14**, 192–195.

Lang C. (1987) Mortality of perch, *Perca fluviatilis* L., estimated from the size and abundance of egg strands. *Journal of Fish Biology* **31**, 717–720.

McQueen D.J., Post J.R. & Mills E.L. (1986) Trophic relationships in freshwater pelagic ecosystems. *Canadian Journal of Fisheries and Aquatic Sciences* **43**, 1571–1581.

Nash K.T., Hendry K. & Cragg-Hine D. (1999) The use of brushwood bundles as fish spawning media. *Fisheries Management and Ecology* **6**, 349–355.

Persson L., Andersson G., Hamrin S.F. & Johansson L. (1988) Predator regulation and primary production along the productive gradient of temperate lake ecosystems. In S.R. Carpenter (ed.) *Complex Interactions in Lake Communities*. New York: Springer, pp. 45–65.

Ploskey G.R. (1985) Impacts of terrestrial vegetation and preimpoundment clearing on reservoir ecology and fisheries in the United States and Canada. Rome: *FAO Fisheries Technical Paper* **258**, 35 pp.

Shapiro J. & Wright D.J. (1984) Lake restoration by biomanipulation, Round Lake, Minnesota, the first two years. *Freshwater Biology* **14**, 371–383.

Treasurer J.W. (1981) Some aspects of the reproductive biology of *perch Perca fluviatilis* L. fecundity, maturation and spawning behaviour. *Journal of Fish Biology* **18**, 729–740.

Zalewski M., Brewinska-Zaras B. & Frankiewicz P. (1990a) Fry communities as a biomanipulating tool in a temperate lowland reservoir. *Archiv für Hydrobiologie* **33**, 763–774.

Zalewski M., Brewinska-Zaras B., Frankiewicz P. & Kalinowski S. (1990b) The potential for biomanipulation using fry communities in a lowland reservoir: concordance between water quality and optimal recruitment. *Hydrobiologia* **200/201**, 549–556.

Zeh M., Ritter E. & Ribi G. (1989) Spawning and egg development of *Perca fluviatilis* in Lake Zurich. *Aquatic Sciences* **51**, 100–107.

Zuromska H. (1967) Mortality estimation of roach (*Rutilus rutilus* L.) eggs and larvae on lacustrine spawning grounds. *Roczniki Nauk Rolniczych* Seria **H 90**, 539–556 (in Polish).

Chapter 18
Fish ecology and conservation in Lake Banyoles (Spain): the neglected problem of exotic species

E. GARCÍA-BERTHOU* and R. MORENO-AMICH
Department of Environmental Sciences and Institute of Aquatic Ecology, University of Girona, Girona, Catalonia, Spain

Abstract

Lake Banyoles is the second largest lake in the Iberian Peninsula. It is usually considered oligotrophic because of the low nutrient concentration and phytoplankton biomass, although it is rather mesotrophic based on its primary production and its benthic community. Knowledge of the fish assemblage of the lake and the conservation problems due to a long history of fish introductions is reviewed. Before 1910, only five or six native species were present. Since that time 12 alien species have been introduced, leading to the apparent loss of two native species (three-spined stickleback *Gasterosteus aculeatus* L. and, possibly introduced several centuries previously, tench, *Tinca tinca* (L.)), and the decline of three others (eel *Anguilla anguilla* (L.), chub *Leuciscus cephalus* (L.), and barbel *Barbus meridionalis* Risso). The current fish assemblage is dominated by alien species, particularly the largemouth bass, *Micropterus salmoides* (Lacépède), and the pumpkinseed sunfish, *Lepomis gibbosus* (L.), in the littoral zone, and roach, *Rutilus rutilus* (L.), in the limnetic zone. Management strategies for the conservation and rehabilitation of the lake are discussed.

Keywords: conservation, endemic species, fisheries management, introductions.

18.1 Introduction

The introduction of fish species is a worldwide phenomenon (Welcomme 1988; Cowx 1998). Exotic inland freshwater fish species have been particularly damaging, causing the extinction of native fish species mainly through predation among other mechanisms (Cowx 1997; 1998).

Lake Banyoles, the second largest lake of the Iberian Peninsula, is typical of this situation and has a long history of introductions, with up to 12 species introduced since 1910. To date, only studies on the feeding ecology of the fish assemblage have been undertaken (García-Berthou 1994; 1999a, b; García-Berthou & Moreno-Amich 2000b, c)

*Correspondence: Department of Environmental Sciences and Institute of Aquatic Ecology, University of Girona, Girona, Catalonia, Spain (email: caegb@fc.udg.es).

to examine the trophic status of the fish populations, particularly in relation to the introduced species. This chapter:

(1) reviews the current ecological knowledge of the fish assemblage of the lake;
(2) assesses the conservation status of the native species; and
(3) proposes possible management solutions.

18.2 Limnology of Lake Banyoles

Lake Banyoles, situated at 42°7′N, 2°45′E and 172 m above sea level in Catalonia, Spain (Fig. 18.1), is of mixed tectonic–karstic origin. The mainly subterranean water sources and high calcium concentration restrict its productivity. Although usually considered oligotrophic because of the low nutrient concentration and phytoplankton biomass, it is rather mesotrophic based on its primary production and its benthic community. A number of studies have been carried out on its morphometry (Moreno-Amich & García-Berthou 1989), hydrology (Casamitjana & Roget 1993), bacterioplankton (Garcia-Gil, Casamitjana & Abella 1996), phytoplankton (Planas 1973), zooplankton (Miracle 1976) and non-littoral zoobenthos (Prat & Rieradevall 1995). Selected features of the lake are surface area, 111.8 ha; mean depth, 14.8 m; water temperature, 7–26°C; and conductivity, 900–2000 μS. Three different littoral habitats were distinguished for

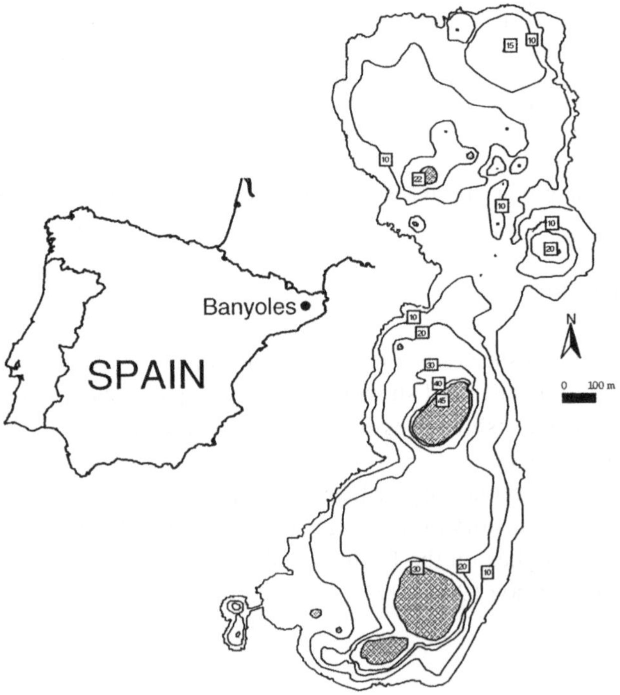

Figure 18.1 Location of Lake Banyoles in the Iberian Peninsula (left) and bathymetric map (right)

fish sampling, corresponding to the plant associations described by Bolòs & Masalles (1983) and characterised by:

(1) reed, *Phragmites australis* ssp. *australis*, dominating the western shore;
(2) cattail, *Typha angustifolia* ssp. *australis* and *Typha latifolia*; and
(3) rush, *Schoenoplectus littoralis*, dominating the zones 1–2 m deep and more common around the eastern shore.

18.3 Fish ecology of Lake Banyoles

Historical fish records from Lake Banyoles have been described elsewhere (García-Berthou & Moreno-Amich 2000a). The fish assemblage before 1910 comprised chub, *Leuciscus cephalus* (L.), Mediterranean barbel, *Barbus meridionalis* Risso, freshwater blenny, *Salaria fluviatilis* (Asso), eel, *Anguilla anguilla* (L.), and three-spined stickleback, *Gasterosteus aculeatus* L. (Fig. 18.2). Up to 12 fish species have been introduced into the lake over the past 100 years, resulting in a decline of five or six native species (Fig. 18.2); the tench, *Tinca tinca* (L.), being an anomaly having possibly been introduced several centuries ago (see García-Berthou & Moreno-Amich 2000a). Three-spined stickleback and tench disappeared around 1960; eel and barbel are now very rare.

The lake is nowadays dominated by alien species. In the littoral zone, the most common fish species are largemouth bass, *Micropterus salmoides* (Lacépède), and pumpkinseed sunfish, *Lepomis gibbosus* (L.), (Fig. 18.3). Pumpkinseed were particularly abundant in rush beds, whereas largemouth bass were more abundant in shallower, more vegetated habitats. This distribution pattern matches their main prey, amphipods (García-Berthou & Moreno-Amich 2000b) and *Atyaephyra desmaresti* (García-Berthou 1994), respectively.

The limnetic zone is dominated by roach, *Rutilus rutilus* (L.), and also inhabited by common carp, *Cyprinus carpio* L., chub and eel (Figs 18.3 and 18.4). The size structure is dominated by large fish, particularly for carp, chub and eel, and there is a lack of younger size groups suggesting recruitment problems (García-Berthou & Moreno-Amich 2000a). Roach are the main zooplanktivorous fish in the lake, mostly preying on the cladoceran, *Daphnia longispina* (García-Berthou, 1999a), whereas carp are benthic feeders (García-Berthou 1994). Although eel and chub were mostly captured in deep waters (Fig. 18.4), their gut contents consisted of large, littoral prey (crayfish and amphibians, respectively) (García-Berthou 1994), suggesting that they fed in the littoral zone at night.

18.4 Conservation of native species

The conservation status of the fish species native to the lake is summarised in Table 18.1. In Spain the five native lake species are considered either endangered (blenny) or vulnerable (other species), a conservation status that is similar or worse than in Europe.

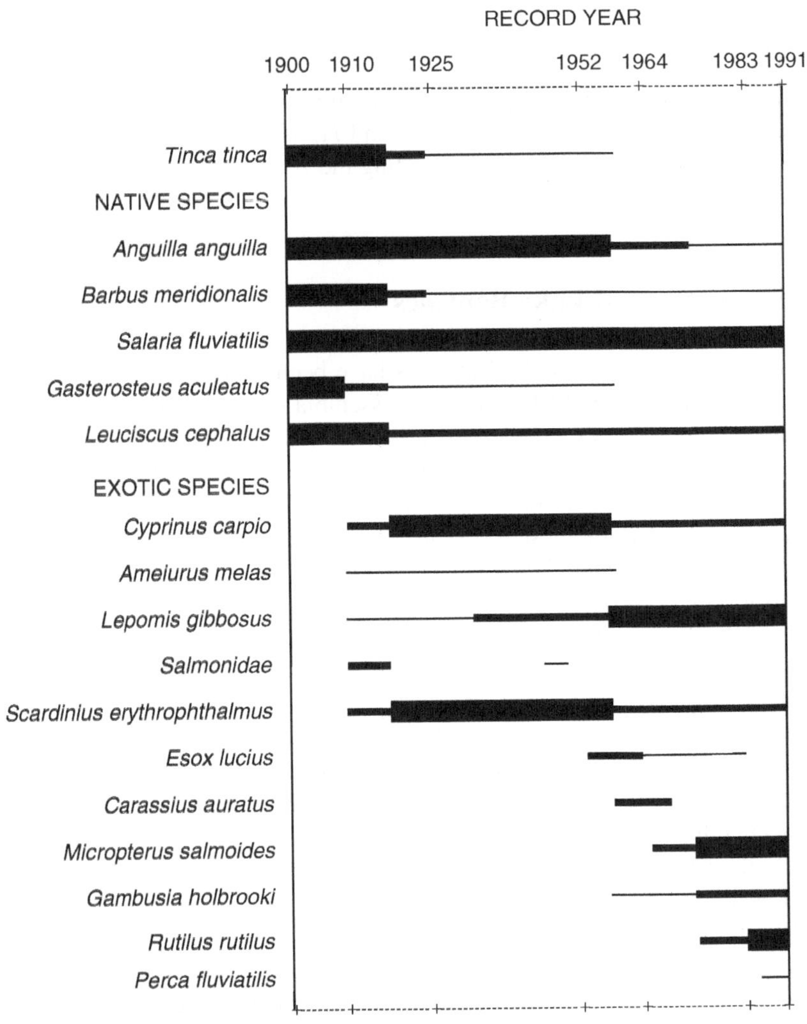

Figure 18.2 Population abundance of the fish assemblage of Lake Banyoles according to the literature comments and its present state. The thickness of the lines is proportional to the species abundance, classified as abundant, common or rare

Despite their precarious situation in Spain, only one species, the blenny, is protected by Spanish law and three of them are not protected by European law. These data suggest that the five fish species should, at least, be protected under Spanish law and, more importantly, effective mechanisms for their conservation should be implemented.

The freshwater populations of the three-spined stickleback are threatened in Europe in contrast to its brackish water counterparts (Lelek 1987). The loss of the population in Banyoles is of concern because it was probably one of the most important freshwater populations in the Iberian Peninsula, where both lakes and sticklebacks are rare (Doadrio, Elvira & Bernat 1991).

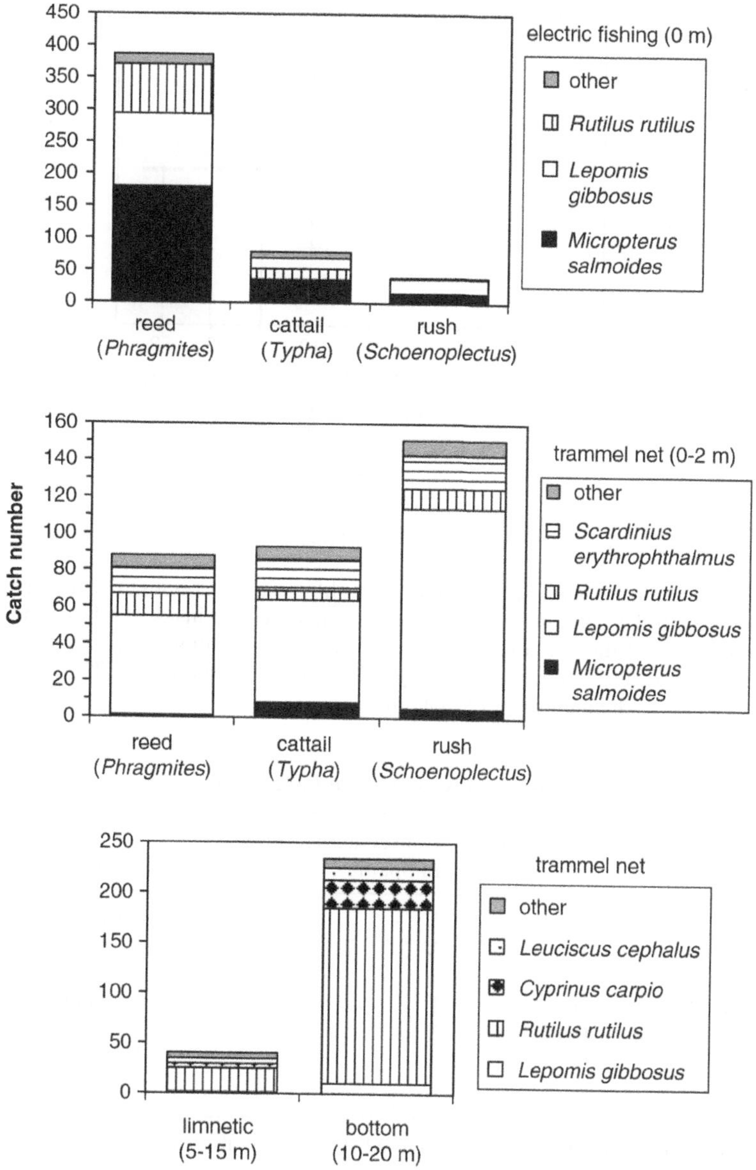

Figure 18.3 Catch number by species and habitat in Lake Banyoles. Catch number is the total number of individuals captured and is equivalent to catch per unit effort because constant sampling effort was applied

The freshwater blenny is the only native species still common in the lake (García-Berthou & Moreno-Amich 2000a), despite the presence of alien predators. This status is fortuitous and should be maintained because this species is among the few Spanish fish considered to be 'endangered' (Blanco & González 1992; Elvira 1995a).

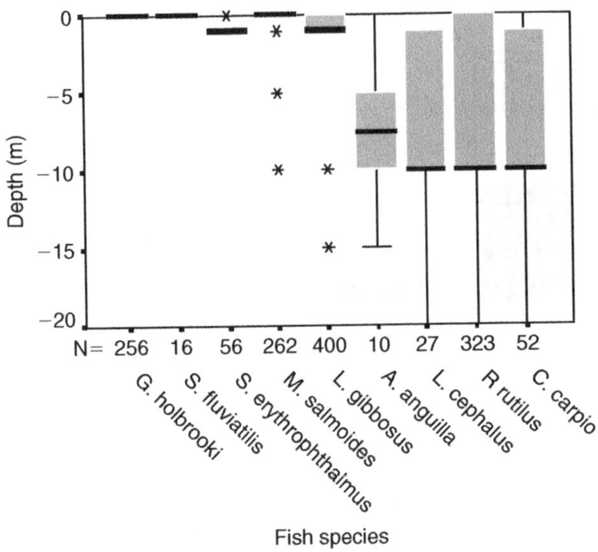

Figure 18.4 Box plots of the capture depth of the different fish species. The box (grey rectangle) corresponds to the 25th and 75th percentiles, the dark line inside the box represents the median depth (50% of the individuals), error bars are the minimum and maximum except for outliers (asterisks corresponding to values beyond 1.5 boxes from the box). The figures (beside the species name) are the number of fish captured

Table 18.1 Conservation status of native species in Lake Banyoles

Species	Conservation in Europe[a,b]	Conservation in Spain[a,c]	Protection[c,d]	Local abundance
Anguilla anguilla	I–V	V	No	Very rare
Barbus meridionalis	V	V	EU	Very rare in the lake, common in tributaries
Salaria fluviatilis	O	E	EU, E	Common
Gasterosteus aculeatus	I–R–V	V	No	Extinct?
Leuciscus cephalus	O	V	No	Very rare in the lake, common in tributaries

[a]IUCN conservation categories: E = endangered, V = vulnerable, R = rare, I = intermediate, O = out of danger.
[b]*Source*: Lelek (1987).
[c]*Sources*: Blanco & González (1992), Elvira (1995a).
[d]EU = protected by European Union laws, E = protected by Spanish laws (excluding regional laws).

The abundance of the eel has diminished throughout Europe during the last 20 years. The falling recruitment of eel is due to a number of factors, including changes in oceanic currents and temperatures, overfishing, migration barriers, pollution and introduction of parasites (Bruslé 1989; Knights, White & Naismith 1996). The decline in eel in Lake Banyoles, however, is because of pollution outflow in the River Terri, which prevents migration. Ongoing restoration of the Terri River should allow reproductive migration of adult eels and arrival of juveniles.

18.5 Management recommendations

The situation in Lake Banyoles is similar to many other freshwaters in Spain. The percentage of alien species ranges from 29% to 59% of the total number of fish species in the different Spanish river basins (Elvira 1995b). As a consequence, Elvira (1995b) pointed out that knowledge of the disruption caused by alien species was insufficient and preservation of native fish fauna ought to be a priority for the Spanish and European authorities. Although some important initiatives have been recently undertaken for the most endangered Iberian endemic species, such as *Valencia hispanica* Val. (Planelles-Gomis 1999) and *Anaecypris hispanica* (Steindachner) (Collares-Pereira, Cowx, Rodrigues, Rogado & Moreira da Costa 1999), the general situation is precarious and may be summarised as follows.

There is no management or effective control by the Spanish authorities for the introduction and translocation of fish, nor regular monitoring of most freshwater fish populations. As in Lake Banyoles, most Spanish freshwater fisheries are recreational and most alien fishes are introduced and translocated by anglers, in frequent uncontrolled, but passively tolerated, attempts. It is therefore suggested that the Spanish authorities should implement the following management actions.

- Implement an educational programme directed at the general public and specifically anglers to change their unawareness about the potential ecological impact of introducing alien species.
- Regulate and restrict the present common practice by anglers of transporting and stocking fish from one river basin to another.
- Control the introduction of alien fish species into the Iberian Peninsula and carefully evaluate new introduction proposals following existing codes of practice (Kohler & Stanley 1984; Turner 1988; Li & Moyle 1993; Coates 1998).
- Establish long-term monitoring programmes of selected river basins to assess the size and evolution of native and alien fish populations (e.g. Maitland & Lyle 1992).

In summary, fisheries management should be performed or supervised by the authorities and should be based in scientific evidence and criteria (see e.g. Kohler & Hubert 1993; Cowx 1994; 1996; 1998), which is currently not the case.

Acknowledgements

We would like to thank everybody who assisted in field and laboratory work, especially Q. Paredes and D. Boix. L. Zamora provided the figure of the bathymetric map. This study was financially supported by the Town Council of Banyoles (grant to RMA) and the Autonomous Government of Catalonia (FI fellowship and CIRIT grant AR89 to EGB).

References

Blanco J.C. & González J.L. (eds) (1992) *Libro rojo de los vertebrados de España*. Madrid: ICONA, 714 pp. (in Spanish).

Bolòs O. de & Masalles R.M. (1983) Mapa de la vegetació de Catalunya. Escala 1 : 50000. Memòria del full núm. 33. Banyoles. Barcelona: Generalitat de Catalunya. (in Spanish).

Bruslé J. (1989) L'anguille européenne (*Anguilla anguilla*), une espèce jugée commune jusqu'à la dernière décénnie, mérite-t-elle d'acquérir aujourd'hui le statut d'espèce menacée? *Bulletin de la Société Zoologique de France* **114**, 61–73 (in French).

Casamitjana X. & Roget E. (1993) Resuspension of sediment by focused groundwater in Lake Banyoles. *Limnology and Oceanography* **38**, 643–656.

Coates D. (1998) Codes of practice for the stocking and introduction of fish. In I.G. Cowx (ed.) *Stocking and Introduction of Fish*. Oxford: Fishing News Books, Blackwell Scientific Publications, pp. 383–396.

Collares-Pereira M.J., Cowx I.G., Rodrigues J.A., Rogado L. & Moreira da Costa L. (1999) The status of Anaecypris hispanica in Portugal: problems of conserving a highly endangered Iberian fish. *Biological Conservation* **88**, 207–212.

Cowx I.G. (1994) Stocking strategies. *Fisheries Management and Ecology* **1**, 15–31.

Cowx I.G. (1996) The integration of fish stock assessment into fisheries management. In I.G. Cowx (ed.) *Stock Aassessment in Inland Fisheries*. Oxford: Fishing News Books, Blackwell Scientific Publications, pp. 495–506.

Cowx I.G. (1997) L'introduction d'espèces éstrangères de poissons dans les eaux douces européennes: succès économiques ou désastre écologique? *Bulletin Français de la Pêche et de la Pisciculture* **344/345**, 57–78 (in French).

Cowx I.G. (1998) Stocking strategies: issues and options for future enhancement programmes. In I.G. Cowx (ed.) *Stocking and Introduction of Fish*. Oxford: Fishing News Books, Blackwell Scientific Publications, pp. 3–13.

Doadrio I., Elvira B. & Bernat Y. (eds) (1991) *Peces Continentales Españoles. Inventario y Clasificación de Zonas Fluviales*. Madrid: ICONA, 221 pp. (in Spanish).

Elvira B. (1995a) Conservation status of endemic freshwater fish in Spain. *Biological Conservation* **72**, 129–136.

Elvira B. (1995b) Native and exotic freshwater fishes in Spanish river basins. *Freshwater Biology* **33**, 103–108.

García-Berthou E. (1994) *Ecologia alimentària de la comunitat de peixos de l'Estany de Banyoles*. PhD dissertation, University of Girona, Girona, Spain, 288 pp. (in Spanish).

García-Berthou E. (1999a) Spatial heterogeneity in roach (*Rutilus rutilus*) diet among contrasting basins within a lake. *Archiv für Hydrobiologie* **146**, 239–256.

García-Berthou E. (1999b) Food of introduced mosquitofish: ontogenetic diet shift and prey selection. *Journal of Fish Biology* **55**, 135–147.

García-Berthou E. & Moreno-Amich R. (2000a) Introduction of exotic fish into a Mediterranean lake over a 90-year period. *Archiv für Hydrobiologie* **149**, 271–284.

García-Berthou E. & Moreno-Amich R. (2000b) Food of introduced pumpkinseed sunfish: ontogenetic diet shift and seasonal variation. *Journal of Fish Biology* **57**, 29–40.

García-Berthou E. & Moreno-Amich R. (2000c) Rudd (*Scardinius erythrophthalmus*) introduced to the Iberian peninsula: feeding ecology in Lake Banyoles. *Hydrobiologia* **436**, 159–164.

Garcia-Gil L.J., Casamitjana X. & Abella C.A. (1996) Comparative study of two meromictic basins of Lake Banyoles (Spain) with sulphur phototrophic bacteria. *Hydrobiologia* **319**, 203–211.

Knights B., White E. & Naismith I.A. (1996) Stock assessment of European eel *Anguilla anguilla* L. In I.G. Cowx (ed.) *Stock Assessment in Inland Fisheries*. Oxford: Fishing News Books, Blackwell Scientific Publications, pp. 431–447.

Kohler C.C. & Hubert W.A. (eds) (1993) *Inland Fisheries Management in North America.* Bethesda, Maryland: American Fisheries Society, 594 pp.

Kohler C.C. & Stanley J.G. (1984) A suggested protocol for evaluation of proposed exotic fish introductions in the United States. In W.R. Courtenay Jr & J.R. Stauffer Jr (eds) *Distribution, Biology and Management of Exotic Fishes.* Baltimore, Maryland: John Hopkins University Press, pp. 387–406.

Lelek A. (1987) *The Freshwater Fishes of Europe. Vol. 9. Threatened Fishes of Europe.* Wiesbaden: AULA-Verlag, 343 pp.

Li H.W. & Moyle P.B. (1993) Management of introduced fishes. In C.C. Kohler & W.A. Hubert (eds) *Inland Fisheries Management in North America.* Bethesda, Maryland: American Fisheries Society, pp. 287–307.

Maitland P.S. & Lyle A.A. (1992) Conservation of freshwater fish in the British Isles: proposals for management. *Aquatic Conservation: Marine and Freshwater Ecosystems* **2**, 165–183.

Miracle M.R. (1976) Distribución en el espacio y en el tiempo de las especies del zooplancton del lago de Banyoles. *Monografías ICONA* 5, 270 pp. (in Spanish).

Moreno-Amich R. & García-Berthou E. (1989) A new bathymetric map based on echo-sounding and morphometrical characterization of the Lake of Banyoles (NE-Spain). *Hydrobiologia* **185**, 83–90.

Planas M.D. (1973) Composición, ciclo y productividad del fitoplancton del lago de Banyoles. *Oecologia aquatica* **1**, 1–106 (in Spanish).

Planelles-Gomis M. (ed.) (1999) *Peces Ciprinodóntidos Ibéricos. Fartet y Samaruc.* Generalitat Valenciana, València, 357 pp. (in Spanish).

Prat N. & Rieradevall M. (1995) Life cycle and production of Chironomidae (Diptera) from Lake Banyoles (NE Spain). *Freshwater Biology* **33**, 511–524.

Turner G.E. (1988) Codes of practice and manual of procedures for consideration of introductions and transfers of marine and freshwater organisms. *FAO, EIFAC/CECPI Occasional Paper* 23, 44 pp.

Welcomme R.L. (1988) International introductions of inland aquatic species. *FAO Fisheries Technical Paper* 294, 318 pp.

Chapter 19
Conservation of the endangered whitefish, *Coregonus lavaretus*, population of Haweswater, UK

I.J. WINFIELD*, J.M. FLETCHER and D.K. WINFIELD
CEH Windermere, The Ferry House, Ambleside, Cumbria, UK

Abstract

Within the UK, the whitefish, *Coregonus lavaretus* (L.), occurs in just seven lakes, and is accordingly protected under nature conservation legislation. The status of one population in Haweswater, a reservoir in the English Lake District, has declined since the 1960s and previous studies have linked this deterioration with increased variations in lake level. In 1992, an additional potential threat developed with the establishment of a breeding colony of cormorants, *Phalacrocorax carbo* L., at the lake. Here, the status of the whitefish population through the 1990s, as monitored by the examination of entrapped specimens and echo sounding, is related to more recent data on lake levels. A successful spawning ground improvement feasibility study and assessment of the potential impact by cormorants are also presented. Finally, resulting conservation measures relating to lake levels, the cormorant population, and the establishment of refuge populations are described.

Keywords: conservation, cormorant, *Coregonus lavaretus*, Haweswater, lake level, reservoir, whitefish.

19.1 Introduction

The whitefish, *Coregonus lavaretus* (L.), occurs in just seven UK lakes and is accordingly protected under Schedule 5 of the Wildlife and Countryside Act 1981. The present investigation is concerned with the population of Haweswater, a large reservoir in Cumbria, north-west England. Before the 1990s, the only publications concerned with the population biology of this population were Bagenal (1970) and Maitland (1985) describing specimens from the mid-1960s and early-1980s, respectively.

In the early-1990s, an extensive gill-netting survey of all whitefish populations in England and Wales revealed the Haweswater population to be of poor status, with individuals ranging from 160 to 350 mm in length and from 2 to 13 years in age, and exhibiting inconsistent recruitment (Winfield, Cragg-Hine, Fletcher & Cubby 1996). Moreover, subsequent comparisons of whitefish entrapped between 1992 and 1994 with those examined by Bagenal (1970), who described the population biology of 188

*Correspondence: CEH Windermere, The Ferry House, Far Sawrey, Ambleside, Cumbria LA22 0LP, UK (email: ijw@ceh.ac.uk).

whitefish entrapped in the water abstraction system between 1965 and 1967 for which length ranged from about 70 to 380 mm and age from 1 to 9 years, showed that there had been a decline in age class equitability and individual growth, which can be associated with an increase in lake level variability (Winfield, Fletcher & Cubby 1998a). An analysis of the numbers of whitefish entrapped from 1973 to 1994 (Winfield *et al.* 1998a) revealed a marked decrease in the early 1980s (see also Fig. 19.1(a)) and it is notable that Maitland (1985) found only individuals between 230 and 350 mm in length among 14 un-aged whitefish entrapped in 1983. In 1992, an additional potential threat developed with the establishment of a breeding colony of piscivorous cormorants, *Phalacrocorax carbo* L., at the lake.

Here, the status of the whitefish population through the 1990s, including data from long-term entrapment records, is related to more recent data on lake levels. A successful spawning ground improvement feasibility study and assessment of the potential impact by cormorants are also described. Finally, resulting conservation measures relating to lake levels, the cormorant population, and the establishment of refuge populations are documented.

19.2 Study site

Situated in the north-west of England (National Grid Reference NY 480 140), Haweswater lies at an altitude of 240 m, with a surface area of 391 ha and a maximum depth of 57 m. The present surface area and maximum depth were increased by 252 ha and 29 m, respectively, over those of the original lake by the completion of a dam in 1941 which facilitates the lake's current operation as a potable water reservoir by North West Water (NWW). In addition to whitefish, the fish community of this oligotrophic lake comprises Arctic charr, *Salvelinus alpinus* (L.), brown trout, *Salmo trutta* L., eel, *Anguilla anguilla* (L.), perch, *Perca fluviatilis* L., minnow, *Phoxinus phoxinus* (L.), and three-spined stickleback, *Gasterosteus aculeatus* L. (Winfield *et al.* 1996). The only fishery activity is very limited recreational angling for trout.

19.3 Status through the 1990s

Between 1992 and 1994, whitefish entrapped in the water abstraction system of Haweswater ranged from 225 to 350 mm in fork length and from 2 to 11 years old (Winfield *et al.* 1998a). These age data, together with age frequency distributions of entrapped whitefish from 1995 to 1999 are shown in Fig. 19.1(b). In addition to reinforcing the decline in the numbers of whitefish entrapped during the 1990s (as shown in Fig. 19.1(a)), these data also revealed the rarity of old and young whitefish since 1994. The only exception to this failure to recruit was the relatively strong 1995 year class sampled as 2-year-old individuals in 1997. However, these fish have not persisted in subsequent years.

An extensive netting survey carried out on 16 May 1996, when five bottom-set survey gill nets of a multifilament design (each measuring approximately 1.5 m deep and 60 m

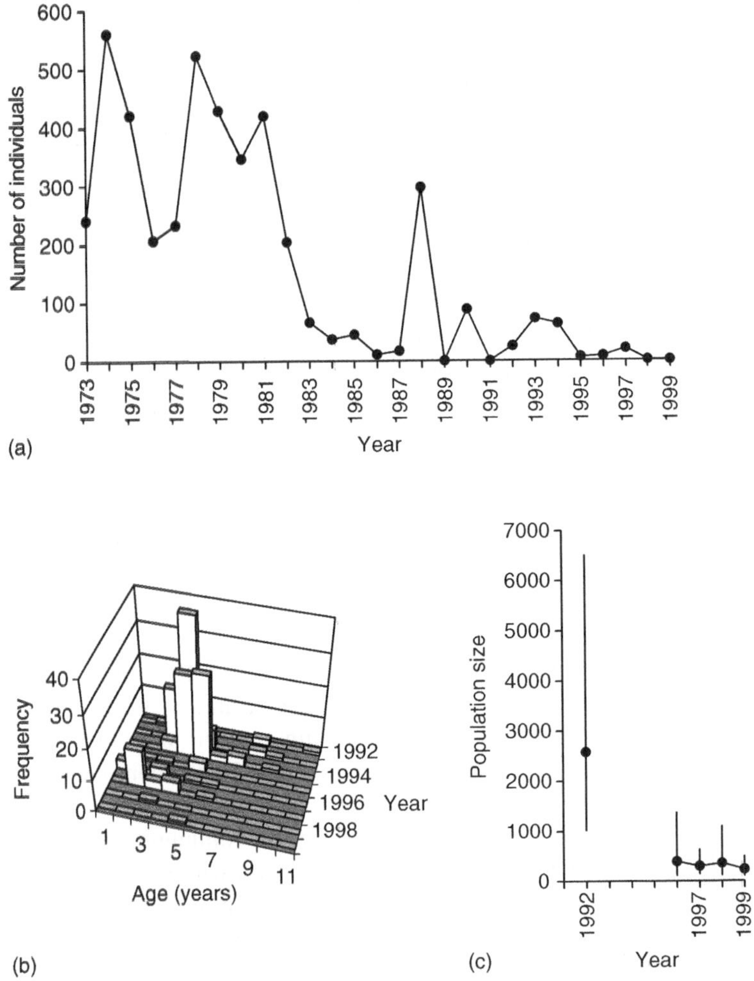

Figure 19.1 (a) Numbers of whitefish entrapped from 1973 to 1999, (b) age frequency distributions of entrapped whitefish from 1992 to 1999, and (c) estimated population sizes (geometric means with 95% confidence limits) of adult whitefish in May 1992, 1996, 1997, 1998 and 1999

long with 10 panels of equal length of bar mesh sizes 8, 10, 12, 16, 22, 25, 30, 33, 38 and 43 mm) were set for *ca.* 24 h, caught just two whitefish (110 and 296 mm in length, 1 and 7 years in age, of year classes 1995 and 1989, respectively) from a site near the lake's deepest point. Note that one of the two individuals was a member of the 1995 year class which had also appeared relatively strongly in the entrapment samples. Growth achieved by these two fish was relatively slow when compared with individuals from the 1960s (see Bagenal 1970). Clearly, the abundance of whitefish in Haweswater is very low and it is unlikely that survey gill netting will produce significant sample sizes unless, and until, the population recovers. Consequently, further netting surveys are not recommended and the population is now monitored by entrapment analysis and echo sounding.

Figure 19.1(a) shows the annual numbers of whitefish entrapped from 1973 to 1999, although biological data are unavailable from 1973 to 1991. Entrapment levels have remained low through the 1990s, even though the volume of water abstracted has not decreased (Winfield *et al.* 1998a). Although entrapment data have considerable sampling and other biases (Maitland 1985), they do provide the only, and thus invaluable, long-term record of whitefish relative abundance in Haweswater.

19.4 Monitoring by echo sounding through the 1990s

In addition to continuing to record the number of individuals entrapped and examining their biological characteristics, the whitefish population of Haweswater is now also monitored by quantitative echo sounding. Following isolated surveys in May 1992 and May 1996, since 1997 the fish community of Haweswater has been monitored in May, July and September by echo-sounding surveys using a Simrad EY200P portable echo sounder with a 200 kHz single-beam transducer, followed by data analysis using HADAS (for full details see Winfield *et al.* 1998a).

At Haweswater, echoes from fish above 250 mm in length can be assumed to be almost exclusively adult whitefish (see Winfield *et al.* 1998a), and data from May 1992 and 1996 to 1999 are given in Fig. 19.1(c) as estimated population sizes of adult whitefish. Mean population estimates of adult whitefish have varied significantly (ANOVA, $F_{4,45} = 5.390$, $P < 0.01$), with *post-hoc* comparisons by the Tukey HSD test revealing that the 1992 densities were significantly higher ($P < 0.05$) than equivalent figures from all other years, with no other significant differences present ($P > 0.10$). In addition to this decrease in abundance during the 1990s, these data are also alarming because of the extremely small geometric mean estimate of 2574 adults in 1992 falling to between 225 and 383 adults in the late-1990s. In terms of corresponding population densities, the observed range for adults in Haweswater between May 1996 and September 1999 from 1 to 8 individuals ha^{-1} may be compared with, for example, a range of 10–70 adult whitefish ha^{-1} found in Lake Osensjoeen, Norway (Linloekken 1995). Thus the population density of adult whitefish in Haweswater in the late-1990s was very low when compared with populations elsewhere in Europe.

19.5 Lake levels in the 1980s and 1990s

A detailed analysis of daily lake levels from 1 January 1961 to 8 June 1996, including the specific periods of January to March, when whitefish eggs are spawned and incubated inshore, and June to October, when whitefish young and adults in unregulated lakes use the inshore as a feeding area, was given by Winfield *et al.* (1998a). Both periods of the year showed significant increases in the range of lake levels and Fig. 19.2(a) shows the typical pattern observed for many years, in which the lake level fell during the incubation seasons of 1985–1987 to the extent that eggs are likely to have become

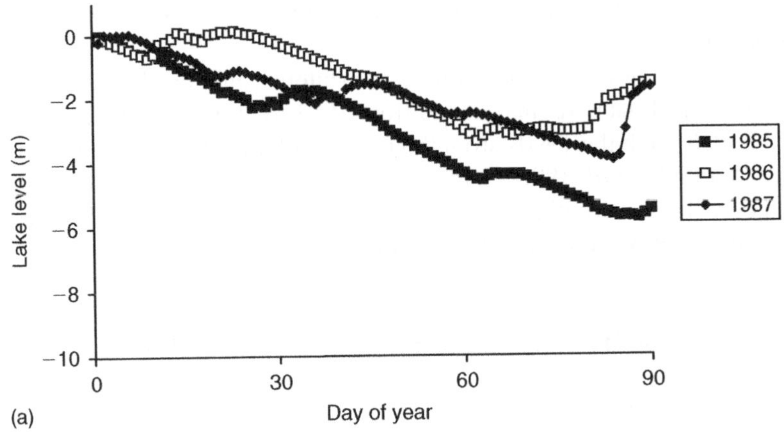

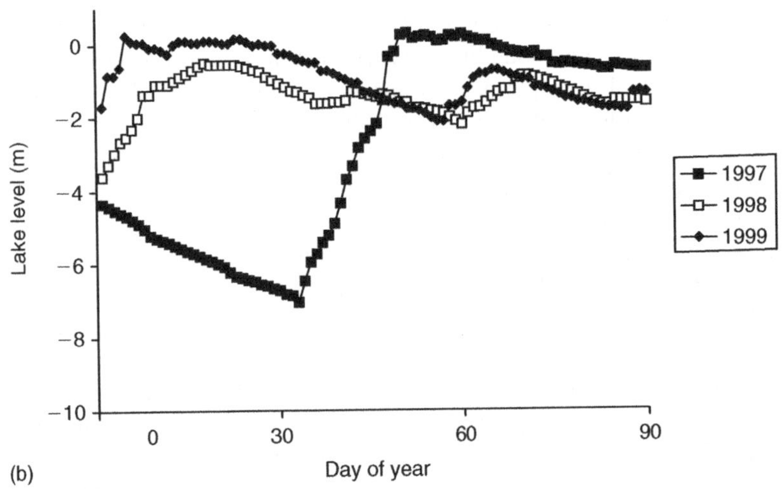

Figure 19.2 (a) Lake levels during the first 90 days of the year from 1985 to 1987, and (b) lake levels during the first 90 days of the year from 1997 to 1999. Whitefish spawning probably occurs largely within days 20–40

exposed. Such effects have the potential to kill entire year classes of whitefish, and thus be responsible for the inconsistent recruitment of the 1980s noted by Winfield *et al.* (1996). Lake levels in the late-1990s (Fig. 19.2(b)) are considered below.

19.6 Spawning ground improvement feasibility study

The present poor status of whitefish probably arises from the extreme lake level fluctuations of Haweswater during the 1970s to mid-1990s, particularly those during the egg incubation period (Winfield *et al.* 1998a). Such impacts of reservoir operations on whitefish populations have been reported for numerous other lakes in Europe, for example, Lake Kemijarvi in Finland (Heikinheimo-Schmid & Huusko 1988). However, the main concern of such studies has been decreased commercial catches. Rather than directly addressing the environmental damage, catches have typically been increased by stocking (Anon. 1994). Although this approach may increase catches, it is also likely to have effects on the genetic composition of the impacted population which are highly undesirable when biodiversity considerations are paramount.

An alternative approach at Haweswater, given that the lake level regime is now operated as sympathetically as possible (see below), is to attempt to ameliorate the environmental effects of the drawdown. One promising option is the use of artificial spawning substrata, such as Astroturf or a similar plastic grass as an artificial 'gravel', which can be moved further into the lake basin in response to falling water levels, thus keeping incubating whitefish eggs submerged. Although such materials have been used for other fish species in streams (e.g. Gray & Cameron 1987), despite extensive searches the authors are unaware of any such work specific to whitefish or any other *Coregonus* species in lakes.

Consequently, a pilot study was undertaken using an artificial grass (Model RA1) made by Nordon Enterprises Limited (Altham, UK). On 30 January 1997 when the lake was 6.65 m below full, five weighted mats of artificial grass each measuring 1.8 by 4.3 m were deployed on the bottom of Haweswater at a depth of 3–4 m on the east shore at the only known spawning ground of whitefish.

The mats were subsequently inspected using an underwater video camera deployed from a boat on 4 February 1997, by which time whitefish eggs were observed on them. Strong winds and storms precluded any observations during the subsequent few weeks, resulting in the next inspection being made on 12 March 1997 when whitefish eggs were again observed.

However, although incubation on the grass mats was apparently successful, unusually high rainfall during the above storms resulted in the lake level increasing rather than decreasing over the incubation period, such that the lake reached overflow by 19 February 1997. Subsequently, Haweswater remained full or near full during the remainder of the incubation period (see Fig. 19.2(b)). Although this precluded a full testing of the spawning ground improvement system, the resulting lake level regime of 1997 was good for natural whitefish incubation. Advantage was later taken of low lake levels of 8.54 m below full on 17 September 1997 to make a visual survey of the entire exposed shore of the lake to identify additional potential spawning grounds and suitable places for future deployments of full scale systems, which will be installed by NWW if the need arises. A total of 21 such locations was identified.

19.7 Impact of cormorants

During the May 1996 echo-sounding survey, 26 cormorant nests were observed on the only island of Haweswater, which was thought to represent a considerable increase in the local breeding population size of this piscivorous bird during recent years. This suspicion was confirmed by more accurate data kindly made accessible by the Royal

Society for the Protection of Birds (RSPB). Given the concerns currently being expressed over the potential effects of increased cormorant populations on freshwater fish populations elsewhere in the UK, this development at Haweswater warranted further investigation in the context of its potential impact on the whitefish population. Consequently, a project was undertaken to determine the numbers and local feeding behaviour of cormorants from November 1996 to December 1997 (Winfield, Winfield & Fletcher 1998b).

Minimum and maximum counts of cormorants recorded at approximately 2-week intervals over this period (Fig. 19.3(a)), which show that although wintering numbers were typically less than 10 individuals, numbers increased with the arrival of nesting birds in April and peaked in July at 84 individuals, including young. Numbers subsequently decreased as young and adults dispersed from the lake.

Using data on the fish standing stock from the May 1997 echo-sounding survey, the above cormorant counts, and a daily consumption rate of 0.6 kg per cormorant per day, the total annual consumption by cormorants was expressed as a percentage of the fish community standing stock and ranged from 943% to 3867% for the minimum and maximum daily cormorant counts, respectively (Fig. 19.3(b)). It was surmised that such a local predation rate cannot be sustained by the Haweswater ecosystem. The solution to this apparent paradox is that although cormorants do feed at Haweswater, where their feeding efficiency is very low (Winfield *et al.* 1998b), they also make extensive movements off the lake to feed elsewhere (Fig. 19.3(c)). However, even the limited amount of feeding undertaken at Haweswater itself has the potential to impact markedly on the scarce whitefish population and so, adopting the precautionary principle, management of the cormorant population is warranted.

19.8 Resulting conservation measures

19.8.1 *Sympathetic lake level management*

Given the present extremely poor status of the Haweswater whitefish population and the role in its decline probably played by the lake level variations of this strategically-important reservoir, NWW has undertaken, as far as is practicable and within the constraints of its abstraction licences, to operate the reservoir to ensure that it is full at the end of March each year. A wider implementation of programmes to reduce pipeline leakage and promote the efficient use of water by customers also serve to reduce the need for level fluctuations at Haweswater, although an additional major factor is the local pattern of rainfall. Fortunately, these measures and the rainfall of recent winters from 1997 to 1999 have been such that no significant falls in levels have been experienced during the corresponding incubation periods (Fig. 19.2(b)).

19.8.2 *Management of cormorants*

Given the concerns over the potential impact of cormorants on the whitefish population, following a meeting in 1998 attended by a variety of organisations including

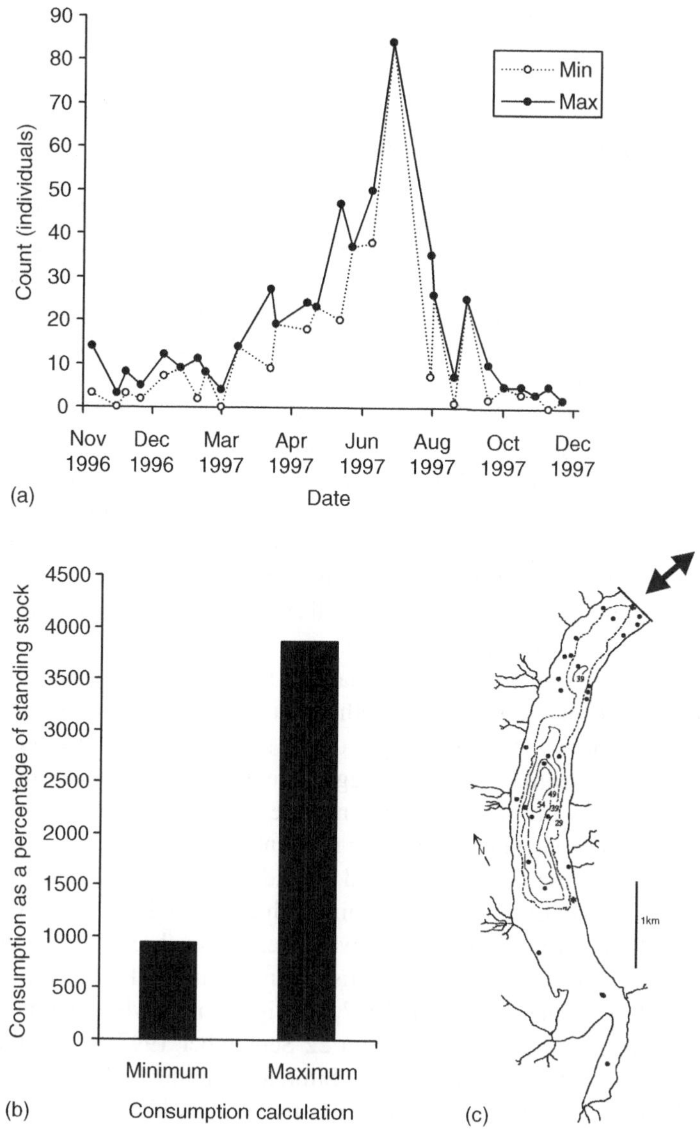

Figure 19.3 (a) Within-day minimum and maximum counts of cormorants between November 1996 and December 1997, (b) food consumption of cormorants in 1997 expressed as a percentage of the fish community standing stock, and (c) locations of cormorants (closed circles) observed feeding and their major departure and arrival point (double-headed arrow)

English Nature (EN), the Environment Agency and the RSPB, steps were taken by NWW to discourage cormorants from nesting at Haweswater in 1999. This was attempted by deploying a variety of scaring methods on the nesting island from April to July, although the only effective technique was found to be frequent disturbance by humans. In this way, nesting was successfully prevented, although adult cormorants still used the lake for roosting, loafing and some feeding. Nesting will again be prevented in 2000.

19.8.3 *Establishment of refuge populations*

Although true conservation of the Haweswater whitefish requires the solution of environmental problems at the lake itself, as a precautionary measure, steps have also been taken to establish two refuge populations. Following exploratory surveys and discussions with EN, the high altitude Blea Water (483 m, National Grid Reference NY 449 109) and Small Water (452 m, NY 454 100) within the Haweswater catchment were selected as recipient bodies.

On 5 February 1997, about 48 000 eggs and milt were obtained from five female (size range, 295–341 mm, age range, 4–9 years) and 21 male (size range, 285–317 mm, age range, 4–8 years) whitefish gill netted on the only known spawning ground of Haweswater. Although these numbers of donor fish were relatively small, they should be considered against the May 1996 estimate of an adult population of only 383 individuals (Winfield *et al.* 1998a). Stripping was performed with eggs from each female being kept separate until after they had been fertilised by milt from three to five males. Males were not reused in subsequent fertilisations.

Because there was insufficient daylight remaining to allow safe transport of the fertilised eggs, they were kept overnight at a depth of about 0.7 m in Haweswater as equal volumes in two containers lined with a layer of the synthetic grass described above. The next morning, i.e. 6 February 1997, the eggs were transported in water in stainless steel vacuum flasks on foot to the two recipient water bodies, where they were introduced to six containers in each lake lined with the synthetic grass.

The containers were subsequently inspected by underwater video camera and some temporary egg retrieval six times before hatching, which began in early-May and was complete by mid-June. Very high egg survival rates were observed at both lakes. Subsequent visual searches for young whitefish were unsuccessful at both lakes on several occasions through the summer of 1997, but eight young whitefish in excellent condition were sampled from Small Water on 22 September 1997 in a 10 mm mesh size monofilament gill net of a gang of nets. Similar sampling at Blea Water on 23 September 1997 did not result in the capture of any young whitefish, although sampling effort was relatively smaller in this considerably larger water body and so such fish are not necessarily absent.

19.9 Concluding remarks

The UK whitefish populations are of great national and international conservation value, in part because their gene pools have not been subjected to the effects of translocations, stockings or intense fisheries as is common elsewhere in Europe (see Beaumont, Bray, Murphy & Winfield 1995). Nevertheless, Haweswater is a strategically-important supply of potable water for north-west England and the conservation management of its whitefish population must operate within this context. The most damaging aspects of lake level variations are now minimised as far as possible, with an additional amelioration system in reserve, and significant progress has been made towards establishing refuge populations. Assuming that the population of adult whitefish has remained above the minimum viable level over the last decade, then, if impacts from cormorants are controlled to sustainable levels, the recent improvement in incubation conditions should lead to a major future recruitment to the spawning stock.

Acknowledgements

We thank Craig Denny, Susannah Moss and colleagues at North West Water Limited for their provision of entrapment specimens and lake level data. We are also grateful to our diving colleagues David Abel, Roland Fleck, Paul Hodgson and Ben James for their assistance with the installation, inspection and retrieval of the artificial spawning substratum. Trevor Furnass, Charles Paxton, Kate Tobin, Stephen Walker and Millie Winfield also provided welcome support in the field, while Bill Kenmir of the Royal Society for the Protection of Birds kindly made available unpublished data and background information on cormorants. This work was funded by North West Water Limited and the Environment Agency.

References

Anon. (1994) Guidelines for stocking coregonids. *EIFAC Occasional Paper No. 31*. Rome: FAO, 18 pp.

Bagenal T.B. (1970) Notes on the biology of the schelly *Coregonus lavaretus* (L.) in Haweswater and Ullswater. *Journal of Fish Biology* **2**, 137–154.

Beaumont A.R., Bray J., Murphy J.M. & Winfield I.J. (1995) Genetics of whitefish and vendace in England and Wales. *Journal of Fish Biology* **46**, 880–890.

Gray R.W. & Cameron J.D. (1987) A deep-substrate streamside incubation box for Atlantic salmon eggs. *Progressive Fish Culturist* **49**, 124–129.

Heikinheimo-Schmid O. & Huusko A. (1988) Management of coregonids in the heavily modified Lake Kemijaervi, northern Finland. *Finnish Fisheries Research* **9**, 435–445.

Linloekken A. (1995) Monitoring pelagic whitefish (*Coregonus lavaretus*) and vendace (*Coregonus albula*) in a hydroelectric reservoir using hydroacoustics. *Regulated Rivers: Research and Management* **10**, 315–328.

Maitland P.S. (1985) Monitoring Arctic charr using a water supply screening system. *Information Series International Society of Arctic Char Fanatics No. 3*, pp. 83–88.

Winfield I.J., Cragg-Hine D., Fletcher J.M. & Cubby P.R. (1996) The conservation ecology of *Coregonus albula* and *C. lavaretus* in England and Wales, U.K. In D. Hefti & A. Kirchhofer (eds) *Conservation of Endangered Freshwater Fish in Europe*. Basel: Birkhauser Verlag, pp. 213–223.

Winfield I.J., Fletcher J.M. & Cubby P.R. (1998a) The impact on the whitefish (*Coregonus lavaretus* (L.)) of reservoir operations at Haweswater, U.K. *Advances in Limnology* **50**, 185–195.

Winfield I.J., Winfield D.K. & Fletcher J.M. (1998b) The feeding behaviour of cormorants at Haweswater. Final Report. Unpublished Report by the Institute of Freshwater Ecology to North West Water Limited. WI/T11068J7/2, 46 pp.

Chapter 20
Fish community and habitat changes in the artificially stocked fishery of Lake Naivasha, Kenya

P. HICKLEY* and D. HARPER
Department of Biology, University of Leicester, Leicester LE1 7RH, UK

Abstract

Lake Naivasha is a freshwater lake with 'Ramsar' status lying in a chain of lakes situated in the eastern rift valley of Kenya. Only five species of fish are present, all of which have been introduced. Three species, largemouth bass, *Micropterus salmoides* Lacépède, *Oreochromis leucostictus* (Trewewas) and *Tilapia zillii* (Gervais), form the basis of an important gill net fishery and the bass is also taken by rod and line for sport. Also introduced and exploited commercially is *Procambarus clarkii* (Girard). Some of the mechanisms underlying the principal ecological changes that have been observed in recent times are described. Fish and crayfish catches are examined in relation to changes in water level and aquatic macrophyte abundance. Catches of *O. leucostictus* increased with rise in water level whereas those of *T. zillii* decreased. Bass appeared not to be influenced by water level but had a positive relation with crayfish catch. Submerged macrophytes disappeared between 1987 and 1999 and possible causes are considered, the hypothesis being that grazing by crayfish was responsible. Using the importance of food items in bass stomachs as an indicator of prey species availability, an inverse relationship was demonstrated for crayfish abundance against area of submerged water plant cover.

Keywords: aquatic macrophytes, commercial fisheries, introduced species.

20.1 Introduction

Lake Naivasha is a shallow (3–6 m deep), freshwater lake, approximately 160 km^2 in area, situated in the eastern rift valley of Kenya about 100 km north of Nairobi. It lies in a closed basin at an altitude of 1890 m above sea level, receives 90% of its water from the perennial River Malewa and is subject to considerable fluctuations in water level. Dominant vegetation types are marginal papyrus, *Cyperus papyrus* L., submerged *Najas horrida* Magnus, and floating *Salvinia molesta* Mitch. and *Eichhornia crassipes* (Mart.). The abundance of all these aquatic macrophytes is, however, in a continual state of flux. An overview of the lake and its changing ecology can be found in Harper, Mavuti & Muchiri (1990).

---

*Correspondence: National Coarse Fisheries Centre, Environment Agency, Hoo Farm Industrial Estate, Worcester Road, Kidderminster, DY11 7RA, UK (e-mail: phil.hickley@environment-agency.gov.uk).

When Lake Naivasha was first studied (*ca* 1900) only one species of fish was present, the endemic *Aplocheilichthys antinorii* (Vinc.) which was last recorded in 1962 (Elder, Garrod & Whitehead 1971). Since 1925 various fish introductions have been made, some successful and some not (Muchiri & Hickley 1991). Present today are *Oreochromis leucostictus* (Trewewas), *Tilapia zillii* (Gervais), the largemouth bass, *Micropterus salmoides* Lacépède, *Barbus amphigramma* Blgr. and the guppy, *Poecilia reticulata* Peters. Since 1959 the two tilapias and bass have formed an important fishery (Muchiri & Hickley 1991) with all three species being commercially exploited using gill nets, and bass also being taken by rod and line for sport.

The Louisiana red swamp crayfish, *Procambarus clarkii* (Girard), was introduced into Lake Naivasha in 1970 with a view to establishing a commercial fishery (Parker 1974). When Lowery & Mendes (1977) recorded its increasing numbers and distribution from 1974 to 1976 it was an important organism in the eastern basin, but since then has spread throughout the lake. Initially opened in 1975, the crayfish fishery at the outset produced several hundred metric tonnes annually for export, mainly to Europe, but since 1983 much lower catches have supplied only the local, mostly tourist, market (Harper *et al.* 1990).

Hickley, Bailey, Harper, Kundu, Muchiri, North & Taylor (in press) examined the status and future of the Lake Naivasha fishery. They:

(1) identified three phases in its development – an initial 'boom and bust' (1959–1973), a period of stability (1974–1988) and, currently, a poorly performing fishery with low reported catches;
(2) concluded that the annual yield could be enhanced – a prospect based upon the mean theoretical yield (about 900 t year^{-1}) being considerably higher than the overall maximum sustainable yield (about 650 t year^{-1}) as computed from catch per unit effort (CPUE) figures; and
(3) proposed that the introduction of additional, appropriate fish species should be seriously considered, albeit only in association with improved enforcement and stock conservation regimes.

This chapter describes some of the mechanisms underlying the principal ecological changes that have been observed in recent times (since 1987) by Harper, Adams & Mavuti (1995) and Hickley *et al.* (in press). Various key inter-relationships are considered, in particular those of fish catches and the influence of water level, largemouth bass catches compared with those of crayfish and, finally, the impact of crayfish density upon submerged macrophytes. The supposed improvement in understanding of these components of the Lake Naivasha ecosystem are used to draft selected recommendations for the better management of the aquatic resources.

20.2 Methods

Catch data were obtained from the Fisheries Department of the Kenya Government. Finfish are taken with multifilament gill nets (>50 mm bar mesh) set from canoes and the crayfish are caught with traps. All fish must be landed at approved locations where

Fisheries Department personnel record the catches. Data collection commenced in 1963. From 1987 annual sampling of the fish population was carried out using survey gill nets. The CPUE for the commercial fishery was calculated as kg per canoe. The CPUE measure for the survey nets was the catch converted to numbers of fish per standard gill net per hour, where a standard gill net comprised five 30-m sections, being one each of 11, 15, 20, 24 and 35 mm bar mesh.

Bass stomach contents were analysed as described by Hickley, North, Muchiri & Harper (1994), i.e. the number of guts in which each food item occurred was recorded and expressed as % Occurrence, and the % Abundance of different foods in individual guts was calculated as the count of a food item expressed as a percentage of the total count for all food items. To assess the relative importance of each dietary item consumed, a prominence value (PV) (Wilhm 1967 as adopted by Hickley & Bailey 1987) was calculated from the product of its % Abundance and the square root of its % Occurrence.

20.3 Results

20.3.1 *Fish catches*

It appears that annual catches of *O. leucostictus* follow a similar trend to the fluctuations in water level, with increases in catch associated with an increase in water level (Fig. 20.1). For the bass, *T. zillii* and crayfish such subjective appraisal is more difficult. Annual reported catches in the second half of the period were very low, the exception being the 1998–1999 increase following the 1997–1998 rise in water level.

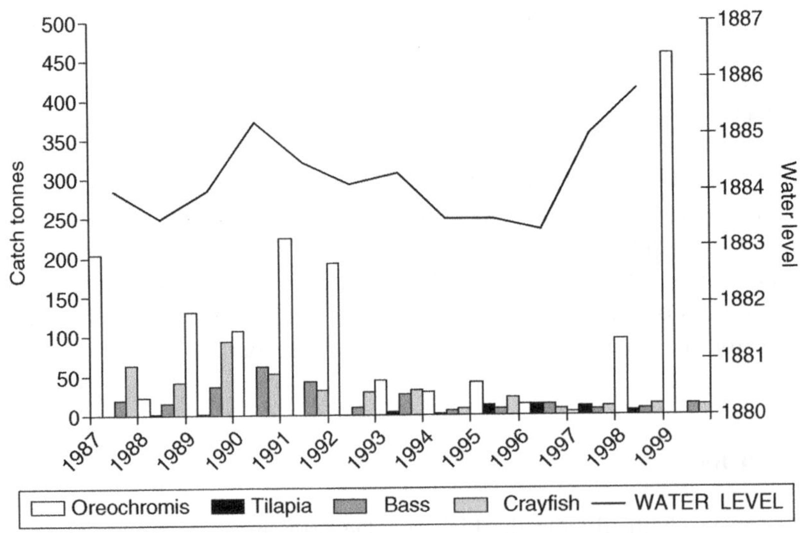

Figure 20.1 Annual fish catches taken by commercial gill nets, annual crayfish catches taken by commercial traps and mean annual water level since 1987

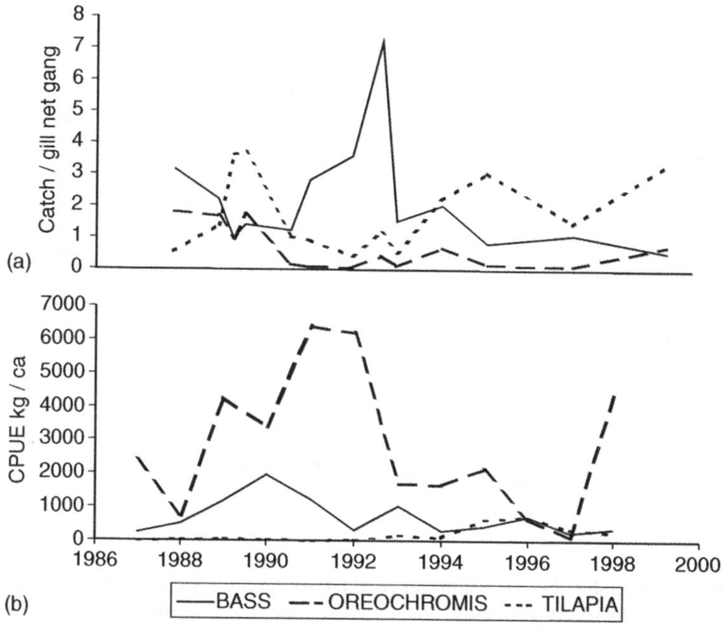

Figure 20.2 Annual CPUE for: (a) survey gill nets, expressed as number of fish per standard gill net gang per h; (b) commercial gill nets, expressed as kg total annual catch per canoe

The overall species composition differed noticeably between the commercial nets, which took adult fish, and the survey nets, which caught juveniles (Fig. 20.2). In the commercial catches *O. leucostictus* averaged 67% of the catch, bass 22.2% and *T. zillii* 10.2%. In contrast, survey catch composition was *O. leucostictus* 12.4%, bass 46.5%, *T. zillii* 38.7% and *B. amphigramma* 2.4%. Notwithstanding these percentages being gravimetric for the commercial nets and numeric for the survey nets, the differences in catch composition probably reflect capture efficiency of the two types of gear relative to the target species. Nevertheless, there is a tendency for the tilapia and bass catches to deviate in opposite directions to each other, or at least not to follow the same pattern against time (Fig. 20.2).

20.3.2 *Fish catch and water level*

Figure 20.3 shows catches against lake water level for all species combined, *O. leucostictus*, *T. zillii* and bass, respectively. Water level was plotted as the average for the calendar year preceding that during which the catches were made, so as to guarantee that all catches used in determining the mean value were made subsequent to, rather than prior to, the water level measurements. Fish catch for all species combined increased with increase in water level ($P < 0.01$). Split into the component species, the relationships with increasing water level were direct for *O. leucostictus* ($P < 0.01$), inverse for *T. zillii* ($P < 0.05$) and absent for largemouth bass. It is assumed that the

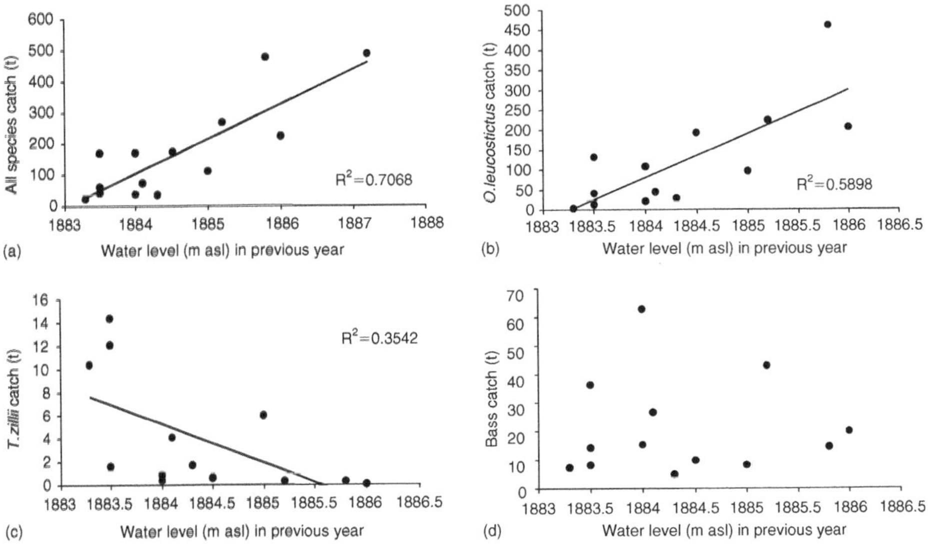

Figure 20.3 Relationship between annual mean fish catches for the period 1987–1999 and annual mean water level in previous year: (a) all species combined; (b) *O. leucostictus*, (c) *T. zillii*, (d) *M. salmoides*

predominance of *O. leucostictus* in the overall catch influenced the similarity of the respective regression lines.

20.3.3 *Bass and crayfish*

A simplified food web was presented by Muchiri, Hickley, Harper & North (1994) in which crayfish were placed both as the principal consumer of submerged macrophytes and also as the main prey of the bass. Hickley *et al.* (1994) described largemouth bass in Lake Naivasha as a generalised macro-predator. The most important invertebrate prey organisms for the juvenile bass (<260 mm fork length) were *Micronecta scutellaris* (Stal.) and dipteran pupae. For larger bass (>260 mm fork length), the crayfish, *P. clarkii*, was the preferred food.

The analysis of bass stomach contents ($n=1991$) showed noticeable changes from year to year in the prominence of the different food items (Table 20.1). For example, for large bass, *Xenopus* toads ranked first in 1987 compared with *P. clarkii* from 1990 onwards. For small bass, whilst *M. scutellaris* ranked first in most years, the status of some other invertebrates changed dramatically. A case in point was the appearance of conchostracans in the 1991/1992 samples.

Figure 20.4 shows annual catch of bass from the commercial gill nets plotted against annual catch of crayfish from the crayfish trap fishery. The resultant regression line ($P < 0.01$) suggests that, in keeping with the seeming importance of crayfish as bass prey, there could be a possible relationship between the bass and crayfish population sizes.

Table 20.1 The ranked importance of food items found in stomachs of *M. salmoides* based on PV for the years 1987–1998 inclusive for all habitats combined where the food item ranked fifth or higher in any year

Food item	Rank of food item in given year									
	1987	1988	1989	1990	1991	1992	1993	1994	1996	1998
Small bass										
M. scutellaris	1	1	1	1	1	1	1	3	1	2
Chironomid pupae	2	3	2	2	2	2	2	1	2	
P. clarkii	3		5	3			5	2	3	1
Cladocera	4		3	4	4	3	4			
Chironomid larvae		4	4	5	5	5	3		4	
Conchostraca					3	4				
Fish	5							4		5
Anisops		2						5		3
Trichopteran larvae		5							5	
Ostracoda										4
Large bass										
P. clarkii	2	2	2	1	1	1	1	1	1	1
M. scutellaris		1	1	2		2				3
Xenopus	1									
Fish	3	3		3			2	2		2
Chironomid pupae			3							
Frogs		4	5	4						
Chironomid larvae			4							
Coleoptera										4
Lepidoptera								3		
Laccocorris		5								

Section (a) is small bass, 60–260 mm fork length ($n = 1799$) and section (b) is large bass, 260–500 mm fork length ($n = 192$).

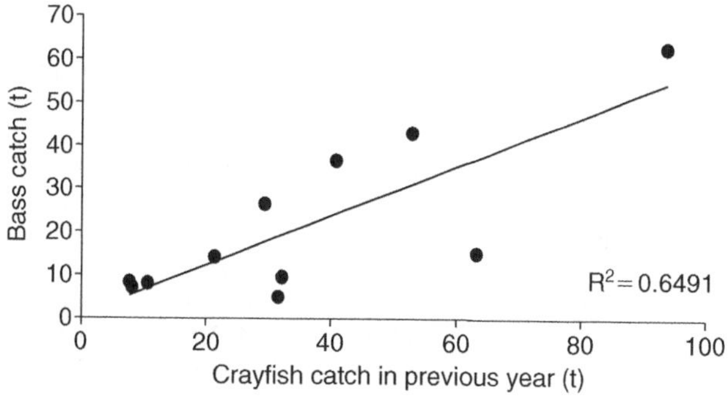

Figure 20.4 Relationship between annual catch of bass taken by commercial gill nets and annual catch of crayfish taken by commercial traps in the previous year

20.3.4 *Aquatic plants and crayfish*

Maps published by Gouder de Beauregard, Harper, Malaisse & Symoens (1998) showed the changes that have taken place in the submerged macrophyte beds of Lake Naivasha. Since 1987, the trend was one of reduction in overall coverage. Although suspecting such reduction could be due to grazing by crayfish, Harper *et al.* (1995) concluded that there was no clear coincidence of high crayfish catch returns and low plant cover, believing that the catch returns did not adequately reflect the crayfish population in the lake but rather the performance of the trap fishery.

The variation in importance of bass food types in different years (Table 20.1) suggests a similar opportunism of bass feeding habit in Lake Naivasha, as has been observed elsewhere (Hodgson & Kitchell 1987). Accordingly, it is feasible to adopt the PV of prey species in bass stomachs as an indicator of the abundance and availability of such species. Table 20.2 shows approximate areas of Lake Naivasha occupied by submerged aquatic macrophytes in different years (after Gouder de Beauregard *et al.* 1998; Harper *et al.* 1995) compared with the abundance of crayfish as represented by their PV in bass stomach contents. Submerged macrophyte abundance was significantly ($P < 0.001$) inversely correlated against the PV of crayfish (Fig. 20.5).

20.4 Discussion

20.4.1 *Fish catches*

Muchiri & Hickley (1991) attributed some of the observed fluctuations in fish catches to fishing pressure, changing lake levels and the loss of submerged macrophytes.

Table 20.2 Approximate areas of Lake Naivasha occupied by submerged aquatic macrophytes in different years (after Gouder de Beauregard *et al.* 1998; Harper *et al.* 1995) compared with the abundance of crayfish as represented by their importance in bass stomachs (assessed as PV)

Year	Submerged macrophyte (ha)	Importance of crayfish in bass diet PV			Sample size *n*	
		Small bass	Large bass	All bass	Small bass	Large bass
1987	1900	80.1	257.0	168.5	127	21
1988	2600	0.0	130.1	65.0	139	36
1989	2300	28.8	183.4	106.1	235	24
1990	900	68.1	612.6	340.3	321	16
1991	650	8.5	1000.0	504.3	264	37
1992		9.8	844.0	426.9	244	29
1993		11.9	875.0	443.5	110	5
1994		198.1	670.0	434.0	126	10
1995	350					
1996	80	32.0	1000.0	516.0	154	2
1997						
1998	(0)	758.4	527.0	642.7	79	12

The nominal zero value for 1998 represents the fact that no submerged macrophytes could be located during the survey period of that year.

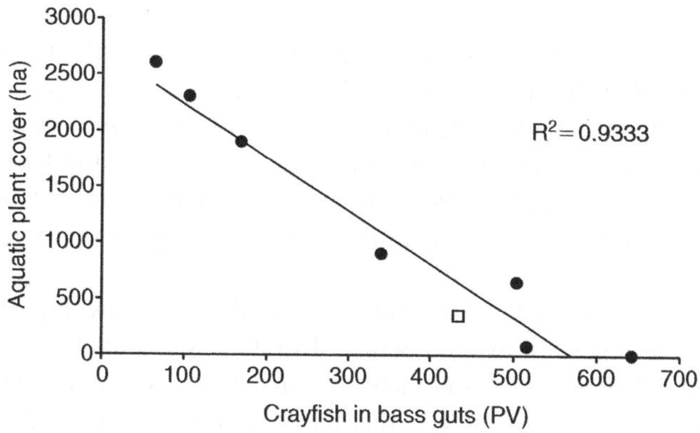

Figure 20.5 Relationship between area covered by submerged macrophytes and PV of crayfish in bass guts 1987–1998. Closed circles are for the same year; open square is 1995 macrophyte cover (ha) against 1994 crayfish PV

Overall fish catches appeared to be related to trends in water level changes with a rise in lake level generally followed by increased catches and a fall in level followed by a corresponding decline in fish catch. This supposition was confirmed for *O. leucostictus* and *T. zillii* (Fig. 20.3). The reasons why the two tilapias responded differently is not clear. *Tilapia zillii* is found in deeper water than *O. leucostictus,* which shows a greater tendency to frequent the margins (Muchiri 1990). Thus, for *O. leucostictus* (increased catch with increased water level), it could be that as the water rises the areas close to the marginal vegetation become more accessible to the commercial nets, whereas for *T. zillii* (increased capture with reduced water level) it is possible that as the water level goes down the deeper water can be sampled more efficiently by the nets.

For bass, the catches of which did not seem to be influenced by water level, it would appear that crayfish abundance is the more likely influencing factor (Fig. 20.4). Such bass and crayfish interactions have been recognised elsewhere. In achieving a 98% reduction in a population of papershell crayfish, *Orconectes immunis* (Hagen), in a fish rearing impoundment in Wisconsin, using traps and stocking largemouth bass, Rach & Bills (1989) found that it was predation by the bass which was the major factor. Similarly, Taub (1972) reported that largemouth bass reduced populations of *Cambarus diogenes* Girard to the extent that insufficient crayfish could be caught to make population estimates.

A reduced presence of submerged macrophytes could also have affected recruitment of new individuals to sustain the bass fishery by disrupting the availability of shelter from predators, including birds such as cormorants, *Phalacrocorax* spp., grebes, *Podiceps* spp., gulls, Laridae, and terns, Sternidae (Henderson & Harper 1992). Bettolli, Maceina, Noble & Betsill (1992) showed that habitat complexity, as mediated by vegetation abundance, was the principal factor regulating piscivory by largemouth bass in the littoral zone of Lake Conroe, Texas, and Miranda & Hubbard (1994) found that 0-group bass depend on vegetation for cover. Bass may also be influenced in other ways by loss of macrophytes. In Lake Baldwin, Florida, grass carp, *Ctenopharyngodon idella* (Val.), were used to eradicate all submerged macrophytes whereupon one component of the bass population sought locations with other forms of underwater structures and another moved to open waters devoid of structures and made little use of the inshore regions (Pothoven & Vondracek 1999). Colle, Cailteux & Shireman (1989) showed that following mechanical harvesting used to reduce *Myriophyllum spicatum* L. in Fish Lake, Wisconsin, growth rates for younger bass increased and for older bass declined. In two Minnesota lakes, after herbicide treatment, enhanced growth of largemouth bass was observed, probably related to improved opportunity for piscivory (Unmuth & Hansen 1999).

Absence of vegetation would be unlikely to affect directly spawning site availability for the finfish because nest excavation in the substratum is used by both bass (McClane 1978) and *T. zillii* (Greenwood 1966), and *O. leucostictus* is a female mouth-brooder (Greenwood 1966). Also, it is unlikely that the loss of macrophyte beds as a feeding area or as a food, would have a noticeable effect. Both *T. zillii* and *O. leucostictus* have high proportions of detritus in their diet. Detritus is an unlimited food item which has allowed the two tilapia species to co-exist (Muchiri, Hart & Harper 1995).

For the crayfish, reduced lake depth could increase reproductive opportunity. Olouch (1990) described *P. clarkii* in Lake Naivasha as breeding not only inside burrows but also on the sediment in shallow water at a depth between 0.5 and 4 m and concluded that reproduction is very much a lake edge activity. In recent times, however, much more of the lake has become <4 m deep. In 1991 the average depth was 3.35 m (Hickley *et al.*, in press), 1.75 m less than in 1983 (Åse, Sernbo & Syren 1986).

20.4.2 *Aquatic plant abundance*

With regard to the fluctuations in aquatic plant abundance (Gouder de Beauregard *et al.* 1998; Harper *et al.* 1995), several factors need to be taken into account, namely water depth, water quality and consumption. Change in water depth seems to have more influence on aquatic species present than total macrophyte abundance (Gouder de Beauregard *et al.* 1998). Water quality changes are likely to have an impact but relationships with aquatic plant density could be complex and reciprocal. Furthermore submerged plant beds have been shown to influence their own light regime, creating a micro-environment of calm water with reduced phytoplankton (Harper 1992), whereas during the recent period of reduced macrophyte occurrence, very high chlorophyll a values ($178\,\mu g\,L^{-1}$) were recorded (Mbogo, personal communication). It should be noted, however, that although the overall demise of the submerged vegetation might or might not be linked to the detrimental change in lake water quality, the situation has undoubtedly been exacerbated by a parallel decline in the fringing *C. papyrus* stands (Boar, Harper & Adams 1999). In addition, Smart (1990) suggested that macrophytes can provide alternative grazing for hippopotamuses, *Hippopotamus amphibius* L. This is on the basis that the changing lake environment and increasing agriculture around the lake has resulted in both habitat loss and restricted access to nocturnal grazing areas due to electric fences. Hippopotamuses were observed eating in the water, taking both *Najas* and *Potamogeton*.

The crayfish, *P. clarkii*, is generally assumed to be herbivorous (Penn 1943). In this context, according to Harper *et al.* (1990), it is likely that the disappearance of blue water lilies, *Nymphaea nouchali* Burm.f., which was more or less complete by the end of the 1970s, was a result of the combined grazing pressure of coypu, *Myocastor coypus* (Molina), and crayfish. Once the lilies had gone the crayfish would have had a more substantial grazing effect upon submerged vegetation, which in turn had disappeared from the lake by 1983. With no macrophytes the crayfish could only subsist as detritivores and the population, as observed subjectively by local residents, seemingly declined. Between 1984 and 1987 the aquatic vegetation returned during a period of water level decline. Harper *et al.* (1990) concluded that the reappearance of submerged plants after a 10-year period of absence from the lake was almost certainly a result of the disappearance of coypu and the population decline of crayfish.

For the second extirpation of submerged macrophytes, which occurred from 1987 onwards, Gouder de Beauregard *et al.* (1998) concluded that, although more than one factor should be considered, crayfish were primarily responsible for the reduction in submerged macrophytes, similar to this study (Fig. 20.5).

This argument is supported by other studies where crayfish, *P. clarkii*, were shown to exert a considerable ecological impact, affecting species composition, diversity and biomass of plants (Andres 1977; Feminella & Resh 1986; Ilheu & Bernado 1995; Gutierrez-Gurrita, Sancho, Bravo, Baltanas & Montes 1998).

20.4.3 *Future management of Lake Naivasha*

Implementation of management proposals based upon scientific findings can sometimes prove difficult, so care in communication is paramount if integration and acceptability of actions is to be achieved (Hickley & Aprahamian 2000). Accordingly, the riparian owners of Lake Naivasha have developed a comprehensive management plan to assist this process (LNRA 1999). With respect to the fishery, it has already been recommended by Hickley *et al.* (in press) that the overall lake management package should include conservation measures based on sound ecology (such as refuge areas and close times), appropriate legislation and the enforcement thereof (such as minimum mesh sizes and a ban on trading in undersized fish), education of all parties and the addressing of associated social issues. In addition, from the results presented here, there are two principal conclusions that should be taken into account.

First, water level fluctuations affect fish catches and thus the stability and predictability of the commercial fishery performance. A number of water usage measures are identified in the Lake Naivasha Management Plan (LNRA 1999) which, if implemented, could ameliorate some of the fluctuations in water level.

Recommendation 1: Water usage should be managed with a view to stabilising lake level fluctuations.

Secondly, the crayfish, *P. clarkii,* has a detrimental effect upon the submerged macrophyte beds of Lake Naivasha. The first time that the crayfish induced the disappearance of these water plants a decline in the crayfish population followed, thus facilitating the return of the vegetation. With the present absence of submerged plants it remains to be seen whether the crayfish population will decline as in the past, thus enabling the macrophytes to recover. Already in December 2000 (2 years after the disappearance of the plants), however, there has been a noticeable reduction in crayfish numbers (Foster & Mbogo, personal communication). Nonetheless, given that crayfish are not currently exploited to the same degree as previously (Fig. 20.1), and taking into account the environmental perturbations that are superimposed this time around, measures to control crayfish density should be incorporated into the overall management package to assist the re-establishment of submerged plants.

Recommendation 2: Attempts should be made to control the crayfish population by:

- encouraging the reinstatement of a high level of commercial trapping,
- protecting the bass population using net mesh size enforcement in the commercial fishery and the release of catch in the recreational fishery.

Acknowledgements

This study is part of a wider research project at Naivasha undertaken by the Universities of Leicester, UK, and of Nairobi, Kenya. We are grateful to the Office of the President of the Government of Kenya for research permission. We are indebted to the Fisheries Department in Naivasha for access to catch data. Work in the field would not have been possible without the support of the Earthwatch Institute and the assistance of their volunteers. The principal author is particularly grateful to Dr R.R. Boar for her editorial rigour. The views expressed are those of the authors and not necessarily those of their parent organisations.

References

Andres L.A. (1977) The economics of biological control of weeds. *Aquatic Botany* **3**, 111–123.

Åse L.E., Sernbo K. & Syren P. (1986) Studies of Lake Naivasha, Kenya, and its drainage area. *Naturgeografiska Institutionen Stockholms Universitet, Forsningsrapport* **63**, 1–75.

Bettolli P.W., Maceina M.J., Noble R.L. & Betsill R.K. (1992) Piscivory in largemouth bass as a function of aquatic vegetation abundance. *North American Journal of Fisheries Management* **12**, 509–516.

Boar R.R., Harper D.M. & Adams C.S. (1999) Biomass allocation in *Cyperus papyrus* in a tropical wetland, Lake Naivasha, Kenya. *Biotropica* **31**, 411–421.

Colle D.E., Cailteux R.L. & Shireman J.V. (1989) Distribution of Florida largemouth bass in a lake after elimination of all submerged aquatic vegetation. *North American Journal of Fisheries Management* **9**, 213–218.

Elder H.Y., Garrod D.T. & Whitehead P.J.P. (1971) Natural hybrids of the African cichlid fishes *Tilapia spirulus nigra* and *T. leucosticta*, a case of hybrid introgression. *Biological Journal of the Linnean Society* **3**, 103–146.

Feminella J.W. & Resh V.H. (1986) Effects of crayfish grazing on mosquito habitat at Coyote Hills Marsh. *Proceedings and Papers of the Annual Conference of the California Mosquito and Vector Control Association* **54**, 101–104.

Gouder de Beauregard A.-C., Harper D.M., Malaisse F. & Symoens J.-J. (1998) Dynamique récente et cartographie de la végétation aquatique (1960–1996) du lac Naivasha (Rift Valley, Kenya).

Mededelingen der Zittingen, Koninklijke Acdemie voor Overzeese Wetenscaappen, Bulletin des Seances, Academie Royale des Sciences d'Outre-Mer **44**, 373–389.

Greenwood P.H. (1966) *The Fishes of Uganda*. Kampala: The Uganda Society, 131 pp.

Gutierrez-Gurrita P.J., Sancho G., Bravo M.A., Baltanas A. & Montes C. (1998) Diet of the red swamp crayfish *Procambarus clarkii* in natural ecosystems of the Donana National Park temporary fresh-water marsh (Spain). *Journal of Crustacean Biology* **18**, 120–127.

Harper D.M. (1992) The ecological relationships of aquatic plants at Lake Naivasha, Kenya. *Hydrobiologia* **232**, 65–71.

Harper D.M., Adams C. & Mavuti K. (1995) The aquatic plant communities of the Lake Naivasha wetland, Kenya: pattern, dynamics and conservation. *Wetlands Ecology and Management* **3**, 111–123.

Harper D.M., Mavuti K.M. & Muchiri S.M. (1990) Ecology and management of Lake Naivasha, Kenya, in relation to climatic change, alien species introductions and agricultural development. *Environmental Conservation* **17**, 328–336.

Henderson I.G. & Harper D.M. (1992) Bird distribution and habitat structure on Lake Naivasha, Kenya. *African Journal of Ecology* **30**, 223–232.

Hickley P. & Bailey R.G. (1987) Food and feeding relationships of fish in the Sudd swamps (River Nile, southern Sudan). *Journal of Fish Biology* **30**, 147–159.

Hickley P., Bailey R.G., Harper D.M., Kundu R., Muchiri S.M., North R. & Taylor A. The status and future of the Lake Naivasha fishery, Kenya. *Hydrobiologia* (in press).

Hickley P., North E., Muchiri S.M. & Harper D.M. (1994) The diet of largemouth bass, *Micropterus salmoides*, in Lake Naivasha, Kenya. *Journal of Fish Biology* **44**, 607–619.

Hickley P. & Aprahamian M. (2000) Fisheries science and the managerial imperative. In I.G. Cowx (ed.) *Management and Ecology of River Fisheries*. Oxford: Fishing News Books, Blackwell Science, pp. 346–360.

Hodgson J.R. & Kitchell J.F. (1987) Opportunistic foraging by largemouth bass (*Micropterus salmoides*). *American Midland Naturalist* **118**, 323–335.

Ilheu M. & Bernado J.M. (1995) Trophic ecology of red swamp crayfish *Procambarus clarkii* (Giraud) – preferences and digestibility of plant foods. In *Freshwater Crayfish 10. Tenth International Symposium of Astacology*. Baton Rouge, LA (USA): Louisiana State University, pp. 132–139.

LNRA (1999) *Lake Naivasha Management Plan*. Lake Naivasha Riparian Association, PO Box 1011 Naivasha, Kenya, 77 pp.

Lowery R.S. & Mendes A.J. (1977) *Procambarus clarkii* in Lake Naivasha, Kenya, and its effects on established and potential fisheries. *Aquaculture* **11**, 111–121.

McClane A.J. (1978) *Field Guide to Freshwater Fishes of North America*. New York: Henry Holt, 212 pp.

Miranda L.E. & Hubbard W.D. (1994) Winter survival of age-0 largemouth bass relative to size, predators, and shelter. *North American Journal of Fisheries Management* **14**, 790–796.

Muchiri S.M. (1990) *The feeding ecology of tilapias and the fishery of Lake Naivasha, Kenya*. PhD Thesis (unpublished), University of Leicester, UK, 233 pp.

Muchiri S.M. & Hickley P. (1991) The fishery of Lake Naivasha, Kenya. In I.G. Cowx (ed.) *Catch Effort Sampling Strategies: Their Application in Freshwater Fisheries Management*. Oxford: Fishing News Books, Blackwell Science, pp. 382–392.

Muchiri S.M., Hickley P., Harper D.M. & North E. (1994) The potential for enhancing the fishery of Lake Naivasha, Kenya. In I.G. Cowx (ed.) *Rehabilitation of Freshwater Fisheries*. Oxford: Fishing News Books, Blackwell Science, pp. 348–357.

Muchiri S.M., Hart J.B. & Harper D.M. (1995) The persistence of two introduced species in Lake Naivasha, Kenya, in the face of environmental variability and fishing pressure. In T.J. Pitcher and P.J.B. Hart (eds) *Impact of Species Change in African Lakes*. London: Chapman & Hall, pp. 299–319.

Olouch A.O. (1990) Breeding biology of the Louisiana red swamp crayfish *Procambarus clarkii* Giraud in Lake Naivasha Kenya. *Hydrobiologia* **208**, 85–92.

Parker I.S.C. (1974) *The Status of the Louisiana Red Swamp Crayfish in Lake Naivasha.* Nairobi, Kenya: I.S.C. Parker Wildlife Services, 45 pp.

Penn G.H. (1943) A study of the life history of the Louisiana Red Crayfish *Cambarus clarkii* Girard. *Ecology* **24**, 1–18.

Pothoven S.A. & Vondracek B. (1999) Effects of vegetation removal on bluegill and largemouth bass in two Minnesota lakes. *North American Journal of Fisheries Management* **19**, 748–757.

Rach J.J. & Bills T.D. (1989) Crayfish control with traps and largemouth bass. *The Progressive Fish-Culturist* **51**, 157–160.

Smart A.C. (1990) The density of *Hippopotamus amphibius* Linnaeus at Lake Naivasha, Kenya. *Tropical Freshwater Biology* **2**, 241–247.

Taub S.H. (1972) Exploitation of crayfish by largemouth bass in a small Ohio pond. *Progressive Fish-Culturalist* **34**, 55–58.

Unmuth J.M.L. & Hansen J.M. (1999) Effects of mechanical harvesting of Eurasian water milfoil on largemouth bass and bluegill populations in Fish Lake, Wisconsin. *North American Journal of Fisheries Management* **19**, 1089.

Wilhm J.T. (1967) Comparison of some diversity indices applied to populations of benthic invertebrates in a stream receiving organic wastes. *Journal of Water Pollution Control Federation* **39**, 1673–1683.

Chapter 21
Is the invasion of water hyacinth, *Eichhornia crassipes* Solms (Mart.), a blessing to Lake Victoria fisheries?

M. NJIRU*, A.N. OTHINA, A. GETABU
Kenya Marine and Fisheries Research Institute, Kisumu, Kenya

D. TWEDDLE
JLB Smith Institute, Grahamstown, South Africa

I.G. COWX
University of Hull International Fisheries Institute, Hull, UK

Abstract

Following infestation of Lake Victoria by water hyacinth, *Eichhornia crassipes* Solms (Mart.) in the 1990s, condemnation of the weed came from many stakeholders. It was blamed for reduction in fish catches to spread of diseases like malaria and bilharzia, but the advantages the plant had on the fishery were overlooked. Catches of *Clarias gariepinus* (Burchell) and *Protopterus aethiopicus* (Heckel), which had declined in the 1980s, increased markedly following hyacinth infestation. In particular, the Nyanza Gulf, Kenya, with many sheltered bays that harboured extensive hyacinth mats, supported greater catches of *C. gariepinus* and *P. aethiopicus* than open water beaches, with the exception of stations with swampy areas or those occasionally covered by hyacinth. The increase in *C. gariepinus* and *P. aethiopicus* catches was attributed to increased food availability and breeding areas due to water hyacinth, fishers subsequently targeting the species and reduced competition from declining stocks of *Lates niloticus* (L.).

Keywords: *Clarias gariepinus*, Lake Victoria, *Protopterus aethiopicus*, water hyacinth.

21.1 Introduction

Water hyacinth, *Eichhornia crassipes* Solms (Mart.) was first observed in the Ugandan sector of Lake Victoria in 1988 (LVEMP 1995). Infestation was suspected to have occurred through the Kagera River in Uganda. The rapid expansion of the weed's coverage in the lake was reported to inflict economic, social, health and environmental impacts. Important users associated with the fisheries, transport and water quality were affected by the spread of the weed. Large mats of water hyacinth posed major obstruction to transport and fisheries operations. Delays in processing fish catches resulted in deterioration of fish quality and to avoid spoilage operators had to carry ice which increased operational costs.

*Correspondence: Kenya Marine and Fisheries Research Institute, PO Box 1881 Kisumu, Kenya (email: kmfkisu@net2000ke.com).

Water hyacinth was observed to impact negatively on the health of lakeside communities. It provided habitat for agents of malaria and bilharzia and harboured snakes. It turned water green and dirty making the supply unsuitable for drinking and other domestic use. Reduction of oxygen levels in the water created an environment unsuitable for fish survival, subsequently reducing species diversity. Lake Victoria, which supports a thriving fishing industry, supplying affordable protein, income and employment to more than 10 million people was adversely affected by the infestation. Water hyacinth was perceived to affect fisheries through reduced levels of production, a reduction in species diversity, poor quality fish, rising cost of operation resulting in lower income to fishers and higher prices to consumers (LVEMP 1995).

Exploratory bottom trawling and commercial catches recorded between 1958 and 1970 showed that *Clarias gariepinus* (Burchell) and *Protopterus aethiopicus* (Heckel) were important commercial species in Lake Victoria after haplochromines, *Bagrus docmak* (Forsskål) and *Oreochromis esculentus* (Graham) (Kudhongania & Cordone 1974). There was a decline of *C. gariepinus* and *P. aethiopicus* in the late 1970s and 1980s (Ogutu-Ohwayo 1990) attributed to loss of habitat, pollution, fishing pressure and establishment of *Lates niloticus* (L.) (Hecky 1993; Muggide 1993; Ochumba & Kibaara 1989). Since then the fishery has been dominated by *L. niloticus*, *Oreochromis niloticus* (L.) and a native cyprinid, *Rastrineobola argentea* (Pellegrin) (Witte & Dansen 1995).

Recent commercial and trawl catches in the Nyanza Gulf of Lake Victoria showed an increase of both *C. gariepinus* and *P. aethiopicus*, which were attributed mainly to lake infestation by water hyacinth (Getabu & Nyaundi 1999). Although hyacinth has brought problems to the lake community, the weed also appears to have had a positive effect on the fishery. This paper highlights the need to look at the positive aspects of the hyacinth and discusses factors leading to the increase of *C. gariepinus* and *P. aethiopicus*.

21.2 Materials and methods

21.2.1 *Study area*

The Kenyan sector of Lake Victoria comprises the main lake and the semi-enclosed Nyanza Gulf, also known as the Kavirondo or Winam Gulf. The lake receives its main flow from several rivers originating from the East Africa Rift Valley. The main geographical, hydrological and physical characteristics of Lake Victoria and Nyanza Gulf have been summarised by Burgis, Mavuti, Moreau & Moreau (1987), Mavuti and Litterick (1991) and Crul (1995). The Nyanza Gulf, which comprises the major part of the lake in Kenyan waters, lies within the equatorial region. The water temperature and solar radiation are relatively constant throughout the year (mean values of 22 ± 3°C and 1200 ± 140 MEM^{-1} S^{-1}). The Nyanza Gulf has an area of approximately 1920 km^2, with a length of 60 km and a width varying from 6 to 30 km. The gulf is shallow with a mean depth of 6 m and lies at an altitude of 1134 m above sea level. The study area was divided into four zones (Fig. 21.1).

- *Zone I* – Inshore gulf, shallow with an average depth of 5 m with several sheltered bays. The region is influenced by the drainage of several major rivers (Sondu and

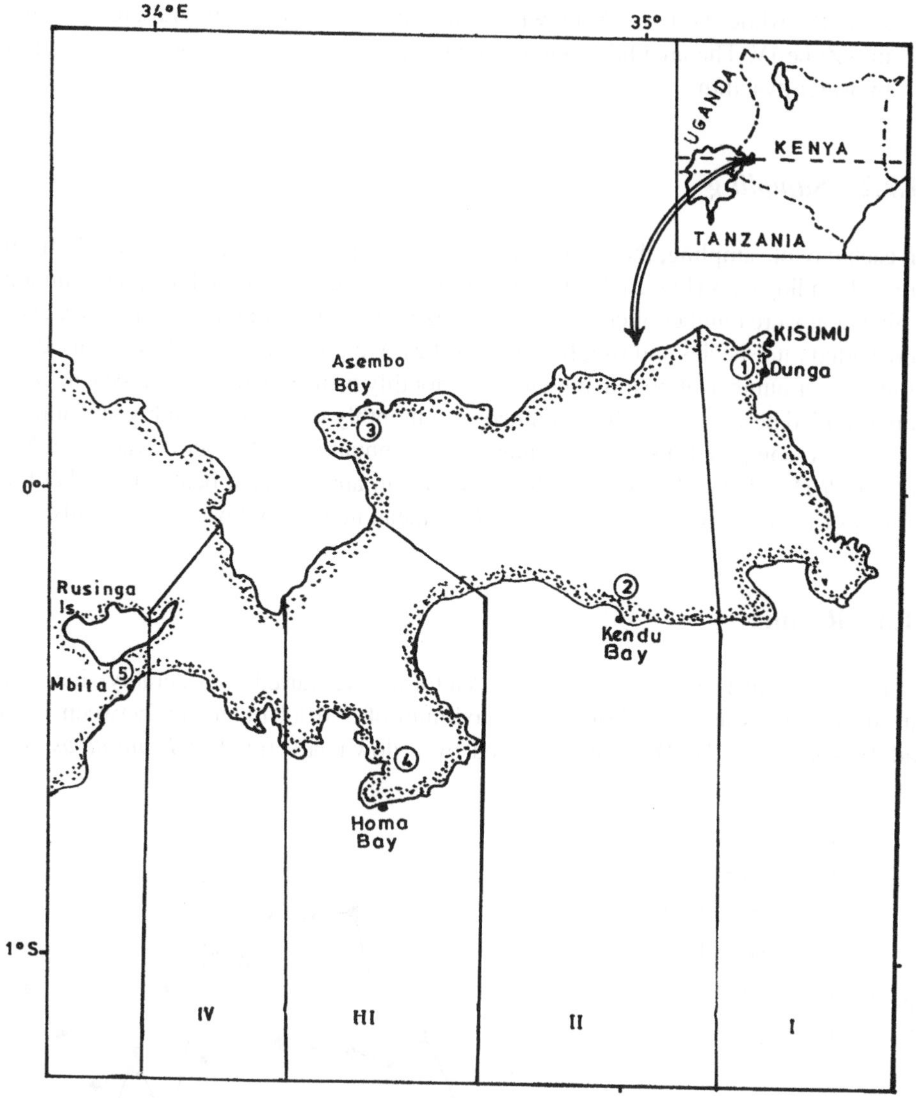

Figure 21.1 Map of Nyanza Gulf showing sampling areas

Nyando) which pass through rich agricultural areas. The area also receives municipal effluents. The site had the heaviest water hyacinth coverage, with several regions with permanent cover.

- *Zone II* – Inshore gulf, with an average depth of 20 m and few sheltered bays. The region is influenced by the drainage of the River Oluoch that passes through a rich agricultural area. The area receives municipal effluents and had occasional water hyacinth coverage.
- *Zone III and IV* – Offshore open waters, shallow with an average depth of 5 m with the shores exposed to heavy wind and wave action. The River Kuja drains into

Zone III, while the River Yala, which pass through rich agricultural areas, drains into Zone IV. The area has no major municipal influence and was rarely covered by water hyacinth.

21.2.2 *Sampling*

On each day of sampling, the catches of from five to 10 boats, selected randomly from targeted landing sites (Fig. 21.1), were recorded. Parameters recorded for each sampled fishing boat were number of crew, gear type, size, time and duration of fishing. Catches were sorted into species and weighed (nearest kg). Weight recorded daily was raised by a function of number of boat fishing to give monthly estimates for a given beach. Data from sampled beaches were then summed on a monthly basis, and monthly and annual means were computed. Estimate of annual total annual catch (AC in t) was given by: $AC = CPBD \times FD \times B$, where $CPBD$ is the mean annual catch boat^{-1} (kg), FD the number of fishing days (365 or 366) and B the mean number of boats fishing daily.

21.3 Results

Total landings in Kenya waters of Lake Victoria have varied considerably since the introduction of Nile perch, reaching a maximum of 200 000 t annually between 1988 and 1994 (Fig. 21.2). This was followed by a downward trend in *L. niloticus* (L.)

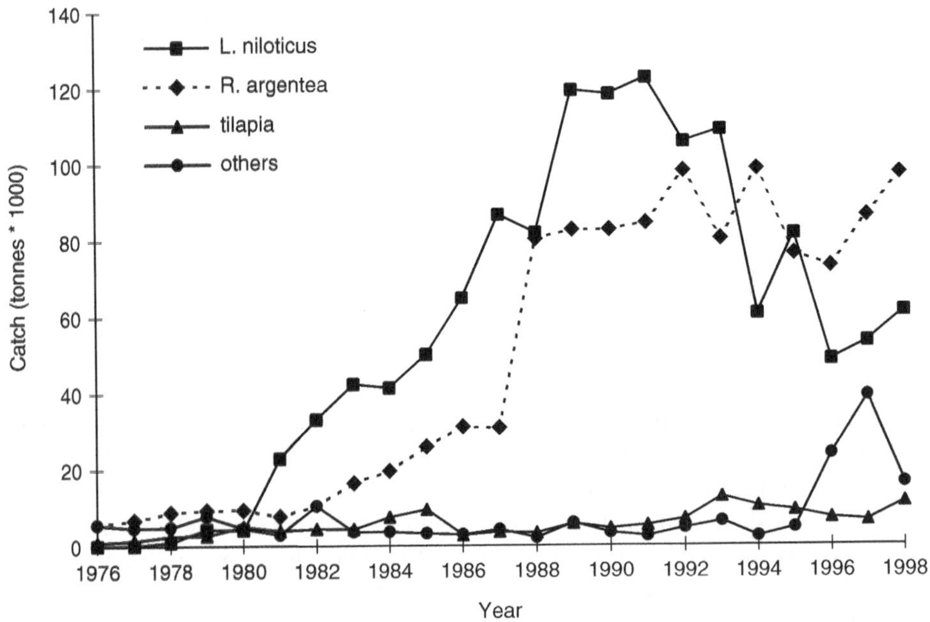

Figure 21.2 Annual catches from Kenyan waters of Lake Victoria 1976–1998

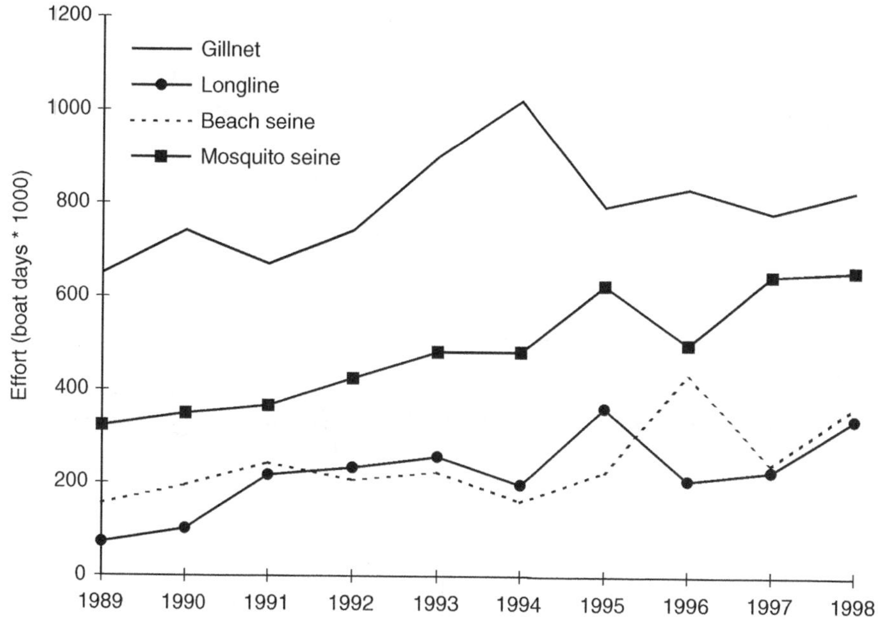

Figure 21.3 Trends in effort for the major fishing gears

catches, while those of *R. argentea* stabilised. Tilapiines and haplochromines showed a small increase in recent years. Other species, mainly *P. aethiopicus* and *C. gariepinus* showed a sharp increase after 1995, declining again in 1998, coupled with a slight increase in effort, expressed as boat days (Fig. 21.3).

Highest catches of *C. gariepinus* and *P. aethiopicus* were from Zone I and the least in Zone IV (Figs 21.4 and 21.5). No catches were recorded when the landing sites were completely covered by water hyacinth. Trawl surveys in Kenyan waters of Lake Victoria between 1997 and 1999 showed higher catches of *C. gariepinus* and *P. aethiopicus* at 5–9 m and 0–4 m, respectively (Fig. 21.6).

21.4 Discussion

There has been a general, and well documented, decline in the stocks of indigenous fish species in Lake Victoria between the 1950s and 1970s, and an increase of *L. niloticus* and *R. argentea* catches since the mid-1980s (see Ogutu-Ohwayo, 1990 and Crul 1995 for reviews). The increases in *L. niloticus* and *R. argentea* were attributed to an increase in fishing effort (Othina 1999). CPUE also increased annually from 82 kg boat day^{-1} in 1986 to 180 kg boat day^{-1} in 1989, which was mainly due to the explosion of *L. niloticus* (Othina 1999). The decline of native species was mainly due to overfishing, predation and competition by *L. niloticus* (Ogutu-Ohwayo 1990; Witte & Densen 1995) and ecological changes occurring in the lake (Crul 1995).

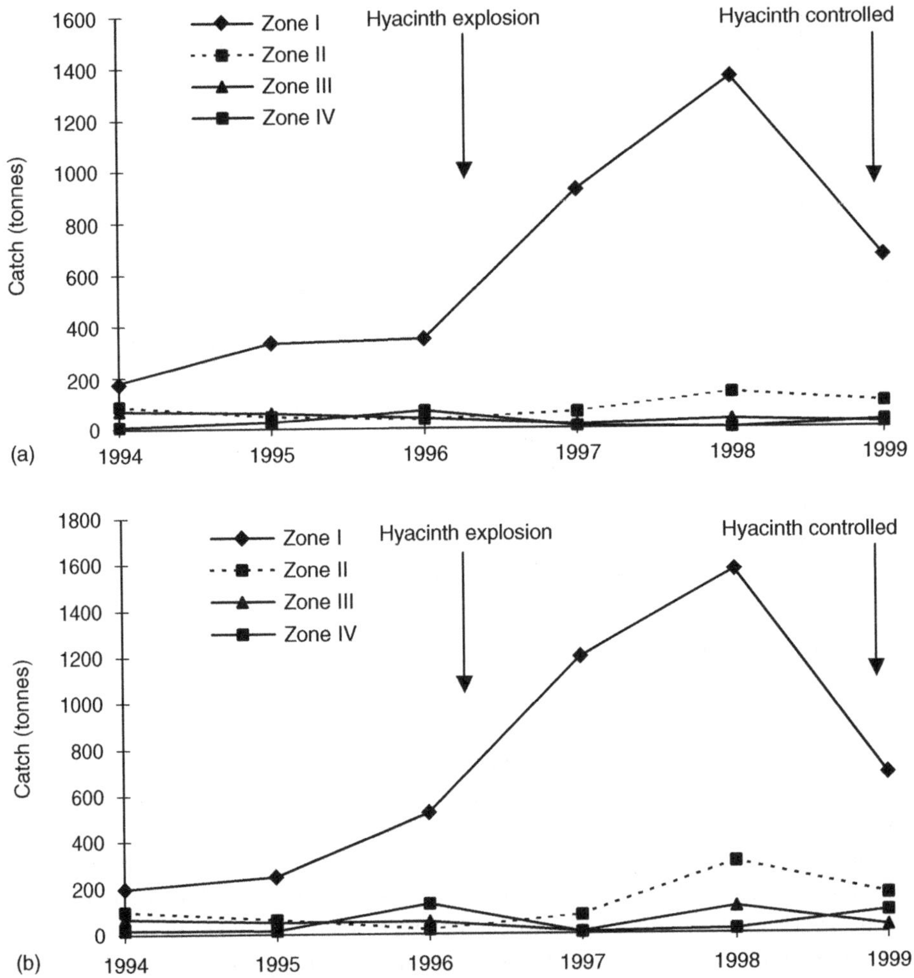

Figure 21.4 Annual catches of (a) *Clarias gariepinus* and (b) *Protopterus aethiopicus* from Zones I, II, III and IV of the Nyanza Gulf

The increased prevalence of *C. gariepinus*, *P. aethiopicus* and haplochromines between 1997 and 1999, compared to the 1980s was attributed to a number of factors. Firstly, the reduction in *L. niloticus* stocks reduced predation pressure on *C. gariepinus*, *P. aethiopicus* and the haplochromines, or their prey, and removed competition. Secondly, the infestation of the lake by water hyacinth has provided optimal conditions for breeding and growth of *C. gariepinus*, *P. aethiopicus* and haplochromines. *C. gariepinus* is found mainly in shallow inshore waters, swamps and major affluent rivers (C. Rabuor, personal communication). Infestation of Lake Victoria by the water hyacinth has increased the swampy area, offering more feeding and breeding areas. Water hyacinth has introduced considerable organic matter on the bottom and experimental trawling caught a wide variety of invertebrates, especially Ephemeroptera and

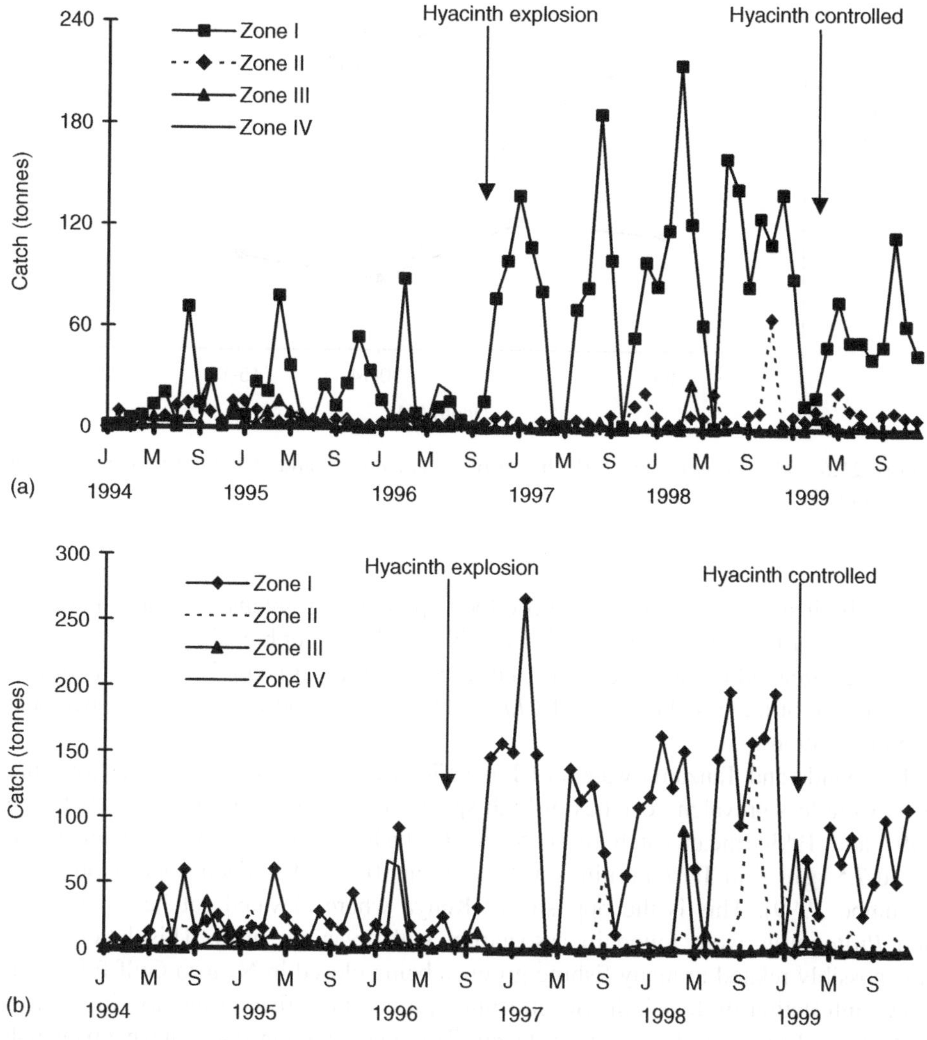

Figure 21.5 Monthly catches of (a) *C. gariepinus* and (b) *P. aethiopicus* from Zones I, II, III and IV of the Nyanza Gulf

Gastropoda, and highlighted an increased biomass of haplochromines. Insects and haplochromines are major dietary components of *Clarias* and *Protopterus*, while gastropods are preferred by *Protopterus*.

Coverage of water hyacinth also causes deoxygenation of water, and at times anoxia below the dense mats. *P. aethiopicus* is an obligatory air breather while *C. gariepinus* can survive in foul stagnant water due to accessory organs, enabling them to utilise atmospheric oxygen. The localised deoxygenation could have thus played a role in the increase in *C. gariepinus* and *P. aethiopicus* because they could occupy areas that Nile perch could not because of its intolerance to low oxygen levels.

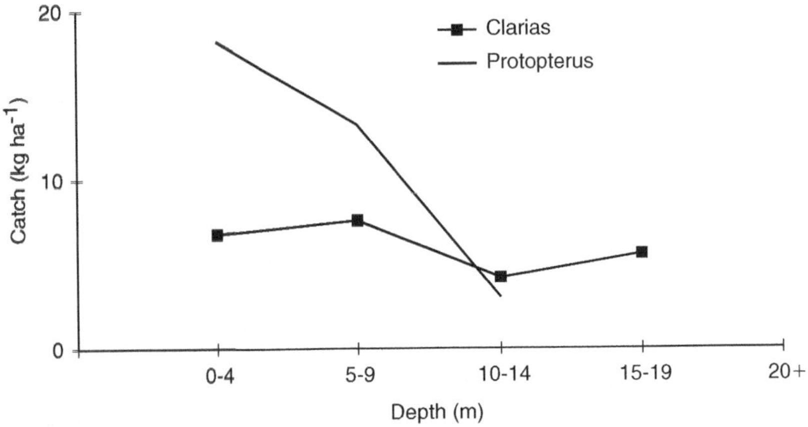

Figure 21.6 Mean trawl catches (kg ha^{-1}) of *C. gariepinus* and *P. aethiopicus* in different depths in Lake Victoria

Finally, it is possible that El Niño effects, particularly intense rainfall, increased flows in the affluent rivers and provided suitable conditions for breeding of *C. gariepinus* in flooded areas adjacent to the rivers, thus increasing recruitment. However, there was no evidence of increased catches of *Clarias* by riverine fishermen (J. Mugo, personal communication).

In Uganda and Tanzania waters of Lake Victoria, where extensive water hyacinth mats were less prevalent, catches of both species remainded low, especially between 1997 and 1999 (mean catch of 0.05 and 0.2 kg hr^{-1} respectively in Uganda and 0 and 7.5 kg hr^{-1} in Tanzania) in waters between 10 and 19 m deep (Okaronon 1999; Mkumbo 1999). This is the opposite of Kenya where elevated catches of *Clarias* and *Protopterus* were found in association with hyacinth mats. This latter effect was possibly related to many fishing grounds being closed in Nyanza Gulf following heavy infestation by hyacinth. Beach seining, a common fishing method in shallow part of the lake, was drastically reduced. This phenomenon could have given fish time to reproduce and grow. The downward trends of catches in 1999 were attributed to reduction in water hyacinth and overfishing. The hyacinth has been substantially reduced in the lake by biological control using weevils and the fishers who targeted the fish have overexploited the stocks due to the high value of the species on the local markets.

The upshot of studies on the relationship between water hyacinth presence and status of the fisheries is to debate whether the remaining water hyacinth in the lake should be removed. Most fishers indicate that the mats should be left in areas where they do not impede other economic activities. However, biological control cannot be contained and it is unlikely that selective removal will be possible. Whatever the outcome, it is imperative that the effect of water hyacinth on fisheries production is monitored in the long term to provide an understanding of the interactions between the fish population dynamics and hyacinth prevalence.

Acknowledgements

We wish to acknowledge the Kenya Marine and Fisheries Research Institute staff involved in data collection and compilation. The financial support for the project was provided by European Union under the Lake Victoria Fisheries Research Project (Ref.: ACP-RPR 227).

References

Burgis M.J., Mavuti K.M., Moreau J. & Moreau I. (1987) The central plateau. In M.J. Burgis & J.J. Symonens (eds) *African Wetlands and Shallow water Bodies: Directory (Zones Humides et Lacs Peu Profonds d' Afrique: Répertoire)*, Collections de Travaux Document 211, Paris, pp. 359–388.

Crul R.C.M. (1995) *Limnology and Hydrology of Lake Victoria*. UNESCO/IHP-IV Project M-5.1. UNSECO. Publishing, Paris, 78 pp.

Getabu A. & Nyaundi J. (1999) Relative abundance and distribution of fish in Kenyan waters of Lake Victoria. In I.G. Cowx & D. Tweddle (eds) *Report on Fourth FIDAWOG Workshop*, Kisumu 19–20 August 1999. LVFRP/TECH/99/07, pp. 46–62.

Hecky R.E. (1993) The eutrophication of Lake Victoria. *Verhandlungen der Internationale Vereinigung für Theoretische Angewandte Limnologie* **25**, 39–48.

Kudhongania A.W & Cordone A.J (1974) Batho-spatial distribution patterns and biomass estimates of the major demersal fishes in Lake Victoria. *African Journal of Tropical Hydrobiology and Fisheries* **3**, 15–31.

LVEMP (1995) Lake Victoria Environmental Management programme. Report submitted by Kenya, Uganda and Tanzania to World Bank. Washington, World Bank, 321 pp.

Mavuti K.M. & Litterick M.R. (1991) Composition, distribution and ecological site of zooplankton in Lake Victoria Kenyan waters. *Verhandlungen der Internationale Vereinigung für Theoretische Angewandte Limnologie* **24**, 1117–1122.

Mkumbo O.C. (1999) Recent trends in distribution patterns and catch rates from trawl surveys in Lake Victoria, Tanzania. In D. Tweddle & I.G. Cowx (eds) *Report on the Third FIDAWOG Workshop*, Kisumu 19–20 August 1999. LVFRP/TECH/99/06, pp. 18–29.

Muggide R. (1993) The increase in phytoplankton primary productivity and biomass in Lake Victoria. *Verhandlungen der Internationale Vereinigung für Theoretische Angewandte Limnologie* **25**, 846–849.

Ochumba P.B.O. & Kibaara D.I. (1989) Observations on blue–green algal blooms in the open waters of Lake Victoria. *African Journal of Ecology* **27**, 23–34.

Ogutu-Ohwayo R. (1990) The decline of the native fishes of lakes Victoria and Kyoga (East Africa) and the impact of introduced species, especially the Nile perch, *Lates niloticus* and the Nile tilapia, *Oreochromis niloticus*. *Environmental Biology of Fishes* **27**, 81–96.

Okaronon J.O. (1999) The fish stocks of Lake Victoria (Uganda). In D. Tweddle & I.G. Cowx (eds) *Report on the Third FIDAWOG Workshop*, Kisumu 19–20 August 1999 LVFRP/TECH/99/06, pp. 30–47.

Othina A. (1999) The status of the artisanal fishery of Lake Victoria, Kenya, with notes on improvements to the catch data collection. In I.G. Cowx & D. Tweddle (eds) *Report on Fourth FIDAWOG Workshop*, Kisumu 19–20 August 1999. LVFRP/TECH/99/07, pp. 78–79.

Witte F. & Densen, W.L.T. (1995) *Fish Stocks and Fisheries of Lake Victoria. A Handbook for Field Observations*. Cardigan, Britain: Samara Publishing, 404 pp.

Section III
Management

Chapter 22
Managing the exploitation of brook trout, *Salvelinus fontinalis* (Mitchill), populations in Newfoundland lakes

M.C. van ZYLL de JONG*
Wildlife Research and Science Division, Government of Newfoundland and Labrador, Corner Brook, Newfoundland, Canada

N.P. LESTER, R.M. KORVER
Aquatic Ecosystem Science Section, Ontario Ministry of Natural Resources, Peterborough, Ontario, Canada

W. NORRIS and B.L. WICKS
Indian Bay Ecosystem Research Project, Indian Bay, Newfoundland, Canada

Abstract

In response to anglers' allegation of a declining fishery, a monitoring and research programme was initiated in Indian Bay Brook Newfoundland, Canada. Creel and index-fishing data from lakes were used to construct a model for managing the exploitation of brook trout, *Salvelinus fontinalis* (Mitchill). The model describes the relationship between angling effort (angler-h ha^{-1} $year^{-1}$) and fishing yield (kg ha^{-1} $year^{-1}$). Based on model predictions, current fishing effort is nearing the sustainable limits of the system (3 angler-h ha^{-1} $year^{-1}$) and maintenance of a quality fishery demand regulations that restrict fishing effort or reduce its impact. A dynamic simulation model (calibrated for the study population) was used to compare the effectiveness of various types of management regulations (e.g. creel limits, size-based restrictions on harvest). Simulation results indicated that creel limits will not prevent over-fishing and that size-based management is needed to offer a sustainable high-quality fishery. Management guidelines and further data requirements are discussed.

Keywords: brook trout, modelling, management, population dynamics, yield.

22.1 Introduction

The brook trout, *Salvelinus fontinalis* (Mitchill), is the most common freshwater sport fish in Newfoundland with respect to both its distribution and its importance with anglers. The species exhibits huge variability in its life history (e.g. Macfadden 1964; Scott & Crossman 1973; Power 1980). It exhibits a ubiquitous distribution in Newfoundland and Labrador covering a wide range of habitats and aquatic productivities, thus rendering the management of this species highly data dependent. Despite

*Correspondence: Wildlife Research and Science Division, Government of Newfoundland and Labrador, Corner Brook, Newfoundland, PO Box 2323, Canada A2H 6J8 (email: mdejong@wild.dnr.gov.nf.ca).

this, current management regulations are determined stochastically in the absence of data. The absence of data-driven management, coupled with historically high participation rates, has resulted in the commonly held view by anglers that brook trout is declining in individual size and stock abundance. Despite angler's allegations of a declining fishery, few studies have been conducted on brook trout population dynamics (Ryan 1984; O'Connell & Dempson 1996; Lester, Korver, van Zyll de Jong, Norris & Wicks 1999; van Zyll de Jong, Lester, Korver, Norris & Wicks 1999) or exploitation patterns (e.g. Fowlow, Hoenig & van Zyll de Jong 1996) in Newfoundland lakes. In addition, comparative studies and simulation exercises on the effect of management strategies on Newfoundland brook trout are absent. This paucity of fundamental data limits an ability to provide scientifically defensible management actions or plans.

In an effort to provide information for brook trout stock assessment, a multi-year research project in Indian Bay Brook, Newfoundland was initiated in 1993. The objectives of the project were:

(1) to provide the scientific data needed to evaluate the status of fish populations;
(2) to provide a model for managing the exploitation of brook trout fisheries; and
(3) to offer informed advice for management decisions.

22.2 Methods

22.2.1 *Study area*

Indian Bay Brook, Newfoundland (Fig. 22.1) flows north-east from the upper section of the Bonavista Peninsula into Bonavista Bay (51°10′10″N; 56°01′25″W). Between 1995 and 1999, 15 lakes were sampled using a standard stock assessment programme (Table 22.1). The watershed lies in the central Newfoundland eco-region (Damman 1983). Pure black spruce forests and Aspen stands dominate the area due to the prevalence of fire in the natural history of the region. The topography is rolling to undulating and is characterised by sandy loam soils. All ponds sampled support populations of brook trout, Atlantic salmon, *Salmo salar* L., ouananiche, *Salmo salar ouananiche* (McCarthy), rainbow smelt, *Omerus mordax* (Mitchill), American eel, *Anguilla rostrata* (LeSueur.), and three-spined stickleback, *Gasterosteus aculeatus* L. In addition, Big Wings Pond supports a population of Arctic charr, *Salvelinus alpinus* (L), and in Third Pond, Back-up Pond and Second Pond, banded killifish, *Fundulus heterocleatus* (LeSueur), have been observed in net catches.

22.2.2 *Standardised stock assessment programme*

Standardised index netting surveys were conducted during the spring of 1995–1999. Capture methodology standards and guidelines are outlined in Table 22.2. To assess year-to-year changes in abundance, sampling was allocated to one season, the spring. In addition, a reproductive survey was carried in the autumn each year. Reproductive surveys followed the same methodological standards and guidelines as the spring

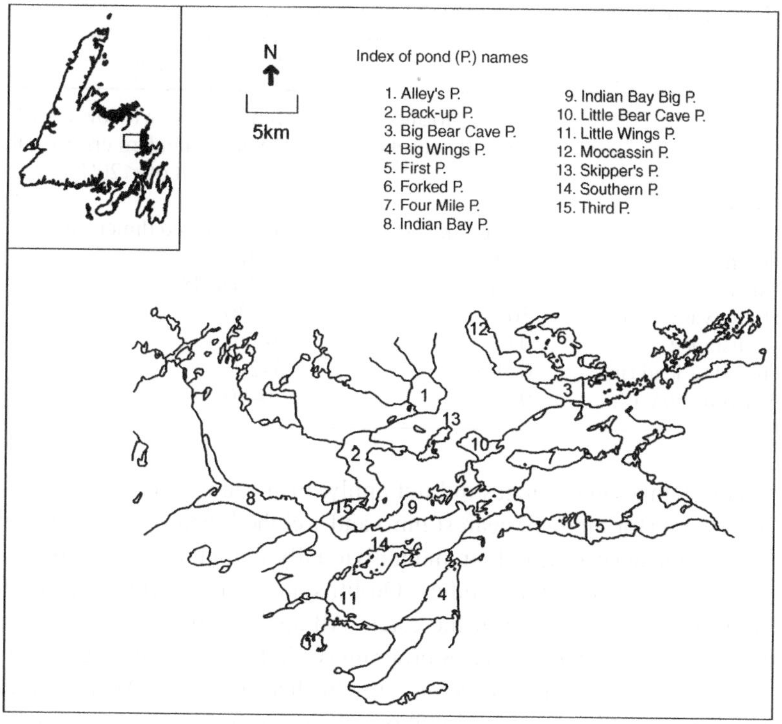

Figure 22.1 Indian Bay watershed including study lakes

Table 22.1 Location, physical and chemical variables measured in the Indian Bay Brook ponds

Pond	Latitude	Longitude	SA	MD	SD	EL	PER	DC	CON	TDS
Alley's Pond	49°07′04″	54°07′48″	408	4.7	1.05	45	7.5	21.0	34.2	31.6
Back-up Pond	49°04′00″	54°10′28″	964	3.4	1.55	45	17.0	21.4	30.3	28.8
Big Bear Cave Pond	49°07′00″	53°59′00″	512	9.3	1.5	60	12.0	12.4	35.9	32.8
Big Wings Pond	49°00′30″	54°07′15″	1088	10.4	1.84	45	21.5	14.2	42.8	37.8
First Pond	49°02′45″	54°00′13″	559	5.5	2.03	30	17.0	9.0	40.0	35.8
Forked Pond	49°08′28″	54°01′22″	570	6.7	1.66	76	14.0	15.6	55.0	46.6
Four Mile Pond	49°08′21″	54°01′25″	417	6.4	1.72	60	12.5	11.7	31.7	29.8
Indian Bay Pond	49°05′40″	54°18′25″	967	5.6	2.16	45	23.8	25.7	46.4	40.4
Indian Bay Big Pond	49°04′00″	54°07′00″	1094	5.0	2.35	45	27.6	17.6	33.0	30.8
Little Bear Cave Pond	49°05′30″	54°05′00″	200	4.3	1.40	45	7.0	15.5	32.8	30.6
Little Wings Pond	49°00′30″	54°12′00″	138	3.1	1.84	60	7.8	18.4	55.0	46.6
Mocassin Pond	49°08′20″	54°04′28″	500	6.3	1.51	76	12.0	18.3	35.6	32.7
Skipper's Pond	49°05′30″	54°07′30″	116	2.91	1.75	45	6.7	18.1	37.8	34.2
Southern Pond	49°01′43″	54°10′15″	365	4.5	1.57	60	10.6	18.7	52.5	44.8
Third Pond	49°03′00″	54°12′00″	272	3	1.54	45	9.0	22.2	41.0	36.5

SA, surface area (ha), MD, mean depth (m), SD, shoreline development (no units), EL, elevation (m), PER, perimeter (km), DC, distance to the coastline (km), CON, conductivity (μS) and TDS, total dissolved solids (μS).

Table 22.2 Capture methodological standards and guidelines

Parameter	Target	Acceptable range
Season	May 15 to July 15	Temperature window (9–19°C)
Sampling size	10–140 net sets	Relative standard error for the catch per net set of <20%
Set duration (h)	24	23–25
Gear	1.5 m fyke net	Standardised dimensions
Net separation (m)	200	>200
Reuse of sites	None	At least 48 h
Lead length in water (m)	30	25–30
Lead-shore distance	0.0	0–3 m
Depth with end of lead (m)	0–1.5	0–2
Lead angle from shore (°)	90	70–90

surveys, except temperature ranges did not apply. Fish caught were identified, sexed, counted, measured (fork length, nearest mm) and weighed (nearest 0.1 g). Brook trout were tagged with lake specific colour and individually-numbered, fingerling floy tags, and released for mark–recapture studies. Otoliths and scales were collected for age interpretation from a sub-sample of the catch. All ageing structures were interpreted for age and growth using the CSAGES program (Calcified Structure Age and Growth Extraction Software) (Casselman & Scott, unpublished software). At each netting site general weather conditions (i.e. precipitation, cloud cover, wind direction and speed) and site-specific habitat attributes (i.e. substrate, macrophyte cover, depth and site temperature) were recorded. During reproductive surveys all of the above information was collected plus maturity and fecundity data.

Three reproductive surveys were carried out during the autumn from 1996 to 1998. The length at maturity was expressed as the length when 50% of fish were mature (L_m). Knife-edge transition models were developed using probit analysis. Estimates of the number of eggs per female and the relative fecundity of brook trout were derived. All egg counts were absolute.

A one-stage progressive count roving creel survey was conducted during the winter fishing season for each year between 1993 and 1999. The summer fishery was monitored through access point creel surveys performed in 1992, 1993, 1997–1999. In both seasonal surveys catch and effort were recorded. Individual fish attributes were measured during winter interviews. A separate roving survey was conducted during the summer to sample the catch. All fish were treated as previously. An otolith and scales were taken for a sub-sample of the catch for ageing purposes. All analysis is described in detail in Lester, Petzold, Dunlop, Monroe, Orsatti, Schaner & Wood (1991), van Zyll de Jong *et al.* (1999) and Lester *et al.* (1999).

22.2.3 *Derivation of input parameters*

All relative abundance estimates (mean catch per net set or CPUE) were calculated in FISHNET 2.0L (OMNR, unpublished software). The Schnabel (1938) multiple mark–recapture method was used to calculate population size. Biomass was the sum of density-at-age for each pond multiplied by the mean weight-at-age for each pond. An estimate of natural mortality was obtained from the pooled catch-at-age curves from the fyke net catches. Since young fish (<age 2) were not vulnerable to angling, survival from ages 1 to 2 can be used to estimate natural mortality as $N_2/N_1 = S = e_m$.

To apply this method, indices of abundance were adjusted to account for age-specific differences in fyke net vulnerability. Recapture–mark ratios indicated that vulnerability was constant for ages $\geqslant 2$ and reached 70% of the maximum by age 1. Indices of abundance were calculated as $N_i = C_i/V_i$ where N_i is the index for age i, C_i is the age-specific mean catch per net set, and V_i is the age-specific vulnerability.

The most common method of expressing age at length is the von Bertalanffy growth model:

$$L_t = L_\infty (1 - \exp(-K(t - t_0))),$$

where L_t is the mean length of the fish at age t, L_∞ is the asymptotic length at infinite age, K is the growth coefficient (or rate at which the fish grows towards L_∞ each year (units year^{-1})), and t_0 is the extrapolated age at which L_t is zero. Due to sampling problems which included the under representation of younger ages, von Bertalanffy parameters could not be fitted by traditional non-empirical methods. The parameters would be unrealistic. The following method was used to allow an approximation of von Bertalanffy parameters (Payne, Korver, MacLennan, Nepszy, Shuter & Thomas 1990). If L_∞, is constrained to an empirical measure of the maximum length of fish caught, L'_∞, and t_0 constrained to zero, and K ascribed as, K', the von Bertalanffy equation can be rearranged into the form of a linear regression:

$$-\ln(1 - (L_t/L'_\infty)) = Kt - Kt_0.$$

The slope of the line when forced through the origin (equivalent to setting t_0 to zero) gives an estimate of K'. This empirical method was applied to the pond data set. Early growth rate (ω) was calculated as the product of L'_∞ and K'.

The relationship between weight and length was defined by $W = aL^b$; where W is weight (g), L the fork length (mm), and a and b are derived constants estimated for each lake for each year by least squares regression of $\log_{10}$ transformed mean weight and length at age.

Female spawning biomass (FSB) was determined as the product of the ratio of female to males in the population, the biomass at each age (B_n) and the per cent maturity at each age ($\%M_n$), and summed for a total estimate of FSB for the population:

$$\text{FSB} = \Sigma Bf \times SR \times \%M_n.$$

Total egg production (TEP) for a pond is TEP = FSB $\times f$, where f is the relative fecundity expressed as mean number of eggs per kilogram. Egg survival (α) to age 1 was estimated by dividing the population estimate at age 1 ($N1_n$) in year n by the TEP of spawning females in the pond in year $n-1$: $\alpha = N1_n/\text{TEP}_{n-1}$.

A versatile model was used to describe the expected relationship between parental stock size and recruitment (Shepherd 1982). The carrying capacity parameter (B_0) and the shape parameter (β) were estimated using the methods described in Shepherd (1982).

All fisheries' parameters were calculated according to methods outlined in Fowlow *et al.* (1996) and Pollock, Hoenig, Jones & Greene (1997). Length at first capture was estimated with probit analysis. Release and skunk rates were also calculated according to methods outlined in van Zyll de Jong *et al.* (1999).

22.2.4 *Brook trout exploitation model*

The exploitation model predicts the effect of angling on the yield and fishing quality of a brook trout fishery and addresses several key questions.

(1) What is the maximum potential yield ($kg\,ha^{-1}$)?
(2) How much fishing effort will contribute to this maximum yield?
(3) How much fishing effort can be sustained?

Key reference points are also provided (e.g. effort at maximum yield, effort at extinction) for evaluating the capacity of the fishery. An age-structured equilibrium model was used to describe the relationship between yield and fishing mortality rate. This model was derived by combining a generalised stock–recruitment relationship with conventional yield-per-recruit and biomass-per-recruit functions (Shepherd 1982; Shuter, Jones, Kover & Lester 1998). The algebraic solution is fairly complex, but can be summarised as

$$\text{Yield} = f(W_\infty\ \omega\ t_m\ f_{max}\ \alpha_{max}\ M\ \beta\ B_0\ t_c\ F).$$

In this expression, the parameters can be grouped as follows.

(1) *Life history parameters*, defining rates of growth, natural mortality and reproduction in the absence of intraspecific competition; where W_∞ is the asymptotic weight of an adult fish (kg); ω is the growth rate (cm per year) early in life; t_m the age at maturity (knife-edge transition assumed); f_{max} fecundity (the number of eggs per kilogram for a mature female at low population density); α_{max} survival from egg to age 1 at low population density; M the instantaneous natural mortality rate (per year) for fish aged 1 year and over.
(2) *Habitat quantity parameters*, defining the rates at which early survival or fecundity are reduced from their maximum values as the population fills available habitats and approaches its carrying capacity, where β is a parameter that sets the rate of decline in survival or fecundity; and B_0 a scaling parameter ($kg\,ha^{-1}$) that reflects the amount of habitat available to the population and hence is directly related to both the carrying capacity of the population and its maximum sustainable yield.
(3) *Fishery parameters*, defining what part of the population is exploited and the intensity of exploitation, where t_c is the age at first capture (knife-edged transition to full vulnerability assumed); and F is the instantaneous fishing mortality rate (per year) applied to all fish aged $>t_c$.

22.2.5 *Management simulation model*

The Fisheries Management Support System (FMSS), developed by R. Korver of the Ontario Ministry of Natural Resources was used to simulate the effect of the following regulations:

(1) no regulations;
(2) bag limit = 6 fish or 2 lb + 1 fish;
(3) minimum size = 22.5 cm;
(4) minimum size = 28 cm;
(5) minimum size = 35 cm;
(6) protected slot = 22.5–28 cm; and
(7) maximum size = 35 cm.

These size-based regulations used criteria related to the growth and maturation of Indian Bay brook trout (Fig. 22.2). The first minimum size regulation (3) implies all fish less than the expected size at maturity (i.e. 22.5 cm) must be released. The next minimum size regulation (4; 28 cm) implies that the average fish would reproduce once before reaching a size that could be harvested by anglers. The last minimum size regulation (5; 35 cm) implies the average fish would reproduce three times before reaching a vulnerable size. The protected slot (22.5–28 cm) implies fish could not be harvested during their first year of reproduction. The maximum size regulation (35 cm) implies that large and relatively old fish would be protected.

For each scenario, effort was set at a fixed level (7 angler-h ha^{-1}) and a brook trout population was simulated for 50 years. The response during the last 25 years was used to compare the effects of each regulation. The responses measure included: adult abundance – number of fish ha^{-1}; CPUE – number of fish caught per angler-h; CPUE 35 – number of fish >35 cm caught per angler-h; and HUEW – kilograms of

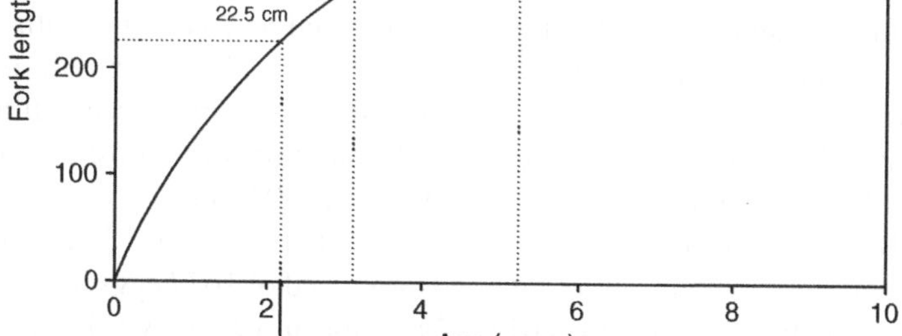

Figure 22.2 Growth and age at maturation information for Indian Bay brook trout

fish harvested per angler-h. These fishing quality measures reflect different aspects of the fishing experience. CPUE indicates whether the angler is catching fish, CPUE 35 indicates the chance of catching a large fish, and HUEW indicates whether the angler can expect to bring fish home. The criteria, 7 angler-h ha^{-1}, was chosen as a basis for comparing these regulations because it represents a level of fishing effort that is barely sustainable when there are no regulations. Thus, the question being asked was whether it is possible to have a quality fishery at high effort by employing one of these regulations. The second question addressed was how much effort could be sustained under each regulation (at what effort level did the population crash). This was determined by running simulations at progressively higher effort levels until the fish population became extinct.

22.3 Results

22.3.1 *Model input parameters*

An empirical method was used to estimate von Bertalanffy parameters from a 13-pond data set consisting of length-at-age. L'_∞ for all ponds combined based on 39 fish was determined as 405 mm, and K', based on mean lengths between the ages 2 and 7, was set at 0.37 (Fig. 22.2). The length–weight relationship for fish caught in fyke nets surveyed from 1995 to 1998 was $\log W = 0.009 \times 3.05 \log L$.

Relative abundance was expressed as mean catch-at-age per net set (i.e. CPUE) of brook trout from the spring fyke net catches. Schnabel population estimates were completed on nine lakes some with multiple-year estimates. Estimates of population density ranged from 3.6 to 179 fish ha^{-1} on Little Bear Cave Pond and Skipper's Pond, respectively. Biomass ranged from 0.43 to 15.55 kg ha^{-1} in Indian Bay Pond and Skipper's Pond, respectively.

An estimate of natural mortality was obtained from the pooled catch curve. Since young fish (i.e. <age 2) were not vulnerable to angling, the survival from ages 1 to 2 can be used to estimate natural mortality. Indices of abundance were first adjusted to account for age-specific differences in fyke net vulnerability. Recapture–mark ratios indicated that vulnerability was constant for ages $\geqslant$2 and reached 70% of the maximum by age 1. An estimate of 0.67 for survival from ages 1 to 2 was obtained, implying a natural mortality rate, $M = 0.45$.

Mean relative fecundity for all ponds was estimated as 2540 eggs with a range of 2398–2634 eggs kg^{-1}. Autumn fyke netting surveys revealed a sex ratio of approximately equality. Length at maturity was calculated from autumn fyke netting surveys in ponds from a 1998 survey. The length at 50% maturity was 22.5 cm, length at 5% maturity was 15.0 cm and length at 95% maturity was 30.0 cm.

Mean FSB was 0.66 kg ha^{-1} with a range of 0.21–1.30 kg ha^{-1} TEP ranged from 355 to 3292 eggs ha^{-1} with mean for all ponds surveyed of 1677 eggs ha^{-1}. Egg to age 1 survival ranged from 0.0004 to 0.0130. Variation in egg survival was negatively correlated with spawner biomass. Although the result was not significant ($r^2 = 0.34$, $P = 0.13$, $n = 8$) statistical power was very low given the small sample size. The

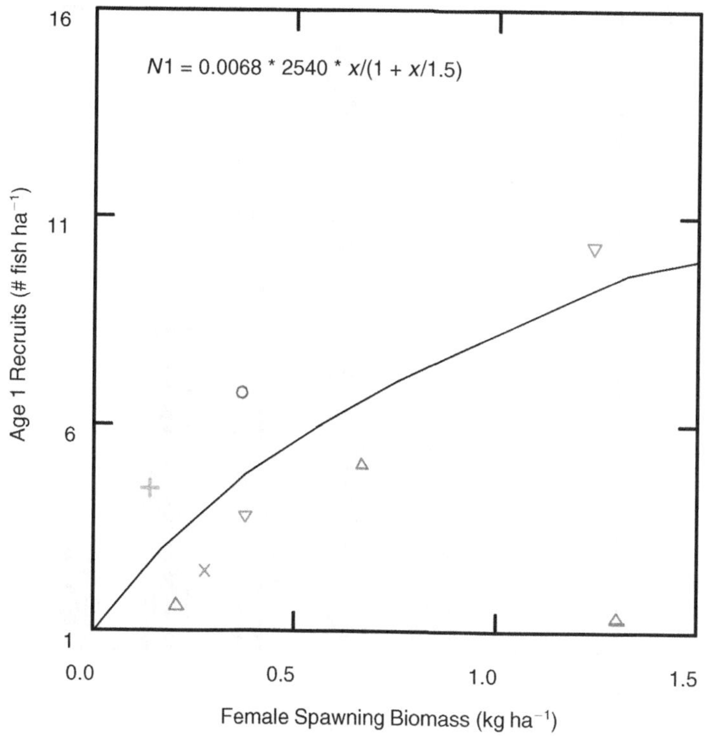

Figure 22.3 Relation between abundance at age 1 and the FSB. The solid line is a Shepherd stock–recruitment relation for Indian Bay brook trout. The equation in the upper left-hand corner describes the relationship where $N1$ is age 1 recruits and x is the FSB

regression implies that maximum egg survival (i.e. at low densities) was 0.007; i.e. 7 in 1000 eggs survive to become age 1 fish. These estimates of maximum egg survival and average fecundity (2540 eggs kg^{-1}) were used to constrain the initial slope of the stock–recruitment relationship (Fig. 22.3).

Combined figures for both the winter and summer fisheries gave a mean annual yield of 0.31 kg ha^{-1} and a mean annual effort of 2.67 angler-h ha^{-1}. The mean annual CPUE was 1.33 fish h^{-1} and HWUE 0.33 kg h^{-1}. The pooled angled catch curve which was used to estimate the age at full vulnerability is shown in Fig. 22.4, with corresponding length set at 20 cm, 15 cm and 25 cm for 5%, 50% and 95% vulnerability, respectively.

22.3.2 *Equilibrium yield model output*

The model predicted that yield increases while fishing mortality rate remained <0.22 and reached a maximum level of 0.4 kg ha^{-1} (Fig. 22.5). As mortality rate increased further, yield decreased and expired when mortality equalled 0.56 (Fig 22.5). This expiration of the yield was due to the decrease in fish abundance that resulted from higher fishing mortality. To convert fishing effort into a fishing mortality rate (F) one

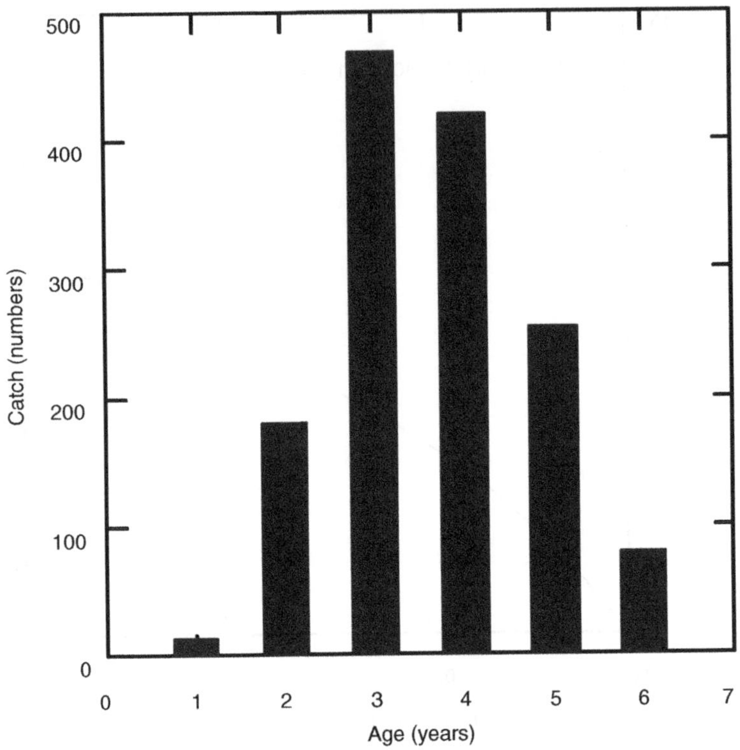

Figure 22.4 Age composition of harvested brook trout for all ponds for the period 1994–1998

needs an estimate of the catchability coefficient of angling (q) or $F = qE$, where E is effort intensity (angler-h ha^{-1}) measured on an annual basis. Catchability can be calculated from the relationship between CPUE and fish abundance. Given that catch is equal to the product of F and abundance, then,

$$\text{Catch} = q \times E \times \text{Abundance} \quad \text{and} \quad \text{CPUE} = q \times \text{Abundance}.$$

Thus, catchability is the slope of the relationship between CPUE and fish abundance. This interpretation assumes that q is a constant (i.e. independent of abundance) and therefore CPUE is linearly related to abundance.

For the Indian Bay fishery, angling CPUE was plotted against estimated fish density for eight ponds where mark–recapture studies supplied estimates of abundance (Fig. 22.6). Abundance estimates refer to fish that are at least 2 years old. Younger fish are excluded because they are not vulnerable to angling. Most of the points fall near a regression line forced through the origin whose slope implies $q = 0.072$. A larger set of observations from fyke net index fishing also yielded indices of abundance. Mean catch per net set of fish that were at least 2 years old were used to measure the relative abundance of fish vulnerable to angling. The relationship between angling CPUE and relative abundance was asymptotic (Fig. 22.6). When abundance is high, there is no relationship between CPUE and abundance. At lower levels of abundance (i.e. <6),

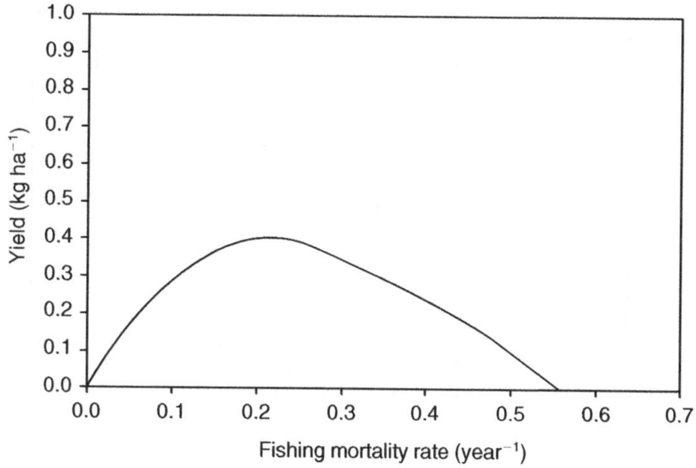

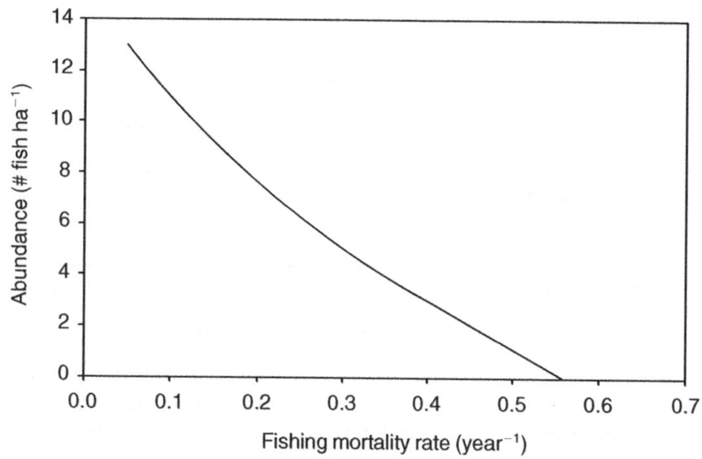

Figure 22.5 The first model (upper graph) shows the expected relationship between fishing mortality rate and fish yield, and the second model (lower graph) depicts the expected relationship between fish abundance and fishing mortality rate

CPUE declines roughly in a linear fashion, indicating that catchability is constant. It was concluded that the relationship between CPUE and abundance was not linear when considering the full range of abundance that may exist. Once a critical level is reached, CPUE decreases roughly in proportion to abundance. Substantial reduction in fish abundance may occur before angling CPUE is impacted. Based on results shown in Fig. 22.5, it was argued that the critical abundance level was approximately 12 fish ha^{-1} and that $q = 0.072$ when density was less than this criterion. This implies $F = 0.072E$ or $E = 14.3F$, or by increasing effort by 1 angler-h ha^{-1} increases fishing mortality by 0.072.

To generate a yield–effort relation, the x-axis of the yield–F relation was transformed by setting *effort* $= 14.3F$. The result showed that the yield peaks when effort equals 3.0 angler-h ha^{-1} and expires at an effort of approximately 8 angler-h ha^{-1}

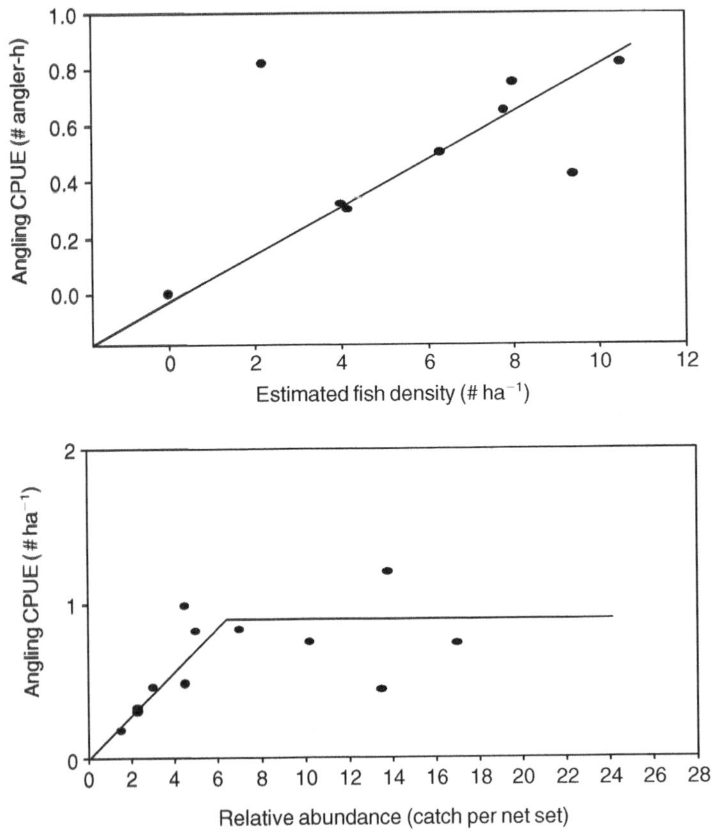

Figure 22.6 The following describes the catchability relationship. The upper graph shows the relationship between angling CPUE and fish density estimated from mark–recapture experiments. The lower graph depicts the relationship between angling CPUE and relative abundance from index-netting survey catch rates (catch per net set)

(Fig. 22.7). CPUE decreased as effort increased (Fig. 22.7). At the effort level that resulted in maximum yield, CPUE equalled 0.52 fish per angler-h.

Observed yield deviated somewhat from the predicted relation (Fig. 22.7). These deviations should not be interpreted as an indication that the model is invalid. The model describes an expected equilibrium response, whereas the data describe a transient response. The model addresses the question: if effort was maintained at a specified level for many years, what stable level of yield (and CPUE) would eventually result? It may take many years before the system becomes stable. (In the real world, the system may never become stable because environmental factors are never constant.) Thus, differences between predicted and observed yields are expected. On some ponds, the observed yield was much higher than the predicted maximum sustainable level. This difference is expected when exploiting a previously unexploited resource. A time lag (of at least one generation) exists before the effects of exploitation are

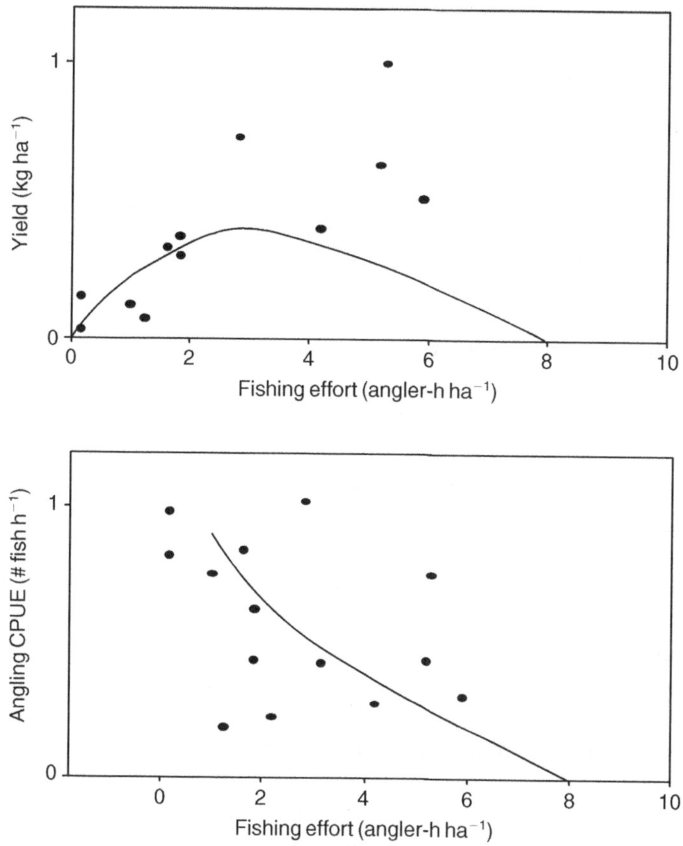

Figure 22.7 The first graph (upper) describes the yield–effort relationship and the second graph (lower) describes the angling CPUE–fishing effort relationship. The solid lines represent the equilibrium model and the solid dots represent observed lake values

manifested in terms of future production. Yield measured over a short time interval is not a good indicator of a sustainable yield.

22.3.3 *Diagnosis of the fishery*

Is the Indian Bay trout fishery overexploited? Using effort at maximum yield (i.e. 3 angler-h ha^{-1}) as a reference point, effort exceeded the criterion on 5 of 13 ponds (Fig. 22.7). Thus, it can be concluded that some ponds were being overexploited. If the mean effort is examined for the set of ponds (=2.6 h ha^{-1}) it can be concluded that the system, as a whole, is operating near full capacity. Although some ponds are over-exploited, these excesses are compensated by under utilisation of other ponds. Given open access to all ponds, it might be expected that angler migration between ponds would be driven by differences in CPUEs and would eventually result in a distribution of fishing effort that equalises CPUE. Thus, over the long term all ponds would be exploited near the optimum level.

Can the Indian Bay fishery sustain an increase in fishing effort? The simple answer is yes. Mean effort is well below the extinction value (i.e. 8 angler-h ha^{-1}) so increases could occur without endangering the fish stocks. The complex answer is 'yes, but at a cost'. The costs are: (1) further degradation of fishing quality; and (2) increased likelihood of endangering fish stocks. Fishing regulations that dampen the impact of fishing effort are needed to avoid these costs.

22.3.4 *Which management scenario works?*

Results indicated that the creel limit was not effective (Fig. 22.8). When fishing effort is high, CPUE is low and most anglers do not catch their limit. Consequently, the effect of a creel limit is the same as no regulation at all. Size-based regulations varied in terms of their effectiveness. Protecting small fish (<22.5 cm, immature) and protecting large fish (>35 cm) was marginally effective. Protecting fish during the first year of reproduction (22.5–28 cm) was more effective, but did not compare with the success of the remaining two scenarios. Minimum sizes of 28 and 35 cm were by far the most effective regulations. The 28-cm regulation resulted in a higher HUEW, implying a better food fishery, but in all other measures the 35-cm minimum size excelled.

The success of a 35-cm minimum size is not surprising. Given the growth and natural mortality rates used in the model, the probability that a fish survives to reach this size is only 0.10. Thus, this size regulation will result in a mainly catch-and-release fishery. Although a minimum size of 28 cm results in a higher kill rate, the difference in terms of kilograms retained per hour of fishing (HUEW) was marginal (Fig. 22.8). For both scenarios, HUEW was in the order of 0.6 kg h^{-1}, implying approximately 16 angler-h of fishing would be needed to bring home 1 kg of fish. For other regulations, the take home rate was even smaller. None of these regulations will support a food fishery when effort is as high as 7 h ha^{-1}.

22.4 Discussion

Brook trout population dynamics for other lentic brook trout populations are very limited. Quinn, Korver, Hicks, Monroe & Hawkins (1994) tested the effect of various management scenarios on brook trout population in Algonquin Park in Ontario, Canada. They showed that minimum size limits tended to show increased abundance and decreased harvest, and population size structures improved as minimum size limits were raised. There is considerable literature on lotic brook trout populations (Hunt, Brynildson & Mc Fadden 1962; Shetter & Alexander 1966; Hunt 1970; Clarke, Alexander & Gowing 1980; Power & Power 1996). The general conclusion from comparative and simulation studies on lotic populations was that a varying combination of size-based limits in association with low harvest limits was needed to be effective in

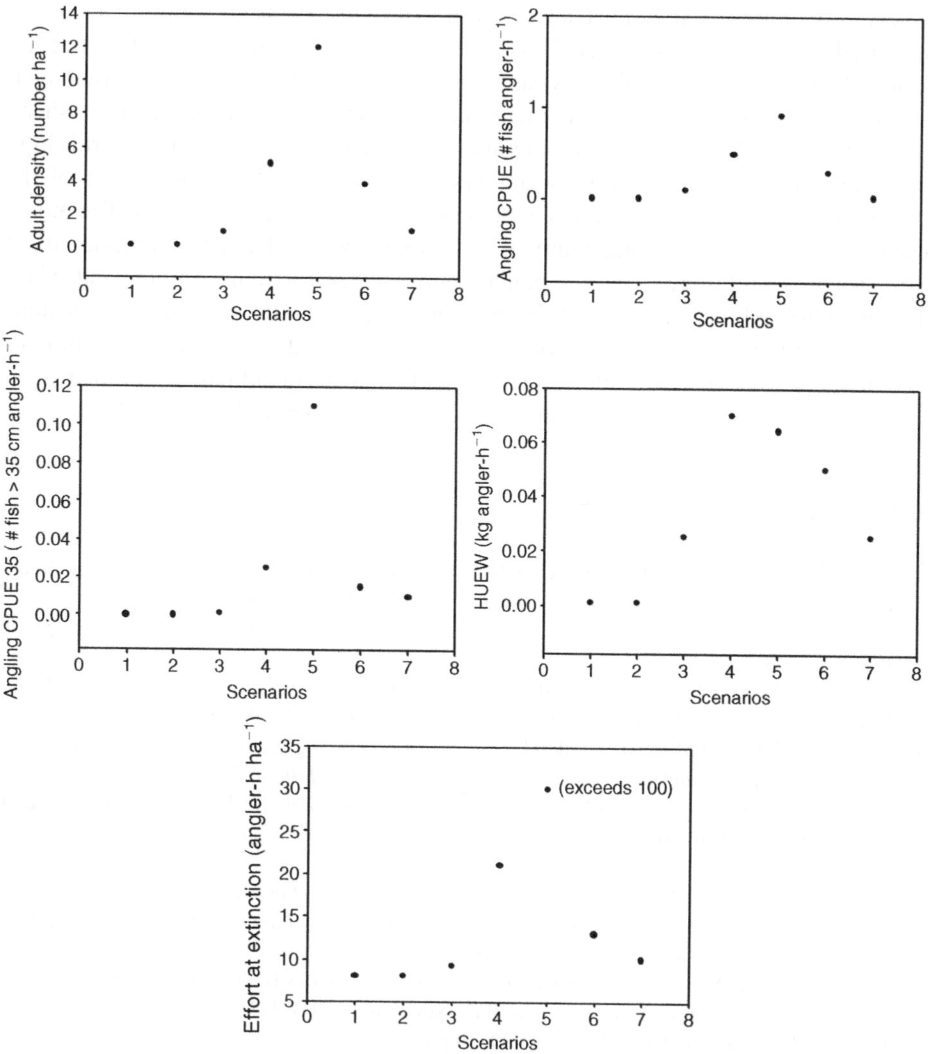

Figure 22.8 Results of management simulations: 1. no regulations; 2. bag limit; 3. <22.5 cm minimum size limit; 4. <28 cm minimum size limit; 5. <35 cm minimum size limit; 6. 22–28 cm slot limit; and 7. >35 cm maximum size limit (indicators variable responses from left to right descending; adult density, angling CPUE, angling CPUE 35, HUEW and effort at extinction)

preventing excessive angler harvest of brook trout. These previous studies support the predictions of the model presented.

The model predictions imply that the maximum sustainable yield of brook trout on Indian Bay Ponds was approximately 0.4 kg ha^{-1}, and the maximum yield occurs when angling effort is approximately 3 angler-h ha^{-1}. The model predicted effort at extinction of approximately 8 angler-h ha^{-1}. The model suggested that the Indian Bay fishery is operating near full capacity in terms of the brook trout yield it can supply. Increased fishing effort is expected to result in a decreased yield and further degradation of fishing quality. Simulations based on the model indicated that size-based regulations are needed to accommodate increased effort and sustain a high quality fishery. Creel limits that allow six fish (or 2 lb plus 1 fish) are not effective when effort is high (i.e. 7 angler-h ha^{-1}) because most anglers will not catch their limit. None of the regulations evaluated will provide a good food fishery when effort is high. The model has weaknesses due to the uncertainty of several parameters. These uncertainties affect estimates of key reference values (e.g. maximum yield, effort at maximum yield, effort at extinction) and the diagnosis of fishery status. They should not affect conclusions regarding the relative effectiveness of different management actions. The real utility of these models lies is that they allow the manager to analyse quantitatively a full array of management options before selecting one as optimal (Power & Power 1996).

References

Clarke Jr R.D., Alexander G.R. & Gowing H. (1980) Mathematical descriptions of trout-stream fisheries. *Transactions of the American Fisheries Society* **109**, 587–602.

Damman A.H.W. (1983) An ecological subdivision of the island of Newfoundland. *Monographiae Biologicae* **48**, 163–206.

Fowlow P.S., Hoenig J.M. & van Zyll de Jong M.C. (1996) Creel Estimates Indian Bay Brook Fishery (1994–1997). *Atlantic Fisheries Research Document 96/11*, St John's: DFO, 26 pp.

Hunt R.L., Brynildson O.M. & Mc Fadden J.T. (1962) Effects of angling regulations on a wild brook trout fishery. *Wisconsin Conservation Department Technical Bulletin 26*, 154 pp.

Hunt R.L. (1970) A compendium of research on angling regulations for brook trout conducted at Lawrence creek Wisconsin. *Wisconsin Department of Natural Resources, Research Report 54*, Madison, 124 pp.

Lester N.P., Petzold M.M., Dunlop W.I., Monroe B.P., Orsatti S.D., Schaner T. & Wood D.R. (1991) *Sampling Ontario Lake Trout Stocks: Issues and Standards: Lake Trout Synthesis*. Toronto: Ontario Ministry of Natural Resoures, 117 pp.

Lester N.P., Korver R.M., van Zyll de Jong M.C., Norris W. & Wicks B. (1999) A model for managing exploitation of brook trout in Indian Bay, Newfoundland. *DFO Atlantic Stock Assessment Research Document 99/99*. St John's: DFO, 19 pp.

Macfadden J.T. (1964) A population study of the brook trout (*Salvelinus fontinalis*) *Wildlife Monographs* 7, 73 pp.

O'Connell M.F. & Dempson J.B. (1996) Spatial and temporal distributions of salmonids in two ponds in Newfoundland, Canada. *Journal of Fish Biology* **48**, 738–757.

Payne N.R., Korver R.M., MacLennan D.S., Nepszy S.J., Shuter B.J. & Thomas E.R. (1990) The harvest potential and dynamics of lake trout populations in Ontario. *Lake Trout Synthesis Population Dynamics Working Group Report*. Toronto, Ontario: Ontario Ministry of Natural Resources, 72 pp.

Pollock K.H., Hoenig J.M., Jones C.M. & Greene C. (1997) Catch rates estimation for roving and access point surveys of anglers. *North American Journal of Fisheries Management* **17**, 11–20.

Power G. (1980) The brook char, *Salvelinus fontinalis* . In E.K. Balon (ed.) *Charrs: Salmonid Fishes of the Genus Salvelinus*. The Hague: Junk, pp. 141–203.

Power M. & Power G. (1996) A modelling framework for analyzing anthropogenic stresses on brook trout (*Salvelinus fontinalis*) populations. *Ecological Modelling* **80**, 171–185.

Quinn N.W.S., Korver R.M., Hicks F.J., Monroe B.P. & Hawkins R.R. (1994) An empirical model for lentic brook trout. *North American Journal of Fisheries Management* **14**, 692–709.
Ryan P.M. (1984) Fyke net catches as indices of abundance of brook trout, *Salvelinus fontinalis* and Atlantic salmon *Salmo salar*. *Canadian Journal of Fisheries and Aquatic Sciences* **41**, 377–380.
Schnabel Z.E. (1938) The estimation of the total fish population of a lake. *American Mathematical Monographs* **45**, 348–352.
Scott W.B. & Crossman E.J. (1973) Freshwater fishes of Canada. *Fisheries Research Board of Canada Bulletin* **184**, 966 pp.
Shepherd J.G. (1982) A versatile new stock recruitment relationship for fisheries and the construction of sustainable yield curves. *Journal du Conseil International pour l'Exploration de la Mer* **40**, 67–75.
Shetter & Alexander (1966) Angling and trout populations of the north branch of the au sable river, Crawford and Otsego counties, Michigan, under special regulations, 1958–63. *Transactions of the American Fisheries Society* **95**, 85–91.
Shuter B.J., Jones M.L., Korver R.M. & Lester N.P. (1998) A general life history model for regional management of fish stocks: the inland lake trout (*Salvelinus namaycush*) fisheries of Ontario. *Canadian Journal of Fisheries and Aquatic Sciences* **55**, 2161–2177.
van Zyll de Jong M.C, Lester N.P., Korver R.M., Norris W & Wicks B.L. (1999) Population dynamics and recreational fisheries of brook trout (*Salvelinus fontinalis*) in Indian Bay Brook Newfoundland. *DFO Atlantic Stock Assessment Secretariat Research Documents 99/98*. St John's, 41 pp.

Chapter 23
Factors affecting the performance of stillwater coarse fisheries in England and Wales

R. NORTH*
Environment Agency, National Coarse Fisheries Centre, Kidderminster, Worcestershire, UK

Abstract

Angling for coarse (non-salmonid) fish involves over one million active participants in England and Wales. Performance of stillwater coarse fisheries, measured as angler catch ($g\,man^{-1}\,h^{-1}$) was influenced primarily by fish abundance but was considerably modified by seasonal factors, especially temperature. Fish population data from 280 fisheries revealed a negatively skewed distribution of fish biomass (4.5–14 280 kg ha^{-1}). Fisheries management strategies strongly influenced fish biomass and species composition, with intensively managed waters being dominated by carp, *Cyprinus carpio* L., and roach, *Rutilus rutilus* (L.), and most likely to contain alien or riverine fish species.

Anglers' catch rates in case study fisheries were ranked in order of fish biomass, with commercially run, intensively managed fisheries consistently producing much higher catches than have previously been published for stillwaters or rivers. It appears that, given appropriate management of habitat, fish and anglers, stillwater fisheries can sustain unnaturally high fish stocks, produce consistently excellent catch rates and withstand exceptionally high angling pressures.

Keywords: cyprinids, fisheries management, recreational fisheries, stocking.

23.1 Introduction

Angling is one of the most popular participant sports in the United Kingdom. A national survey of angling (National Rivers Authority 1995a) reported that over three million people went fishing in 1993 and 1994. This figure is most likely an overestimate of the number of anglers who fished regularly. Their numbers are probably better gauged by annual fishing licence sales which, for anglers aged 12 years and over in England and Wales, were 1.002 million in 1994 (National Rivers Authority 1995b) and 1.144 million in 1998 (Environment Agency 1999).

In the 1994 NRA survey, a large majority of anglers (86%) fished at some time for non-salmonid (coarse) fish and 67% fished exclusively for coarse fish. Over half of coarse fish anglers (52%) showed a preference for angling in stillwaters rather than rivers or canals. Since 1994, the proportion of anglers favouring stillwaters is likely to

*Correspondence: Environment Agency, National Coarse Fisheries Centre, Arthur Drive, Hoo Farm Industrial Estate, Worcester Road, Kidderminster, Worcestershire DY11 7RA, UK (email: rick.north@environment-agency.gov.uk).

have increased in response to the growing number of purpose-built stillwater fisheries that have been created, often near to centres of population, offering ease of access and a high likelihood of good catches.

Practically all coarse fishing in England and Wales is purely recreational in that the catch of fish is returned alive to the water either immediately after capture or at the end of the fishing session. With the exception of fish that may be killed for food, pike, *Esox lucius* (L.), pikeperch, *Stizostedion lucioperca* (L.), eel, *Anguilla anguilla* L., and occasionally perch, *Perca fluviatilis* (L.), fishing mortality in its normally accepted sense is minimal and generally not a significant factor in population dynamics.

The Environment Agency has a legal responsibility not only to regulate freshwater fisheries but also to maintain, develop and improve them (Salmon and Freshwater Fisheries Act 1975). Recognising the present and future importance of stillwater fisheries to coarse angling, the Environment Agency investigated factors affecting their performance. Improved understanding of the links between fish biomass, species composition, fisheries management strategies and fisheries performance measured as angler catch should lead to a better-informed implementation of fisheries legislation relating to the movement and introduction of fish. It will also permit the Environment Agency to actively encourage fishery owners and managers to adopt sustainable management strategies to the ultimate benefit of fish and anglers alike.

This chapter presents preliminary data on fish biomass, density, species composition and growth in stillwaters, together with angling catch data from case study fisheries.

23.2 Methods

Data on fish populations in stillwaters were collected from survey reports from the various fisheries departments of the Environment Agency and its predecessor organisations across England and Wales. Most surveys were by seine netting with a small number by electric fishing or hydroacoustic techniques. Some data were derived from dewatering of stillwaters.

Surveys were considered suitable for inclusion if the author of the report had calculated a population estimate by an appropriate method, e.g. mark-recapture, catch depletion, area-based minimum. The majority of surveys included species-specific biomass and density estimates but a small proportion recorded only total values. A proportion of the surveys also provided growth rate information, where growth was expressed relative to the standard growth rates of Hickley & Dexter (1979), but sufficient data were available only for roach, *Rutilus rutilus* (L.) and bream, *Abramis brama* (L.).

Three categories of stillwater fishery were recognised according to the prevailing management strategy. Natural fisheries were those where there had been no known fisheries management activity of any kind for at least 25 years. Fish populations in natural fisheries are maintained entirely by recruitment from existing stocks. Improved fisheries were those where fisheries management of some kind had taken place within the last 25 years with the objective of improving the quality of angling. Usually the management activities involved the stocking or removal of moderate quantities of fish or minor habitat manipulation. Fish stocks in improved fisheries are maintained principally by natural recruitment, albeit partly from recently stocked fish, and remain within limits set by environmental parameters. Intensive fisheries exist where the management objective is to maximise the fish stock by extensive manipulations and are operated exclusively for angling. Natural recruitment in intensive fisheries rarely plays a part in determining the total population size.

Angler catch data were obtained directly from stillwater fisheries in the north-west and midlands areas of England. Choice of fisheries was limited as many fishery owners keep scant or no records of angler catch. The three fisheries used as case studies were venues for regular angling competitions and comprehensive records of catches were kept. Catch data consisted of results from organised competitions by individual anglers from permanently numbered fishing positions and comprised weights of fish caught, date and duration of angling. Typical length of fishing competitions was 5 h but a few were fished over 6, 3 or 2.5 h. Catch rates (g $man^{-1} h^{-1}$) were calculated for all anglers whose catches were weighed. A small and variable proportion of anglers returned their fish to the water without weighing them if they thought that they could not win a prize, thus catch data are minima. Species composition of the catch was often not formally recorded. Because of concerns over commercial confidentiality of some information, especially concerning stocking densities, case study fisheries are not identified.

23.2.1 *Case study fisheries*

Fishery 1 is a lake of some antiquity located on an estate in the English Midlands. It has an area of 30.4 ha with 95 fishing stations. In 1998, coarse fish biomass estimated by hydroacoustic survey was 550–600 kg ha^{-1}. Rainbow trout, *Oncorhynchus mykiss* (Walbaum), are stocked in the lake and support a separate fishery in spring and summer with the coarse fishery operating from September to March each year. Catch data were available from 1993/1994 to 1998/1999.

Fishery 2 is a relatively recently created intensively managed venue with three pools in the north-west of England. One of these pools, a fishery of 0.9 ha accommodating 61 anglers, was selected as the study site. It supports a biomass of approximately 2000 kg ha^{-1}. Permanently installed blown-air aeration facilities are present and used at times of low dissolved oxygen concentration. Catch data were available for the whole of the first year of operation, September 1998 to August 1999.

Fishery 3, created in 1991 in the English Midlands, is one of the earlier examples of an intensive fishery in the UK. It is a complex of seven pools and the largest (0.8 ha), which supports 60 anglers, was selected for study. Fish biomass in the pool was estimated to be at least 4000 kg ha^{-1}, based on stocking history. Growth and natural recruitment is likely to have increased the biomass beyond that given but no subsequent estimate is available. By comparison, the mean biomass at harvest of fish in rearing ponds of the Environment Agency's coarse fish farms is 3720 kg ha^{-1}. Removable paddle-wheel aerators are installed and used during periods of low oxygen concentration or potential ice-cover. Catch data were available for 45 weeks of 1999; data for the period 18 April to 7 June had been lost. Fishery 3 was the only site where more

Table 23.1 Species of fish present and weight range of individual fish captured by anglers in the three case study fisheries

	Species presence and weight range (g)		
Species	Fishery 1	Fishery 2	Fishery 3
Carp[a]	>5000	100–250	1000–4000
Crucian carp	–	100–200	200–500
Goldfish[b]	–	50–200	100–450
Gudgeon	–	15–30	–
Tench	–	100–250	250–750
Bream	100–1500	–	25–1500
Rudd	–	25–150	–
Roach	50–450	25–150	10–500
Chub	–	–	450–650
Orfe[b]	–	25–100	–
Dace	–	–	50–200
Perch	15–150	–	25–900
Pike	Up to 15 000	–	–

[a]Includes common, mirror, leather, ghost and koi.
[b]Includes 'wild' and ornamental colour variants.

than one competition was held on the same day. Here it was common practice during the summer months for there to be one competition in the day and a second, on the same fishing stations and often involving many of the same anglers, in the evening.

Table 23.1 shows the species of fish present and the size range of fish caught by anglers in each of the case study fisheries.

Water temperature was not recorded on a regular basis at any of the fisheries. For Fisheries 1 and 3 an inferred temperature was derived from continuously recorded temperature of the River Severn at Bewdley (NGR SO 782 762), which was within 25 km of the fisheries. It was considered that monthly mean river temperature gave a reasonable approximation of that of the stillwaters.

Supplementary data were provided by the Environment Agency's coarse fish farms, fishery owners and managers.

23.3 Results

23.3.1 *Fish stocks*

Survey information on fish stocks was collected from 280 stillwaters in England and Wales. The size of waters varied from 0.025 to 15 ha, with a mean size of 1.1 ha. Eight waters were considered to be natural fisheries, 199 improved and 52 intensive.

The distribution of fish density was negatively skewed, with a range of 10–126 200 fish ha^{-1} (Fig. 23.1). Fifty per cent of waters contained fewer than 5000 fish ha^{-1}.

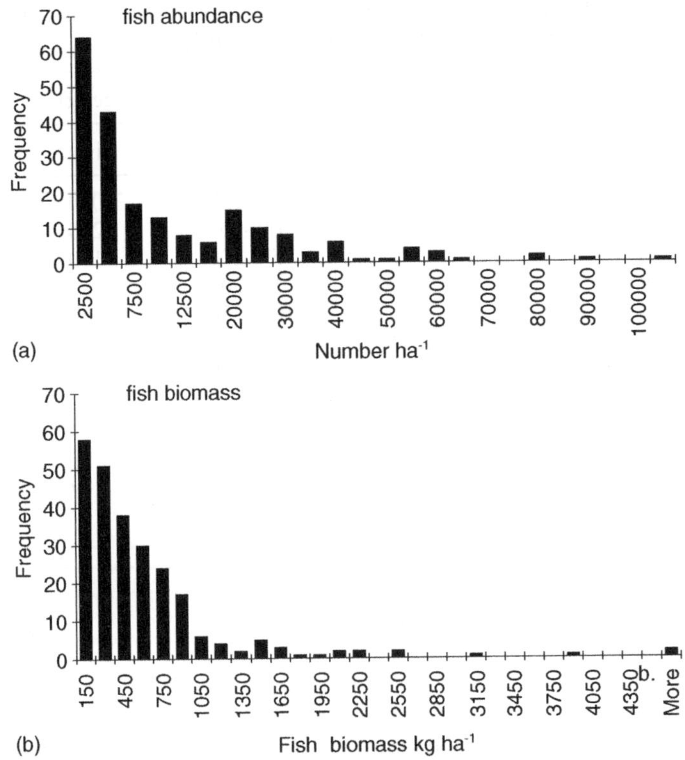

Figure 23.1 Frequency distribution of (a) fish abundance (n ha^{-1}) and (b) fish biomass (kg ha^{-1}) in 280 stillwaters in England and Wales

The biomass distribution (Fig. 23.1(b)) was similar to that for density, with a range of 4.5–14 280 kg ha^{-1} and a mean of 592 kg ha^{-1}. Individual species' biomass distributions closely resembled those of total biomass with the majority of fisheries containing relatively low standing stocks.

Natural fisheries contained a maximum biomass of <500 kg ha^{-1} and 75% were <250 kg ha^{-1} (Fig. 23.2). The robustness of these data may be compromised by the small sample size. Improved fisheries showed the typical, negatively skewed distribution, with up to 2250 kg ha^{-1} (Fig. 23.2) in some waters and a median value of 325 kg ha^{-1}. Intensive fisheries exhibited the highest biomass of the fisheries surveyed, with a maximum of 14 280 kg ha^{-1} and a median of 750 kg ha^{-1} (Fig. 23.2). Intensive fisheries that operated as high-profile commercial waters typically held 1500–3500 kg ha^{-1} and sometimes more. Figure 23.3 shows the median, 40–60 and 25–75 percentile biomass ranges for the three types of fishery.

A total of 20 species of fish was recorded. The number of species in any fishery was 11 or less and followed a normal distribution, with a mode of five. Very few fisheries contained fewer than three or more than eight species. Roach was the most frequently occurring species, found in 83% of waters surveyed. Carp, bream, perch

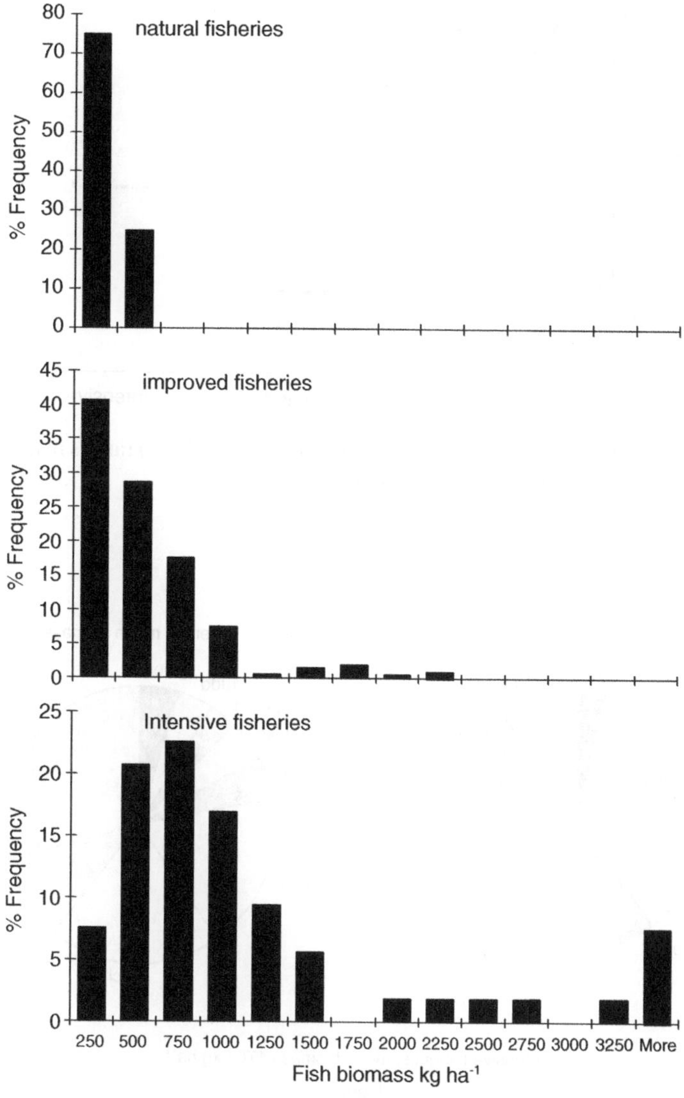

Figure 23.2 Percentage frequency of fish biomass (kg ha^{-1}) in three categories of stillwater coarse fishery in England and Wales

and tench were also common and present in 54–59% of waters. Rudd, *Scardinius erythrophthalmus* (L.), were found in 46%, crucian carp, *Carassius carassius* (L.), in 36% and pike in 33% of fisheries. Occurrence of crucian carp may have been overestimated because of confusion with brown goldfish, *Carassius auratus* (L.). Other species were present in 45% of fisheries and included minor species (gudgeon, *Gobio gobio* (L.), three-spined stickleback, *Gasterosteus aculeatus* (L.)), native riverine species (chub, *Leuciscus cephalus* (L.), barbel, *Barbus barbus* (L.), dace, *Leuciscus leuciscus* (L.)) and alien species (brown goldfish, ornamental goldfish, *Carassius auratus* (L.),

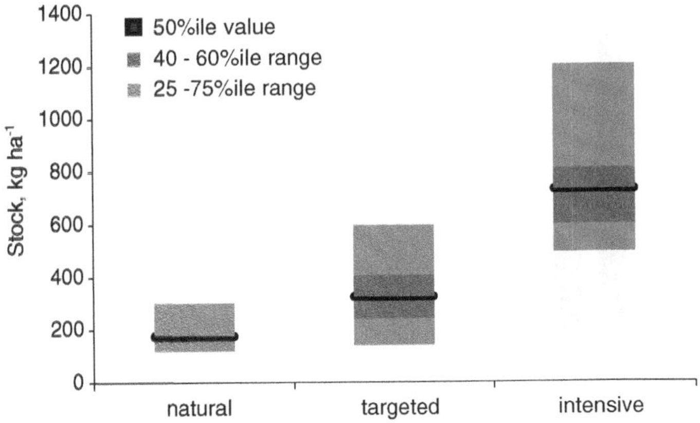

Figure 23.3 Median, 25–75 and 40–60 percentile biomass (kg ha^{-1}) ranges in three categories of stillwater coarse fishery in England and Wales

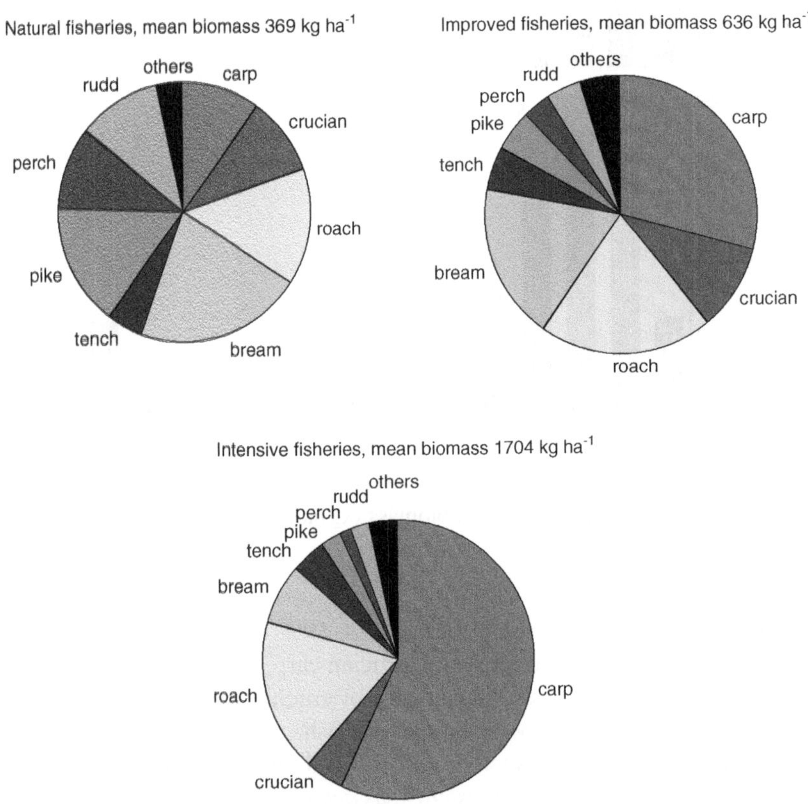

Figure 23.4 Relative abundance by biomass (kg ha^{-1}) of fish species in three categories of stillwater coarse fishery in England and Wales

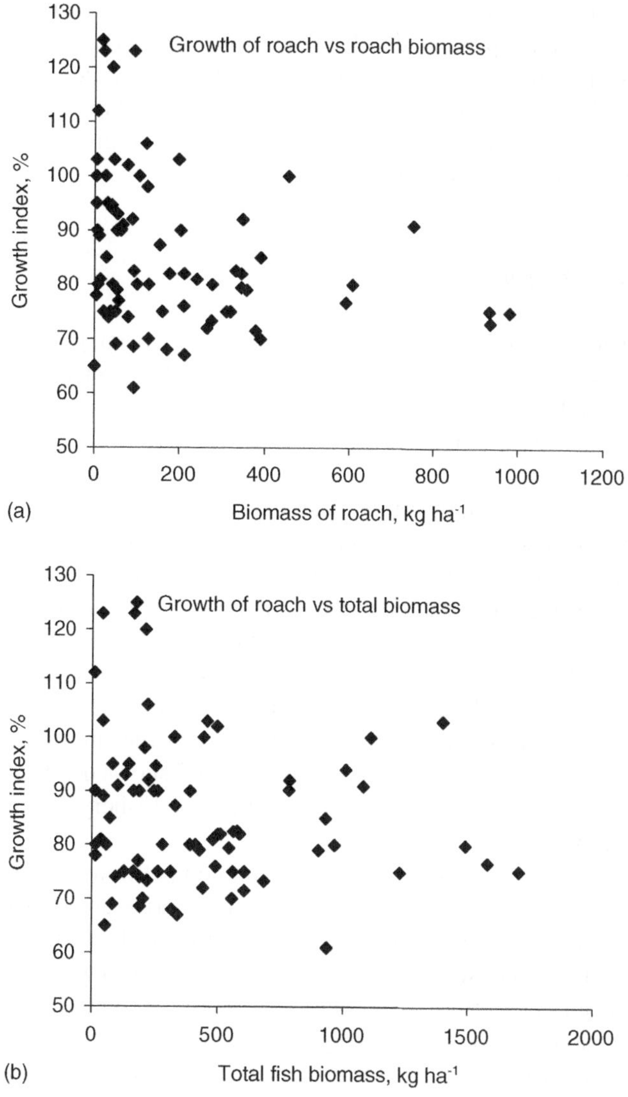

Figure 23.5 Observed growth indices (Hickley & Dexter 1979) for roach against biomass (kg ha^{-1}) of (a) roach and (b) all fish present

koi carp, ghost carp, golden and silver orfe and grass carp, *Ctenopharyngodon idellus* (Val.)).

The relative abundance by weight of the different species varied with the category of fishery (Fig. 23.4). Natural fisheries contained a similar biomass of roach, bream, pike, perch and rudd. Improved fisheries were dominated by carp, roach, bream and crucian carp. These four species accounted for over 75% of the biomass. In intensive fisheries, carp predominated, contributing approximately 60% of the total biomass, and together with roach accounted for over 80% of the total. Intensive fisheries contained the lowest proportions of pike, perch and rudd. Tench, *Tinca tinca* (L.) was the only species that made up a similar proportion of the biomass in all fishery types.

Growth rates of roach and bream appeared to be influenced by the total and species biomass (Fig. 23.5(a) and (b)) for roach; maximum observed growth indices declined with increasing total or roach biomass.

23.3.2 *Fishery performance*

Results from a total of 680 angling competitions (386 from Fishery 1, 110 from Fishery 2 and 184 from Fishery 3), representing over 60 000 man h of angling effort and producing a total catch of over 86 500 kg, were collected. Angling pressure varied among the three fisheries, ranging from 188 man $h^{-1}ha^{-1}$ $year^{-1}$ for Fishery 1, to 14 744 and 17 742 man $h^{-1}ha^{-1}$ $year^{-1}$ for Fisheries 2 and 3, respectively. Mean catch rates of anglers weighing in at the three fisheries were: 671 g $man^{-1}h^{-1}$ for Fishery 1, 1029 g $man^{-1}h^{-1}$ for Fishery 2 and 3447 g $man^{-1}h^{-1}$ for Fishery 3. As Fishery 1 operated a restricted fishing season (September–March) a more reasonable comparison may be drawn from the catch rates for the same period in the other fisheries, which were 672 g $man^{-1}h^{-1}$ for Fishery 2 and 2753 g $man^{-1}h^{-1}$ for Fishery 3.

A strong seasonality was evident in the performance of all fisheries (Fig. 23.6). The most obvious environmental variable influencing metabolic activity in fish is water temperature and there was a close correlation between the inferred monthly mean temperature and monthly mean catch rates (e.g. Fig. 23.7 shows the data for Fishery 1). Although it was not possible to infer temperatures for Fishery 2 it is likely that a similar relationship existed.

Mean individual catch rates for anglers winning competitions and in each position down to 20th for Fishery 3 and 25th for the other fisheries are shown in Fig. 23.8. Catch rates in the fisheries were ranked in the same order as the fish biomass. For individual fishing competitions, winning catches varied with season but followed a similar pattern to the mean catch rates. At Fishery 3 winning weights ranged from 3.8 to 80.7 kg with a mean of 34.6 kg; for Fishery 2 the winning weights varied from 2.67 to 27.7 kg with a mean of 11.69 kg; and for Fishery 1 the winning weights were from 1.53 to 20.7 kg with a mean of 9.28 kg.

Overall exploitation rates, expressed as total annual catch as a percentage of estimated total stock, were 24% for Fishery 1, 737% for Fishery 2 and 1540% for Fishery 3. Exploitation from an effort of 100 man h of angling was 0.38% for Fishery 1, 5.7% for Fishery 2 and 10.8% for Fishery 3.

23.4 Discussion

In stillwaters in England and Wales fishery managers intending to create 'better fisheries' have over recent decades sought to influence fish populations and hence fishery performance in various ways, mostly by direct manipulation of fish stocks.

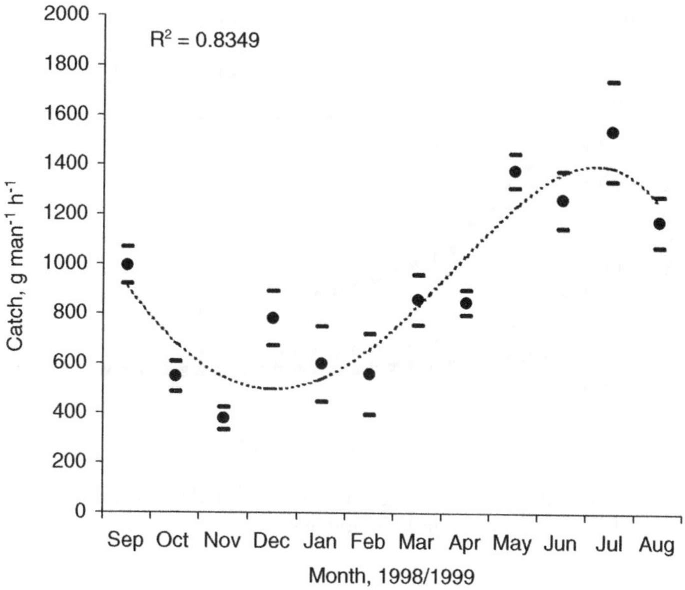

Figure 23.6 Seasonal variation in mean monthly angler catch rate (g man^{-1}h^{-1} + 1 SE) in Fishery 2 over a 12-month period

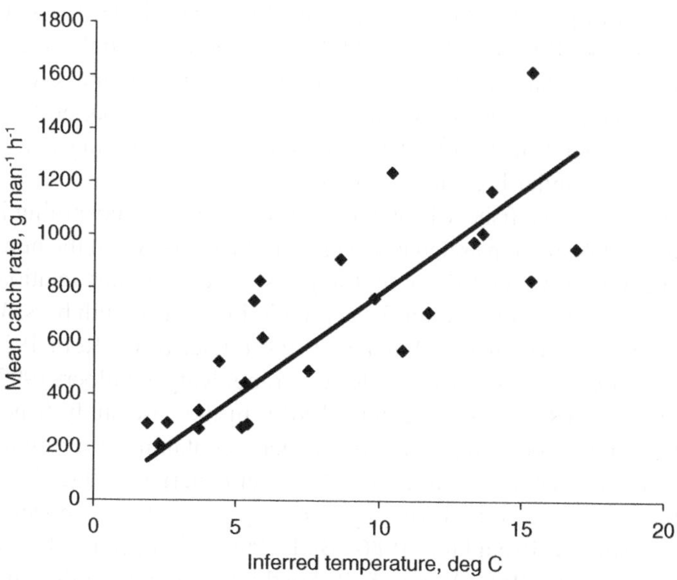

Figure 23.7 Monthly catch per unit effort (g man^{-1}h^{-1}) in Fishery 1 in 1995–1999, in relation to inferred mean monthly water temperature ($r = 0.8104$, $n = 24$, $P < 0.01$)

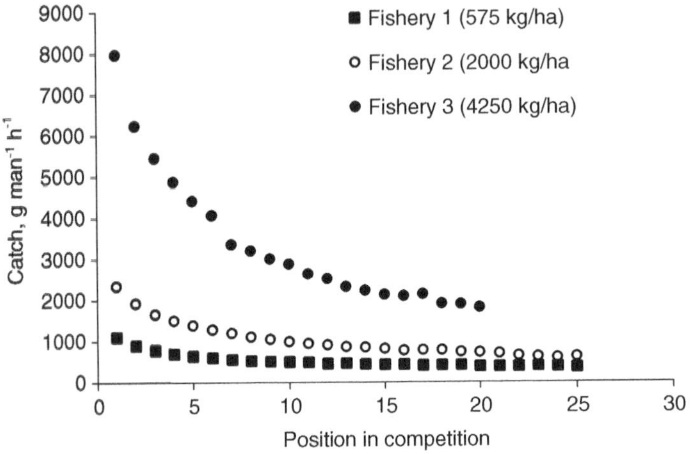

Figure 23.8 Mean individual catch rates (g man^{-1}h^{-1}) of anglers' fishing competitions in three stillwater coarse fisheries with different fish biomass

As a result of such direct management, improved and intensively managed fisheries contain fish biomasses higher than that in natural fisheries, with distorted species compositions dominated by carp, roach and bream.

A large number of fisheries supported fish biomasses in excess of those published for eutrophic lakes in the UK and Europe where populations over 350 kg ha^{-1} and occasionally as low as 150 kg ha^{-1} have been suggested as ecologically damaging (e.g. Wright & Phillips 1992; Giles, Wright & Shoesmith 1995; Prejs, Pijanowska, Koperski, Martyniak, Boron & Hliwa 1997). Frequently, natural recruitment from originally stocked fish has resulted over time in very high biomasses, in excess of 2000 kg ha^{-1} in one case. Indeed several intensive fisheries supported fish biomasses higher than those found in the Environment Agency's fish rearing units and which were maintained by similar life-support facilities.

The quantity of fish in a fishery is one of the primary factors governing its performance. Beukema (1969) demonstrated a strong relationship between the number of carp in an experimental stillwater fishery and anglers' catch. Similar relationships have been found in other types of stillwater fishery, including largemouth bass, *Micropterus salmoides* (Lacepède), (Buynak & Mitchell 1993; Lagler & De Roth 1953) walleye, *Stizostedion vitreum* (Mitchill), (Beard, Hewett, Yang, King & Gilbert 1997) and some marine reef fisheries (Richards & Schnute 1986). In the case study fisheries, angler catch was related to stock size, with the fishery containing the highest biomass producing the highest catch rates. Intensive stillwater fisheries produced catch rates far in excess of those published for UK river systems (e.g. North 1980; Cooper & Wheatley 1981; Cowx, Fisher & Broughton 1986; Ball 1996), European stillwater fisheries (Gerard & Timmermans 1991; Wolos & Piskorski 1991) and North American sport fisheries (e.g. Quertemus 1991).

Performance of all three types of fishery was strongly influenced by other factors in the short term. Water temperature is likely to have been the single most important abiotic factor but others probably included dissolved oxygen concentration, pH and ammonia concentration. In fisheries that are used throughout the year, which is the great majority in England and Wales, seasonal metabolic changes in the fish such as reproductive status might also influence angling performance. There are many anecdotal reports of several species becoming virtually uncatchable shortly before and during spawning.

A fish's previous experience of capture may affect fishery performance in two ways. Some species of fish learn to some degree to avoid a second or subsequent capture. Beukema (1970a, b) and Raat (1985) demonstrated acquired hook avoidance in different strains of carp and pike. Hook avoidance acquired by naïve carp lasted for at least 1 year after first capture but vulnerability to angling did not decrease further if a fish was captured more than once. A one-step learning process could account for the consistent performance of intensively fished waters such as Fishery 3 where individual fish are probably caught up to 15 times per year but do not become progressively more difficult to catch. The observed recapture frequency exceeds published figures for carp, e.g. three tagged fish caught seven times each in 3 years (Linfield 1980).

Learned hook avoidance is one element of fish capture affecting fishery performance over a relatively long term but it is often assumed by anglers that a fish's experience of being caught also has short-term physiological effects that reduce the probability of its recapture. There is evidence to suggest that the impact of such physiological effects may be overestimated. Pottinger (1998) reported that blood chemistry of carp returned to pre-capture norms within 24 h of being caught and that recovery was not affected by retention in keepnets. Hackney & Linkous (1978) reported that a tagged largemouth bass was caught by artificial lures three times in just over 1 h and I. Talbot (personal communication) reported a tagged tench being caught from the same place and on the same bait on 3 consecutive days. Longer-term physiological impact of captures is also unlikely as carp that were caught more than once by anglers showed better growth than those caught once or not at all (Raat 1985).

Angler-related factors exert an important influence on fishery performance. Accessibility of the fish to the anglers varies according to the size and configuration of the stillwater in question. At Fishery 1, for example, anglers in a typical 25-entrant competition would be able to fish approximately 21% of the total lake area, assuming an effective maximum fishing range of 40 m. The same number of anglers would theoretically be able to fish almost eight times the area of Fishery 3 although this was not possible. The process of angling has mechanical limits that determine the largest possible catch from any given water. These limits come into operation if anglers are fishing for either very large fish, e.g. 3–6 kg carp which may take 10–15 min each to land, or very small fish where only very few skilled anglers are capable of exceeding a catch rate of >100 fish h^{-1}.

Stillwater coarse fisheries in the UK seem able to support very high angling effort without apparent detriment. Shuter, Jones, Korver & Lester (1998) considered that 4.0–6.6 man $h^{-1} ha^{-1} year^{-1}$ was the fishing effort that would produce the maximum sustainable yield of lake trout, *Salvelinus namaycush*, Walbaum, in Ontario lakes.

More comparable examples of fishing effort are available for European cyprinid fisheries. Wolos (1991) reported fishing effort of 118–1225 man $h^{-1}ha^{-1}year^{-1}$ in Polish stillwaters and Gerard (1998) 3965 man $h^{-1}ha^{-1}year^{-1}$ in a Belgian ship canal. Fishing effort in natural and improved fisheries were within the published range but at intensive fisheries it was 5–15 times higher.

Management of anglers with respect to fish welfare plays a crucial role in maintenance of fish stocks in intensive fisheries but is also an increasingly significant aspect of the management of other types of fishery. Fishery rules concerning hook size and type (barbless) are almost universal and are widely believed to reduce damage to fish and consequent mortality, although some authors consider benefits of barbless hooks to be exaggerated (Schill & Scarpella 1997). Restrictions governing the type and quantity of bait that may be used are frequently applied with a view to avoiding water quality and fish nutrition problems.

Perhaps the most effective rules apply to the retention of captured fish in nets and their subsequent weighing. Many fisheries require small and large fish to be held in separate nets and limit the amount of fish that may be held in one net. While this practice may not be essential for water quality reasons (Pottinger 1997), it does reduce physical damage at the time when fish are removed from the net for weighing.

The present study has identified and partially quantified several of the more important factors affecting the performance of stillwater coarse fisheries. Fish population size, as biomass, appears to be the primary determinant of angler catch but the relationship is modified strongly by seasonally varying environmental factors, especially temperature. Notwithstanding the variation among fisheries in absolute catch rates of anglers, similarities in the overall pattern of performance were observed in different fishery types suggesting that a common set of factors is probably involved in determining the angling performance of all stillwater coarse fisheries. Evidence from intensively managed fisheries indicated that very high stock densities, and consequently angler catch rates, can be established and maintained by suitable management of fish and anglers.

Acknowledgements

The author would like to thank the Environment Agency Area Fisheries Scientists, who provided the data on fish population surveys and all the fishery owners, managers and anglers who freely gave of their opinion, expertise and catch information. The opinions expressed are those of the author and not of the Environment Agency.

References

Ball M. (1996) An analysis of coarse fishing match data from the Reading and District Angling Association, River Kennet 1981–1985. *Internal Report of National Rivers Authority, Thames Region, Oxford* 1, 40 pp.

Beard D.T., Hewett S.W., Yang Q., King R.M. & Gilbert S.J. (1997) Prediction of angler catch rates based on walleye population density. *North American Journal of Fisheries Management* **17**, 621–627.

Beukema J.J. (1969) Angling experiments with carp (*Cyprinus carpio*) I: Differences between wild, domesticated and hybrid strains. *Netherlands Journal of Zoology* **19**, 596–609.

Beukema J.J. (1970a) Acquired hook avoidance in the pike (*Esox lucius* L.) fished with artificial and natural baits. *Journal of Fish Biology* **2**, 155–160.

Beukema J.J. (1970b) Angling experiments with carp, *Cyprinus carpio* II: Decreasing catchability through one trial learning. *Netherlands Journal of Zoology* **20**, 81–92.

Buynak G.L. & Mitchell B. (1993) Electric fishing catch per effort as a predictor of largemouth bass abundance and angler catch in Taylorsville Lake, Kentucky. *North American Journal of Fisheries Management* **13**, 630–633.

Cooper M.J. & Wheatley J.A. (1981) An examination of the fish population in the River Trent, Notingham, using angler catches. *Journal of Fish Biology* **19**, 539–556.

Cowx I.G., Fisher K.A.M. & Broughton N.M. (1986) The use of angler's catches to monitor fish populations in large water bodies with particular reference to the river Derwent, Derbyshire, England, UK. *Aquaculture and Fisheries Management* **17**, 95–104.

Environment Agency (1999) *Salmonid and Freshwater Fisheries Statistics for England and Wales 1998*. Bristol: Environment Agency, 50 pp.

Gerard P. (1998) The impact of angling on the population dynamics of roach (*Rutilus rutilus*, L.) in a former ship canal. In P. Hickley & H. Tompkins (eds) *Recreational Fisheries: Social, Economic and Management Aspects*. Oxford: Fishing News Books, Blackwell Science, pp. 235–244.

Gerard P. & Timmermans J.A. (1991) Comparison of fish population estimates and angling catch estimates in a narrow ship canal in Belgium In I.G. Cowx (ed.) *Catch Effort and Sampling Strategies: Their Application in Freshwater Fisheries Management*. Oxford: Fishing News Books, Blackwell Science, pp.127–133.

Giles N., Wright R.M. & Shoesmith E.A. (1995) The effects of perch, *Perca fluviatilis* L., and bronze bream, *Abramis brama* L., on insect emergence and benthic invertebrate abundance in experimental ponds. *Fisheries Management and Ecology* **2**, 17–25.

Hackney P.A. & Linkous T.E. (1978) Striking behaviour of the largemouth bass and the use of the binomial distribution for its analysis. *Transactions of the American Fisheries Society* **107**, 682–688.

Hickley P. & Dexter K.F. (1979) A comparative index for quantifying growth in length of fish. *Fisheries Management* **10**, 147–151.

Lagler K.F. & De Roth G.C. (1953) Populations and yield to anglers in fishery for largemouth bass. *Paper of the Michigan Academy of Science* **38**, 235–253.

Linfield R.S.J. (1980) Catchability and stock density of common carp, *Cyprinus carpio* L. in a lake fishery. *Fisheries Management* **11**, 11–22.

National Rivers Authority (1995a) National Angling Survey 1994. *Fisheries Technical Report 5*. Bristol: National Rivers Authority, 30 pp.

National Rivers Authority (1995b) *Salmonid and Freshwater Fisheries Statistics for England and Wales*. Bristol: National Rivers Authority 1994, 57 pp.

North E. (1980) The effects of water temperature and flow upon angling success in the River Severn. *Fisheries Management* **11**, 1–9.

Pottinger T.G. (1997) Changes in water quality within anglers' keepnets during the confinement of fish. *Fisheries Management and Ecology* **4**, 341–354.

Pottinger T.G. (1998) Changes in blood cortisol, glucose and lactate in carp retained in anglers' keepnets. *Journal of Fish Biology* **53**, 728–742.

Prejs A., Pijanowska J., Koperski P., Martyniak A., Boron S. & Hliwa P. (1997) Food web manipulation in a small, eutrophic lake Wirbel, Poland: long-term changes in fish biomass and basic measures of water quality. A case study. *Hydrobiologia* **342/343**, 383–386.

Quertemus C.J. (1991) The use of Bass club tournament results to evaluate relative abundance and fishing quality. *American Fisheries Society Symposium* **12**, 515–519.

Raat A.J.P. (1985) Analysis of angling vulnerability of common carp, *Cyprinus carpio* L., in catch-and-release angling in ponds. *Aquaculture and Fisheries Management* **16**, 171–187.

Richards L.J. & Schnute J.T. (1986) An experimental and statistical approach to the question: is catch per unit effort an index of abundance? *Canadian Journal of Fisheries and Aquatic Sciences* **43**, 1214–1227.

Schill D.J. & Scarpella R.L. (1997) Barbed hook restrictions in catch-and-release trout fisheries: a social issue. *North American Journal of Fisheries Management* **17**, 873–881.

Shuter B.J., Jones M.L., Korver R.M. & Lester N.P. (1998) A general life history based model for regional management of fish stocks: the inland lake trout (*Salvelinus namaycush*) fisheries of Ontario. *Canadian Journal of Fisheries and Aquatic Sciences* **55**, 2161–2177.

Wolos A. (1991) Angler's opinions as to the quality of the fishing and the fishery management in selected Polish waters. In I.G. Cowx (ed.) *Catch Effort Sampling Strategies: Their Application in Freshwater Fisheries Management.* Oxford: Fishing News Books, Blackwell Science, pp. 134–142.

Wolos A. & Piskorski P. (1991) Anglers' catches as an illustration of the community structures, angling pressure and angling regulations, based on inland waters in the Krosno region, Poland. In I.G. Cowx (ed.) *Catch Effort Sampling Strategies: Their Application in Freshwater Fisheries Management.* Oxford: Fishing News Books, Blackwell Science, pp. 166–176.

Wright R.M. & Phillips V.E. (1992) Changes in the aquatic vegetation of two gravel pit lakes after reducing the fish population density. *Aquatic Botany* **43**, 43–49.

Chapter 24
Management of fisheries in a large lake – for fish and fishermen

T. VEHANEN*, P. HYVÄRINEN
Finnish Game and Fisheries Research Institute, Manamansalontie 90, FIN-88 300 Paltamo, Finland

M. MARTTUNEN, P. KYLMÄLÄ
Finnish Environment Institute, Helsinki, Finland

H. TERVO
University of Oulu, Thule-institute, Oulu, Finland

Abstract

Fishermen exploiting fish stocks, for example, for professional or recreational purposes, form a heterogeneous group. Combining the aims and views of these different groups operating in the same lake is an important task for managers. Therefore, in addition to the biological effects of management activities, the aims of the fishermen and economic impacts should be considered. Guidelines from a project conducted in Lake Oulujärvi, one of the largest lakes in Finland are described. The study combined biological information of the lake's fish stocks and the objectives of people exploiting the lake. Possible interactions between fish species were modelled. The economic value of Lake Oulujärvi's fisheries and its direct and indirect impacts on employment in the area were also measured.

Keywords: fisheries management, modelling, stocking, socio-economic factors, *Salmo trutta*, whitefish.

24.1 Introduction

The manipulation of fish stocks and fishing, alongside the physical rehabilitation of the environment, are usually the main tasks for fisheries management in lakes. In addition to biological aspects, socio-economic factors are also important when management decisions are made (Hickley & Tompkins 1998). Fishermen exploiting fish stocks form a heterogeneous group, for example, professional fishermen and the associated fishing sector often stress economic values, whereas many recreational fishermen point out the importance of leisure values. Combining the aims and views of these different stakeholders operating in the same lake is currently an important task for managers.

*Correspondence: Finnish Game and Fisheries Research Institute, Manamansalontie 90, FIN-88 300 Paltamo, Finland (email: vehanen@tp.rktl.fi).

Fish stocking and introductions are widely used management methods, annually involving billions of individual stocked fish in managed fisheries (Cowx & Welcomme 1998). When a stocking programme is planned, careful examination of the possible effects of stocking should be carried out (Cowx 1994; Hickley 1994). In addition to measuring the success of the stocking operation, information on the possible adverse effects and interactions with other fish species should be evaluated. Here, effective models are needed to help managers in the planning of stocking operations (e.g. Jørgensen 1988; Stefansson & Palsson 1998).

In this chapter the current status of Lake Oulujärvi's fish stocks and fisheries were examined, and possibilities for their future development in relation to management and socio-economic aspects were assessed.

24.2 Materials and methods

24.2.1 *Study lake*

Lake Oulujärvi (27°10′N, 64°20′E) has an area of 928 km^2, with a mean depth of 7.6 m and a maximum depth of 36 m. The lake has been regulated since 1951 for hydropower purposes, and the average annual amplitude of the water level is 1.9 m. The colour of the water is below 80 Pt mg L^{-1} and the total phosphorus below 20 $\mu g\ L^{-1}$.

There are about 50 professional fishermen working in the lake annually. Vendace, *Coregonus albula* (L.), and whitefish, *Coregonus lavaretus* (L.) sl., are the main target species for professional fishermen. The largest part of their catch comes from trawling, although gill nets and fyke nets are also used. Annually, slightly less than 4000 households use mainly gill nets and angling in the recreational fishery. In addition to vendace and whitefish, recreational fishermen target predatory species, like brown trout, *Salmo trutta* L., and pike, *Esox lucius* L.

24.2.2 *Population estimates*

Catch statistics from Lake Oulujärvi have been collected since 1973 (Salojärvi 1991; 1992). Virtual population analysis (VPA) was used to estimate the number of vendace, brown trout, pikeperch, *Stizostedion lucioperca* (L.), and whitefish from 1973 to 1995. Two commercially-important types of whitefish exist in the lake: the indigenous blue whitefish (number of gillrakers [mean ± SD] 33.5 ± 3.2) and the stocked northern densely-rakered whitefish (number of gillrakers 52.5 ± 4.2) (Salojärvi 1992). These whitefish were analysed separately. Pikeperch stock in Lake Oulujärvi plummeted in the 1950–1960s due to overfishing and a slight lowering in water temperature (Colby & Lehtonen 1994). The stocking of pikeperch fingerlings was started on a large scale in the mid-1980s, and pikeperch standing stock in 1995 rose due to this stocking. Due to damming and dredging of the spawning rivers, the brown trout catch in Lake Oulujärvi is almost totally based on the stocking of 2–3-year-old fish.

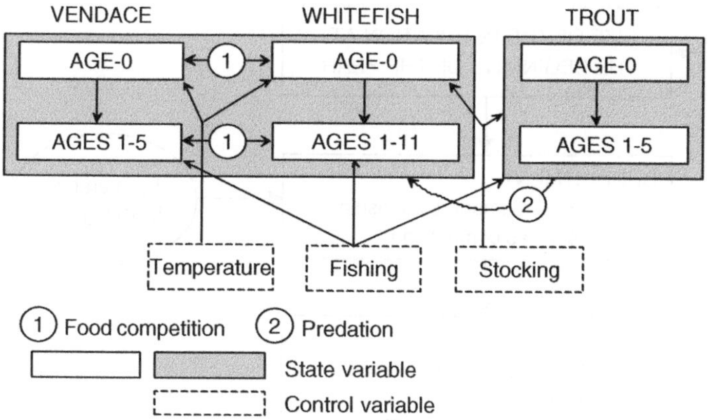

Figure 24.1 A diagram of the most important interactions in the computer model used to model the interactions between fish species in Lake Oulujärvi

24.2.3 *Modelling*

A computer model was used to simulate vendace, whitefish and brown trout stocks and their interactions in Lake Oulujärvi. Pikeperch stock was excluded from the simulations because it only started to recover at the end of the observation period (1973–1995). The model used was designed to work with 1–3 species at a time, and it was executed by Excel® and Visual Basic® (Marttunen & Kylmälä 1997).

The basic processes simulated in the model are fish growth, reproduction and mortality (Fig. 24.1). These can be guided by fish stocking, fishing and water temperature. The deterministic model (Fig. 24.2) requires data on the fish populations in the beginning of simulation, parameters for different interactions (e.g. stock–recruitment curve and density dependent growth, etc.), and information on guiding factors (fish stocking, fishing, temperature). Some of the relationships are optional, but most are obligatory. The model was calibrated using fish data from Lake Oulujärvi. A detailed description is given by Marttunen & Kylmälä (1997).

24.2.4 *Fishermen's opinion inquiry*

The opinions and views of the fishermen were collected by personal interviews and postal questionnaires. A questionnaire was mailed to all the professional fishermen (50) and a random sample of 300 people were selected from the recreational fishermen. Another questionnaire was sent to those who did not reply to the first one. The questionnaire included 26 questions concerning three main themes:

(1) the current status of fish stocks and fisheries in the lake;
(2) conflicts between different groups exploiting the lake; and
(3) views and means for further development of the lake.

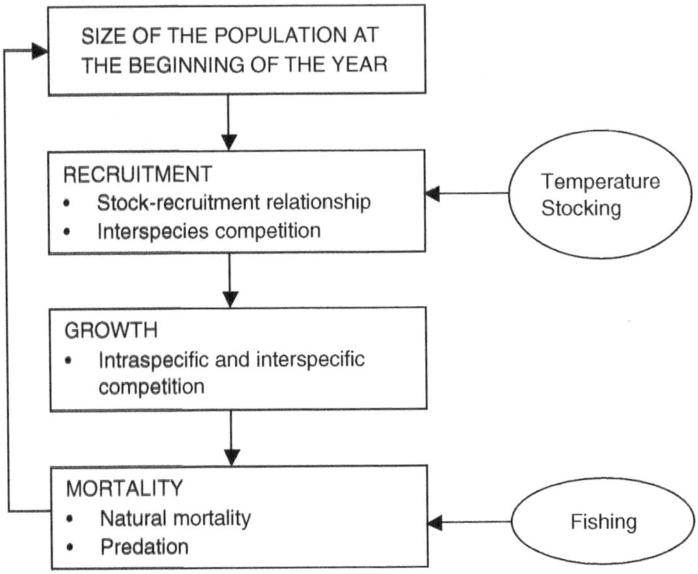

Figure 24.2 Diagrammatic representation of the computer model used to determine the interactions between fish species in Lake Oulujärvi

The answers were analysed in three different categories:

(1) professional fishermen;
(2) local recreational fishermen; and
(3) recreational fishermen not living in the area, but having a country cottage by the lake (non-local recreational fishermen).

To widen the information received from the postal questionnaire, 21 people were interviewed individually. The interviewees were stakeholders in the lake's fisheries: commercial entrepreneurs related to fishing, local and national authorities, professional fishermen, researchers and representatives of fishing organisations. The interview followed the structure of the mailed questionnaire, the themes being the same.

24.2.5 *Economic analysis*

The economic value of Lake Oulujärvi's fisheries in 1995, and its direct and indirect impacts on employment in the area, were evaluated. There are several stakeholders involved in Lake Oulujärvi's fisheries and the economy related to it. Professional fishermen and recreational fishermen exploit the fish stocks in the lake. The State of Finland and municipalities in the area have invested in the infrastructure. Other groups related to fisheries are tourist entrepreneurs, fish industries and trade, fish hatcheries and farms, the power company regulating the lake, and industries and households in the area. The economic analysis was based on the postal questionnaire sent to the fishermen and on interviews with representatives of the above mentioned groups. When conducting the interviews, field data from which an evaluation of the cost of the manipulation of fish stocks in 1995 and the annual cost of investing in the physical rehabilitation of the environment were also obtained.

The aim of the economic analysis was to estimate the total annual cash flow of the rehabilitation of the environment, manipulation of fish stocks and fishing activities in the lake area. In addition to the cash flows, the impact on employment (direct and indirect) was evaluated. This was conducted by combining the results of the questionnaires and interviews with the results of some former studies dealing with employment impacts of fisheries and tourism in the Lake Oulujärvi area and in the Lake Inari area in Finland (Tervo & Mäenpää 1996).

Cash flows of the fisheries were estimated for different groups of fishermen. The contribution of professional fishermen to the total cash flow was estimated according to the value of their annual catches, based on wholesale market prices. The difference between retail and wholesale prices was taken into account when evaluating the fishmongers' contribution to the total monetary flow. Annual fishing costs were evaluated on the basis of the postal inquiry, according to the value of the equipment owned by the fishermen, how often they renewed it and the fixed fishing costs. This was done to estimate the employment impact on the trade sector (and also to check if their fishing activities were profitable or not). Fishing costs for the recreational fishermen were evaluated in the same way. Because recreational fishermen do not sell their catches, their contribution to the total cash flow was estimated according to their fishing costs. More sophisticated methods to evaluate the recreational values of the fishing activities were not applied, because the approach was one of cash flow, not a proper benefit-cost analysis (Tervo & Mäenpää 1996).

24.3 Results

24.3.1 *Fish populations in Lake Oulujärvi*

Between 1973 and 1995 the annual catch of indigenous blue whitefish varied between 20 and 60 t. This whitefish stock peaked three times during the study period: at the beginning of 1973, in the mid-1980s and at the beginning of the 1990s (Fig. 24.3). After the stocking of northern densely-rakered whitefish began, the catch rose to 50 t in the middle of the 1980s. Despite higher numbers of fish being stocked annually, catches declined and have not recovered since the mid-1990s.

Vendace catches ranged between 50 and 350 t. Fluctuations in vendace stock were much larger than those in blue whitefish stock, with peaks in 1973, 1977, 1987 and 1994–1995 (Fig. 24.3).

Brown trout catch in Lake Oulujärvi varied from a couple of 1000 kg to slightly less than 50 t. The maximum catchable stock was 50 000 fish (Fig. 24.3), consisting mainly of 2–3-year-old stocked brown trout. Pikeperch catch was 10 t in 1996, but rose rapidly due to recruits from increased stockings.

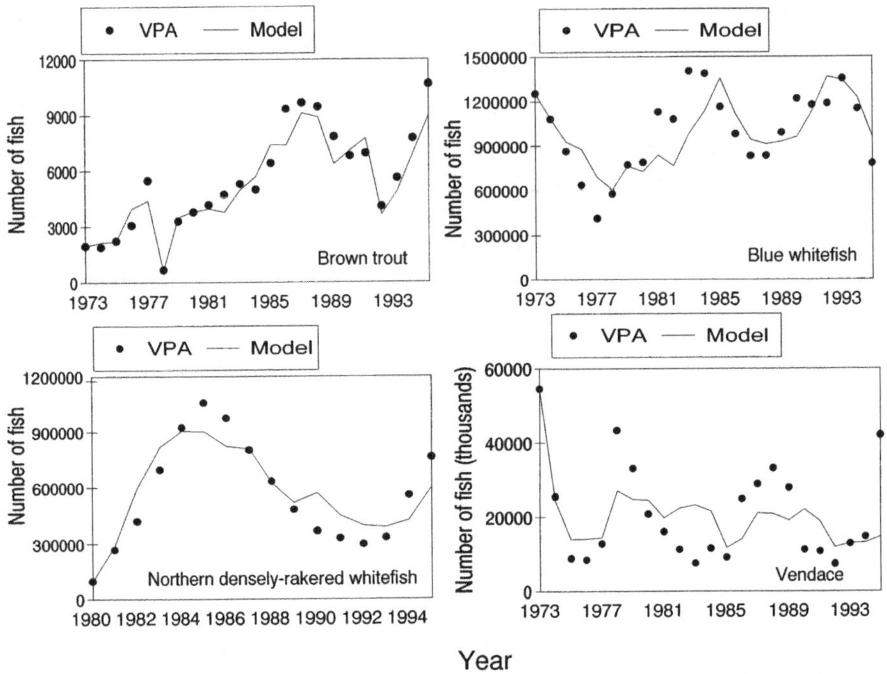

Figure 24.3 Population size of brown trout (1–5 lake years old), blue whitefish (2–10 years old), northern densely-rakered whitefish (2–8 years old) and vendace (1–5 years old) in Lake Oulujärvi estimated by VPA (•) and by the deterministic computer model (—)

24.3.2 *Modelling*

The calibration of the model indicated that in addition to the stock–recruitment curve, the recruitment of vendace and whitefish was affected by the number of competing species (either vendace or whitefish) present in the lake, i.e. the model indicated a slight competitive interaction between the species. Although natural mortality of whitefish and vendace depended on predation by brown trout, predation only slightly increased the predictability of the model and did not explain the fluctuations in whitefish and vendace stocks. Natural mortality of brown trout depends mainly on the amount of their primary prey, vendace, in the lake.

The model could successfully simulate the fluctuations in whitefish and brown trout stocks (Fig. 24.3). Although the level of vendace stock could be simulated within the given relationships, the fluctuations in stock were not satisfactorily modelled, suggesting that factors other than those included in the model act on the stock.

Approximately 40 000 brown trout have been stocked annually in Lake Oulujärvi and this was used to assess the effects of either halving the number of brown trout stocked or increasing the stocking rate by 50%. The model indicated that halving the number of stocked trout would result in a 28% decline in the average catches of brown tout, but would raise the annual vendace catch by 14%, and also have a slight positive effect on the whitefish catch (Table 24.1). Raising the annual stocking to 60 000 fish

Table 24.1 The effect of the different number of brown trout stocked annually in Lake Oulujärvi on brown trout, vendace and whitefish according to modelling

	Number of 1-year-old fish	Population size	Mean weight of fish	Total catch
Annual stocking 20 000 brown trout				
Brown trout	−18	−18	20	−28
Vendace	14	12	0	14
Blue whitefish	8	7	−2	5
Northern densely-rakered whitefish	6	14	−2	12
Annual stocking 60 000 brown trout				
Brown trout	1	22	−14	16
Vendace	−19	−25	0	−17
Blue whitefish	−10	−14	3	−6
Northern densely-rakered whitefish	−8	−25	3	−13

The numbers indicate the percentage change to actual average stocking density from 1973 to 1995 (*ca.* 40 000 brown trout).

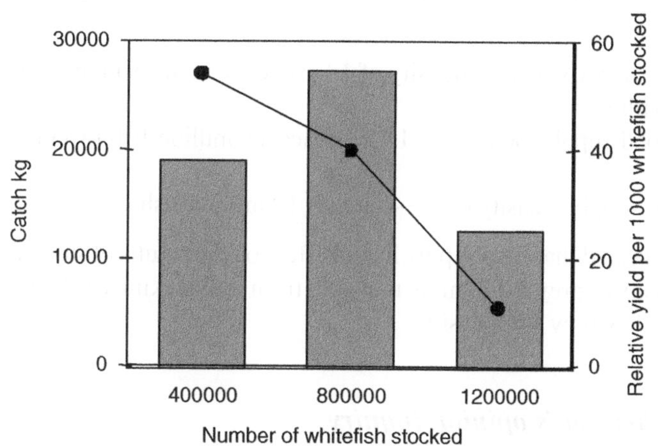

Figure 24.4 The average annual whitefish catch from three different stocking densities and the relative yield (yield per 1000 fingerlings stocked) from stocking simulated by the deterministic model

would increase the trout catch by 20%, but have a negative effect on the size of harvested fish, and on vendace and whitefish catches (Table 24.1).

The number of stocked northern densely-rakered whitefish varied widely. Simulation of different stocking densities suggested that stocking at high densities (close to 1 million juveniles annually, Fig. 24.4) is wasteful. Both the total catch and relative yield per thousand stocked whitefish juveniles declined rapidly at high densities.

The model suggested that food supply, i.e. the amount of the primary prey fish vendace (Vehanen, Hyvärinen & Huusko 1998), is an important factor that affects the

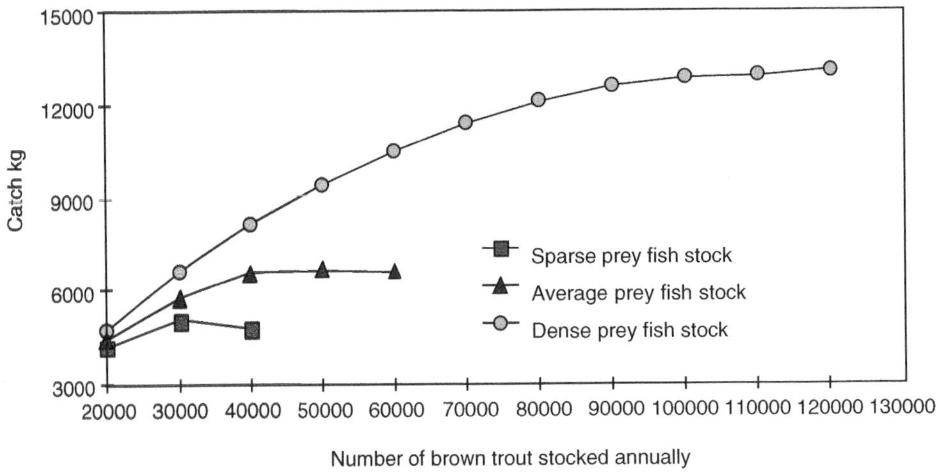

Figure 24.5 Relationship between brown trout yield from stocking and annual number of brown trout stocked in three different prey fish scenarios. See text for details

results of brown trout stocking. The results of brown trout stocking were modelled for three scenarios:

(1) poor food supply when the density of 1+-vendace in the beginning of the year was about 7 million fish;
(2) average food supply, density of 1+-vendace 11 million fish (average in the lake); and
(3) good food supply, density of 1+-vendace 19 million fish.

The results indicated that food supply has an effect on the results of stocking (Fig. 24.5). When the density of prey fish is high, the yield from the stocking can be nearly doubled, compared to a low prey fish density.

24.3.3 *Fishermen's opinion inquiry*

Altogether, 275 people (78.6%) replied to the questionnaire. There were marked differences in the views of the different groups exploiting the lake. The strongest motive for fishing was income among the professional fishermen, and relaxation and outdoor activity for recreational fishing. Professional fishermen were more content with the current catches of the lake, but dissatisfaction was higher among the recreational groups (non-local recreational fishermen, Fig. 24.6). Dissatisfaction was more prevalent with respect to catches of the most valued sport fishes – brown trout and pikeperch.

Professional fishermen targeted vendace (the most sought after species in more than 80% of the answers), whereas the most sought after fish species among recreational fishermen were predatory salmonids (more than 50% of the answers), brown trout and landlocked salmon, *Salmo salar* L. The difference between local and non-local

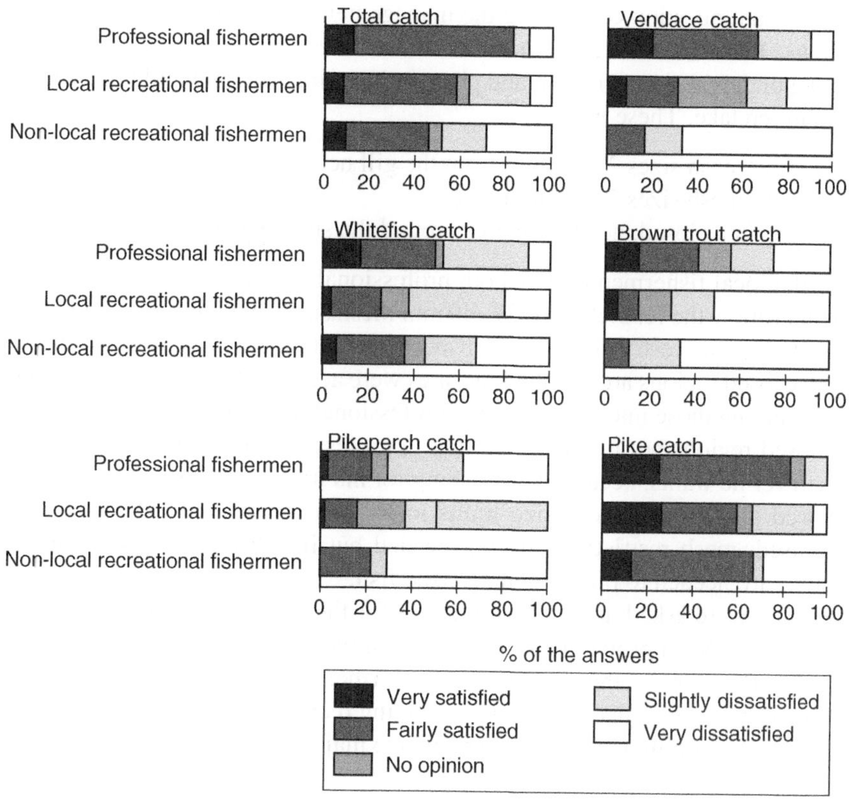

Figure 24.6 Distribution of answers from the opinion survey when fishermen were asked about their satisfaction about their total catch or catch of different fish species in Lake Oulujärvi in 1995

recreational fishermen was evident in the attitudes towards rainbow trout, *Oncorhynchys mykiss* (Walbaum), a non-native species which is usually stocked as table-sized fish. More than half (53.8%) of the non-local recreational fishermen classified rainbow trout as their most desired species, whereas the percentage among local recreational fishermen was 37.4% and only 17.6% among professional fishermen. In addition to targeting different species, there were also differences in the desired mean sizes of harvested fish among the different fishing groups. In general, all groups wanted to catch larger fish than they do at present (brown trout, pikeperch and whitefish), but the desired size of fish was largest among professional fishermen and smallest among non-local recreational fishermen.

Fishermen were asked if any fishing method used in the lake conflicts with other methods used. In the majority of the answers (50% among professional fishermen, 67% among local and 87% among non-local recreational fishermen) trawling was named as a fishing method that conflicts with the use of other fishing methods (e.g. gill nets, pound nets) in the lake.

An important question concerning the management of fisheries is response to gear restrictions. Fishermen were given three different restrictions, specifically aimed at preventing undersized brown trout and pikeperch getting caught in gill nets, which is a problem in the lake. These were:

(1) excluding mesh sizes 27–40 mm from the gill nets;
(2) excluding mesh sizes 27–55 mm; and
(3) to establish a closed, no-fishing area around the stocking sites.

Most of the local fishermen (60%), both professional and recreational, were ready to accept and follow the restriction of 27–40 mm mesh sizes, but the non-local fishermen resisted (60%) this restriction. A wider mesh restriction was opposed by all groups (>60%). Closed regions around stocking sites were accepted by all fishermen (>60%). However, among those interviewed, some professional fishermen had a strong aversion to the closed regions because they felt that 'a professional fishermen needs access to fish wherever he wants, because he is fishing for his livelihood'. A number of persons interviewed also related mesh size limits to the size of stocked fish; if table-sized fish are used, mesh restrictions are not needed but if fish are stocked as juveniles, restrictions are needed to preserve undersized fish.

Fishermen were asked about the guidelines for the future development of the fisheries in the lake. Most professional fishermen felt the current situation was acceptable. Slightly less than 40% of the local recreational fishermen shared the same opinion, but even more (42.7%) felt that action towards aiding recreational fishing should be taken in the future. This opinion was shared among the majority of the non-local recreational fishermen.

24.3.4 *Economic analysis*

The annual economic value of Lake Oulujärvi fisheries was estimated at 20 million FIM (3.36 million Euro) annually (Table 24.2). The largest individual money flow, about half of the total value, comes from the fees and expenditure of the recreational fishermen.

Table 24.2 Summary of the estimated cash flow in Lake Oulujärvi's fisheries in 1995

	Million Euro
Production of hatcheries and fish stocking	0.25
Annual costs of investments	0.08
Professional fish catch	0.62
Fish trade and industry	0.34
Expenses of the recreational fishermen	1.77
Expenses of fishing tourism	0.10
Wages not mentioned elsewhere	0.03
Indirect incomes	0.17
Cash flow altogether	3.36

Table 24.3 Summary of the estimated effect of Lake Oulujärvi's fisheries on employment in the area in 1995

	Man years
Professional fishing	16.8
Hatcheries and fish stocking	3.2
Fishing authorities	0.3
Researchers	0.5
Fish trade	3.7
Fish industry	0.5
Expenses of the professional fishermen	1.8
Expenses of the recreational fishermen	3.9
Expenses of the fishing tourism	1.0
Total direct effect	31.9
Total indirect effect	8.0
Total man years	40.0

Another economically-important component was the value of the catch of the professional fishermen. The effect of the fisheries on employment in the area was estimated at 40 man years annually. Most of the jobs come from professional fishing (Table 24.3), but fish trade and fish hatcheries in the area also have a notable effect on employment.

24.4 Discussion

The study in Lake Oulujärvi illustrated the importance of combining information from several sources when management decisions are made. In many cases the primary goal of fisheries management is to manage fish stocks effectively according to principles of sustainable development (Cowx 1998). The present results show that different groups of fishermen have different objectives for their activities in the lake, which need to be taken into account when management actions are considered (e.g. Hickley & Tompkins 1998). In addition, management actions may also influence the economic value of the lake's fisheries, when, for example, balancing the exploitation of fish stocks between professional and recreational use.

Fish stocks in Lake Oulujärvi fluctuate naturally but are also heavily influenced by fish stocking. The large, natural fluctuations of vendace have particularly important effect on the total catches of the lake. Due to the damming of large rivers in the area, the stocks of migratory salmonids have decreased and the role of fish stocking as a management tool is especially important.

Both the intraspecific and interspecific interactions may have an effect on the results of fish stocking (Gunn, McMurtry, Bowlby, Casselman & Liimatainen 1987; Lachance & Magnan 1990). However, these effects are difficult to outline and quantify without suitable data, and a model to calculate and visualise the interactions is important for management. For example, the stocking of northern densely-rakered whitefish increased the catches of this whitefish form, but results varied. Output from the model indicated that high stocking densities result in poorer catches. In addition, modelling suggested that the yield from the stocking of brown trout was highly dependable on the amount of suitable prey fish species, vendace, indicating strong interspecific relationships (e.g. Stewart, Kitchell & Crowder 1981; Stewart & Ibarra 1991). The best results from trout stocking could be obtained when the stocking density is adjusted according to the fluctuations in vendace stock. However, a model is always a simplification of nature and many factors may remain unaccounted. In Lake Oulujärvi, uncertainty in fisheries management comes mainly from the effect of the recent increase in the pikeperch stock, together with the lack of knowledge of the amount of its primary prey, smelt, *Osmerus eperlanus* (L.) (Vehanen *et al.* 1998). Also, the long-term effects of increased fishing mortality on vendace after the start of trawling in 1987 remains to be followed up.

Assessment of the outcome of management actions on the different group of fishermen is important (Salmi & Auvinen 1998). In many cases, as in Lake Oulujärvi, recreational fishermen favour predatory species like brown trout, whereas professional fishermen consider brown trout as a competitor of their primary target fish, vendace. When the purpose is to ensure the best conditions for both professional and recreational fishing, a balance in the amount of stocked predatory species should be found which ensures that predation against vendace does not rise to too high a level, but also satisfies the catch expectations of recreational fishermen. In the current situation of high vendace stock, professional fishermen tended to be satisfied with their catches, but the recreational fishermen were especially dissatisfied with the catches of predatory brown trout and pikeperch. Therefore, increased stocking of these species is often demanded. However, it is often forgotten that gear restrictions to protect the undersized fish from exploitation are often more effective at increasing catches than increased stocking (e.g. Gigliotti & Taylor 1990). To be effective, gear restrictions should be approved by all groups of fishermen. However, according to this study this is not currently the case.

Although the economic value of the lake's fishery is sometimes forgotten by management, it is of special interest to the local authorities. Currently both the economic value and the effect on employment of Lake Oulujärvi's fisheries in the area are considerable. The tendency in the area is to try to increase further the economic income by investing in fishery-related tourism. In many cases this approach has been effective (Møller & Petersen 1998). However, to attract a considerably larger number of fishery tourists, larger catches of sport fish are needed, as are larger investments in fish stocking and infrastructure, but considerable uncertainty in the profitability of the investment remains.

In the development of the lake's fisheries, the different views of the fishermen have to be balanced in an ecologically and economically viable way. It is important that information on the basis of management action is discussed and shared with all the groups exploiting the lake. For example, through informing the fishermen about the effectiveness of gear restrictions it is possible to gain their approval. It is more sensible to increase catches of predatory species by increasing the size of harvested fish rather than increasing the stocking volume. When the needs of the different groups of fishermen are taken into account, both recreational and professional fishermen can work effectively in the same lake.

References

Colby J. & Lehtonen H. (1994) Suggested causes for the collapse of zander, *Stizostedion lucioperca* (L.), populations in northern and central Finland trough comparisons with North American walleye, *Stizostedion vitreum*, (Mitchill). *Aqua Fennica* **24**, 9–20.

Cowx I.G. (1994) Stocking strategies. *Fisheries Management and Ecology* **1**, 15–30.

Cowx I.G. (1998) (ed.) *Stocking and Introduction of Fish*. Oxford: Fishing News Books, Blackwell Science, 456 pp.

Cowx I.G. & Welcomme R.L. (eds) (1998) *Rehabilitation of Rivers for Fish*. Oxford: Fishing News Books, Blackwell Science, 260 pp.

Gigliotti L.M. & Taylor W.W. (1990) The effect of illegal harvest on recreational fisheries. *North American Journal of Fisheries Management* **10**, 106–110.

Gunn J.M., McMurtry M.J., Bowlby J.N., Casselman J.M. & Liimatainen V.A. (1987) Survival and growth of stocked lake trout in relation to body size, stocking season, lake acidity, and biomass of competitors. *Transactions of the American Fisheries Society* **116**, 618–627.

Hickley P. (1994) Stocking and introduction of fish – a synthesis. In I.G. Cowx (ed.) *Rehabilitation of Freshwater Fisheries*. Oxford: Fishing News Books, Blackwell Science, pp. 247–254.

Hickley P. & Tompkins H (1998) (eds) *Recreational Fisheries: Social, Economic and Management Aspects*. Oxford: Fishing News Books, Blackwell Science, 310 pp.

Jørgensen S.E. (1988) *Fundamentals of Ecological Modelling*. Amsterdam: Elsevier, 391 pp.

Lachance S. & Magnan P. (1990) Performance of domestic, hybrid, and wild strains of brook trout, *Salvelinus fontinalis*, after stocking: the impact of intra- and interspecific competition. *Canadian Journal of Fisheries and Aquatic Sciences* **47**, 2278–2284.

Marttunen M. & Kylmälä P. (1997) Kalakantojen hoitomalli Inarijärven kalaistutusten vaikutusten arvioinnissa. *Suomen ympäristö* **117**. (in Finnish with English abstract).

Møller M.M. & Petersen J.D. (1998) The Funen sea trout project, Denmark – a development project for the environment and tourism. In P. Hickley & H. Tompkins (eds) *Recreational Fisheries: Social, Economic and Management Aspects*. Oxford: Fishing News Books, Blackwell Science, pp. 70–79.

Salmi P. & Auvinen H. (1998) Local conflicts in Finnish Lake Fisheries. In P. Hickley & H. Tompkins (eds) *Recreational Fisheries: Social, Economic and Management Aspects*. Oxford: Fishing News Books, Blackwell Science, pp. 116–128.

Salojärvi K. (1991) Recruitment mechanisms of the vendace (*Coregonus albula* (L.)) in Lake Oulujärvi, northern Finland. *Aqua Fennica* **21**, 163–173.

Salojärvi K. (1992) Compensation in whitefish (*Coregonus lavaretus* L. s.l.) populations in Lake Oulujärvi, northern Finland. *Finnish Fisheries Research* **13**, 31–48.

Stefansson G. & Palsson O.K. (1998) A framework for multispecies modelling of Arcto-boreal systems. *Reviews in Fish Biology and Fisheries* **8**, 101–104.

Stewart D.J. & Ibarra M. (1991) Predation and production by salmonine fishes in Lake Michigan, 1978–88. *Canadian Journal of Fisheries and Aquatic Sciences* **48**, 909–922.

Stewart D.J., Kitchell J.F. & Crowder J.B. (1981) Forage fishes and their salmonid predators in Lake Michigan. *Transactions of the American Fisheries Society* **110**, 751–763.

Tervo H. & Mäenpää I. (1996) Inarijärven kalataloudellinen kustannus-hyötyanalyysi. *Suomen ympäristökeskus monisteita* **2** (in Finnish).

Vehanen T., Hyvärinen P. & Huusko A. (1998) Food consumption and prey orientation of piscivorous brown trout (*Salmo trutta*) and pikeperch (*Stizostedion lucioperca*) in a large regulated lake. *Journal of Applied Ichthyology* **14**, 15–22.

Chapter 25
Fishery management practices in French lakes

C. ARGILLIER*, O. PRONIER
Cemagref, Unité Ressources Ichtyologiques en Plans d'Eau, Montpellier, France

T. CHANGEUX
Conseil Supérieur de la Pêche, 134 Avenue de Malakoff, 75116 Paris, France

Abstract

A survey on local fishery management methods applied in French lakes, ponds and reservoirs carried out in 1998 showed that fish stocking was the main management measure. These operations relied upon the experience of field staff rather than a standard decision process. An analysis of the relationship between stocking effort and fish communities proved the poor efficiency of these programmes apart from in mountain lakes. The absence of fisheries monitoring is considered the main factor limiting the efficiency of fisheries management practices. Since 1995, fisheries management plans were set up throughout France. Lakes and reservoirs included in this process must take into account anglers' demand, as well as environmental preservation.

Keywords: fishery management, fish stocking, fish community, gravel pit, lake, reservoir.

25.1 Introduction

Due to its geological and climatic situation, France is poor in natural lakes. Apart from small glacier lakes in mountainous regions, most wetland zones were drained during the Industrial Revolution. However, an increasing demand for drinking water, hydropower, flow regulation and irrigation simultaneously lead to the building of many large reservoirs.

There is now growing public interest in these sites related to an increasing demand from the outdoor leisure sector and socio-environmental concerns (Anonymous 1992). Recreational fisheries also greatly contribute to the tourist appeal of these water bodies.

Apart from landlocked waters, permanently isolated from the hydrographic network in France, water resources are legally considered as common property and are therefore collectively managed. Fishing rights belong to the owners of the river banks and generally are granted to anglers' associations. Recent legislation requires management authorities to plan and co-ordinate their actions. This framework led them to co-operate with the Conseil Supérieur de la Pêche (the French National Council of Fishing) to set up sustainable actions plans taking conservation needs into consideration. This

*Correspondence: Cemagref, Unité Ressources Ichtyologiques en Plans d'Eau, 361 rue J.F. Breton, BP 5095, 34033 Montpellier Cedex 1, France (email: christine.argillier@cemagref.fr).

resulted in two documents summarising regional policy; the first, called Plan Départemental de Protection du Milieu Aquatique et de Gestion des Ressources Piscicoles (PDPG; Departemental Plan for Freshwater Environment Protection and Fish Resources Management) (Holl, Auxiètre & Bordes 1994), selects operational sites and remedial measures necessary for sustainable fishery management; the second, called the Plan Départemental de Promotion du Loisir-pêche (PDPL; Departemental Plan for the Promotion of the Recreational Fishery), includes actions concerning recreational fisheries (Changeux, Bonnieux & Armand 2001).

This chapter presents the results of an inquiry aimed at anglers' federations, to investigate their management practices and expectations for lentic systems. The relationship between fish stocking and standing stocks was analysed in relation to recent policy concerning sustainable management strategies.

25.2 Materials and methods

25.2.1 *Study of fisheries management practices and managers concerns*

A questionnaire was sent to the 82 regional unions overseeing the angling associations. The purpose was to collect information concerning fisheries management practises in lakes, reservoirs and gravel pits between 1994 and 1997. Managers had to provide information about species manipulations (stocking and culling), fishery surveys and spawning ground improvements, water body characteristics and target angling species, dates and purposes of the implemented practices. Two open questions were formulated to identify other management operations and difficulties encountered in their application. Thereafter managers were asked to grade, in order of interest, the following actions: policy, habitat improvement, fish stocking, catch survey and populations studies. The last part was dedicated to free comments about water body management. The response rate was 74%, and results were considered to provide a national perspective of the issues.

25.2.2 *Relationship between fish stocking and fish community*

This analysis was carried out on 45 natural lakes and reservoirs. Fisheries data were collected during either gillnetting surveys or impoundment drain down for dam investigation. Fish stocking information was provided by managers in the decade prior to fish community assessment. Relative density of each species was encoded into three categories corresponding to absent, low and high abundances. Stocking effort was expressed in numbers of fish per hectare per year and encoded similar to the abundances. Water bodies located above and below 1500 m in altitude were analysed separately. This distinction was guided by a previous study showing a great difference between their fish community structure and habitat characteristics (Pronier & Irz 1999). Investigations on mountain sites focused on salmonid species: brown trout, *Salmo trutta* (L.), rainbow trout, *Oncorhynchus mykiss* (Walbaum), brook trout, *Salvelinus fontinalis* (Mitchell), lake trout, *Salvelinus namaycush* (Walbaum) and Arctic charr, *Salvelinus alpinus* (L.). Analyses on lowland water bodies were performed on pike, *Esox lucius* L., common carp, *Cyprinus carpio* L., roach, *Rutilus rutilus* (L.), tench, *Tinca tinca* (L.), brown trout and rainbow trout. These species were selected on the basis of the survey results. Two matrices were constructed with species as variables and water bodies as the samples. The two matrices were submitted to separate standardised principal components analysis (PCA). The relationship between the fish community and fish stocking was then investigated using a co-inertia analysis (CoA) (Dolédec & Chessel 1994). This symmetric method is reliable in the case of data sets comprising large number of variables with regard to the number of samples. The axes are designed to maximise the co-variability of the two matrices. A Monte-Carlo random permutation test was performed to assess the significance of the co-structure.

25.3 Results

25.3.1 *Fisheries management survey*

Within the diversity of management actions carried out in French water bodies, stocking was the most frequent (68%), followed by fish monitoring (11%), spawning ground improvement (9%), species eradication (6%), macrophyte enhancement (2%) and water quality improvement (1%). The most common habitat restoration practice was littoral zone restoration to preserve or to develop spawning grounds, but artificial spawning structures were also often used. More than 95% of these works were performed in reservoirs with fluctuating water levels and in gravel pits with steep shores. The purpose was to sustain pike, roach, and to a lesser extent perch, *Perca fluviatilis* L., populations. Furthermore, a community-level improvement was expected in 20% of the cases. Turbidity control was seldom practised (only 2% of the total actions) and attempted by chemical treatment. Chemical or mechanical control of macrophytes was rarely performed.

Removal and enhancement of fish stocks, recorded on more than 700 sites, was the most common management action. Eradication of undesirable fish species appeared to be rather inefficient. The most common target species was black bullhead, *Ictalurus melas* (Rich.), frequently invasive in artificial water bodies. More than 2100 fish stocking operations, defined as the input of one species at one site on a single occasion, were recorded during the study period. An average of three species was annually stocked in each water body, but this number was lower in natural lakes than in artificial systems. The number of species stocked was greatest in reservoirs (34), gravel pits (22) and lakes (17). Nevertheless, six species account for 70% of these operations (Fig. 25.1). Pike was the most intensively sustained species in artificial sites, frequently associated with roach, carp and tench. Most stocking effort was directed towards brown trout in natural lakes. Rainbow trout accounted for 10% and 15% of stocking in artificial water bodies and in lakes respectively. The introduction of new species was carried out less frequently.

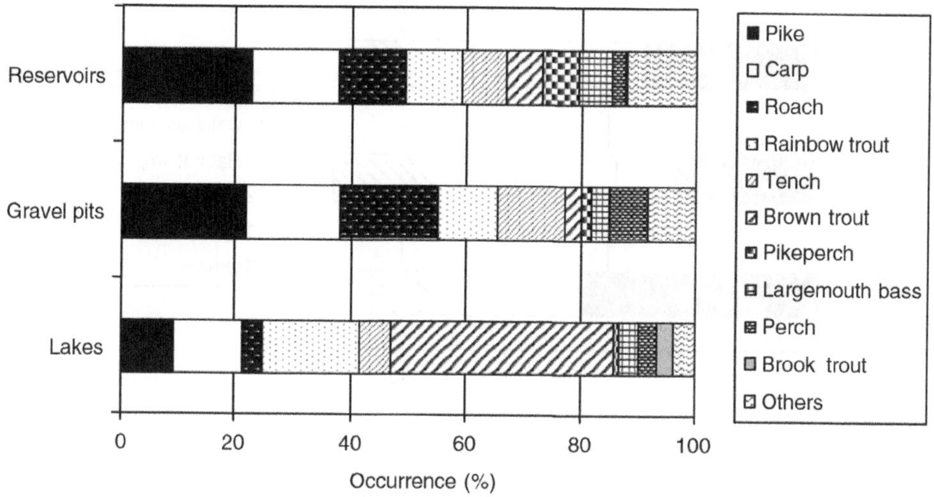

Figure 25.1 Percentage occurrence of fish species in the overall stocking programmes (ratio between the species occurrences in stocking programmes and the total number of stocking programmes recorded)

The aim of these programmes was generally not clearly documented. Stock preservation or improvement was the main objective when a limiting factor was identified. Pike stocking, for example, was often carried out to overcome poor natural recruitment in reservoirs where water level fluctuations are problematic. Nevertheless, in other cases, the need to satisfy anglers' demands was often the implied purpose.

Comparison between sites surveyed and those stocked showed that this practise was generally applied without any prior knowledge of the fish community. Fish sampling mainly targeted small natural lakes or when reservoirs were drained for dam examination. Moreover, catches were generally not recorded. Therefore the stocking strategy was usually based on the experiences of local management staff. This empirical approach was not considered satisfactory by managers but was reported in the free comments of the survey form to be a consequence of lack of basic information on fish resource and management guidelines. Direct or indirect monitoring of the fish populations was the major concern of anglers' federations (Fig. 25.2), particularly in natural lakes. Numerous local sampling and catch surveys are now planned to partially answer this demand.

25.3.2 *Fish stocking and fish community*

In mountain lakes, the preliminary PCA carried out on stocking data showed that simultaneous stocking of brown trout and Arctic char occurred in some lakes, but that stocking of other species did not correspond to any general pattern (Fig. 25.3). The same analysis run on the species relative densities showed the absence of clear species associations. The CoA indicated a strong relationship between fish population

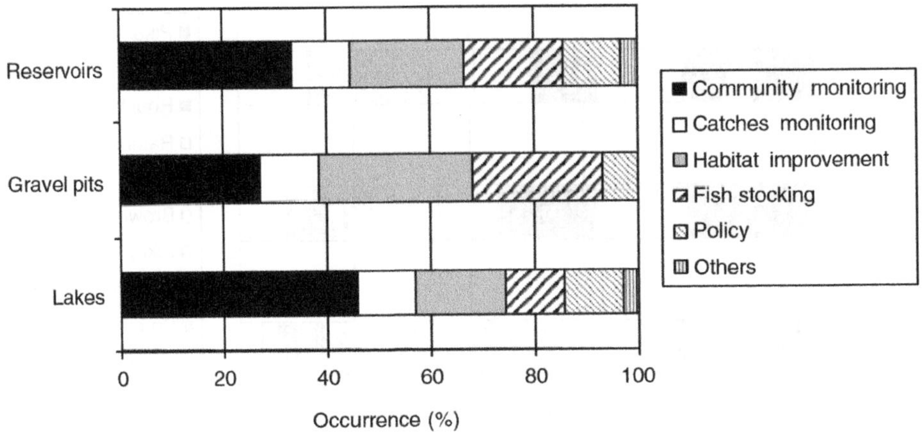

Figure 25.2 Percentage occurrence of managers' priority concerns (ratio between the number of cases when the management action was considered as first priority and the total number of received answer forms)

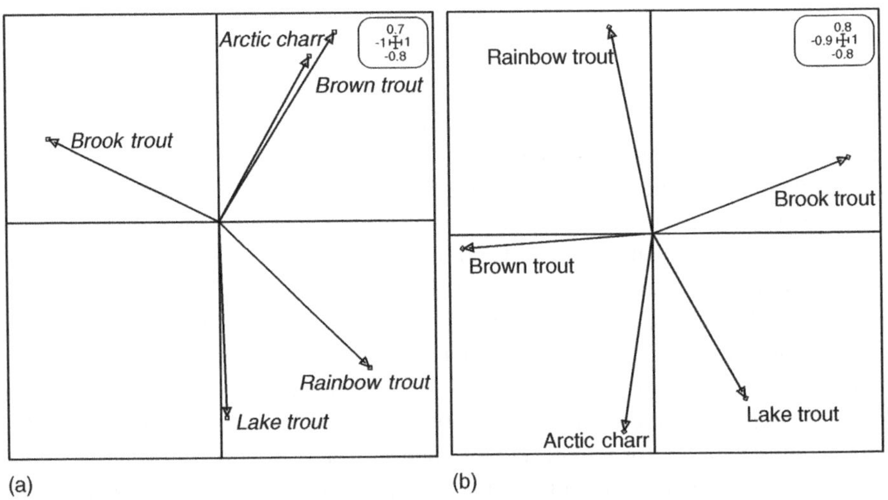

Figure 25.3 PCA carried out on stocking effort (a) and species numerical relative abundance (b) for mountain water bodies

structure and fish stocking with a significant Monte-Carlo test ($P < 0.001$, 1000 permutations) (Fig. 25.4). Each species was well separated in the two representations.

Conversely, in lowland lakes, PCA showed antagonism between brown trout and pike relative density (Fig. 25.5(b)). Furthermore, brown trout and rainbow trout were often associated in stocking programmes and opposed to pike stocking acitivies (Fig. 25.5(a)). The CoA revealed the absence of common structure between the two tables. This was confirmed by plotting the relative abundance of the six main species in the community

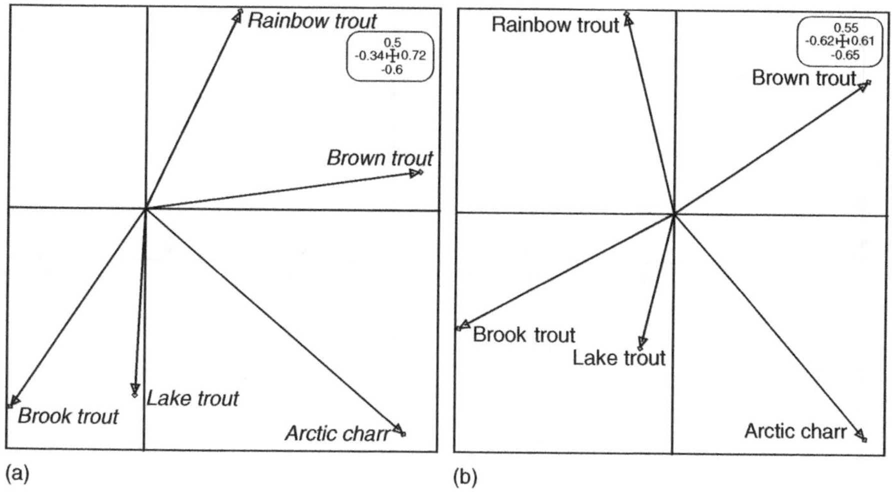

Figure 25.4 CoA of the relationship between stocking effort (a) and species numerical relative abundance (b) in mountain water bodies

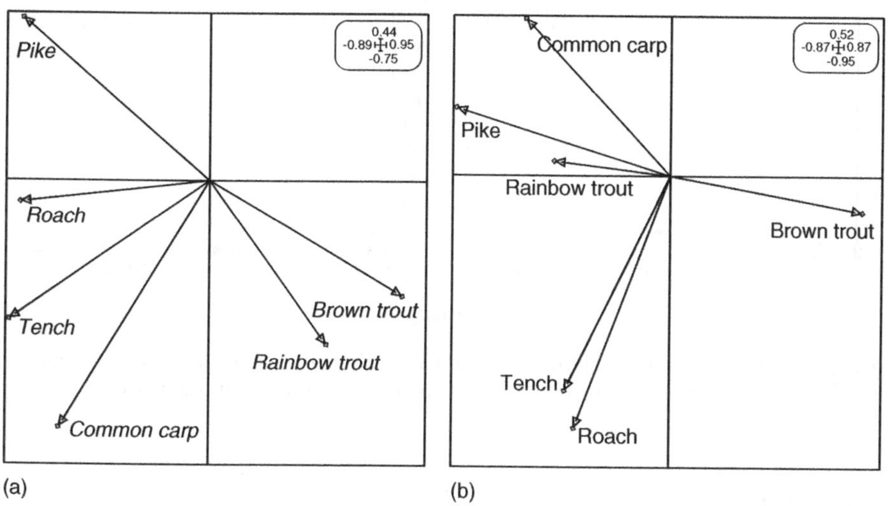

Figure 25.5 PCA carried out on stocking effort (a) and species numerical relative abundance (b) for lowland water bodies

against the importance of fish stocking (Fig. 25.6). No general trend was observed, even when pike and roach, the most commonly manipulated fish, were included.

25.4 Discussion

The inquiry revealed that management actions simultaneously aimed to manipulate the fish community structure and respond leisure fishery demands.

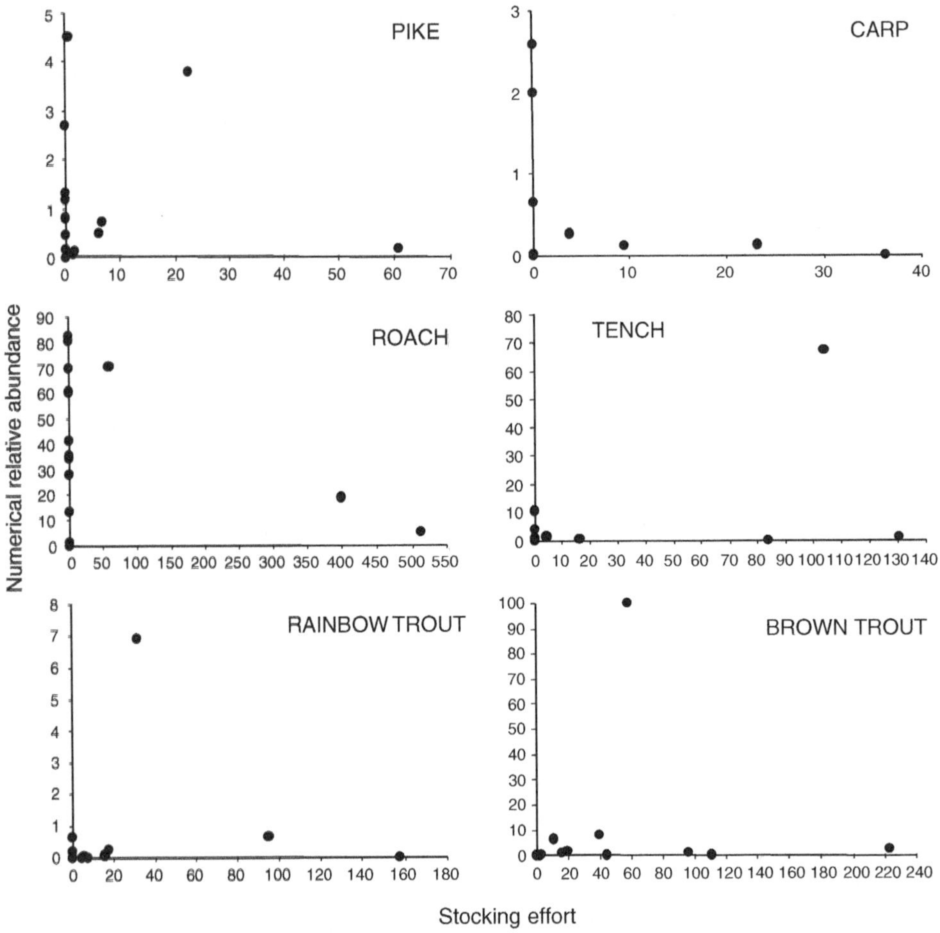

Figure 25.6 Graphical representation of the relationship between stocking effort (number of fish ha^{-1} $year^{-1}$) and species numerical relative abundance for lowland water bodies

25.4.1 *Fish community control*

Fish standing stock resulted from three major influences that can be categorised into colonisation opportunities affected by human manipulations, condition of the habitat and interspecific relationships. These factors determine which species develop, regress or disappear.

The few practices concerning habitat improvement suggested this action is not a priority in lacustrine water bodies, despite being of great concern for the management of lotic systems (CSP 1996). Nevertheless managers are conscious of the limitations induced by a weak shoreline diversity in artificial water bodies (gravel pits and reservoirs). In practice, the functional rehabilitation of the systems mainly involves enhancement of spawning grounds associated with stocking. No real biomanipulation operations were recorded apart from the unsuccessful attempts to eradicate black bullhead. Similar failures are reported in river systems (Jestin 1985).

Most fish stocking programmes were carried out without prior consideration of the context nor was success monitored. Few of the strategic steps recommended by Cowx (1999) are taken into consideration. The potential ecological (Hansson, Annadotter, Bergman, Hamrin, Jeppesen, Kairesalo, Luokkanen, Nilsson, Sondergaard & Strand 1998; Whittier & Kincaid 1999) and socio-economic consequences of such actions are seldom assessed in the decision process. This was partially justified by a lack of knowledge necessary for a sustainable planned strategy and explained by the complexity and cost of fish surveys in deep water bodies.

The response of fish populations to management actions was satisfactory on mountain water bodies that are ecologically simple and well documented. This confirms the acclimation of salmonids in these sites as observed in Pyrenean lakes where introductions have been carried out since the 1940s (Delacoste, Baran, Lascaux, Abad & Besson 1997). Moreover, stocking is now limited to sustaining these acclimated species. These management programmes, followed by species competition, have resulted in the dominance of a single species (Fraser & Power 1984; Heggberget 1984).

In lowland water bodies, there is discrimination between management practices in salmonid-dominated water bodies and those at other sites, but no relationship was found between fish relative abundance and stocking. This apparent inefficiency of stocking could result from habitat bottlenecks at the population level, for example the absence of spawning grounds for brown trout in the lowland tributaries. Welcomme, Kohler & Courtenay Jr (1983) also highlighted the problem for pike and recommended habitat improvement rather than stocking to maximise the species density. Results from this study argue for the development of an efficient habitat improvement engineering policy. Habitat complexity and species diversity in lowland systems do not permit efficient control of fish population size without combining stocking with other management measures.

An analysis of site potential is now provided by the PDPG (Holl *et al.* 1994), which bases its diagnosis on the situation of target species; generally brown trout in salmonid-dominated reaches and pike in lowland waters. The condition of these species is considered representative of the condition of the environment. The water body and the adjacent riverine systems are considered as a whole and should allow the target species to complete its life cycle by ensuring suitable conditions for reproduction, hatching and growth. This is considered successful if no limiting factors prevent the population reaching the carrying capacity (Potential Number of Adult Wild Fish see Changeux *et al.* 2001 for more details) of the water body, and managers are recommended not to intervene. Where limiting factors affect some vital functions, resulting in a standing stock which has declined by more than 20% of the site's potential, the system is considered 'perturbed' (20–80%) or 'damaged' (more than 80%). Here, the PDPG proposes to managers a series of habitat improvement measures aiming at restoring a level corresponding to a minimum 20% of the carrying capacity. This action plan should therefore avoid inefficient enhancements. In these cases, the interest of anglers in the target species is used to monitor their catches and provide a useful tool to evaluate the efficiency of a management programme (Cooper & Wheatley 1981; Cowx, Fisher & Broughton 1986).

25.4.2 *Response to fisheries demand*

The priority given to fish stocking shows that the main concern of decision makers is to satisfy anglers' demand. Stocking generates ecological and disease risks for the standing stock. When the existing stock is in a good condition, the risk is important compared to the expected benefit. If a stock exhibits problems, this can make the risk acceptable with regard to the expected results of stocking. The long-term goal remains a self-sufficient community but manipulations can provide temporary improvements. Evaluation of the social and economic benefits of these actions is provided in the PDPL. The potential number of anglers is calculated on the basis of the human distribution and the local accommodation infrastructure. Fisheries can be categorised into three types: general public fishing characterised by a demand for easy catches, sport fishing and natural fishing targeting wild fish. The distinction between sport and natural fishing is based on stock abundance.

To satisfy the demand for general public fishing near to cities, put-and-take fisheries can be recommended in some of the reservoirs identified as 'perturbed' or 'damaged' in the PDPG.

In conclusion, the two plans (PDPG and PDPL) provide a strategy combining ecological and socio-economic approaches to the fisheries. Proper application of these schemes should limit enhancement of small reservoirs in the headwaters, which have no bottlenecks to development, and enhance anglers' interests in gravel pits and reservoirs surrounding urban centres.

Acknowledgements

This work was supported by the Conseil Supérieur de la Pêche, project number 97577 and funded by the Conseil Régional Languedoc-Roussillon. We would like to express our gratitude to Pascal Irz for his comments and assistance on this paper.

References

Anonymous (1992) La pêche en France – Enquête nationale: toute en paradoxes. *Eaux libres* 4–18.

Changeux T., Bonnieux F. & Armand C. (2001) Cost benefit analysis of fisheries management plans. *Fisheries Management and Ecology* **8**, 425–434.

Cooper M.J. & Wheatley G.A. (1981) An examination of the fish population in the River Trent, Nottinghamshire using angler catches. *Journal of Fish Biology* **19**, 539–556.

Cowx I.G. (1999) An appraisal of stocking strategies in the light of developing country constraints. *Fisheries Management and Ecology* **6**, 21–34.

Cowx I.G., Fisher A.M. & Broughton N.M. (1986) The use of anglers' catches to monitor fish populations in large water bodies, with particular reference to the River Derwent, Derbyshire, England. *Aquaculture and Fisheries Management* **17**, 95–103.

CSP (1996) Enquête auprès des gestionnaires sur les domaines d'intérêt pour une meilleure orientation des actions techniques et scientifiques de l'établissement. Conseil Supérieur de la Pêche, Le Paraclet, Internal report, 11 pp. (in French).

Delacoste M., Baran P., Lascaux J.M., Abad N. & Besson J.P. (1997) Bilan des introductions de salmonidés dans les lacs et ruisseaux d'altitude des Hautes-Pyrénées. *Bulletin Français de Pêche et de Pisciculture* **344/355**, 205–219.

Dolédec S. & Chessel D. (1994) Co-inertia analysis: an alternative method for studying species-environment relationships. *Freshwater Biology* **31**, 277–294.

Fraser N.C. & Power G. (1984) The interactive segregation of landlocked Arctic charr (*Salvelinus alpinus*) from lake charr (*S. namaycush*) and brook charr (*S. fontinalis*) in two lakes of subarctic Quebec, Canada. In L. Johnson & B. Burn (eds) *International Symposium on Arctic Charr.* Winnipeg Man. (Canada), 4 May 1981, pp. 163–181.

Hansson L.-A., Annadotter H., Bergman E., Hamrin S.F., Jeppesen E., Kairesalo T., Luokkanen E., Nilsson P.-A., Sondergaard M. & Strand J. (1998) Biomanipulation as an application of food-chain theory: contrainsts, synthesis, and recommendations for temperate lakes. *Ecosystems* **1**, 558–574.

Heggberget T.G. (1984) Habitat selection and segregation of parr of Arctic charr (*Salvelinus alpinus*), brown trout (*Salmo trutta*) and Atlantic salmon (*Salmo salar* L.) in two streams in north Norway. In L. Johnson & B. Burn (eds) *International Symposium on Arctic Charr.* Winnipeg Man. (Canada), 4 May 1981, pp. 217–231.

Holl M., Auxiètre J.P. & Bordes G. (1994) *Gestion Piscicole et Plans de Gestion. Conception et Pratique*. Paris: Conseil Supérieur de la Pêche, 230 pp.

Jestin J.M. (1985) Gestion de la ressource aquatique. Le point de vue halieutique. In D. Gerdeaux & R. Billard (eds) *Gestion Piscicole des Lacs et Retenues Artificielles.* Aups (France), 15–16 Nov. 1983, Institut National de la Recherche Agronomique, pp. 159–165.

Pronier O. & Irz P. (1999) Relationship between fish populations and physical characteristics of French lakes and reservoirs. In *Sustainable Lake Management – 8th International Conference on the Conservation and Management of lakes.* Copenhagen (Denmark), 17–21 May 1999, ILEC, pp. S11A-9.

Welcomme R.L., Kohler C.C. & Courtenay Jr W.R. (1983) Stock enhancement in the management of freshwater fisheries: a European perspective. *North American Journal of Fisheries Management* **3**, 265–275.

Whittier T.R. & Kincaid T.M. (1999) Introduced fish in north eastern USA lakes: regional extent, dominance, and effect on native species richness. *Transactions of the American Fisheries Society* **128**, 769–783.

Chapter 26
Fisheries and fisheries management on Lake Peipsi–Pihkva

T. SAAT*, V. VAINO
University of Tartu, Tartu, Estonia

E. AFANASJEV and N. KONCEVAJA
Pskov Branch of GosNIORH, Pskov, Russia

Abstract

This chapter focuses on the historical trends in the commercial fisheries of Lake Peipsi–Pihkva. The fish resources of this cross border lake are very important both for Russia and Estonia. Recent changes in the Estonian and Russian economies, and the appearance of the state border dividing the lake, have resulted in shifts in the exploitation patterns. Changes in the management of the fisheries are described.

Keywords: cross border water-body, Estonia, Lake Peipsi–Pihkva, management, Russia.

26.1 Characterisation of Lake Peipsi–Pihkva

Lake Peipsi–Pihkva is the fourth largest lake in Europe lying in the territories of Estonia and Russia. It is a lowland (average 30.1 m above Baltic Sea level), shallow, non-stratified, eutrophic lake consisting of three parts (Fig. 26.1) with a total area of 3555 km^2 (Table 26.1). The drainage area of Lake Peipsi–Pihkva is 44 240 km^2 (16 323 km^2 in Estonia, 27 917 km^2 in Russia). The main inflows are the rivers Emajõgi (Estonia) and Velikaja (Russia). The only outflow is through the River Narva (discharge 12.6 $km^3 year^{-1}$) into the Gulf of Finland. The River Narva was dammed in 1959, which blocked the route of migratory fish (eel, *Anguilla anguilla* (L.), salmon, *Salmo salar* L., river lamprey, *Lampetra fluviatilis* (L.)) to their reproduction or nursery areas in the river and lakes.

The average phosphorus concentration is 46 $\mu g L^{-1}$, total nitrogen 876 $\mu g L^{-1}$ (higher in Lake Pihkva, lower in Lake Peipsi). Mean Secchi depth is 1.63 m, and mean pH 8.14. Primary production is estimated at 203.5 g C $m^{-2} year^{-1}$, zooplankton production 22 g C $year^{-1}$. Lake Peipsi is very rich in macrozoobenthos, its abundance (without big molluscs) is 2617 individuals m^{-2}, biomass 12.34 g m^{-2} (52.2 kJ m^{-2}). Biomass of large molluscs (mainly *Dreissena polymorpha*) is 238 g m^{-2} (Nõges, Haberman, Jaani, Laugaste, Lokk, Mäemets, Nõges, Pihu, Starast, Timm & Virro 1996).

*Correspondence: Estonian Marine Institute, 18b Viljandi Road, 11216 Tallin, Estonia (email: tsaat@sea.ee).

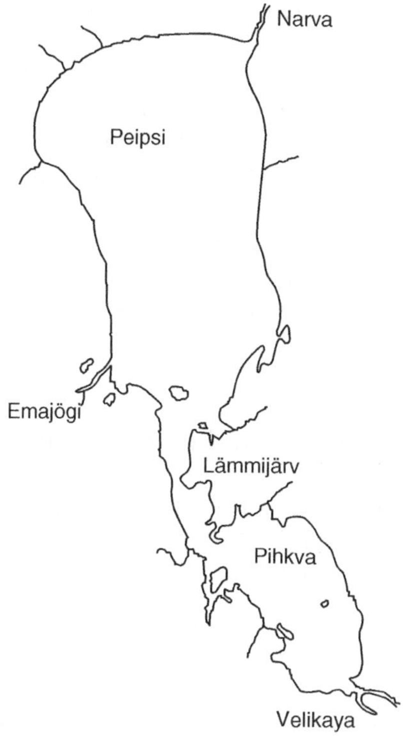

Figure 26.1 Lake Peipsi–Pihkva

Table 26.1 General characteristics of Lake Peipsi–Pihkva

Part of lake	Area (km^2)	Mean depth (m)	Maximum depth (m)	Volume (km^3)	Features
Peipsi	2611	8.3	12.9	21.8	Non-stratified eutrophic with mesotrophic features
Lämmijärv	236	2.5	15.3	0.6	Non-stratified eutrophic with dystrophic features
Pihkva	708	3.8	5.3	2.7	Non-stratified eutrophic

26.2 Fish fauna and main commercial species

Fish fauna of Lake Peipsi–Pihkva and the lower reaches of its inflow and outflowing rivers comprises 37 species. The main commercial species (annual commercial catch usually >500 t) are smelt, *Osmerus eperlanus m. spirinchus* (Pallas), perch, *Perca fluviatilis* L., pikeperch, *Stizostedion lucioperca* (L.), roach, *Rutilus rutilus* (L.), bream, *Abramis brama* (L.), and ruffe, *Gymnocephalus cernuus* (L.), followed by pike,

Esox lucius L., vendace, *Coregonus albula* (L.), whitefish, *Coregonus lavaretus maraene* (Bloch), burbot, *Lota lota* (L.) (Shirkova & Pihu 1966). However, the annual commercial catch in recent years has fallen to 20–25 kg ha^{-1} $year^{-1}$.

There have been several stockings of alien species in Lake Peipsi, mostly carp, *Cyprinus carpio* L., but also other species. They include introduction of salmonids (*Salmo salar* L., *Salmo trutta* L.) in 1852 (Investigations on the State of Fisheries 1860), and, in the Soviet times, even sturgeons (*Acipenser* spp.) and Chinese carps. None of these species have established self-reproducing populations. The only alien species regularly inhabiting lakes is *Carassius auratus gibelio* (Bloch). Artificial propagation and stocking of local species, mostly whitefish, has been performed since 1930s, but has now been curtailed.

26.3 History of stocks, fishery and management

Historically, Lake Peipsi–Pihkva was a smelt-bream lake. In the eighteenth and the beginning of the nineteenth century the most important commercial species was bream. However, catches of bream declined because of:

(1) use of small mesh size beach seines and, subsequently, high catch of juveniles, especially on the Russian side of the lakes;
(2) too many fishermen;
(3) intensive catch of bream and other species during spawning migration and on the spawning grounds.

This lead to a decline in the stocks of bream, as well as asp *Aspius aspius* (L.), *Abramis ballerus* (L.) (a species absent in the lakes and in the whole of Estonia nowadays) and vendace *Coregonus albula*. All these species spawn in bays or in the littoral zone, which are also the nursery areas. By contrast smelt, the juveniles of which live in deeper areas, became more abundant (Baer 1851; Investigations on the State Fisheries 1860), estimated commercial catch reached 70–100 kg ha^{-1} $year^{-1}$ in the 1910s (Tjurin 1974). It was around this time that fishing regulations for Lake Peipsi–Pihkva were put into practice (Investigations of the State Fisheries 1860).

Before World War II, commercial fishing on lakes was concentrated mostly on the Russian side (83.6% of the total catch; Table 26.2), but there were considerable differences in the traditional fisheries between countries. In Russia the fishery concentrated on smelt (47.1% of the catch). For this reason, gear with very small mesh sizes (6–12 mm) were used intensively, which lead to a huge by-catch of small fish (standard length, SL, <12 cm; accounting for almost one-third of the catch), including juveniles of bream, pikeperch, perch and other species. The amount of larger fish (mainly bream, pikeperch, pike, burbot) was only 2.4%. By contrast, larger fish were preferred by Estonian fishermen (37.3% of the total catch) while lake smelt and the by-catch of small fish amounted to only 23.2% and 3% of the total catch, respectively (Table 26.2).

During World War II, commercial fishing almost collapsed. This and favourable environmental conditions for fish reproduction from 1945 to 1952 improved the stocks of the main commercial species. However, fishing intensity increased rapidly after the

Table 26.2 Mean annual catch (t) of fish in Lake Peipsi–Pihkva in 1931–1940, 1980–1989 and 1990–1998

	1931–1940		1980–1989		1990–1998	
Species	Estonia	Russia	Estonia	Russia	Estonia	Russia
Smelt	443	4597	1125	1462	676	1102
Vendace	24	240	709	1117	60	26
Whitefish	61	14	30	38	35	23
Pike	298	100	104	223	65	117
Bream	55	27	76	85	172	317
Roach	153	~586	241	413	114	475
Pikeperch	198	57	68	100	431	570
Perch	290	~781	680	470	605	248
Burbot	55	27	76	85	18	26
Other valuable[a]	44	11	3	1	21	7
Inferior species[b]	58	~2852	835	1381	~168	435
Total	1910	9767	4046	5530	2365	3346

[a]Eel, asp, vimba bream, etc.
[b]In recent years, >90% ruffe, also small roach, perch. In earlier years, substantial part consisted of juveniles of valuable species, e.g. perch, pikeperch, bream.

war, including trawl fishing (up to 30 trawlers were used). As the result, stocks of pikeperch as well as other valuable species declined rapidly. Trawling (except for smelt which is performed by Russian fishermen to the present day) was prohibited in late 1950s. However, a new gear type was introduced: Danish seines, which were believed to be effective in suppressing perch, ruffe and roach. These species were considered food competitors for valuable species (Tjurin 1957). However, the Danish seine fishery also affected pikeperch and other valuable species. In the 1970s the number of Danish seines was reduced to 40 (the same number as nowadays), and codend mesh size was increased from 10–12 to 20–40 mm. Also burbot was considered a harmful predator and attempts were made to suppress its stocks (Efimova 1963); this species was rehabilitated in the late 1960s (Pihu & Pihu 1968).

By the end of the Soviet period, the absolute and relative amount of fish taken by the Estonian fishermen increased significantly. Identical goals (maximum catch regardless of its quality) and identical fishery regulatory measures all over the lakes lead to equality of the catch composition. The amount of lake smelt (27.8%), vendace (17.5%) and small fish (20.6%) in the catches of the Estonian fishermen increased. By this time, due to intensive trawling in the 1950s, commercial resources of bream and pikeperch were low, and the amount of large fish was only about 10% of the total catch (Table 26.2).

In the 1990s, after re-establishment of independency, catch composition by the Estonian commercial fishermen rapidly returned to the pattern of the 1930s (Table 26.2). The main interest of fishermen is concentrated on perch and pikeperch, most of which is exported to Western Europe. In recent years, due to the opening of export markets, Russian fishermen have also become more interested in these species.

26.4 Recent changes in stocks and fishery

Recent changes in the fish fauna include, first of all, increase in the pikeperch stock and collapse of the vendace stock in early 1990s. The smelt stock in the more eutrophicated Lake Pihkva has collapsed, and there has been some increase in the bream stock. These changes are related both to environmental factors and changes in fishery regulations (limiting of small mesh size active gears and their codend mesh size) (Table 26.3).

Abundance of vendace started to increase in the 1970s, and landings peaked at 3271 t in 1987. The sharp decrease in the stock abundance thereafter is believed to be due to high embryonic mortality during mild winters in the late 1980s and early 1990s associated with the absence of permanent ice cover (Pihu 1996). However, this decrease was also related to an increase in fishing intensity in the 1980s. A small recovery of the stock occurred in the mid-1990s, but it remains at a low level, partly due to high abundance of pikeperch. There are also suggestions that the abundance of vendance depends on the abundance of smelt, which is probably due to competition for food between these two planktivorous fishes.

Pikeperch and bream abundance in the early twentieth century was suppressed by the use of small mesh size active gears. Limiting the number of Danish seines and increasing the codend mesh size, as well as the present level of eutrophication and favourable climatic conditions, have been beneficial to these species. Strong year-classes of pikeperch appeared in 1989, 1991, 1995 and 1997. However, the level of exploitation of pikeperch is high and in recent years the fishery was based on a few strong year-classes (Table 26.4). The appearance of a few strong year-classes of pikeperch in the 1990s has caused problems for management. In 1997, the Russian side unilaterally allowed commercial exploitation of the strong 1995 year-class (SL around 30 cm). For 1998 and 2000, both Russia and Estonia agreed to allow the commercial fishing of four-summer-old pikeperch (1995 and 1997 strong year-classes, respectively; SL slightly less than minimum legal size 40 cm) during the autumn and winter fishery.

Commercial fishing is prevalent on the lakes all year around, but recreational fishing mainly occurs in winter (angling for perch, whitefish). Subsistence fishing, using gill nets, is limited both in Estonia and Russia to shallow coastal areas (up to 1 km from the shoreline). This fishery exploits mostly roach and perch. Illegal fishing and misreporting of commercial catches were common in the early 1990s, but the situation is improving. However, for some species, e.g. bream, illegal fishing, including fishing on spawning areas, remains a serious problem. There are no accurate data on the catches of recreational, subsistence or illegal fishing.

The most effective and profitable gear on Lake Peipsi are Danish seines which can be used in spring (from the ice break until the end of April) and again from 1 September (occasionally earlier in Russia). This gear is mostly used to catch pikeperch and perch. The share of Danish seines in the total catch of Estonia was 27% in 1998 (Table 26.5). Fyke and trap nets are also widely used (their number is not limited except in the summer fishery for vendace), and all smelt and vendace in the Estonian part of lake is taken with these gear. There is also a limited trawl fishery (mostly for smelt) in Russia. Gill nets are mostly used during winter. Until 1999 there

Table 26.3 The main factors affecting recent stock changes of some species in Lake Peipsi–Pihkva. +, positive impact on the stock size, −, negative impact

Indice	Pikeperch	Bream	Vendace	Smelt
Stock size	Increasing	Increasing	Decreasing	Decreasing
Current eutrophication level	+	+	−	−
High summer temperatures	+	+		
Lack of permanent ice cover			−	
Food availability	+	+		
Predation by pikeperch			(−)	−
Overfishing	−	−	−	
Fishery regulations (limitation of small mesh size gear)	+	+		

Table 26.4 Abundance of pikeperch (number per haul) in late summer (August) Danish seine surveys in the Estonian part of Lake Peipsi in 1993–1999

	Age (summers)					
Year	2	3	4	5	>5	Total
1993	11.5	386.8	10.2	11	4	425
1994	0.5	15.4	147.1	4	5	173
1995	6.4	2.3	10.3	12	0	32
1996	152.1	3.4	0.5	2	4	163
1997	10.8	74.4	0.8	0	0	86
1998	153.2	10.8	24.7	0	0	189
1999	13	115	1	1	0	129

Table 26.5 The importance of different gear in the total commercial catch in 1998 (Estonia)

	Gill nets		Danish seines		Fyke and trap nets		Others		
Species	t	%	t	%	t	%	t	%	Total (t)
Bream	139	83	7	4	22	13	1	<1	168
Burbot	8	54	<1	2	6	40	1	4	15
Perch	30	4	517	64	259	32	2	<1	809
Pike	64	66	22	23	8	9	3	3	98
Pikeperch	293	41	385	54	23	3	6	1	707
Roach	36	28	30	24	59	47	0	0	125
Smelt	0	0	0	0	1421	100	0	0	1421
Vendace	0	0	0	0	159	100	0	0	159
Whitefish	53	89	0	0	6	10	0	0	60
Others	1	2	2	5	47	94	0	0	51
Total	624	17	964	27	2011	56	12	<1	3611

Table 26.6 The annual catches in the commercial fishery in 1998 (Estonia)

Species	Catch (t) per month												Total
	I	II	III	IV	V	VI	VII	VIII	IX	X	XI	XII	
Bream	24	30	48	25	17	6	3	1	6	4	1	4	168
Burbot	1	<1	1	2	3	<1	1	1	2	3	1	1	15
Perch	1	3	3	106	228	14	21	16	289	94	33	1	809
Pike	17	12	5	3	4	2	3	2	20	10	3	17	98
Pikeperch	69	26	10	42	16	10	10	5	219	159	54	88	707
Roach	1	1	3	39	15	4	8	12	22	16	5	1	125
Smelt	0	0	0	707	713	0	0	0	0	0	0	0	1421
Vendace	0	0	0	0	0	16	137	6	0	0	0	0	159
Whitefish	4	2	1	1	1	34	11	1	2	<1	<1	1	60
Others	0	0	<1	9	39	1	<1	1	1	1	0	0	51
Total	117	74	70	935	1036	86	193	44	560	286	97	113	3611
%	3.2	2.1	1.9	25.9	28.7	2.4	5.4	1.2	15.5	7.9	2.7	3.1	100

was a specialised summer fishery for whitefish (allowing 55-mm mesh size nets instead of the normal 65 mm). This fishery is now closed as it mostly targeted more valuable pikeperch, because of whitefish quality problems during the high temperature period and to protect the stock of this species (included in the Red Data Book for Estonia).

Fishing on Lake Peipsi–Pihkva is (and has traditionally been) seasonal. Over 50% of fish is taken in April–May. Another increase in catches (about 25% on the annual catch) occurs in September–October (Danish seines). The contribution of the winter fishery (mostly gill net fishing under ice during December–March) is about 10% (Table 26.6).

In the 1990s, fishing conditions in lakes have changed markedly due to natural as well as political and economic reasons. Differences in the strategy and methods of economic reforms between Estonia and Russia (which, among other effects, have resulted in slightly different fishery regulation measures) as well as the boarder political regime have complicated the situation. However, some common tendencies are prevalent.

There was a sharp decrease in registered catches, reaching a minimum in 1994 associated with a decline in fishing during privatization of fishery organizations and misreporting of catches.

In Russia, there has been a substantial increase in the number of companies engaged in fishing and the fish industry, as well as an increase in the number of fishermen from 386 to 692. In Estonia, there are approximately 450 professional fishermen. Former Soviet-time fishing companies have transposed into a few relatively large and strong private companies. Fishing pressure increased in 1995 when local inhabitants were allowed to fish in the 1-km coastal zone with gill nets (subsistence fishing).

Limiting the number of fishing gears (and fees for licences in Estonia) has lead to their more effective use than in the Soviet times. Total catch as well as catch by different types of gear by the Estonian fishermen show similarities between 1962 and 1963

Table 26.7 The number of licences and the catch in 1962–1963 (Pihu 1966) and 1998 in Estonia

Gear	1962–1963, mean	1998
Danish seines		
No. of licences	78	20
Catch (t)	1028	975
Gill nets		
No. of licences	17 055[a]	3000
Catch (t)	341	473
Trap nets		
No. of licences	754	~1200[b]
Catch (t)	2250	2002
Total catch (t)	3623	3611

[a]Probably not all licences were used.
[b]Not all licences were used.

(Pihu 1966) and 1998; despite the number of licences for gill nets and Danish seines being much higher in the 1960s (Table 26.7).

During the last decade, fishing has reoriented towards large (and more valuable) species, first of all, pikeperch but also bream (especially in Russia), pike and perch (especially in Estonia).

26.5 Management in the 1990s

Lake Peipsi–Pihkva is a transboundary water-body, and fisheries are regulated according to the Estonian–Russian Fisheries Agreement from 5 May 1994. An intergovernmental Commission (including members from research institutes, fishermen organizations, producers, control organizations) has two sessions a year. The overall goal of this Commission is to ensure sustainable use of fish resources.

A joint scientific group including researchers from Estonia (University of Tartu) and Russia (Pihkva Branch of the Russian State Research Institute for Lake and River Fisheries, GosNIORH) is attached to the Commission and it is responsible for stock assessment, quota proposals, elaboration of proposals for management measures and other research activities designated by the Commission (e.g. viability of juvenile pikeperch caught by Danish seines, the state of spawning grounds). However, there are problems with regular financing of applied research by both nations.

Decisions and proposals of the Commission are adopted by appropriate governmental bodies. In Estonia, changes in fishing rules usually need a decision of the government. The number of each gear and their distribution between counties is decided by the Minister of the Environment. Licences are further divided by counties on the traditional basis. A special fee is collected for licences; this money is partly allocated (through the State budget) for surveillance, stocking and research. In Russia, management decisions are made by the Fishery Council of the Pskov regional administration.

There are no precautionary reference points determined for stocks in Lake Peipsi–Pihkva. However, formal harvest control rules have been implemented for the main commercial species by the scientific group. The total allowable catch for a fishing year is proposed as a fraction (which depends on species) of the exploitable biomass which is computed as the biomass of fish longer than the minimum legal size. Experimental bottom and (in Russia) pelagic trawl surveys as well as experimental Danish seine surveys are used to calculate the fish biomass in the lakes.

Until 1999, the intergovernmental fishery commission agreed upon recommended maximum catches on both sides and in both lakes (Peipsi and Pihkva). There were quotas for whitefish, as well as spring quotas (April–June) for important spring spawning species, bream, pike and pikeperch. In December 1999 the commission decided to treat the recommended maximum catches proposed by research institutions for all species as quotas and, at the same time, retain quotas for the spring (spawning) period.

Wels, *Silurus glanis* L., grayling, *Thymallus thymallus* (L.), and asp are protected by law in Estonia and their catch is prohibited all year round. Wels is also protected in Russia.

There is a minimum legal size established for the main commercial species, and the by-catch of undersized fish is limited to 5–8% in gill and fyke nets and 15% in Danish Seines in numbers of the total catch of a particular species. The minimum legal size for the main commercial species (total length) are pikeperch 46 cm, bream 35 cm, pike 45 cm, burbot 40 cm, vendace 12 cm, whitefish 35 cm.

Limiting the types of gear and their number is considered very important. Trawl fishing (except for smelt which is performed in the Russian side of both lakes) was prohibited in 1958, and the number of Danish seines was reduced to 40 (20 each side) in 1974; use of Danish seines is normally not allowed on Lake Pihkva. The number of gill nets used in the open water is limited to 3000, and that in the coastal area (up to 1 km from the shoreline) to 1000 on each side in Lake Peipsi. On Lake Peipsi, regardless of its shallowness, at least 3 m high gill nets are traditionally used. One gill net unit is 70 m long. The number of trap nets is established nationally, except the number of traps for vendace which is limited internationally.

Closed areas and close seasons are widely used both in Estonia and Russia, and they are associated with spawning grounds and seasons of major commercial species.

In 1999, minimal distance between commercial gear was enacted in the Estonian fishery regulations.

Minimum mesh sizes are established for the open-water gill net fishery (65 or 55 mm knot to knot, depending on season and species), and for Danish seines (minimum codend mesh size 24 mm in Estonia, 22 mm in Russia). Maximum mesh size for gill nets in the coastal zone is 38 mm in Estonia and 36 mm in Russia.

The number of fishing units and the efficiency of gear are not regulated at present. It is common practice in Estonian waters that a Danish seine licence (issued for a company) is used by two boats and/or crews so that fishing is performed with maximum intensity. One boat leaves for fishing early in the morning and another at around noon, after arrival of the first boat in port. This is the main reason for increased efficiency of this type of fishing (Table 26.7).

References

Baer K.E. (1851) Bericht über einige ichthyologische Nebenbeschäftigungen auf der Reise and den Peipus, vom Ende Aprils bis Anfang Junis. *Bulletin de la Classe physico-mathématique de l'Académie Impériale des Sciences de St. Pétersbourg* **9**, 359–362 (in German).

Efimova A.I. (1963) Fish resources in the Pskov-Chudskoye waterbody and measures towards their increasing. In E.N.Pavlovskij (ed.), *Rybnoye Hozyaistvo Vnutrennih Vodoyomov SSSR*, Moscow, pp. 71–81 (in Russian).

Investigations on the State of Fisheries in Russia (1860) *Vol. 1. Fishery on Lakes Chudskoje and Pskov and in the Baltic Sea.* Sankt Petersburg, 96 pp (in Russian).

Nõges T., Haberman J., Jaani A., Laugaste R., Lokk S., Mäemets A., Nõges P., Pihu E., Starast H., Timm T. & Virro T. (1996) General description of Lake Peipsi–Pihkva. *Hydrobiologia* **338**, 1–9.

Pihu E. (1966) Fishing on Lake Peipsi–Pskov today. *Hydrobiological Researchers Vol. 4. Hydrobiology and Fisheries of Lake Peipsi–Pskov*. Tallinn pp. 270–282 (in Russian, with English summary).

Pihu E. (1996) Fishes, their biology and fisheries management in Lake Peipsi. *Hydrobiologia* **338**, 163–172.

Pihu E. & Pihu E. (1968) On the biology of burbot in the Pskov-Chudskoye waterbody. In G.P. Andrushaitis (ed.) *Ihtiologiya i Ozernoye Rybnoye Hozyaistvo* 1, pp. 109–116. Riga:, Zinatne Publishers (in Russian).

Shirkova A.P. & Pihu E. (1966) The fishes of Lake Peipsi–Pskov and their importance for fisheries. *Hydrobiological Researchers Vol.* 4. *Hydrobiology and Fisheries of Lake Peipsi–Pskov*. Tallinn, pp. 119–127 (in Russian, with English summary).

Tjurin P.V. (1974) Biological justifications of the reconstruction of fish resources of the Pskov-Chudskoje water body. *Izvestiya GosNIORH* **83**, 153–180 (in Russian).

Chapter 27
Role of non-exploited fishery resources in Sri Lankan reservoirs as a source of food for cage aquaculture

U.S. AMARASINGHE*
Department of Zoology, University of Kelaniya, Kelaniya, Sri Lanka

P.A.D. AJITH KUMARA and M.H.S. ARIYARATNE
National Aquatic Resources Agency, Crow Island, Colombo, Sri Lanka

Abstract

The reservoir fisheries of Sri Lanka are almost entirely dependent on the alien cichlid species, most notably *Oreochromis mossambicus* (Peters) and *O. niloticus* (L.). However, small-sized indigenous cyprinid species such as *Amblypharyngodon melettinus* (Val.), *Puntius filamentosus* (Val.), *P. chola* (Hamilton-Buchanan) and *P. dorsalis* (Jerdon), which occur in high abundance in all perennial reservoirs of the country, are not exploited. These small cyprinid species can be differentially exploited using small-mesh gillnets without harming juvenile tilapias. Length frequency data of some small cyprinid species in three Sri Lankan reservoirs, collected between January and December 1999 were analysed using FiSAT. The production per biomass (*P*/*B*) ratios of these unexploited stocks indicate that these fish stocks have potential to withstand heavy fishing pressure and thus sustain productive fisheries.

Community-based cage aquaculture to rear fish fry to fingerling size is a recent development in some perennial reservoirs of Sri Lanka. These fish fingerlings are used to stock in seasonal reservoirs to develop culture-enhanced fisheries. Average daily growth and survival rates of fish fingerlings reared in floating net cages using a feed based on fish meal from local small cyprinids showed better performance than those in the cages where rice bran was used as feed. Their performance was more or less similar to those in cages with commercial feed. The small cyprinid resources in perennial reservoirs of Sri Lanka can thus be used to prepare fish meal as the source of animal protein for aquaculture feeds.

Keywords: cage culture, cichlids, cyprinids, mortality, *P*/*B* ratio, unexploited stocks.

27.1 Introduction

Asia, with $43\,475 \times 10^3 \text{km}^2$ of freshwater lakes and reservoirs (Shiklomanov 1990), has considerable potential for inland fisheries development. However, in most reservoirs in the Asia-Pacific region, the fisheries are based on a few species (Baluyut 1992). In Sri Lanka for example, two exotic Cichlidae species, *Oreochromis mossambicus*

*Correspondence: Department of Zoology, University of Kelaniya, Kelaniya 11600, Sri Lanka (email: zoousa@kln.ac.lk).

(Peters) and *O. niloticus* (L.), contribute the major portion of landings (>90%) from reservoir fisheries (De Silva 1988), despite there being >38 indigenous fish species present, 50% of which are cyprinids (Fernando & Indrasena 1969). These indigenous species are not exploited because of poor consumer acceptability, and gillnet mesh size restrictions targeting tilapias making them difficult to catch because of their small size.

Several studies in Sri Lankan reservoirs (Amarasinghe 1985; 1990; De Silva & Sirisena 1987; 1989; Sirisena & De Silva 1988; Pet & Piet 1993), however, have indicated that these small cyprinids could be exploited using small-mesh (15–52 mm stretched mesh size) gillnets without catching juvenile cichlids. This is because juvenile *O. mossambicus* and *O. niloticus* are only found in shallow (<1.5 m), littoral areas of reservoirs and their adult counterparts, which are targeted by the fisheries, are found in deeper water. As a result, small cyprinids can be differentially exploited in deeper, limnetic areas of reservoirs using small-mesh (18–52 mm) gillnets (Amarasinghe 1985; De Silva & Sirisena 1987). Unfortunately, despite the fishery potential of these small cyprinids having been estimated in Sri Lankan reservoirs by comparing the catch per unit effort (CPUE) values of cichlids and small cyprinids in the experimental gillnets (De Silva & Sirisena 1989; Amarasinghe 1990), there is no evidence to show whether this resource can sustain productive fisheries.

In this chapter, life-history parameters of small cyprinids, major carps and cichlids are compared to understanding whether small cyprinids are capable of supporting productive fisheries. Also the results of cage aquaculture trials in selected Sri Lankan reservoirs are presented with a view to highlighting the performance of different feeds, including farmer-made feeds prepared by using fish-meal based on small cyprinids. Potential synchronization of seasonal exploitation of small cyprinids and the timing of cage aquaculture of fish fingerlings in perennial reservoirs to stock into seasonal ponds is also investigated.

27.2 Materials and methods

Data were gleaned from various sources, including experimental fishing surveys in three Sri Lankan reservoirs to analyse length frequency distributions, published information on demographic parameters of major carps and cichlids (Prakash & Gupta 1986; Amarasinghe 1987; Amarasinghe, De Silva & Moreau 1989; Li, Zhou & Lin 1990; Amarasinghe & De Silva 1992), and results of cage aquaculture trials in three Sri Lankan reservoirs (Ariyaratne 2001). Experimental fishing with mono-filament gillnets of mesh sizes 12.5, 16, 20, 25, 33, 37 mm was carried out in three reservoirs of Sri Lanka, viz. Minneriya (2551 ha; 8°2′N, 80°53′E), Udawalawe (3415 ha; 6°55′N, 80°55′E), and Victoria (2270 ha; 7°20′N, 80°47′E) from January to December 1999. Length frequency distributions of commonly caught small cyprinids were obtained for each mesh size separately. Based on the total fishing effort (gillnet hours), length frequency data for gillnets of each mesh size were adjusted to a constant fishing effort. These adjusted length frequency data in all mesh sizes were then pooled for each sampling period. Data from mesh sizes in which few fishes were caught were not used because length frequency adjustments were invalid.

These adjusted length frequencies were analysed using the FiSAT programs (Gayanilo Jr, Sparre & Pauly 1996) to estimate von Bertalanffy growth parameters. As these small cyprinids are unexploited in the three reservoirs, total mortality of individual fish species was considered equal to natural mortality. Natural mortality (M) of each species was determined from Pauly's empirical equation (Pauly 1980; Gayanilo *et al.* 1996). Allen (1971) and Christensen & Pauly (1992) indicated that when mortality is exponential and growth conforms to von Bertalanffy equation, on average, annual instantaneous total mortality rate equals annual production/mean biomass (P/B ratio) or turnover rate. As such, M values estimated for each of these unexploited small cyprinid species were taken as their P/B ratios. P/B ratios of some Chinese and Indian major carps in different habitats in the world were also determined from the instantaneous rates of total mortality based on Li *et al.* (1990) and Prakash & Gupta (1986). Instantaneous rates of total mortality of commercially-important cichlids in Sri Lankan reservoirs, *O. mossambicus* and *O. niloticus* (Amarasinghe 1987; Amarasinghe *et al.* 1989; Amarasinghe & De Silva 1992), were used to determine their P/B ratios. P/B ratios of small cyprinids, commercially important cichlids and Chinese and Indian major carps were plotted against their asymptotic lengths. These data pairs were then analysed using Bray–Curtis cluster analysis (Bray & Curtis 1957).

Data on proximate composition of three types of feeds (i.e. rice bran, commercial fish feed and farmer-made feed using fish meal prepared from small cyprinids) used in cage aquaculture of fish fingerlings in perennial reservoirs were obtained from Ariyaratne (2001). Performance of cage aquaculture in the form of survival rate and average daily growth (ADG) of various fish species cultured under three different feed types were also obtained from Ariyaratne (2001).

27.3 Results

In the three reservoirs studied, the most commonly caught species in gillnets were *Amblypharyngodon melettinus* (Valenciennes), *Puntius filamentosus* (Valenciennes) and *P. chola* (Hamilton-Buchanan). Growth curves of *A. melettinus*, *P. filamentosus* and *P. chola* superimposed on their length frequency distributions in three reservoirs are shown in Fig. 27.1. However, for Udawalawe reservoir, growth curves are given only for *A. melettinus* and *P. filamentosus* because inadequate data were available for *P. chola*.

Von Bertalanffy growth parameters, instantaneous rates of natural and total mortality and P/B ratios of small cyprinid species in three reservoirs are given in Table 27.1. Also given in Table 27.1 are growth parameters, instantaneous rates of total mortality and P/B ratios of Chinese and Indian carps, which were gleaned from various publications. Growth and mortality parameters, and P/B ratios of *O. mossambicus* and *O. niloticus* in some Sri Lankan reservoirs, which support very profitable fisheries are also presented in Table 27.1. P/B ratios of different species are significantly related to their asymptotic total lengths (L_∞) according to the relationship, $Y = 12.487X^{-0.5429}$ ($r = -0.767$; $P < 0.001$). Results of the Bray–Curtis cluster analysis (Bray & Curtis 1957; Figs 27.2 and 27.3) indicated that small cyprinids which have low L_∞ and high

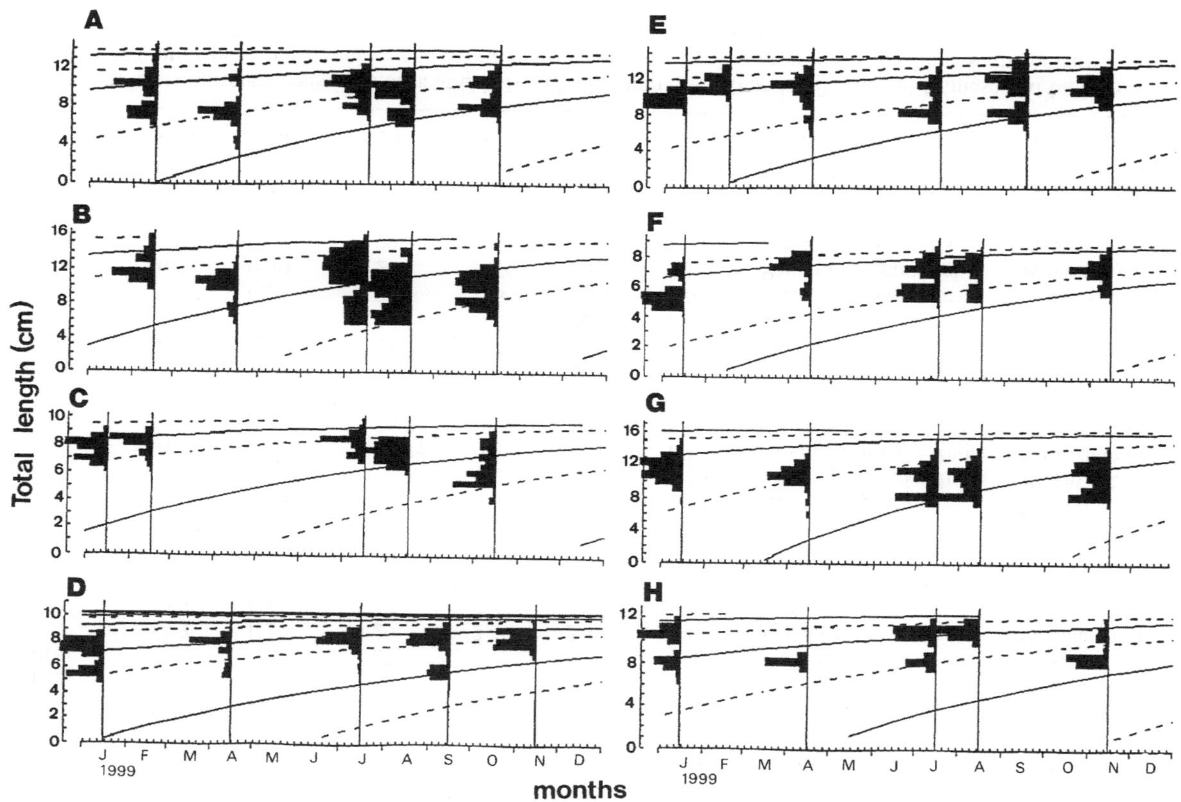

Figure 27.1 Growth curves of *Amblypharyngodon melettinus*, *Puntius filamentosus* and *P. chola* superimposed on their length frequency distributions in three Sri Lankan reservoirs. A: *P. chola* in Minneriya; B: *P. filamentosus* in Minneriya; C: *A. melettinus* in Minneriya; D: *A. melettinus* in Udawalawe; E: *P. filamentosus* in Udawalawe; F: *A. melettinus* in Victoria; G: *P. filamentosus* in Victoria; H: *P. chola* in Victoria

Table 27.1 Von Bertalanffy growth parameters, instantaneous rates of total mortality and *P/B* ratios of small cyprinid species in three reservoirs

Group/species	L_∞ (cm)	K (year^{-1})	T°C	*M*	*Z*	*P/B*	Reservoir	Source
Small cyprinids								
A. melettinus (Am1)	10.2	1.5	29	3.19	3.19	3.19	Minneriya	Present study
A. melettinus (Am2)	10.2	1.2	29	2.76	2.76	2.76	Udawalawe	Pressent study
A. melettinus(Am3)	9.6	1.3	27	2.86	2.86	2.86	Victoria	Pressent study
P. filamentosus (Pf1)	16.6	1.5	29	2.79	2.79	2.79	Minneriya	Present study
P. filamentosus (Pf2)	15.4	1.2	29	2.46	2.46	2.46	Udawalawe	Present study
P. filamentosus (Pf3)	16.6	1.9	27	3.15	3.15	3.15	Victoria	Present study
P. chola (Pc1)	14.7	1.2	29	2.49	2.49	2.49	Minneriya	Present study
P. chola (Pc2)	12.8	1.4	27	2.77	2.77	2.77	Victoria	Present study
Major carps								
C. carpio (Ccp)	129.7	0.107	20	0.24	0.48*	0.48	Xin'anjiang, China	Li *et al.* 1990
A. nobilis (An)	161.4	0.135	20	0.26	0.52*	0.52	Xin'anjiang, China	Li *et al.* 1990
H. molitrix (Hm)	84.3	0.292	20	0.51	1.02*	1.02	Xin'anjiang, China	Li *et al.* 1990
C. mrigala (Cm)	77.1	0.084	26.7†	0.27	0.54*	0.54	Govindgarh Lake, India	Prakash & Gupta 1986
L. rohita (Lr)	91.2	0.382	26.7†	0.68	1.36*	1.36	Govindgarh Lake, India	Prakash & Gupta 1986
C. catla (Cct)	115.0	0.324	26.7†	0.57	1.14*	1.14	Govindgarh Lake, India	Prakash & Gupta 1986
Cichlids								
O. mossambicus (Om1)	39.3	0.34	29	0.82	2.42	2.42	Pimburettewa, Sri Lanka	Amarasinghe 1987
O. mossambicus (Om2)	34.6	0.30	29	0.78	2.22	2.22	Parakrama Samudra	Amarasinghe *et al.* 1989
O. mossambicus (Om3)	43.3	0.32	29	0.77	2.42	2.42	Parakrama Samudra	Amarasinghe *et al.* 1989
O. mossambicus (Om4)	38.8	0.24	29	0.66	1.64	1.64	Parakrama Samudra	Amarasinghe *et al.* 1989
O. mossambicus (Om5)	40.0	0.30	29	0.76	2.14	2.14	Parakrama Samudra	Amarasinghe *et al.* 1989
O. mossambicus (Om6)	40.6	0.22	29	0.61	1.07	1.07	Parakrama Samudra	Amarasinghe *et al.* 1989
O. mossambicus (Om7)	45.0	0.45	29	0.955	3.031	3.031	Minneriya, Sri Lanka	Amarasinghe & De Silva 1992
O. mossambicus (Om8)	43.7	0.52	29	1.005	1.399	1.399	Kaudulla, Sri Lanka	Amarasinghe & De Silva 1992
O. niloticus (On1)	54.5	0.43	29	0.878	3.578	3.578	Minneriya, Sri Lanka	Amarasinghe & De Silva 1992
O. niloticus (On2)	54.5	0.34	29	0.753	1.853	1.853	Kaudulla, Sri Lanka	Amarasinghe & De Silva 1992

Also given are growth parameters, instantaneous rates of natural (*M*) and total mortality (*Z*) and *P/B* ratios of Chinese and Indian carps, plus growth and mortality parameters, and *P/B* ratios of *O. mossambicus* and *O. niloticus* in some Sri Lankan reservoirs. * In Chinese major carps, total mortality (*Z*) was taken as 2**M*; † Data from Sugunan (1995). Abbreviations for species (in parentheses) are as used in Fig. 27.2.

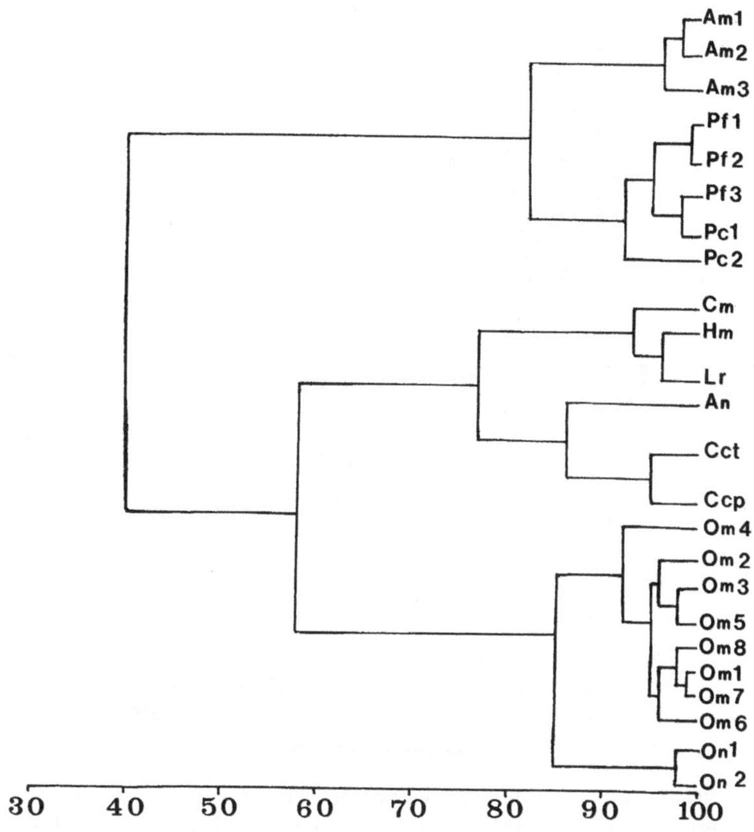

Figure 27.2 Bray–Curtis similarity (Bray & Curtis 1957) of the association between *P*/*B* ratio and asymptotic length of various fish species (small cyprinids, cichlids and major carps). The abbreviations for species are as indicated in Table 27.1

P/*B* ratios, cichlids which support profitable fisheries and major carps, form three different clusters.

These results indicate that small-sized cyprinids have *P*/*B* ratios as high as those of the cichlids which support profitable fisheries in Sri Lankan reservoirs. Conversely major carps, which produce low yields in Chinese and Indian reservoirs have low *P*/*B* ratios.

Proximate composition of three feed types (i.e. rice bran, commercial feed and those based on fish meal prepared from small cyprinids) (Table 27.2; Ariyaratne 2001) had significantly different protein levels from each other (one-way ANOVA and Tukey's test, $P < 0.05$). Feed based on small cyprinids had intermediate levels of % protein between rice bran and commercial feeds. Percentage survival and % ADG of fish fingerlings were also high in net cages where feed based on small cyprinids was used (Table 27.3).

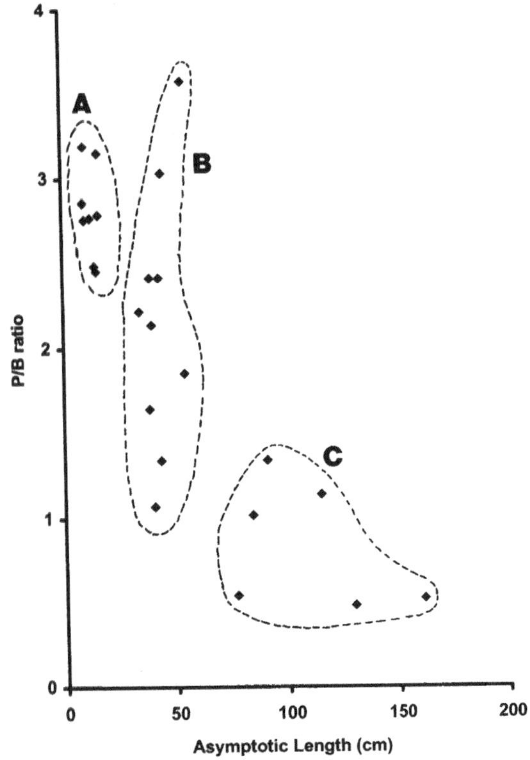

Figure 27.3 Three clusters of the relationship between *P/B* ratio and asymptotic length of various fish species as determined by Bray–Curtis similarity analysis. A: Small cyprinids in three Sri Lankan reservoirs; B: two cichlid species in Sri Lankan reservoir; C: major carps from different localities

27.4 Discussion

It has been established that fish yields are relatively low in reservoirs where large cyprinids form the major proportion of the landings (Fernando & Holcik 1991). On the other hand, Fernando & Holcik (1991) showed that when introduced cichlids are present in tropical reservoirs they support considerably higher yields. When *P/B* ratios of these two groups of fishes are considered, cichlids have higher *P/B* ratios than large cyprinids such as Chinese and Indian major carps. Species with high *P/B* ratios can withstand high levels of fishing mortalities (Evans, Henderson, Bax, Marshall, Oglesby & Christie 1987) so that high yields can be achieved. As such, the differences in life-history patterns of fish species seem to influence colonization success of small-sized cyprinids in Sri Lankan reservoirs.

Small-sized fishes have faster growth rate and higher natural mortality rates than fishes of large maximum length (Pauly 1980; Evans *et al.* 1987). As instantaneous mortality rate is equivalent to the *P/B* ratio (Allen 1971; Christensen & Pauly 1992), small-sized fishes can be expected to have high *P/B* ratios. Small cyprinids that inhabit

Table 27.2 Proximate composition (±SE) of three feed types used in cage aquaculture of fish fingerlings (data from Ariyaratne 2001)

	Rice bran	Commercial feed	Farmer-made feed	*F*	*P*
% Protein ± SE	12.3 ± 0.55[a]	36.0 ± 0.64[a]	27.4 ± 0.98[a]	106.01	<0.001
% Fat ± SE	6.9 ± 0.96[a]	11.3 ± 0.42[a]	6.3 ± 0.84[b]	11.28	<0.001
% Minerals ± SE	10.2 ± 1.18[a]	10.3 ± 0.30[a]	8.0 ± 0.39[b]	4.35	<0.02

Variance ratios of one-way ANOVA (*F*-values) and probability levels (*P*) are also presented here. The values with the same superscripts in each row are not significantly different at the 5% level (Tukey's test).

Table 27.3 Percentage survival and % ADG of different species of fish fingerlings reared in net cages in three Sri Lankan reservoirs (data from Ariyaratne 2001), where applicable ranges are given in parentheses

Culture procedure/ species	No. of culture trials	Feed type	% Survival	% ADG
Advance fry rearing				
C. carpio	4	Rb	47.9 (36.6–78.5)	14.57 (8.90–23.89)
	3	Cf	61.2 (48.5–70.7)	18.53 (6.60–41.74)
	2	Mf	39.2 (36.5–41.5)	11.84 (1.22–22.46)
L. rohita	2	Rb	73.5 (55.1–91.9)	10.47 (0.29–20.60)
	1	Cf	51.4	16.44
	1	Mf	39.9	26.66
Fingerling rearing				
C. catla	1	Rb	23.3	33.20
	1	Cf	46.5	17.13
	1	Mf	25.3	14.49
H. molitrix	1	Rb	19.5	4.95
	1	Cf	40.5	5.03
	1	Mf	34.2	6.67
L. dussumieri	2	Rb	27.2 (12.5–42.0)	33.34 (5.69–60.99)
	2	Cf	71.6 (48.0–95.1)	7.77 (4.98–10.55)
	2	Mf	55.3 (49.0–61.5)	6.76 (6.56–6.96)
O. niloticus	1	Rb	88.4	2.15
	1	Cf	39.9	4.72
	1	Mf	26.6	4.66

Advance fry rearing: rearing of fish fry (2–3 cm) up to large-sized fry (5–6 cm); fingerling rearing: rearing of fish fry (2–3 cm) up to fingerling size (8–9 cm). Rb: rice bran; Cf: commercial fish feed; Mf: farmer-made feed using fish-meal based on small cyprinids.

reservoirs also have high *P*/*B* ratios (Table 27.1). In Sri Lankan reservoirs, *P*/*B* ratios of exotic cichlid species (*O. mossambicus* and *O. niloticus*) which support productive fisheries are also high (Table 27.1). Conversely in major carps, *P*/*B* ratios are considerably lower than those of small-sized cyprinids. Low fish yields of these long-lived, large fish species may therefore be due to their low *P/B* ratios. As small cyprinids are short-lived, and have high *P/B* ratios, they can be expected to be capable of supporting productive fisheries despite being riverine species. Pet, Gevers, Van Densen & Vijverberg (1996) showed that total mortalities of small cypinids (i.e. *A. melettinus*, *P. chola* and *Rasbora daniconius*) and *O. mossambicus* in a Sri Lankan reservoir were very high. According to their estimates, total mortality of small cyprinids were within the range 2.8–4.7 and that of *O. mossambicus* was 6.4–7.4. Pet *et al.* (1996) also showed that out of the total biomass estimated in this reservoir (1829 kg ha^{-1}), only 128 kg ha^{-1} was *O. mossambicus* whereas biomass of small pelagic cyprinid species, mainly *A. melettinus*, was 1098 kg ha^{-1}. Pet *et al.* (1996) recommended that as the utilization of the biological fish production in this Sri Lankan reservoir was only 4.5% in terms of fish yield, to reach a utilization of 10% of the biological fish production from lakes and reservoirs (Blazka, Backiel & Taub 1980), fish yield can be significantly increased by exploiting small pelagic cyprinid species such as *A. melettinus*. Marshall (1993) mentioned that the high *P/B* ratios of sardines (*Limnothrissa miodon*) in Lake Kariba, East Africa contributed to the high yields of the fishery.

Although there is evidence (De Silva & Sirisena 1987; Amarasinghe 1990; Pet & Piet 1993) that there is a considerable potential for exploitation of minor cyprinids, no major fishery has materialized in most reservoirs (Fernando, Gurgel & Moyo 1998). In several Sri Lankan reservoirs (Amarasinghe 1990) the major reason is poor consumer preference for small cyprinids, but small-scale fisheries exist.

As fish meal prepared from small cyprinids is of acceptable nutritive quality to use as supplementary feeds in cage aquaculture in perennial reservoirs, utilization of this fish resource in the form of fish meal appears to be a feasible strategy. In Sri Lanka, an aquaculture development plan has been implemented since 1995. Extensive aquaculture in seasonal reservoirs where water is retained for 7–9 months each year (usually from the inter-monsoonal rainy season in November–December to July–October) is a major component in this development plan. However, in Sri Lanka, there is a paucity of fingerlings available for aquaculture practices. As such, fishing communities are encouraged to rear fish fingerlings in floating net cages in perennial reservoirs. Timing of fingerling rearing is a key factor for the success of the aquaculture development programme in seasonal reservoirs because fingerlings should be made available for stocking in seasonal reservoirs in November–December (De Silva 1988). As such fingerlings should be reared in cages between August and October. Since small cyprinids are caught in large numbers during the periods of low water level in perennial reservoirs, i.e. July–October (Amarasinghe 1990), there is a potential synchronization of seasonal exploitation of small cyprinids and the timing of cage aquaculture of fish fingerlings (Fig. 27.4).

Tacon (1993) mentioned that as most aquafeeds are composed of highly perishable nutrients, it is essential that adequate handling and storage procedures are employed to protect their nutritive quality. However, fishing communities engaged in cage aquaculture in Sri Lankan reservoirs prepare feed dough using fish meal derived from small cyprinids and other ingredients (rice bran and cooked cassava tuber) just before feeding fish. As such the question of storage of feed does not arise in this kind of community-based aquaculture strategy.

Month	M	J	J	A	S	O	N	D	J	F	M	A	M	J	J	A	S
Weather cycle	SW Monsoon & Windy season					NE Monsoon & Calm season						SW Monsoon & Windy season					
Water levels in perennial irrigation reservoirs	Low water level						High water level						Low water level				
Paddy cultivation	Irrigation		Harvest	Fallow period		Field preparation & Sowing			Irrigation		Harvest	Field Prep. & Sowing	Irrigation		Harvest		Fallow Period
Availability of labour	++	+++	+	++	+++	+	+	++	++	++	+	+	++	+++	+	++	+++
Catch efficiency of small cyprinids	Moderate		High				Moderate		Low				Moderate		High		
Fry production in state breeding centres	PL	Fry					PL	Fry					PL	Fry			
Cage culture in perennial irrigation reservoirs			Raising fry upto fingerling size in net cages												Raising fry upto fingerling size in net cages		
Culture-based fisheries in seasonal reservoirs							Stocking of fingerlings										
									Growing period								
																Harvesting & marketing	
Income availability in rural farmer and fisher communities	+	+	-	--	---	---	--	+	--	+	+++	++	+	+	-	--	---

Figure 27.4 Potential synchronisation of seasonal exploitation of small cyprinids in perennial reservoirs and the timing of cage aquaculture of fish fingerlings in perennial reservoirs to stock in seasonal tanks. Note that this cage aquaculture will be useful for increasing income in fishing communities in perennial irrigation reservoirs (+++: greatest amount; ---: lowest amount; PL: post-larval rearing; Fry: fry rearing). This figure was prepared based on the information in De Silva (1988), Amarasinghe (1990) and Murray, Kodithuwakku & Little (2001)

Acknowledgements

This research was funded by European Union INCO-DC Programme (Project Number: IC18-CT-0190) and Australian Centre for International Agricultural Research (Project Number: 9440). Association of Commonwealth Universities (ACU) provided financial support to attend the symposium while the senior author (USA) was in Imperial College of Science, Technology and Medicine, University of London under a post-doctoral fellowship awarded by ACU.

References

Allen K.R. (1971) Relation between production and biomass. *Journal of the Fisheries Research Board of Canada* **28**, 1573–1581.

Amarasinghe U.S. (1985) Studies on the exploitation of minor cyprinids in Parakrama Samudra, a man-made lake in Sri Lanka, using gillnets. *Journal of the National Aquatic Resources Agency (Sri Lanka)* **32**, 11–23.

Amarasinghe U.S. (1987) Status of the fishery of Pimburettewa wewa, a man-made lake in Sri Lanka. *Aquaculture and Fisheries Management* **18**, 375–385.

Amarasinghe U.S. (1990) Minor cyprinid resources in a man-made lake in Sri Lanka: a potential supplementary source of income for fishermen. *Fisheries Research* **9**, 81–89.

Amarasinghe U.S. &. De Silva S.S. (1992) Population dynamics of *Oreochromis mossambicus* and *O. niloticus* (Cichlidae) in two reservoirs of Sri Lanka. *Asian Fisheries Science* **5**, 37–61.

Amarasinghe U.S., De Silva S.S. & Moreau J. (1989) Spatial changes in growth and mortality and effects on the fishery of *Oreochromis mossambicus* (Pisces, Cichlidae) in a man-made lake in Sri Lanka. *Asian Fisheries Science* **3**, 57–68.

Ariyaratne M.H.S. (2001) Performance of fingerlings of commonly cultured fish species to different food types when reared in cages. In S.S. De Silva (ed.) *Reservoir and Culture-Based Fisheries: Biology and Management*. Canberra: ACIAR, pp. 359–366.

Baluyut E.A. (ed.) (1992) Indo-Pacific Fishery Commission, country reports presented at the Fifth Session of the Indo-Pacific Fishery Commission Working Party of Experts on Inland Fisheries, Bogor, Indonesia, 24–29 June 1991 and papers contributed to the Workshop on Tilapia in Capture and Culture-Enhanced Fisheries in the Indo-Pacific Fishery Commission countries, Bogor, Indonesia, 27–29 June 1991. *FAO Fisheries Report No. 458*, Suppl., FAO, Rome, 281 pp.

Blazka P., Backiel T. & Taub F.B. (1980) Trophic relationships and eficiencies. In E.D. LeCren & R.H. Lowe-McConnell (eds) *The Functioning of Freshwater Ecosystems*. London: Cambridge University Press, pp. 293–410.

Bray J.R. & Curtis J.T. (1957) An ordination of the upland forest communities of southern Wisconsin. *Ecological Monograph* **27**, 325–349.

Christensen V. & Pauly D. (1992) *A Guide to the ECOPATH II Software System (Version 2.1)*. ICLARM software 6, Manila: International Centre for Living Aquatic Resources Management, 72 pp.

De Silva S.S. (1988) Reservoirs of Sri Lanka and their fisheries. *FAO Fisheries Technical Paper* 298, 128 pp.

De Silva S.S. & Sirisena H.K.G. (1987) New fish resources of reservoirs in Sri Lanka. Feasibility of introduction of a subsidiary gillnet fishery for minor cyprinids. *Fisheries Research* **6**, 17–34.

De Silva S.S. & Sirisena H.K.G. (1989) New fish resources of reservoirs in Sri Lanka. 3. Results of commercial-scale trials and yield estimates of a gill-net fishery for minor cyprinids. *Fisheries Research* **7**, 279–287.

Evans D.O., Henderson B.A., Bax N.J., Marshall T.R., Oglesby R.T. & Christie W.J. (1987) Concepts and methods of community ecology applied to freshwater fisheries management. *Canadian Journal of Fisheries and Aquatic Sciences* **44** (Suppl. II), 448–470.

Fernando C.H. & Holcik J. (1991) Fish in reservoirs. *International Revue der gesamten Hydrobiologie* **76**, 149–167.

Fernando C.H. & Indrasena H.H.A. (1969) The freshwater fisheries of Ceylon. *Bulletin of the Fisheries Research Station, Ceylon* **20**, 101–134.

Fernando C.H., Gurgel J.J.S. & Moyo N.A.G. (1998) A global view of reservoir fisheries. *International Review of Hydrobiology* **83**, 31–42.

Gayanilo Jr. F.C., Sparre P. & Pauly D. (1996) *The FAO-ICLARM Stock Assessment Tools (FiSAT) User's Guide.* FAO Computerized Information Series (Fisheries), No. 8. Rome: FAO, 126 pp.

Li S., Zhou B. & Lin Q. (1990) The yield and growth of major fish species in a large Chinese reservoir. *Asian Fisheries Science* **3**, 185–196.

Marshall B.E. (1992) *Limnothrissa* in man-made lakes: Do we understand the implications of their small size? In B.E. Marshall & R. Mubamba (eds) Papers presented at the Symposium on Biology, Stock Assessment and Exploitation of Small Pelagic Fish Species in the African Great Lakes Region, Bujumbura, Burundi, 25–28 November 1992. *CIFA Occasional Paper 19,* 270 p., Rome: FAO, pp. 1–17.

Murray F.J., Kodithuwakku S. & Little D.C. (2001) Fisheries marketing systems in Sri Lanka and relevance to development of the local reservoir fishery. In S.S. De Silva (ed.) *Reservoir and Culture-Based Fisheries: Biology and Management.* Canberra: ACIAR, pp. 287–308.

Pauly D. (1980) On the relationship between natural mortality, growth parameters and mean environmental temperature in 175 fish stocks. *Journal du Consiel International Pour l'Exploitation de la Mer* **39**, 175–192.

Pet J.S., Gevers G.J.M., Van Densen W.L.T. & Vijverberg J. (1996) Management options for a more complete utilization of the biological fish production in Sri Lankan reservoirs. *Ecology of Freshwater Fish* **5**, 1–14.

Pet J.S. & Piet G.J. (1993) The consequences of habitat occupation and habitat overlap of the introduced tilapia *Oreochromis mossambicus* (Peters) and indigenous fish species for fishery management in a Sri Lankan reservoir. *Journal of Fish Biology* **43** (Suppl. A), 193–208.

Prakash S. & Gupta R.A. (1986) Studies on the comparative growth rates of three major carps of the Govingarh lake. *Indian Journal of Fisheries* **33**, 45–53.

Shiklomanov L.A. (1990) Global water resources. *Nature and Resources* **26**, 34–43.

Sirisena H.K.G. & De Silva S.S. (1988) New fish resources of reservoirs in Sri Lanka. II. Further studies on a gillnet fishery for minor cyprinids. *Fisheries Research* **7**, 17–29.

Sugunan V.V. (1995) Reservoir fisheries of India. *FAO Fisheries Technical Paper* 345, 423 pp.

Tacon A.G.J. (1993) Feed formulation and on-farm feed management. In: M.B. New, A.G.J. Tacon & I. Csavas (eds) *Farm-Made Aquafeeds. Proceedings of the FAO/AADCP Regional Expert Consultation on Farm-Made Aquafeeds, 14–18 December 1993, Bangkok, Thailand.* Bangkok: FAO/RAPA/AADCP, pp. 61–74.

Chapter 28
Potential species for fishery enhancement in Lake Faé, Côte d'Ivoire

K.S. da COSTA*, K. TRAORE and W. YTE
Centre National de Recherches Agronomiques (CNRA), Station Pisciculture Bouaké, Côte d'Ivoire, France

Abstract

Lake Faé (San Pedro) is an agro-hydroelectric dam located in the south-west of Côte d'Ivoire. It has a surface of about 50 km^2. The ichthyofauna comprises 25 species from 12 families, a species richness that is poorer than other reservoirs in the country, e.g. Ayamé, Buyo, Kossou and Taabo. Many of the economically-important fishes found in these reservoirs (*Oreochromis niloticus* (L.), *Sarotherodon galilaeus* (L.), *Heterobranchus isopterus* Bleeker, *Clarias anguillaris* (L.), *Labeo coubie* Rüppell, *Synodontis* spp., *Heterotis niloticus* (Cuv.)) are not present in Lake Faé, where *Chrysichthys nigrodigitatus* (Lacepède), *Schilbe mandibularis* (Günther) and *Brycinus longipinnis* (Günther) are the main components of the catch. As a consequence, consideration was given to diversify the species richness to enhance the fishery of Lake Faé. It is proposed that *Labeo coubie, Distichodus rostratus, Oreochromis niloticus* and *Heterotis niloticus* are introduced to achieve this objective. This chapter describes assessment of the advantages and ecological risks of these introductions in relation to conservation of the biodiversity of this reservoir.

Keywords: Côte d'Ivoire, fish introductions, fishery management, reservoir, yield.

28.1 Introduction

Côte d'Ivoire has a low national fish production despite the population having a tradition for eating fish, which is the main source of animal protein. Around 42% of the urban population and nearly 100% of the rural communities consume fish, although the intake is spatially unbalanced and varies from 1 to 24 g ind.$^{-1}$ d^{-1} from north to south (Da Costa, Traore & Tito de Morais 1998). This deficit leads to a loss to foreign exchange of about 120 billion francs CFA per year. Consequently, increasing fish production is a priority for Côte d'Ivoire. The construction of hydropower and hydro-agri-pastoral reservoirs has allowed the development of inland fisheries, particularly in the bigger reservoirs (Kossou, Buyo, Taabo and Ayamé). Other more recently constructed reservoirs with a relatively small size (e.g. Lake Faé) or reservoirs with an area less than 1 km^2, present an opportunity to increase fish production further.

*Correspondence: Centre National de Recherches Agronomiques (CNRA), Station Pisciculture 01 B.P. 633 Bouaké O1, Côte d'Ivoire, France (email: dacosta@bouaké.ird.ci).

This study examines the potential for increasing fish production in Lake Faé on the River San Pedro, which, with the exception of the River Sassandra, is the only inland source of fish in the south-west of Côte d'Ivoire. The possibility of diversifying the fish community through species introductions to increase fish production was examined.

28.2 Study area

The hydrographic network of south-west Côte d'Ivoire is characterised by the presence of the rivers Cavally and Sassandra, and by a group of small coastal rivers of which the most important are the Dodo, San Pedro and Néro (Fig. 28.1).

The River San Pedro is in the forest zone. It is 112 km long, has a catchment area of 3310 km^2, a slope of 1.7 $m\,km^{-1}$ and mean annual discharge of 31 $m^3\,s^{-1}$. The San Pedro basin is in an equatorial transition zone with two rainy seasons separated by a short dry period in August and a more pronounced dry season from November to March. The heaviest rains are in June. Annual rainfall in the San Pedro watershed is around 1890 mm $year^{-1}$.

Lake Faé is an agro-hydropower reservoir built in 1980 on the River San Pedro with an area of 50 km^2. This dam was initially intended for crop irrigation, mainly in downstream rice fields, and to serve a paper factory which was never built. Lake Faé was converted (installation of turbines) for the production of electricity to supply the city of Soubré and surrounding areas and supplement supplies from the Buyo dam (western Côte d'Ivoire).

An inventory of the ichthyofauna of the River San Pedro was described by Daget & Oltis (1965), and more recently after construction of Fae dam by Teugels, Lévêque, Paugy & Traoré (1988), and Paugy, Traoré & Diouf (1994) (Table 28.1).

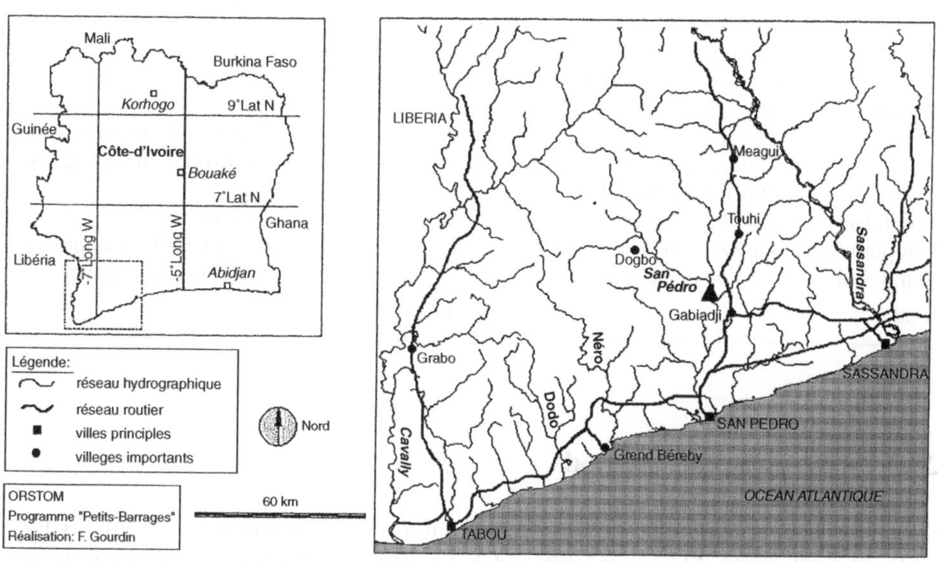

Figure 28.1 Hydrographic net of south-west Côte d'Ivoire

Table 28.1 Inventory fish species in the San Pedro river basin (Teugels *et al.* 1988) and Lake Faé, plus their trophic status (after Welcomme 1985; Doudet 1979; Lauzanne 1988; Welcomme & De Merona 1988; Hugueny 1989; Skelton 1993; Lévêque & Paugy 1999)

Family	Species	San Pedro	Lake Faé	Dietary status
Alestiidae	*Brycinus imberi*	+	+	Omnivore/zooplanktivore
	Brycinus longipinnis	+	+	Omnivore/zooplanktivore
	Brycinus macrolepidotus	+	+	Omnivorous
	Brycinus nurse		+	Omnivore/zooplanktivore
	Lepidarcus adonis	+		
	Micralestes occidentalis	+		
Anabantidae	*Ctenopoma petherici*	+		
Channidae	*Parachanna obscura*	+	+	Generalist carnivore
	Chromidotilapia guntheri	+	+	Aquatic invertebrates
	Hemichromis fasciatus	+	+	Carnivore
Cichlidae	*Sarotherodon melanotheron*	+	+	Herbivore
	Tilapia guineensis		+	Omnivore
	Tilapia mariae	+	+	Omnivore
	Tilapia zillii	+	+	Herbivore
Clariidae	*Clarias ebriensis*	+	+	Omnivore
	Heterobranchus isopterus	+		Omnivore
	Heterobranchus longifilis	+	+	Omnivore
Claroteidae	*Chrysichthys auratus*		+	Aquatic invertebrate
	Chrysichthys maurus	+	+	Aquatic invertebrate
	Chrysichthys nigrodigitatus	+	+	Aquatic invertebrate
Clupeidae	*Pellonula leonensis*	+	+	Omnivore/zooplanktivore
	Pellonula vorax	+		
Cyprinidae	*Barbus ablabes*	+	+	Omnivore/zooplanktivore
	Barbus bigornei	+		
	Barbus macrops	+	+	Omnivore
Eleotridae	*Eleotris vittata*	+		
Gobiidae	*Awaous lateristriga* (*Chonophorus lateristriga*)	+		
Hepsetidae	*Hepsetus odoe*	+	+	Piscivore
Mormyridae	*Marcusenius ussheri*	+	+	Carnivore/benthiphagous
	Petrocephalus bovei	+	+	Carnivore/benthiphagous
	Petrocephalus pelligrini	+		
Mugilidae	*Liza falcipinnis*		+	Detritivore
Notopteridae	*Papyrocranus afer*	+	+	Carnivore/benthiphagous
Schilbeidae	*Schilbe mandibularis*	+	+	Omnivore/insectivore
Total		30	25	

28.3 Materials and methods

Sampling of the fish populations was carried out over three seasons, i.e. 26 June–1 July 1996, 26 September–2 October 1996 and 3–17 February 1997 to account for seasonal changes. No sample was taken in the short dry season (August). Fishing was carried

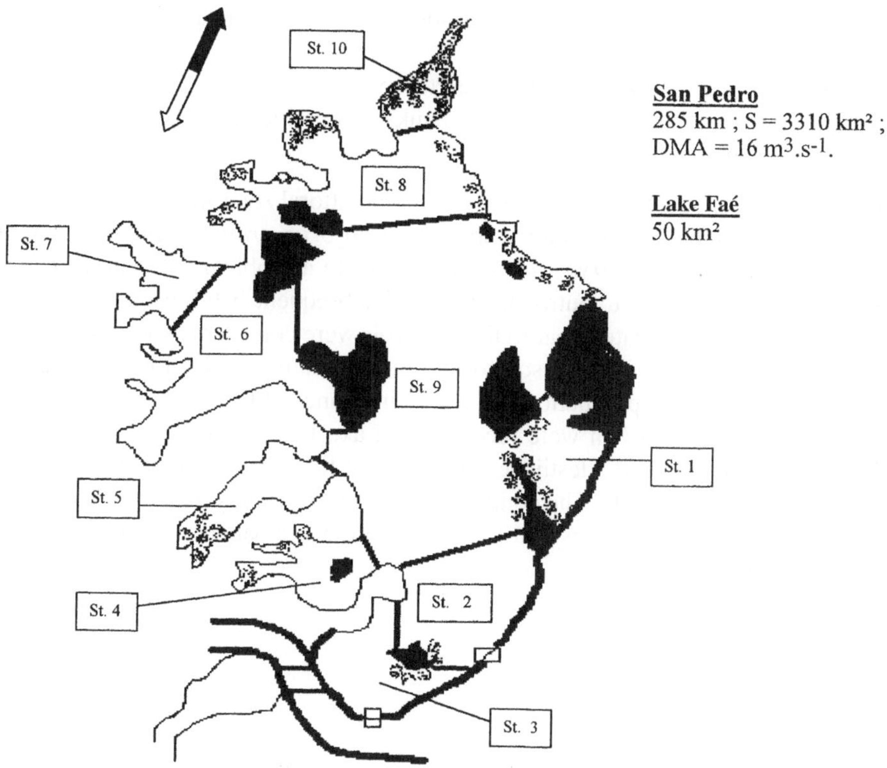

Figure 28.2 Map of Lake Faé showing main sampling stations

out at 10 stations on the lake (Fig. 28.2) using 25 × 2 m gill nets with a range of knot-to-knot mesh sizes between 10 and 80 mm. Fish were also collected from fishermen on the lake or at the lakeside, and at the Faé market. These fishes were generally captured with gill nets, cast nets or bamboo traps.

At each station two identical gangs of gill nets were set over 3–4 nights. The gill nets were set in the evening at around 17:00 h, and lifted around 07:00 h. At the same time as setting and hauling the gill nets, and in the same location, 10 or 15 throws of a cast net, with a mesh size 17.5 mm, were made. Fish caught by the cast net were stored individually for later analysis. Identification of the fishes and their ecological characteristics was based on Lévêque, Paugy & Teugels (1992).

28.4 Results

Five ecological zones were distinguished in Lake Faé (Fig. 28.2):

- the afferent river mouth (station 10), associated with wetlands covered by aquatic plants, especially *Pistia stratiotes*;
- small islands covered by forest and overhanging riparian vegetation associated with both deep and shallow littoral zones;

- the very deep central part of the lake (stations 9 and 10), with large numbers of tree stumps;
- grassy bays (stations 4–10) with some Pterophytes;
- the littoral zone with steep or shallow banks and periodically flooded depending on drawdown.

Fourteen species of aquatic plants were found in the littoral zone and open water of the lake. According to the fishermen three species *Cyclosorus striatus* (Schum.), *Cyperus diffusus* Vahl A. and *Bulbastrylis aplyllaritoides* were a problem when they died off in the flooding period due to eutrophication and subsequent fish kills. Coffee, cocoa and rubber plantations located near to the reservoir were considered sources of potential pollution and could be the reason for the dense floating vegetation on Lake Faé, which restricts fishing, reproduction and the maintenance of the stock of fishes.

Twenty-five species of fish were recorded in Lake Faé, distributed in 12 families (Table 28.1). Cichlidae and Alestiidae dominate the fish community contributing six and four species, respectively. The dominant species were mainly generalist omnivores and zooplanktivorous omnivores (Table 28.1). There was a paucity of detritivores and omnivorous benthivores.

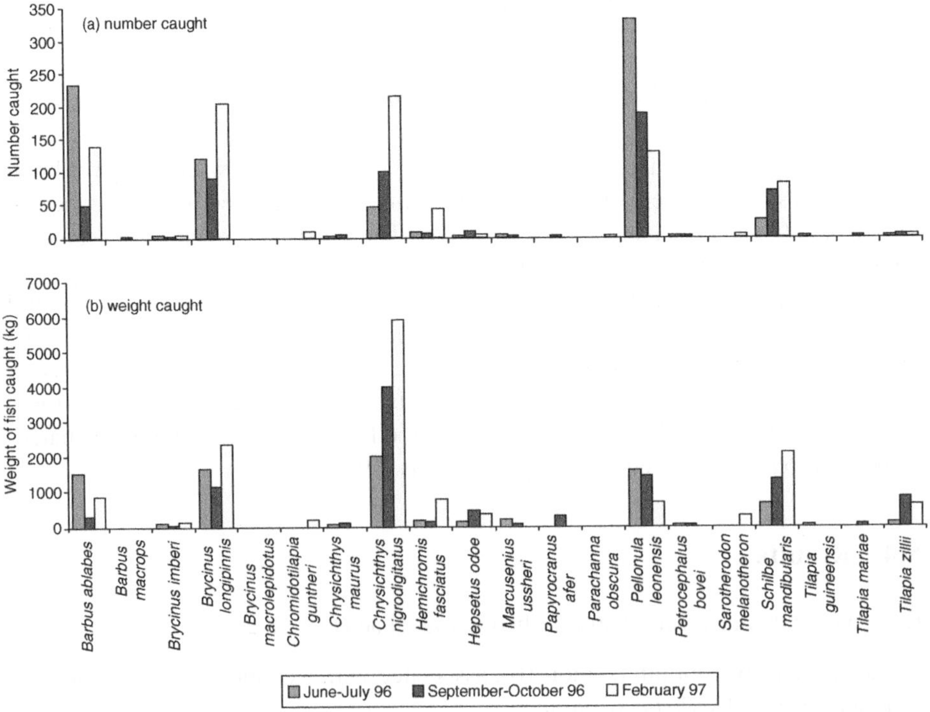

Figure 28.3 Proportion of species captured: (a) by number and (b) by weight with experimental gill nets in Lake Faé

Five species of fish (*Barbus ablabes* (Bleeker), *Brycinus longipinnis* (Günther), *Chrysichthys nigrodigitatus* (Lacepède), *Pellonula leonensis* Boulenger and *Schilbe mandibularis* (Günther)) dominated the catch by number and weight (Fig. 28.3). Mormyridae are represented by small-sized species, such as *Petrocephalus bovei* (Val.) and *Marcusenius ussheri* (Günther), but contribute $<0.8\%$ by weight to the catch. The piscivorous fishes *Hepsetus odoe* (Bloch) and *Hemichromis fasciatus* Peters were low in abundance, contributed less than 2% by weight of the catches (Fig. 28.3).

The capture rate (number of captured fishes per 100 m^2 of gill net per night) varied between 0 and 121 ind. net^{-1} $night^{-1}$, while in weight fluctuated between 0 and 2.01 kg net^{-1} $night^{-1}$. Gill nets with mesh sizes >30 mm did not catch any fish on any occasion.

28.5 Discussion

Most of the species found in the River San Pedro (Teugels *et al.* 1988; Paugy *et al.* 1994) were present in Lake Faé. Thirteen species found by Teugels *et al.* (1988) in the River San Pedro (Table 28.1) were not present in Lake Faé. Their absence was possibly due to their inability to adapt to the lacustrine habitat or overfishing. The poor capture rate and absence of larger individuals in the experimental gill nets points towards overexploitation (Welcomme 1985), and the need to manage the fishery more effectively. However, the situation could also have come about because of the upsurge in standing stock often associated with inundation of reservoirs and the subsequent migration of fishermen to exploit the resources. Once the upsurge has passed, the lack of species able to adapt to the lacustrine conditions, linked to the typical collapse in trophic status, lead to a depauperate fish fauna and standing stock (Cowx & Kapasa 1995), as seems to have happened in Lake Faé.

The fish community had a number of species of brackish-water origin (*Liza falcipinnis* (Val.), *Chrysichthys auratus* (Geoffroy St. Hilaire), *Chrysichthys maurus* (Val.), *Sarotherodon melanotheron* Rüppell, *Brycinus longipinnis*). These species usually migrate upstream in the coastal rivers, and were presumably land-locked by the dam construction, but are able to adapt to an entirely freshwater environment.

Despite being supposedly introduced in 1983 (N'zué, personal communication), *Oreochromis niloticus* (L.) was absent from the catches. Their failure to colonise was possibly because of the poor condition of the fish stocked after transportation to the lake, although competition with *Sarotherodon melanotheron*, heavy fishing pressure or problems with adapting to the draw-dawn regime on the lake (Cowx & Kapasa 1995), may also have contributed.

When compared with other reservoirs in the region (e.g. Ayamé, Buyo, Kossou and Taabo), catches rates were very poor. This may be partially due to the present overexploitation of the indigenous stocks but also poor species diversity and the absence of several highly productive and commercially-valuable species (e.g. *Lates niloticus* (L.), *Oreochromis niloticus* (L.), *Sarotherodon galilaeus* (L.), *Heterobranchus isopterus* Bleeker, *Clarias anguillaris* (L.), *Labeo coubie* Rüppell, *Synodontis* spp.,

Heterotis niloticus (Cuv.)) in other reservoirs in the region. Consideration thus needs to be given to enhancing the fishery through introduction of fish species, which have contributed significantly to fisheries elsewhere in the region. Five species were identified as candidates for introduction based on their economic importance and potential to occupy under exploited feeding niches. These are *Heterotis niloticus* (omnivorous–benthophagous; economically-important; successfully acclimated in other big reservoirs in Côte d'Ivoire); *Labeo coubie* (herbivorous–detritivorous; economically-important); *Distichodus rostratus* Günther (herbivorous; biological control of floating aquatic macrophytes); *Auchenoglanis occidentalis* (Val.) (insectivorous; highly desired by local people); and *Oreochromis niloticus* (herbivorous–phytoplanktivorous; economically-important).

However, before any species is introduced it is imperative the advantages of the introduction are weighed against the potential risks, especially to the endemic fauna (Cowx 1998; Lévêque & Paugy 1999). The introduction of *Heterotis niloticus* and *Labeo coubie* should increase the fish yield by exploiting under-utilised feeding niches. *Heterotis niloticus* is not endemic to Côte d'Ivoire but has adapted to conditions in the reservoirs of the forest zone or savana. It is an omnivorous–benthophagous species with molluscs a predominant component of the diet. As such, it could compete with *Chrysichthys* species which also feed on gasteropods. However, the risk is probably minimal because of the large trophic spectrum of *Chrysichthys* species and presence of abundant food resources in the benthos. Introduction of *Heterotis niloticus* has also been successful in other reservoirs in Côte d'Ivoire with no detectable deleterious effects on the autochtonous fauna. *Labeo coubie* is absent from the River San Pedro basin but is commonly found in most river basins in Côte d'Ivoire. The success of its introduction will be dependent on its ability to adapt to the physical and chemical conditions in Lake Faé.

The introduction of *Oreochromis niloticus* has been successful in many regions of the world (Welcomme & De Merona 1988), including reservoirs in the Côte d'Ivoire (Lessent 1971; Lazard 1990; Da Costa *et al.* 1998) to increase fish yield. The species is very hardy and capable of colonising lacustrine environments (Lévêque & Paugy 1999), thus by using appropriate stocking practices there is no apparent reason why the species should not be successful in Lake Faé, despite failure in the past. However, it does occupy the same feeding niche as other species and a further constraint on its introduction is the presence of *Sarotherodon melanotheron*, which seems to compete successfully with *Oreochromis niloticus* when the former is dominant in the fish community, e.g. in Lake Ayamé (Gourène, Teugels, Hugueny & van den Audenaerde 1999).

The feasibility of introducing *Auchenoglanis occidentalis* is constrained by several factors. Firstly, it does not make a significant contribution to landings in the region, thus it is unlikely to enhance yield in the reservoir. Furthermore, the introduction of *Auchenoglanis occidentalis* in addition to *Heterotis niloticus*, species with the same feeding niche, could increase competition with *Chrysichthys* species, which currently represent the main component of the fish catches. Cosequently, it is not recommended that *Auchenoglanis occidentalis* is introduced into Lake Faé.

Finally, the introduction of *Distichodus rostratus* is potentially favourable because of its value in controlling floating macrophytes which are becoming a problem in Lake

Faé, and are limiting the development of the fisheries through massive fish kills and restricted access.

In terms of priority, it is recommended that *Labeo coubie* and *Distichodus rostratus* are introduced first because they will fill an ecological niche and help control the aquatic macrophytes, respectively. Thereafter further assessment of the feasibility of successfully introducing *Heterotis niloticus* in Lake Faé is needed. Concerning *Oreochromis niloticus*, its re-introduction into Lake Faé is dependent on potential competition with *Sarotherodon melanotheron*, an issue that needs further research.

In view of the poor status of the fishery, the recommendations for enhancing the fisheries through introductions are considered the most appropriate strategy. However, before implementation the full environmental and socio-economic benefits of the introductions should be assessed and Codes of Practice (FAO 1995) for the introduction of fish species should be followed.

References

Cowx I.G. (ed.) (1988) *Stocking and Introduction of Fish.* Oxford: Fishing News Books, Blackwell Science, 456 pp.

Cowx I.G. & Kapasa C. (1995) Changes in reservoir fisheries following impoundment, with special reference to Lake Itezhi-tezhi, Zambia. In T.J. Pitcher & P. Hart (eds) *The Impact of Species Change on African Lakes*. London: Chapman & Hall, pp. 321–332.

Da Costa K.S., Traoré K. & Tito de Morais L. (1998) Effort de pêche et production exploitée dans les petites retenues du Nord de la Côte d'Ivoire. *Bulletin Français Pêche et Pisciculture* **71**, 65–78 (in French).

Daget J. & Oltis A. (1965) Poissons de Côte d'Ivoire (eaux douces et saumâtres). *Mémoires de Institut Français Afrique Noire* **74**, 385 pp.

FAO (1995) Precautionary approach to fisheries: Part 1: Guidelines on the precautionary approach to capture fisheries and species introductions. *FAO Fisheries Technical Paper* 350/1, 41 pp.

Gourène G., Teugels G.G., Hugueny B. & Audenaerde T. Van den (1999) Evaluation de la diversité ichtyologique d'un bassin Ouest-Africain après la construction d'un barrage. *Cybium* **23**, 147–160 (in French).

Hugueny B. (1989) Biogéographie et structure des peuplements de poissons d'eau douce de l'Afrique de l'Ouest: approches quantitatives. Thèse Univ. Paris 7, Paris, 295 pp. (in French).

Lauzanne L. (1988) Les habitudes alimentaires des poissons d'eau douce africains. In C. Lévêque, M. Bruton & G. Ssentongo (eds) *Biologie et Écologie des Poissons d'eau Douce Africains.* Paris: Orstom, pp. 221–242 (in French).

Lazard J. (1990) Transfert des poissons et développement de la production piscicole dans trois pays d'Afrique subsaharienne. *Revue Hydrobiologie Tropicale* **23**, 251–265 (in French).

Lessent P. (1971) Aménagement piscicole des retenues artificielles en Côte d'Ivoire. *Notes et Documents sur la Pêche et La Pisciculture* **2**, 45–56 (in French).

Lévêque C. & Paugy D. (1999) *Les poissons des Eaux Continentales Africaines – Diversité, Écologie, Utilisation par l'homme.* Paris: Editions IRD, 521 pp. (in French).

Lévêque C., Paugy D. & Teugels G.G. (1992) Faune des poissons d'eau douce et saumâtres de l'Afrique de l'Ouest. Tomes 1 et 2. Paris: Editions ORSTOM/MRAC. Collection faune Tropicale n° XXVIII. Tome 1, 384 pp. Tome 2, 902 pp. (in French).

Paugy D., Traoré K. & Diouf P.S. (1994). Faune ichtyologique des eaux douces de l'Afrique de l'Ouest. In G. Teugels (ed.) *Diversité Biologique des Poissons des Eaux Douces et Saumâtres d'Afrique. Synthèses Géographiques.* Annales Museum Tropicale Afrique, Centre de Zoologique **275**, pp. 35–66.

Skelton P. (1993) *A Complete Guide to the Freshwater Fishes of Southern Africa.* South Africa: Southern Books Publishers, 388 pp.

Teugels G.G., Lévêque C., Paugy D. & Traoré K. (1988) Etat des connaissances sur la faune ichtyologique des basins côtiers de Côte d'Ivoire et de l'Ouest du Ghana. *Revue Hydrobiologia Tropicale* **21**, 221–237 (in French).

Welcomme R.L. & Merona B. de (1988) Peuplements ichtyologiques des rivières. In C. Lévêque M. Bruton & G. Ssentongo (eds), *Biologie et Écologie des Poissons d'eau Douce Africains*. Paris: Editions de l'Orstom, pp. 251–276 (in French).

Welcomme R.L. (1985) River fisheries. *FAO Fisheries Technical Paper* 262, 330 pp.

Chapter 29
Pelagic cichlid fishes of Lake Malawi/Nyasa: biology, management and conservation

G.F. TURNER*, R.L. ROBINSON
Biodiversity and Ecology Research Division, School of Biological Sciences, University of Southampton, Southampton, UK

B.P. NGATUNGA
Tanzanian Fisheries Research Institute, Kyela, Tanzania

P.W. SHAW and G.R. CARVALHO
Molecular Ecology and Fisheries Genetics Laboratory, University of Hull, Hull, UK

Abstract

Survey work carried out in the early 1990s indicated that the biomass of cichlid fishes in the pelagic zone of Lake Malawi/Nyasa was around 180 000 t. It appeared that this may represent the last major underexploited fishery resource of Lake Malawi, but several major questions remained unanswered, particularly in relation to species identification, stock movements and nursery areas. In addition to extensive field sampling work, detailed morphological and molecular analyses were carried out. Approximately 21 species of the genera *Rhamphochromis*, *Diplotaxodon* and *Pallidochromis* were recorded, at least nine of which are presently undescribed. All species studied in detail are abundant and widespread in all suitable habitats. Molecular studies indicated minimal population structuring in the three species investigated. The spawning and nursery areas of all known species of *Diplotaxodon* are in deep water, and largely inaccessible to fishing. Spawning areas of *Rhamphochromis* also appear to be in deep water, but the nursery grounds of some species are at least partially in shallow waters. It was concluded that there is minimal risk of species or population extinctions as a result of current patterns of exploitation. Implications of the results for management and conservation of fish stocks are discussed.

Keywords: Cichlidae, exploitation, extinction, genetics.

29.1 Introduction

29.1.1 *The lake and its importance*

Lake Malawi is bordered by three of the world's poorest countries, Malawi, Mozambique and Tanzania. It is one of the largest lakes in the world, with a surface area of 30 800 km^2, and a maximum depth of over 700 m. It probably contains more

*Correspondence: School of Biological Sciences, University of Hull, Hull HU6 7RX, UK (email: g.f.turner@biosci.hull.ac.uk).

endemic fish species than any other lake in the world. At least 650–700 species of cichlid fishes have evolved within the lake basin within the last 700 000 to 2 million years. The production of so many species at such a high rate is surpassed only by the closely-related cichlid fishes of Lake Victoria (Turner 1995). With the devastation of the Victorian cichlids by recent human-induced changes to the lake ecosystem, the Malawian cichlids represent a priceless asset to researchers studying the mechanisms of the evolution of biological diversity. The fish resources of Lake Malawi and its associated water bodies are of tremendous importance as a source of protein and revenue to the people living along the lake shore, particularly in the Malawian sector where the fisheries are most thoroughly developed. Since the beginning of the 1990s, there have been worrying signs of overfishing, particularly in the heavily exploited and highly productive southern areas of the system (Turner 1994a; 1995; Turner, Tweddle & Makwinja 1995; Tweddle, Turner & Seisay 1995). To date, despite a substantial investment in research and development, aquaculture has made little or no contribution to meeting the needs of the rapidly expanding human populations of this region. The initiation of a substantial new capture fishery on the lake would be a major contribution to the development of the region.

29.1.2 *The pelagic zone*

The lake is steep-sided, deep and permanently stratified. Below a depth of 200–250 m there is insufficent oxygen to support fish. At present, both artisanal and mechanised fisheries are largely confined to the shores of the lake, except in the south where there is a large shallow shelf area of high productivity. FAO (1982) concluded that the pelagic zone of the lake was seriously deficient in its fish community and recommended the introduction of a clupeid fish, *Limnothrissa miodon* (Boulenger), from Lake Tanganyika. The reasoning behind this was that large swarms of adult lakeflies, *Chaoborus* sp., were seen on Lake Malawi, but not in Lake Tanganyika. The larvae of these flies are carnivorous and large by the standards of freshwater zooplankton. In the absence of adequate equipment for deep water fishing in the offshore habitat, FAO believed that the principal offshore fish were small shoaling cyprinids, *Engraulicypris sardella* (Günther), known as 'usipa' in the Malawian part of the lake, which they believed were unable to feed on *Chaoborus* larvae. The introduction of the slightly larger clupeids was seen as way of establishing an efficient planktivore which would convert zooplankton into food fish.

In the early 1990s, extensive acoustic and plankton surveys were carried out in the offshore pelagic zone (Menz 1995). Initially, little attention was paid to sampling the fishes, as it was assumed that they would mostly be usipa. However, toward the end of the project, it was realised that cichlid fishes of the genera *Rhamphochromis* and *Diplotaxodon* comprised more than 80% of the fish biomass in the pelagic zone, and that this included a number of species which fed principally on *Chaoborus*. The overall fish biomass was estimated to be between 120 000 and 200 000 t. This suggested that a large fishery could be developed, yielding up to 30 000 t year^{-1} on a sustainable basis (Menz 1995).

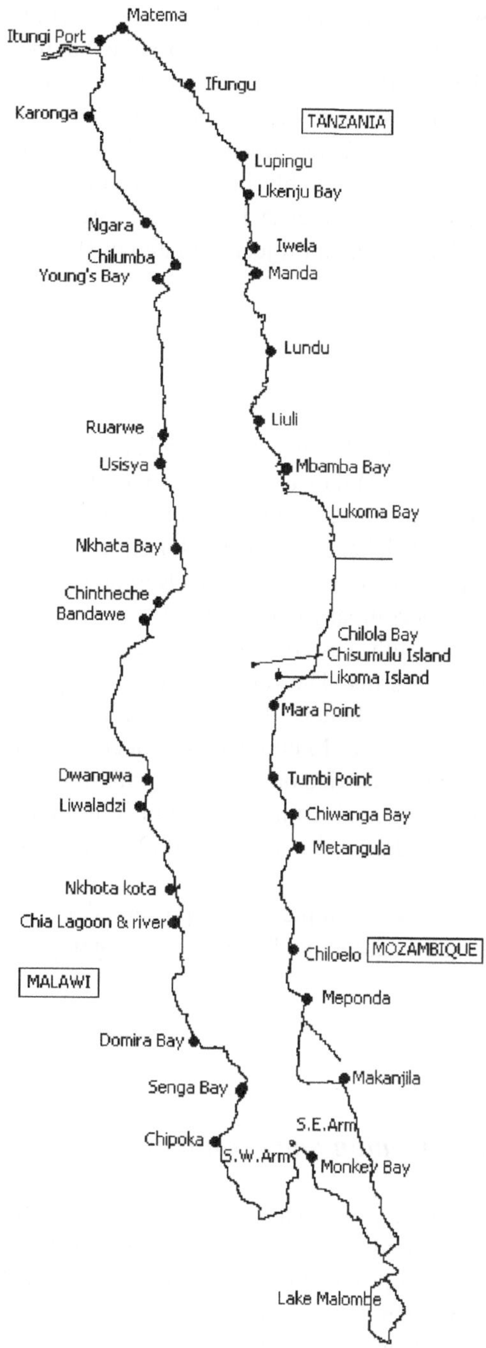

Figure 29.1 Map of Lake Malawi, showing major sampling locations along the shore

This study left several question unanswered. Inshore waters were not investigated, nor were the catches of existing fisheries. It was possible that the offshore stocks were already heavily exploited, perhaps during the breeding season or on their nursery areas. This might mean that a rapid expansion of offshore fishing could lead to a crash in the populations of these fishes. It was difficult to appraise this possibility, as many of the species could not be reliably identified in the field, and it was believed that many were still undescribed (Fryer & Iles 1972). The UK Department for International Development (DfID) Ncheni Project, initiated in 1996, aimed to address these deficiencies.

29.2 Methods

Most of the samples were obtained from catches made by artisanal fisheries or by commercial trawl fisheries. On the Malawian shores, from 1996 to 1998, more than 4500 individually-labelled specimens were collected from more than 4.5 months field work. On the Tanzanian shore of the lake, experimental longline surveys were also conducted by staff from the Tanzanian Institute for Fisheries Research (TAFIRI). It was also possible to send personnel to accompany demersal trawl surveys being carried out by the GEF/SADC project in the south-western part of the lake, and by the Malawi Fisheries Department in the southern arms. The research vessel Usipa was hired for a single lake-wide cruise, from 18 to 27 January 1998. A pelagic trawl was used for 20 hauls, totalling 39 h 40 min fishing time. Material was collected from Mozambican waters during a demersal trawl survey carried out in collaboration with an EU-funded project. Many more specimens were examined in the field or in museum collections made by other projects, notably 1355 *Rhamphochromis* and *Diplotaxodon* specimens collected by the GEF/SADC project.

For many specimens, tissue samples were preserved in alcohol for subsequent molecular analysis, using mitochondrial DNA sequencing or genotyping of microsatellite nuclear DNA.

29.3 Results

29.3.1 *Species identification and taxonomy*

Eleven species of the genus *Diplotaxodon* were recognised (Table 29.1, Fig. 29.2). Of these, five were considered undescribed. Six of the seven previously described species were collected but it was not possible to confirm the identity of *Diplotaxodon ecclesi* Burgess & Axelrod. In general, there were few clear-cut counts or measurements which were useful in the diagnosis of new species. Although a few species had distinctive morphologies, many did not, but they differed in male breeding dress. Several forms were present in small numbers and could not be positively assigned to any of the 11 species. These remain unclassified for the present. A molecular phylogenetic study based on mitochondrial DNA sequences carried out as part of the project demonstrated

Table 29.1 Cichlid species of the genera *Diplotaxodon*, *Pallidochromis* and *Rhamphochromis*

Species	Maximum SL (mm)	Morphological group	Description
Diplotaxodon			
D. limnothrissa	154	Slender small mouth	Turner 1994b
D. 'holochromis'	160	Slender small mouth	Undescribed
D. argenteus	204	Elongate big mouth	Trewavas 1935
D. 'similis'	203	Elongate big mouth	Undescribed
D. aeneus	140	Big-eye	Turner & Stauffer 1998
D. apogon	117	Big-eye	Turner & Stauffer 1998
D. macrops	125	Big-eye	Turner & Stauffer 1998
D. 'offshore'	140	Big-eye	Undescribed
D. 'deep'	224	Deep-body, big-eye	Undescribed
D. greenwoodi	247	Deep-body, big mouth	Stauffer & McKaye 1986
D. 'brevimaxillaris'	212	Deep-body, big mouth	Undescribed
Pallidochromis			
P. tokolosh	280	Big-eye, big mouth	Turner 1994c
Rhamphochromis			
R. esox	420	Cylindrical	Boulenger 1908
R. ferox	209	Slender, small mouth	Regan 1922
R. 'grey'	347	Slender, small mouth	Undescribed
R. longiceps	195	Slender, small mouth	Günther 1864
R. 'longfin'	250	Large mouth & teeth	Undescribed
R. 'maldeco'	312	Large mouth & teeth	Undescribed
R. macrophthalmus	298	Large mouth & teeth	Regan 1922
R. 'stripe'	322	Large mouth & teeth	Undescribed
R. woodi	402	Large mouth & teeth	Regan 1922

that the benthic species *Pallidochromis tokolosh* (Turner) is properly considered a member of the *Diplotaxodon* clade. *Pallidochromis* differs from *Diplotaxodon* in having larger, more widely-spaced teeth and in having a less upwardly-angled mouth.

Nine species of *Rhamphochromis* were tentatively identified, four of which are certainly undescribed. Some of the formally described species will probably have to be synonymised. *Rhamphochromis ferox* Regan has previously been misidentified in the field as a big, large-mouthed predator (Menz 1995). Based on type specimens, it was considered to be small, delicate-looking species, similar to *R. longiceps* (Günther). The big predators were probably *Rhamphochromis woodi* Regan.

29.3.2 *Habitat preferences*

Inshore-living haplochromine cichlids are well-known for their habitat specificity (Fryer & Iles 1972; Ribbink, Marsh, Marsh, Ribbink & Sharp 1983), but prior to the current work, it had not been established whether the same was true of pelagic cichlids.

The eupelagic community was dominated by five species commonly found in this habitat (Table 29.2); the most abundant species were *Diplotaxodon limnothrissa* Turner and *Diplotaxodon* 'offshore', the latter mainly in deeper water. Most of the

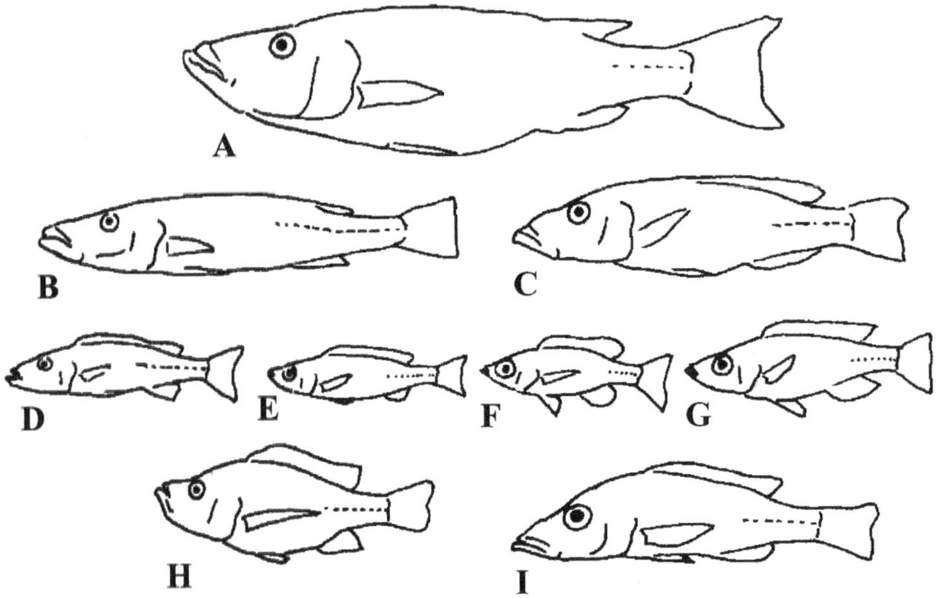

Figure 29.2 Representative pelagic cichlids of Lake Malawi. A. *R. woodi*; B. *R. esox*, C. *R. macrophthalmus*; D. *R. longiceps*, E. *D. limnothrissa*, F. *Diplotaxodon* 'offshore', G. *D. argenteus*, H. *Diplotaxodon* 'brevimaxillaris', I. *P. tokolosh*. Drawings by G.F. Turner, from photographs by the authors

eupelagic forms also occurred in the reef zone, but several species appeared to be confined to this habitat, including *Rhamphochromis* 'grey' and the rare *Diplotaxodon aeneus* Turner & Stauffer. The shelf zone is accessible to bottom trawling and most of this habitat lies in the southern arms of the lake, although there is another large area in the far north. Apart from the deep-water pelagic *Diplotaxodon* 'offshore', all the eupelagic species were encountered in this habitat, but there were also several species which were rarely, if ever, found elsewhere (Table 29.2). *Diplotaxodon* in particular, formed a far greater proportion of the biomass in deeper waters, especially at depths of 100 m or more (Fig. 29.3). Only immature *Rhamphochromis* were collected in the shallower part of the littoral zone, with adult *Rhamphochromis* 'stripe' apparently confined to the deep rocky littoral.

29.3.3 *Distributions and stock structure*

All of the eupelagic species seemed to have lake-wide distributions. Some inshore species seemed not to have very specific habitat requirements, and were also widely distributed. Although several species appeared to have strict habitat requirements, all were collected from several different areas of this habitat, although these areas were sometimes a long distance apart. For example, *Diplotaxodon apogon* Turner & Stauffer was described from the southern arms of the lake, where it is abundant on the

Table 29.2 Habitat preferences of pelagic cichlids in Lake Malawi (all = all developmental stages; a = adults only; f = fry; i = immature)

Habitat	Description	Common species	Others
Eupelagic	Open water above anoxic bottom	*D. limnothrissa* (a, f), *D.* 'offshore' (all), *R. longiceps* (a), *R. ferox* (a), *R. woodi* (a)	
Deep shelf	Soft bottom, 100–220 m	*D. macrops* (all), *D. apogon* (all), *D. limnothrissa* (a, i), *D. greenwoodi* (all), *D.* 'similis' (all), *P. tokolosh* (a, i), *R. woodi* (a, i), *R. macrophthalmus* (all), *R. longiceps* (a)	*D.* 'brevimaxillaris' *R.* 'longfin'
Shallow shelf	Soft bottom, 50–100 m	*D. limnothrissa* (a, f, i), *D. argenteus* (all), *R. ferox* (a), *R. longiceps* (a), *R. esox*(a), *R.* 'maldeco' (all)	*R. woodi*, *D.* 'deep' *D.* 'holochromis' *R. macrophthalmus*
Reefs	Water column near and above rocky shores and submerged reefs	*D.* 'holochromis' (all), *D.* 'similis' (a), *D. limnothrissa* (all), *D. greenwoodi* (all), *R. woodi* (all), *R.* 'grey' (all), *R. longiceps* (all), *R. ferox* (all)	*D. aeneus* *D. argenteus* *R. esox*
Rocky littoral	Benthic habitat on rocky shores	*R.* 'stripe' (all), *R. esox* (f, i), *R. ferox* (f, i)	*R. longiceps*
Margins	Sandy/muddy shores, lagoons, etc.	*R. longiceps* (all)	

shelf area at depths of 100 m or more (Turner & Stauffer 1998). This species then turned up in samples from similar habitats in the South-West Arm, Senga Bay, Domira Bay and off Nkhotakota in the south-western part of the lake, off Metangula on the eastern shore, and in Weissmann Bay in the far north.

Many inshore-living cichlids have high levels of population structure on a small spatial scale, suggesting very limited powers of disperal (van Oppen, Turner, Rico, Deutsch, Ibrahim, Robinson & Hewitt 1997). In contrast, comparing the frequency distributions of allele sizes at six microsatellite DNA loci, little or no genetic differentiation was found between populations of *D. limnothrissa*, *D. macrops* Turner & Stauffer or *D.* 'offshore' collected hundreds of kilometres apart.

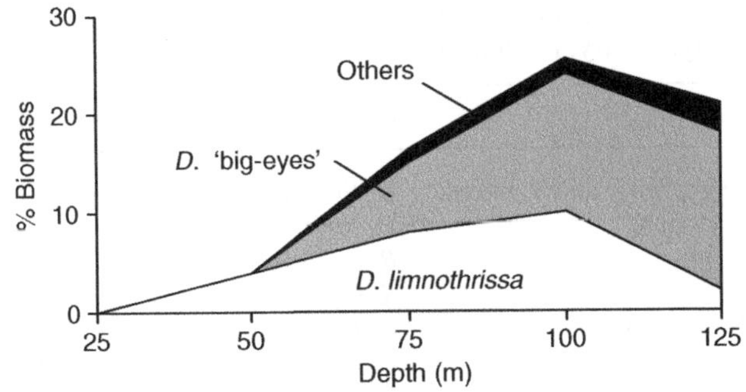

Figure 29.3 *Diplotaxodon* and *Pallidochromis* as percentage biomass of experimental demersal trawl catch in SW Arms (data reanalysed from F. Duponchelle, personal communication). *Diplotaxodon* 'big-eyes' is mainly *D. macrops* and *D. apogon*

29.3.4 *Reproduction*

All haplochromine cichlids previously studied were found to be maternal mouthbrooders. The present studies indicate no reason to doubt that the same is true of the pelagic haplochromines.

Many Malawian cichlids from sandy shores and the shelf habits aggregate in breeding arenas or 'leks'. In trawls, large numbers of ripe males were sometimes collected together with smaller numbers of ripe and spent conspecific females, but relatively few immature fishes. This suggests that the trawl had been fishing over leks. Such aggregations were observed for *D. macrops*, *D. apogon* and *D. greenwoodi* Stauffer & McKaye at around 100–120 m in the southern arms of the lake, and for *Rhamphochromis macrophthalmus* Regan at a depth of 100 m off Metangula in Mozambique.

Females of several species, including *D. limnothrissa*, *D. macrops*, *D. argenteus* and *R. woodi*, were collected carrying eggs, larvae or independently-feeding juveniles in their mouths. Like other mouthbrooding cichlids, females of both of these genera have large, yolky, non-adhesive eggs. Mature eggs were large, generally 4–6 mm in diameter. Some of the largest eggs were found in the smallest *Diplotaxodon* species: *D. limnothrissa, D. macrops* and *D. apogon*. These species reach maximum sizes of only 12–15.5 cm SL and 40–60 g. Consequently, they have very low fecundities (10–40 eggs per ripe female). Such low fecundity is probably unique for a small pelagic zooplanktivore, although it is by no means unusual for a small haplochromine cichlid. *Rhamphochromis* were found to attain rather higher fecundities, 27–680 eggs per ripe female, according to size and species. With such low fecundities, it is unlikely that these species could rapidly recover from population crashes caused by overexploitation.

Breeding seasons were generally long: March–August for *D. limnothrissa* in the SW Arm, all year round for *D.* 'similis', November–April for *D. apogon*, January–September for *D. macrops*, and October–February for *P. tokolosh* (F. Duponchelle, personal communication). *R. ferox* and *R. longiceps* also appear to breed for most of the year, ripe fish being taken from January and February, respectively, until October.

29.3.5 *Breeding and nursery grounds*

Small *Rhamphochromis* of the *ferox/longiceps* group were often seen underwater, and were caught by gears operating near the surface or in shallow waters. These small fish were found to penetrate into swampy areas, including Lake Malombe, the Upper Shire River and Chia Lagoon. Young *Rhamphochomis esox* (Boulenger) were also sometimes found at depths of a few metres or less, mainly on sandy shores, while young *R.* 'stripe' were abundant in shallow rocky areas. Fully ripe or mouthbrooding *D. limnothrissa* were found in all habitats with a bottom depth greater than 20 m, but young of 2–10 cm were most abundant near the surface over shelf areas. Mouthbrooding female *R. woodi* were generally collected over deep bottoms, but immatures of 5–15 cm were mostly collected in the shelf zone near the bottom at depths of 50–100 m. Ripe females and small juveniles of *Diplotaxodon* 'offshore' were often collected far offshore in deep waters, but ripe males were found at depths of 99–184 m from a variety of sites along the mid-western shore from Senga Bay to Nkhata Bay. It seems likely that this species breeds deep-down along the steeper shelving coasts, but that mouthbrooding females rapidly return to the eupelagic zone where they release their fry. All other pelagic cichlids seem to breed in the normal adult habitats, which are mainly in deep waters.

29.3.6 *Diet*

From stomach contents, all *Diplotaxodon* and *Rhamphochromis* specimens examined seem to have fed on prey from the water column (Allison, Irvine, Thompson & Ngatunga 1996; unpublished data). Small individuals fed on crustacean zooplankton, *Chaoborus* larvae and pupae, and young stages of the cyprinid *Engraulicypris*. Larger individuals ate fish, mainly *Engraulicypris* or small pelagic cichlids. Two individual *P. tokolosh* contained remains of benthic cichlids of the genera *Aulonocara* and *Lethrinops* (Turner 1994c).

29.3.7 *Exploitation*

Turner (1995) summarised the mechanised fisheries operating in the south of the lake up to 1991. Four main gears were operating at that time. The ring net (purse seine) fishery employed a 102-mm mesh, too large to catch pelagic cichlids, apart from the occasional *R. woodi*. The other gears all employed 20–38 mm meshes. The midwater trawl was operated with otter boards on the bottom, but the net set to float higher in the water column. In 1990–1991, *D. limnothrissa* comprised 20% of the total catch of the midwater trawl and 53% of the 'small fish' component (Fig. 29.4). *Rhamphochromis* comprised 12% of the 'small fish' element, and other *Diplotaxodon*, mainly *D. argenteus,* about 4%. In the early 1990s, the main target of this fishery was the high-value *Oreochromis* 'chambo' species. The 'small fish', mainly haplochromine element, comprised just 38% of the catch. By 1994, it had risen to 56%, and according to catch

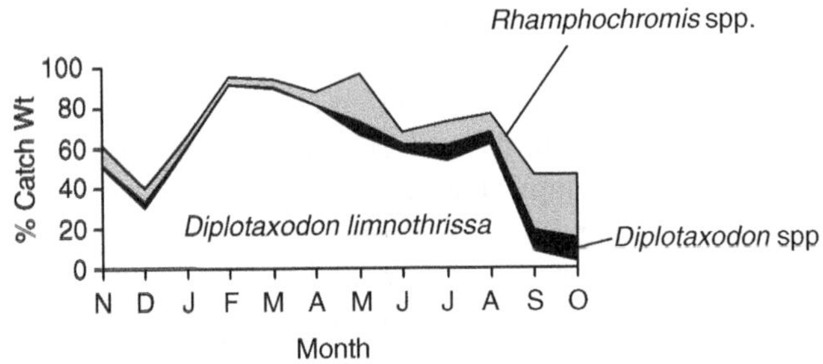

Figure 29.4 Monthly pelagic cichlid catches in midwater trawl, as percentage weight of 'small fish' catch, i.e. excluding large *Oreochromis* and catfish. The category '*Diplotaxodon*' includes all members of this genus, excluding *D. limnothrissa*. The highest catches of *D. limnothrissa* correspond closely to the peak breeding season of that species. Data collected by G.F. Turner in 1990–1991, based on 35 samples, totalling 16 987 fish weighing an aggregate 361 kg

statistics, pelagic cichlids comprised 43% of the total catch. Up to this time the fishery had been operating on the main chambo grounds, in 'Area B' just to the north of Boadzulu Island in SE Arm. With the collapse in chambo stocks in the mid-1990s, the fishery moved its main landing site from Maldeco in the middle of the SE Arm to Monkey Bay in the north of the arm. This permitted the trawlers to exploit the remaining, smaller chambo stocks in the shallower areas around the north of the SE Arm. By 1996–1997, the main by-catch of the fishery was no longer *Diplotaxodon*, but inshore zooplanktivores of the *Copadichromis virginalis* (Iles) complex, along with various benthic cichlids. By August 2000, Maldeco had largely abandoned the semi-pelagic trawl and was operating two fully pelagic trawlers targetting *Diplotaxodon* and *Rhamphochromis* species.

The remaining mechanised fisheries in the early 1990s were demersal trawls. Pair trawlers using illegal double-bagged cod-ends of 20–25 mm mesh fished in shallow waters around the southern arms and Domira Bay, often refusing to submit catch returns to the Malawian Fisheries Department. Despite fishing mainly in shallow waters, pelagic cichlids comprised a surprisingly high proportion of the catch: around 5% by weight of *Diplotaxodon*, mainly juveniles of more surface-living species, such as *D. limnothrissa* and *D. argenteus* (Trewavas), along with around 15% *Rhamphochromis*, mainly small individuals of the *ferox/longiceps* type. Although highly profitable in the SE Arm (Turner & Mdhaili 1992), the pair trawl fishery collapsed in the late 1990s, apparently due to lack of investment in maintenance of the boats and engines by proprietors who had been accustomed to anticipate that the government or development aid programmes would provide them with favourable loans or even free boats and engines.

In the early 1990s, larger single boat demersal trawls operated in deeper water in the SE Arm, catching mainly benthic cichlids, but also a fair proportion of pelagic cichlids (Fig. 29.5); around 18% *D. limnothrissa* and 8% *Rhamphochromis*. In the late 1990s, with loans from the World Bank and the Icelandic Government, this fishery was greatly expanded with two large Icelandic trawlers capable of fishing in deeper waters, one operated for research, but also as a revenue raising activity by the Malawi

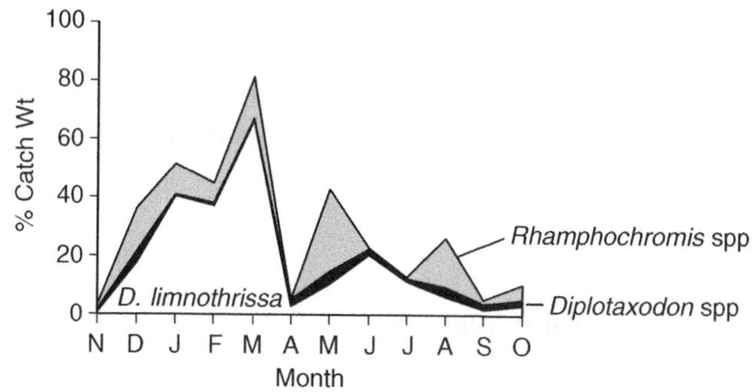

Figure 29.5 Monthly pelagic cichlid catches in commercial bottom trawl, as percentage weight of catch. The category '*Diplotaxodon*' includes all members of this genus, excluding *D. limnothrissa*. Data collected by G.F. Turner from 1990–1991, based on 12 samples, an aggregate 135 kg

Government Fisheries Department. Both vessels covered larger areas than the smaller vessels, frequently fishing in the SW Arm or even further to the north. Perhaps because of their greater power and speed, or because they operated in areas previously not exploited by trawlers, the catches of these vessels frequently included a larger proportion of larger fish, such as big *R. woodi*. Deeper water species, such as *D. macrops* and *D. apogon*, were also landed in large numbers.

Outside of the main chambo grounds in the southern arms, large *Rhamphochromis* are the highest value fish commonly taken in the lake. The rise in tourism along the northern lakeshore of Malawi has led to an increased demand from restaurants for these big 'batala'. Off the steep rocky shores of the Nkhata Bay area, most large *Rhamphochromis* are caught by targetted, size selective angling from dugout canoes fished by day. The main species are *R. woodi* and *R.* 'grey'. Deep-set gillnets tend to land lower value fish, including large numbers of *D. limnothrissa*. The nocturnal light attraction fisheries for *Engraulicypris* land a substantial by-catch of smaller *Rhamphochromis*, mainly *R. ferox* and *R. longiceps*. These species, along with *D. limnothrissa* and *D.* 'holochromis' are also taken in day-fishing 'chirimila' (lift net) catches targetted on reef-living 'utaka', zooplanktivores of the genus *Copadichromis*.

Similar fisheries operate in other rocky areas, including much of the Tanzanian and Mozambican coasts, Likoma and Chisumulu Islands, the Nankumba (Cape Maclear) peninsula and on offshore reefs and islands around Nkhotakota, the southern arms and elsewhere. Off the steep and relatively unproductive coasts around the Livingstone Mountains in Tanzania, fishermen operate large chirimila nets known as 'pajeros'. With great skill and knowledge of fish distributions and currents, fishermen are able to use these nets in a variety of ways, even to exploit the deep-water pelagic species, such as *Diplotaxodon* 'offshore'.

Artisanal fisheries operating in the shallow productive areas exhibited unsustainable tendencies during the 1990s. In 1981, the chambo catch in Lake Malombe exceeded 7700 t (FAO 1993). At that time, there were 72 small-meshed seines operating on the lake. With a rise in the numbers of this gear to over 300 by 1991, the chambo catch fell to 441 t, and by 1994 to 43 t. Although these fisheries have not landed large numbers of pelagic cichlids, apart from some juvenile *Rhamphochromis*, it indicates the potential dangers of unregulated expansion of small-meshed artisanal gears, particularly with regard to species with a spawning/nursery area within range of such gears.

29.4 Discussion

29.4.1 *Conservation of biodiversity*

The widespread populations and lack of stock structure are favourable for the conservation of pelagic cichlids. Extirpation of locally distributed species or genetically unique populations of a species seems a remote possibility. Thus, there seems little cause for concern over the preservation of these species.

29.4.2 *Fisheries management and stock assessment*

If the populations of pelagic cichlid species are comprised of single vast stocks, local overexploitation is likely to be compensated by immigration of fish from elsewhere. However, from the point of view of fisheries management, a note of caution is warranted. The frequencies of selectively neutral alleles (as microsatellites are presumed to be) can be homogenised with a rate of migration between populations as low as five individuals successfully immigrating and breeding per generation. At this rate, a locally overexploited stock might take a very long time to build itself back up to an economically-viable level, particularly given the low fecundity of many of the species. However, given the lack of alternative sources of protein and employment in the region, application of the precautionary principle seems a luxury that cannot be borne by the peoples of the region.

Diplotaxodon limnothrissa: At the height of its exploitation by the midwater trawl in the early 1990s, the total mechanised catch of *D. limnothrissa* was of the order of 700 t year^{-1} (Turner 1996) and it seems likely that the artisanal catch was lower. The total standing stock of this species in the offshore pelagic zone alone was estimated at 87 000 t (Menz 1995). It seems safe to conclude that this species is presently under-exploited and expansion of the fishery is feasible.

The presence of the wide-ranging *D. limnothrissa* in trawl catches in the southern arms poses a problem for stock assessment. If the southern arm stock of this species exhibits a high rate of mixing with the vast offshore populations, then the catch per unit effort (CPUE) is likely to be unaffected by the level of fishing effort, violating the fundamental assumptions of the surplus production models presently used. Ideally, the *Diplotaxodon* catch would be subtracted from the analysis, which would then be performed on the data on resident demersal species alone. If an attempt is made to fully exploit the *D. limnothrissa* stock by expanding inshore trawl fisheries, this is likely to lead to heavy pressure on the demersal species, and it seems probable that the stocks of the demersal species are neither so large, nor so mobile, and could more easily be fished out. It would be better if an alternative method could be found to exploit *D. limnothrissa* in the offshore habitat.

Diplotaxodon 'offshore': As with *D. limnothrissa*, there is a huge eupelagic stock of *D.* 'offshore'. At the beginning of the project, *D. macrops* had not been distinguished from *D.* 'offshore'. As *D. macrops* could be exploited by deep-water demersal trawling, it seemed possible that this might be a way to access this component of the eupelagic stock. This, however, is not the case. At present, it seems that *D.* 'offshore' is only lightly exploited by deep-water artisanal fisheries.

Rhamphochromis ferox and *R. longiceps*: *R. ferox* and *R. longiceps* were not accurately distinguished from each other until late in the project. The offshore stock of these species is large, but they are presently exploited by a wide range of gears, and the current total yield is difficult to estimate. The presence of many immature specimens of these species in catches by small-meshed beach seines and chirimilas is potentially a cause for concern.

Rhamphochromis woodi: Exploitation of the high-value *R. woodi* is likely to be increasing in the north-western part of the lake. At present, the use of size-selective angling methods and the limited range of fishing canoes probably means that there is little risk of overexploitation of this wide-ranging stock. However, large numbers of immature fishes are presently caught by demersal trawls, and it would be useful to estimate the levels of mortality from this cause, and to determine what proportion of the juvenile population lies in the trawlable area.

Other species: Other *Rhamphochromis* and *Diplotaxodon* species, along with *P. tokolosh*, are prinicipally caught as small components of multispecies fisheries. Management of these fisheries must take into account the biology of these other, mainly demersal, cichlid fishes and *Engraulicypris*.

Acknowledgements

This project was funded by the Environment Research Programme of the UK Department for International Development (R6414, ERP 49), and was carried out in collaboration with the Department of Fisheries of Malawi, the Tanzanian Fisheries Research Institute, the Instituto de Investigaçao Pesquiera, Mozambique, and the GEF/SADC Lake Malawi/Nyasa Biodiversity Project. In its latter stages, the project also collaborated with a project funded under the EU Programme on International Cooperation with Developing Countries (INCO-DC), The Trophic Ecology of Demersal Fish Community of Lake Malawi/Niassa, Central Africa. We thank Fabrice Duponchelle for permission to quote his unpublished results.

References

Allison E.H., Irvine K., Thompson A.B. & Ngatunga B.P. (1996) Diets and food consumption rates of pelagic fish in Lake Malawi, Africa. *Freshwater Biology* **35**, 489–515.

Boulenger G.A. (1908) Diagnosis of new fishes discovered by Captain E. C. Rhoades in Lake Nyasa. *Annals and Magazine of Natural History* **8**(2), 238–243.

Burgess W.E. & Axelrod H.R. (1973) New cichlids from Lake Malawi. *Tropical Fish Hobbyist* **1973** (October) 14 & 87–89.

FAO (1982) Fishery Expansion Project, Malawi: Biological Studies on the Pelagic Ecosystem of Lake Malawi. Rome, FAO, *Technical Report 1*, FI: DP/MLW/75/019, 109 pp.

FAO (1993) Fisheries Management in the SE Arm of Lake Malawi, the Upper Shire River and Lake Malombe. Rome, FAO, *CIFA Technical Paper* **17**, 106 pp.

Fryer G. & Iles T.D. (1972) *The Cichlid Fishes of the Great Lakes of Africa.* Edinburgh: Oliver & Boyd, 641 pp.

Günther A (1864) Report on a collection of reptiles and fishes made by Dr. Kirk in the Zambesi and Nyassa regions. *Proceedings of the Zoological Society of London* **1864**, 303–314.

Menz A. (ed.) (1995) *The Fishery Potential and Productivity of the Pelagic Zone of Lake Malawi/Niassa.* Chatham, UK: Natural Resources Institute, 386 pp.

Regan C.T. (1922) The Cichlid fishes of Lake Nyasa. *Proceedings of the Zoological Society of London* **1921**, 675–727.

Ribbink A.J., Marsh A.C., Marsh B.A., Ribbink A.C. & Sharp B.J. (1983) A preliminary survey of the cichlids fishes of rocky habitats of Lake Malawi. *South African Journal of Zoology* **18**, 149–310.

Stauffer J.R. & McKaye K.R. (1986) Description of a paedophagous deep-water cichlid (Teleostei: Cichlidae) from Lake Malawi, Africa. *Proceedings of the Biological Society of Washington* **99**, 29–33.

Trewavas E. (1935) A synopsis of the cichlid fishes of Lake Nyasa. *Annals and Magazine of Natural History Series* **10**(16), 65–118.

Turner G.F. (1994a) Fishing and the conservation of endemic fishes of Lake Malawi. *Archiv für Hydrobiologie* **44**, 481–494.

Turner G.F. (1994b) A description of a commercially important new pelagic species of the genus *Diplotaxodon* (Pisces: Cichlidae) from Lake Malawi, Africa. *Journal of Fish Biology* **44**, 799–807.

Turner G.F. (1994c) *Pallidochromis tokolosh*, a new genus and species of deep water cichlid (Teleostei: Cichlidae) from Lake Malawi, Africa. *Ichthyological Explorations of Freshwaters* **5**, 377–383.

Turner G.F. (1995) Management, conservation and species changes of exploited fish stocks in Lake Malawi. In T.J. Pitcher & P.J.B. Hart (eds) *The Impact of Species Changes in African Lakes.* London: Chapman & Hall, pp. 365–395.

Turner G.F. (1996) *Offshore Cichlids of Lake Malawi.* St. Leon Rot (Germany), Cichlid Press, 240 pp.

Turner G.F. & Mdhaili M. (1992) A Bioeconomic Analysis of the Industrial and Semi-industrial Fisheries of Southern Lake Malawi. FI:DP/MLW/86/013. FAO Field Document 22, 12 pp.

Turner G.F. & Stauffer J.R. (1998) A description of three deep water cichlids of the genus *Diplotaxodon* (Teleostei: Cichlidae) from Lake Malawi, Africa, with a redescription of *Diplotaxodon ecclesi* Burgess & Axelrod. *Ichthyological Explorations of Freshwaters* **8**, 239–252.

Turner G.F., Tweddle D. & Makwinja R.D. (1995) Changes in demersal cichlid communities as a result of trawling in southern Lake Malawi. In T.J. Pitcher & P.J.B. Hart (eds) *The Impact of Species Changes in African Lakes.* London: Chapman & Hall, pp. 387–412.

Tweddle D., Turner G.F. & Seisay M.B.D. (1995) Changes in species composition and abundance as a consequence of fishing pressure in Lake Malombe. In T.J. Pitcher & P.J.B. Hart (eds) *The Impact of Species Changes in African Lakes.* London: Chapman & Hall, pp. 413–424.

van Oppen M.J.H., Turner G.F., Rico C., Deutsch J.C., Ibrahim K.M., Robinson R.L. & Hewitt G.M. (1997) Unusually fine-scale genetic structuring found in rapidly speciating Malawi cichlid fishes. *Proceedings of the Royal Society of London, Series B* **264**, 1803–1812.

Chapter 30
Unsustainable tendencies and the fisheries of Lake Victoria

K. CREAN*
University of Hull International Fisheries Institute, Hull, UK

R. ABILA, C. LWENYA, R. OMWEGA, F. OMWENGA
KMFRI, PO Box 1881, Kisumu, Kenya

A. ATAI, J. GONGA, A. NYAPENDI, K. ODONGKARA
FIRRI, PO Box 343, Jinja, Uganda

M. MEDARD, P. ONYANGO
TAFIRI, PO Box 475, Mwanza, Tanzania

K. GEHEB
UNECIA Ltd, Lake Victoria Fisheries Research Project, Jinja, Uganda

Abstract

During the last 15 years the fisheries of Lake Victoria have been subject to intense fishing effort coupled with deteriorating conditions in the aquatic environment. Efforts to set and maintain ecological limits for the control of fish exploitation have not been successful and the goal of sustainable development for the Lake's fisheries has proved elusive. The factors that contribute to unsustainable tendencies in the Lake's fisheries are explored and the problems that need to be addressed if the adverse trends that affect production and management are to be overcome are investigated.

Keywords: fisheries, Lake Victoria, sustainability.

30.1 Introduction

Lake Victoria is the second largest freshwater lake in the world and the largest in Africa. Yet the lake's formidable size and vast water holding, have not prevented a series of deleterious changes taking place over a relatively recent time scale. Since the 1920s, the lake has been subject to cumulative impacts that have affected not only the physical and chemical quality of its waters, but also the structure and diversity of the lake biota (Coulter, Alanson, Bruton, Greenwood, Hart, Jackson & Ribbink 1986; Craig 1992; Hecky & Bugenyi 1992; Hecky 1993; Mugidde 1993; Lowe-McConnell 1996; Ogutu-Ohwayo, Hecky, Cohen & Kaufman 1997). In the last three decades there has been a steady build-up of fishing effort. This intensified in the 1990s, but possibly

---

*Correspondence: University of Hull International Fisheries Institute, Hull HU6 7RX, UK (email: k.crean@hull.ac.uk).

more important than the amount of fish harvested, was that:

(1) a large proportion of the catch was taken using fishing gears or methods that are injurious to the stock base; and
(2) the introduction of alien species has led to a restructuring of the species composition of the fish community (Ogutu-Ohwayo & Hecky 1991).

These changes in turn have had pronounced social and economic effects on the human communities of the lakeside that depend upon fish for their livelihood (Harris, Wiley & Wilson 1995; Geheb 1999). The decline of Lake Victoria's fish stocks is mirrored in many of the coastal marine fisheries in other parts of the world, and is part of a pervasive crisis where the condition of overfishing is endemic. Indeed, FAO (1995) estimated that some 70% of all the world's fish stocks are now overfished, so that globally the catch of fish has stagnated and we appear to have reached, or even exceeded, the limits of the sustainable harvest in the oceans and large freshwater bodies.

In response to the problems of Lake Victoria, a number of research programmes have been commissioned in an attempt answer to questions that continue to vex the minds of biologists, economists, social scientists and governments concerned with the lake. One of the main tasks is to investigate how it might be possible to establish a sustainable harvesting regime for the lake fisheries. The task is complicated by the debate over what constitutes a 'sustainable' exploitation regime and what the conditions might be for establishing 'sustainable development' over an acceptable time horizon.

The Lake Victoria Fisheries Research Project (LVFRP), initiated in 1997, has sought to address these questions. In essence, the conceptual basis of the project was born out of the need to gain a better understanding of the processes governing the dynamics of the lake fisheries. The intention is to use this information to develop a strategy that can eventually be deployed to help rehabilitate fish stocks and maintain the fishing industry of Lake Victoria. Although not stated directly, understanding the processes that detract from, and might contribute to, 'sustainable development' of the lake's fisheries is an implicit outcome of the research. This chapter sets out to examine the issue of sustainable development in the context of the lake's fisheries.

30.2 Threats to the maintenance of production in the lake fisheries

Lake Victoria has a surface area of 68 000 km^2 and is set in a catchment area of 258 700 km^2. This water mass is distributed within the sovereign boundaries of Kenya (6%), Uganda (43%) and Tanzania (51%). The lake's resources have always been an essential component of the socio-economy of the lakeside communities. Of the 30 million people living in, or around, the catchment area, it is estimated that about 100 000 people are employed directly in the lake fisheries. Another two million are thought to benefit indirectly from the fish industry. This is based on evidence that between 400 000 and 500 000 t of fish are produced each year, with a value of up to US $400 million. Although the figure needs to be treated with caution, as the level of productivity has been subject to spectacular changes over the last three decades (Mkumbo, Ezekiel, Budeba & Cowx, Chapter 8 of this book).

Prior to the introduction of alien fish species the lake's fish fauna was dominated by up to 300 haplochromine species, which accounted for 80% of the fish biomass (Lowe-McConnell 1996). The fish are small and bony and in terms of their socio-economic importance the haplochromines were of most value to local consumers. However, the dominance of haplochromines was challenged in the early 1950s when the alien species Nile perch, *Lates niloticus* (L.) and Nile tilapia, *Oreochromis niloticus* (L.), were introduced to the lake. These species were brought in to boost production, as much sought after food fish, that had the reputation for rapid growth as well as possessing well-established market outlets. Although Nile perch took time to become established, by the early 1980s it had begun to dominate catches. Correspondingly, many of the other species, especially the haplochromines, began to decline. Nile perch became the main predator of haplochromines and certainly contributed to the substantial reduction in the biomass of these species, so much so, that by 1985 Nile perch made up 59% of the total annual catch of lake fish (Mkumbo *et al.*, Chapter 8).

The success of the Nile perch fishery was recognised not only in East Africa but also internationally, and a thriving overseas market for the species was established. Inevitably the success of the fishery attracted other users, which brought with it a subsequent increase in fishing effort. The build-up has been especially noticeable in the last decade, with the Nile perch fishery subject to intense pressure, from not only increased capacity, but also through the use of illegal gears (including poisons). In lakeside fishing communities, partly as result of immigration, the population has increased and with it agricultural and industrial activity. This in turn resulted in the following major changes that, for the most part, have had an effect upon the lake's fisheries.

Over the last two decades there is no doubt that use of illegal fishing methods and gears has compromised and seriously threatened the future of fisheries in each of the riparian states (Owino 1999). Although the threat may have had a relatively low key beginning, the scale of investment in new technology has expanded rapidly, to the point, for example, where some fishing operations are undertaken with beach seines that are hundreds of metres in extent and drift nets that may be kilometres in length. Furthermore, small meshed nets may also be found sewn into the gears that are then operated by powerful trawlers. These methods and gears have given the industry the capacity to make economically efficient fish catches. In many instances the techniques used do not discriminate between juvenile and adult fish, leading to a reduction of the stock base. Although highly undesirable, the adverse impact of unselective and efficient fishing gears and methods was dwarfed by the damaging effects of fish poisons.

Poisoning of fish in Uganda is believed to have begun in 1996 when two fishermen from outside of the country used pesticide to stun and kill lake fish (*The New Vision*, 25 March 1999). The technique was successful in making cheap and easy catches of fish, and thus other fishermen quickly adopted it. The use of pesticides became widespread in the riparian countries (*The East African*, 8–14 February 1999, p. 6) and it was not long before there were deaths of people poisoned by consuming contaminated fish. It seems unthinkable that anyone could be so shortsighted as to opt for poisoning as a harvesting method. Yet there were reasons for it including: a decline in the availability of naturally occurring plant poisons that were used prior to pesticides; the availability of synthetic agro-chemicals that were efficient substitutes; conventional fishing gears were expensive compared to pesticides; and there was, and is, a great consumer demand for fish caught by whatever method.

The incidences of death of people eating poisoned fish caused the authorities to react, although penalties in law were not a strong deterrent. In Uganda, for example, an offender who was caught and convicted would be liable to a fine of only 50 000 Ugandan shillings (≡US $28). In Tanzania, the punishment for first time offenders convicted of poisoning fish was a fine of 300 000 Tanzanian shillings (≡US $378) and the possibility of 3 years' imprisonment. However, in Tanzania, as the enormity of the crime against the lake and its people hit home, it was suggested that fish poisoning should be deemed an offence under the Economic Sabotage and Organised Crimes Act.

The effects of the use of fish poisons were also felt outside of East Africa, as the international community was obliged to act to protect its own consumers, and on 25 March 1999 the European Union (EU) banned the import of fish from East Africa, including Lake Victoria (The East African, 8–14 February 1999). This was a devastating blow to the national economies of those countries bordering the lake. For example, in Uganda, before the ban, the Nile perch fishery was the country's second most important generator of export revenue. The governments and stakeholders in the fisheries around the lake acted together to tackle the fish poisoning threat and their efforts appear to have worked. Nevertheless the international community, and the EU in particular, took time to be convinced, and the international ban was not been completely lifted until February 2001.

Beyond the use of illegal fishing methods and gears there are other threats with which the lake's resources and peoples have had to contend. The early part of the 1990s, for example, was marked by an explosion in the growth of water hyacinth, *Eichhornia crassipes* (Mart.) Solms. This plant is a surface dwelling weed that was introduced to fresh waters in Africa from its native home in South America. Its spread was aided by the rivers that feed into Lake Victoria. As well as congestion and blocking of landing areas, the water hyacinth gave rise to less obvious but more telling problems: reduction in light penetration; reduced levels of dissolved oxygen; and damage to fish breeding areas (Njiru, Othina, Getabu, Tweddle & Cowx, Chapter 21). At its peak it was estimated that there were 1.25 million t of weed dispersed over the lake's surface (Twongo, Bugenyi & Wanda 1995). Mechanical methods and use of herbicides were to prove largely ineffective at controlling the weed. Indeed, it is only since 1998 that there has been progress in finding a way of limiting its spread. Work on the biology of the weed in its native South America showed that two species of weevils were predators of the weed in its original habitat. Specimens of the weevils were imported and cultured in East Africa and after successful laboratory and field trials the insects were released to Lake Victoria. The early results indicate that in many locations in Uganda, and Kenya and Tanzania, the weed is in retreat although it is not clear if this is due to the effects of the introduced biological controls or has resulted from the impact of a yet unidentified cyclic phenomenon (Njiru *et al.*, Chapter 21).

It is easy to understand how these relatively catastrophic and high profile threats afflicting the lake have attracted public interest and alarm. However, well before the introduction of Nile perch and water hyacinth there were processes active in the lake and its catchment area which may well, in the fullness of time, eclipse the magnitude of the other problems. Around the lake margins and further afield in the catchment area, the rapid and inexorable transformation of land utilisation patterns, has surreptitiously entrained a series of far reaching changes that are now influencing the lake environment and its biota. Deforestation, farming, industrial development and population growth (partly as a result of immigration) are all hastening the pace of change in the utilisation and availability of the lakes resources. For the most part these changes have had a detrimental effect upon the quality of the lake waters (*The Guardian*, 8 July 1992, p. 10). The release of nutrients from anthropogenic activities in the catchment has set off a chain of reactions. For instance, there have been substantial increases in chlorophyll concentration and primary productivity, as well as decreases in silica and sulphur concentrations, in sharp contrast to the measurements of these indicators made 30 years ago (Hecky & Bugenyi 1992). Furthermore, an anoxic zone has developed near the lake bottom with a subsequent loss of productivity and restriction of the areas that will sustain fish life. There is also concern that the waters starved of oxygen are expanding and anoxic conditions have been discovered in shallow areas of the lake (Hecky, Bugenyi, Ochumba, Gophen, Mugidde & Kaufman 1994).

In summary, Lake Victoria has been subjected to deteriorative processes that increasingly threaten the viability of its natural resources and especially the industry based on fish. Yet governments, where and when they have intervened, have had little success in stemming the overall decline. In part this may be because of a misjudgement of the complexity and seriousness of the situation, furthermore there is the likelihood that the interventions were not of the right order of magnitude or comprehensiveness to be effective.

30.3 Unsustainable tendencies and the lake fisheries

The parlous situation with respect to the fisheries of Lake Victoria is mirrored elsewhere in the world's fishing industries (McGoodwin 1994; Crean & Symes 1996) and for many researchers, the goal of sustainable development seems as far away as ever. Indeed it is argued that the processes that work against sustaining resource exploitation are not accurately portrayed.

The concept of sustainable development of natural resources has come to prominence during the last 30 years (Norgaard 1988; Newby 1990). Whilst it has had a profound influence on our views of the exploitation of natural resources, authorities charged with balancing the needs of industry with maintenance of the natural biological capital, have discovered that the practical expression of this ideal has proved elusive.

Despite the intuitive appeal of the concept, and its widespread adoption as a policy goal by governments and non-governmental organisations alike, it has come to be viewed by some as an ambiguous ideal with little practical utility (Manning 1992). Not least of all this has been because of the difficulty of defining ecological limits of natural systems where the scientific knowledge is usually partial, imperfect and contestable (Pearse & Walters 1992; Wilson & Kleban 1992). Furthermore, there is the difficulty, or some would argue, impossibility, of seeking to maintain economic growth within defined ecological limits (Lélé 1991). Rather than this narrow perspective, Redclift (1991), for example, perceived the utility of sustainable development as deriving from the multi-dimensional nature of the concept. A concept which would incorporate environmental, economic and, most importantly, social and moral components. Rees (1985) commented that the achievement of sustainable development would depend on being able to balance the complex processes, which create social values and drive political and economic decision-making. However, the author was pessimistic that there seemed little prospect that current economic, social and political processes could be adjusted to sustainable forms within the necessary time scale. Norgaard (1988, p. 607) sounded a more hopeful note by asking whether sustainable development might not be an expression of the 'resource base, environment, technologies and culture evolving over time in a mutually reinforcing manner?'.

Despite problems associated with the practical expression of the sustainable development concept, there is support from all quarters for the institution of processes that would lead to the equilibration of the social, economic and environmental dimensions of resource exploitation. However, there are, as yet, no emerging strategies of how this might be achieved. For the most part, the current institutional and organisational make up of most governments ensures that there is heavy reliance on continual (and usually unsuccessful) interventions, most of which involve adjustment of the regulatory framework.

These broader commentaries on the complexity and deep-rooted nature of problems associated with sustainable development of natural resource systems have been placed in a specific fisheries context. Drummond & Symes (1996) cast doubt on the capacity of governments (in this case the European Commission) to treat the ills of the prevailing fisheries management policies by adopting the classic 'command and control' methods. They argued that this would not be sufficiently all-embracing and penetrative to counter what they termed 'unsustainable tendencies'. They argued that the process of drawing ecological limits with the intention of controlling economic and social institutions is a flawed approach that will not give rise to a sustainable outcome. Thus, interventions invoked to maintain resource exploitation within limits, at best only temporarily suspend the deeply rooted processes that might ultimately bring about 'extinction' of the resource and the dependent exploitation processes.

The Drummond & Symes (1996) model may be taken a stage further and can be extended into the conceptual representation realised by Lélé (1991). The author produced what was considered a 'more realistic' interpretation of sustainable development (Fig. 30.1), which draws together elements that are rooted in the disciplines of biology, ecology, technology, economics, social science and the environmental science. It is essentially a flow diagram but there is a tendency for the inputs to be unidirectional and concentrate on a 'sink' where environmental degradation and poverty are linked together. The processes that establish the sink are driven by, amongst other things, short-term decisions, loss of biological capital and pollution. In the model the variables of access to resources, culture/values and technology via environmental degradation strongly influence the directional flow between affluence and poverty. Interventions (by government) are not shown in the model but may be assumed to have occurred, however, with the net outcome of serving only to reinforce the deteriorative flows already established.

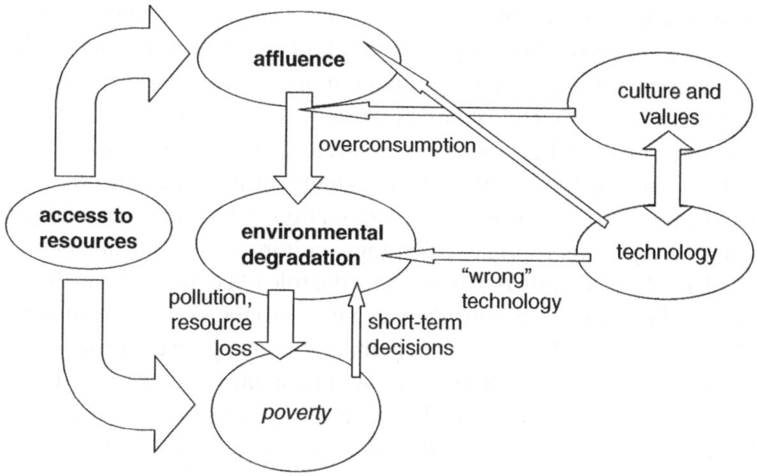

Figure 30.1 A more realistic modelling of sustainable development (*Source*: Lélé 1991)

The concept of unsustainable tendencies readily translates into the context of Lake Victoria's fisheries. Examples of interventions that have only served to reinforce unsustainable tendencies are attempts to control exploitation of undersized fish. Whilst there is in existence a plethora of fisheries ordinances directed at this problem, they appear to have failed. The measures adopted have had little impact because in this case there is a market for undersized fish that cannot be controlled by gear regulations. The consumption of undersized fish is an unsustainable tendency that is a function of culture, social values and demography. It is a growing problem that the current regulatory interventions cannot address. There are indicators that, at least in part, show the damage caused by harvesting of immature individuals of commercially-targeted fish populations. In Uganda, for example, fishermen comment that there is less fish now than 5 years ago; more time has to be spent in the capture of fish; species diversity has been reduced; there are more boats than 5 years ago; the average size of fish is smaller than that landed 5 years ago; illegal fishing techniques are still in use; fishing pays less now than it did 5 years ago. As a result of the widespread theft of passive fishing gears, such as gill nets, fishermen are adopting fishing methods where the gear can be more readily monitored and protected. Thus, many fishing groups are switching to the use of beach seines which are much less vulnerable to theft, but are more damaging to the stock base in terms of their take up of undersized (juvenile) fish and damage to nesting areas (Craig 1992).

30.4 Conclusions

The models described by Drummond & Symes (1996) and Lélé (1991) are useful tools in explaining the complex of problems that confront attempts to establish sustainable development of the fisheries of Lake Victoria. They expose the weakness of interventions by government that are usually technical or fiscal in nature and take little account of trans-sectoral conditions. Overall the processes shown in the Lélé model cannot readily be equilibrated, as the forces involved tend to short-circuit with a tendency to poverty, driven by some or all of the main functions of the model. Whilst the Lélé model is a generic representation of key elements that drive the unsustainable exploitation of resources, it has a relevance to the Lake Victoria situation. The open access status of the lake's fisheries resources coupled with a lack of alternative employment options for the peoples of the catchment area are powerful forces that have fuelled the process of environmental degradation. There are clear examples of the link between poverty–inappropriate technology–culture–environmental degradation.

What the LVFRP has identified to date is that the role of economic growth in fisheries, in relation to the dispersion of benefits between user groups, is not fully understood and further analysis is required. Yet it is apparent that economic growth *per se* is not a means of poverty removal and may well entrain environmental sustainability. The evidence appears to show that affluence associated with the user groups that prosecute the fisheries is skewed towards those that deal with the international trade in Nile perch and there is so far a lack of evidence to show that the benefits are more widely distributed (Jansen, Abila & Owino 1999). Inequitably access to the resource as a result of the distorted accumulation of capital reinforces affluence for a minority. The situation with respect to the distribution of benefits is dynamic and will be responding to the build-up of population in the lakeside communities through 'normal' demographic trends and a shift towards immigration of peoples who have previously not been regarded as stakeholders.

The problems of the lake are deeply embedded in its social fabric, and interact not only at the level of the lakeside communities but also at regional and international levels. The vertical discordance in culture and values between these levels, and also horizontally at a given level, is the source of the unsustainable tendencies that adversely affect the lake's fisheries. What is clear is that sustainable development, if and when attained, will be reached by a route, which starts with the acceptance that there are structural, technological and cultural causes of both poverty and environmental degradation. Thereafter the stakeholders at all levels will have to come to terms with the concept that solutions will require radical change in the structure, function and relationships between the myriad institutions that influence the lake's fisheries economy.

Acknowledgements

The financial support for the project was provided by European Union under the Lake Victoria Fisheries Research Project (Ref.: ACP-RPR 227).

References

Coulter G.W., Alanson B.R., Bruton M.N., Greenwood P.H., Hart R.C., Jackson P.B.N. & Ribbink A.J. (1986) Unique qualities and spatial problems of the African Great Lakes. *Environmental Biology of Fishes* **17**, 161–183.

Craig J.E. (1992) Human induced changes in the composition of fish communities in the African Great Lakes. *Reviews in Fish Biology* **2**, 93–124.

Crean K. & Symes D. (eds) (1996) *Fisheries Management in Crisis: A Social Science Perspective.* Oxford: Fishing News Books, Blackwell Science, 232 pp.

Drummond I. & Symes D. (1996) Rethinking sustainable fisheries: the realist paradigm. *Sociologia Ruralis* **36**, 152–162.

FAO (1995) *Code of Conduct for Responsible Fisheries*. Rome: FAO, 41 pp.

Geheb K. (1999) Small-scale regulatory institutions in Kenya's Lake Victoria fishery: past and present. In H. Kanawabe, G.W. Coulter & A.C. Roosevelt (eds) *Ancient Lakes: Their Biological and Cultural Diversity*. Brussels: Kenobi Productions, pp. 113–121.

Harris C.K., Wiley D.S. & Wilson D.C. (1995) Socio-economic impacts of introduced species in Lake Victoria fisheries. In T.J. Pitcher & P.J.B. Hart (eds) *The Impact of Species Change in African Lakes*. London: Chapman and Hall, pp. 215–242.

Hecky R.E. (1993) The eutrophication of Lake Victoria. *Verhandlungen Internationale Vereinigung für Theorestiche und angewandte Limnologie* **25**, 39–48.

Hecky R.E. & Bugenyi F.W.B. (1992) Hydrology and chemistry of the African Great Lakes and water quality issues: problems and solutions. *Mitteilungen Internationale Vereinigung Theorische Angewandt Limnologie* **23**, 45–54.

Hecky R.E., Bugenyi F.W.B., Ochumba P.B.O., Gophen M., Mugidde R. & Kaufman L. (1994) Deoxygenation of the deep waters of Lake Victoria. *Limnology and Oceanography* **39**, 1476–1480.

Jansen E.G., Abila R.O. & Owino J.P. (1999) Constraints and opportunities for 'community participation' in the management of the Lake Victoria fisheries. *IUCN Eastern Africa Programme, Report No. 6*, Nairobi, Kenya: IUCN, 30 pp.

Lélé S.M. (1991) Sustainable development: a critical review. *World Development* **19**, 607–621.

Lowe-McConnell R.H. (1996) Fish communities in the African Great Lakes. *Environmental Biology of Fishes* **45**, 219–235.

Manning E. (1992) Presidential address: sustainable development, the challenge. *Canadian Geographer* **34**, 290–302.

McGoodwin J.R. (1994) *Crisis in the World's Fisheries: People, Problems and Policies*. Stanford: Stanford University Press, 247 pp.

Mugidde R. (1993) The increase in phytoplankton primary productivity and biomass in Lake Victoria (Uganda). *Verhandlungen Internationale Vereinigung für Theorestiche und angewandte Limnologie* **25**, 826–849.

Newby H. (1990) Ecology, amenity and society – social science and environmental change. *Town Planning Review* **61**, 3–13.

Norgaard R.B. (1988) Sustainable Development. A Co-Evolutionary View. *FUTURES*, December 1988, 606–620.

Ogutu-Ohwayo R. & Hecky R.E. (1991) Fish introductions in Africa and some of their implications. *Canadian Journal of Fisheries and Aquatic Science* **48**(Suppl.1), 8–12.

Ogutu-Ohwayo R., Hecky R.E., Cohen A.S. & Kaufman L. (1997). Human impacts on the African Great Lakes. *Environmental Biology of Fishes* **50**, 117–131.

Owino J.P. (1999) Socio-economics of the Lake Victoria Fisheries: traditional and central management systems of the Lake Victoria Fisheries in Kenya. *IUCN Eastern Africa Programme, Report No. 4*, Nairobi, Kenya: IUCN, 24 pp.

Pearse P.H. & Walters C.J. (1992) Harvesting regulation under quota management systems for ocean fisheries: decision making in the face of natural variability, weak information, risks and conflicting incentives. *Marine Policy* **16**, 167–182.

Redclift M. (1991) The multiple dimensions of sustainable development. *Geography* **76**, 36–42.

Rees J.A. (1985) *Natural Resources: Allocation, Economics and Policy*. London: Routledge, 449 pp.

Twongo T., Bugenyi F.W.B. & Wanda F. (1995) The potential for further proliferation of water hyacinth in lakes Victoria, Kyoga and Kwania and some urgent aspects of research. *African Journal of Tropical Hydrobiology and Fisheries* **6**, 1–10.

Wilson J.A. & Kleban P. (1992) Practical implications of chaos in fisheries: ecologically adapted management. *Maritime Anthropological Studies* **5**, 67–75.

Chapter 31
Principles and approaches to the management of lake and reservoir fisheries

I.G. COWX*
University of Hull International Fisheries Institute, Hull, UK

Abstract

The majority of inland fisheries tend to be heavily exploited or utilised for a variety of reasons, and this generally leads to the need for management of the resources. Management of lake and reservoir fisheries has traditionally orientated around interventions to enhance the fish stocks or regulation of fishing activity, usually under the jurisdiction of central administrators. Increasing demands on water resources and the ever expanding role of society in regulatory processes, coupled with the universally encouraged drive for sustainable development, has shifted this emphasis towards integrated, community-driven management. This chapter examines the principles underlying existing management of lake and reservoir fisheries and reviews the issues affecting the fisheries in an effort to identify limitations in current practices.

Keywords: fisheries, habitat degradation, management, rehabilitation, stock enhancement.

31.1 Introduction

The majority of inland fisheries tend to be heavily exploited or utilised for a variety of reasons, and this generally leads to the need for management of the resources. The objectives are usually associated with the types of use (e.g. commercial or recreational fishing), level of exploitation (for supply of fish protein, brood stock or ornamentals), as well as with socio-economic factors connected with the associated community (e.g. maintenance of employment or conservation value).

Management of lake and reservoir fisheries has traditionally orientated around interventions to enhance the fish stocks or regulation of fishing activity, usually under the jurisdiction of central administrators. Increasing demands on water resources and the ever expanding role of society in regulatory processes, coupled with the universally encouraged drive for sustainable development, has shifted this emphasis towards integrated, community-driven management. This chapter examines the principles underlying existing management of lake and reservoir fisheries and reviews the issues affecting the fisheries in an effort to identify limitations in current practices. Finally,

---

*Correspondence: University of Hull International Fisheries Institute, Hull HU6 7RX, UK (e-mail: i.g.cowx@hull.ac.uk).

opportunities and strategies for future management of fisheries in a multiple resource-user environment are considered.

31.2 Lakes and reservoir ecosystems

Lake and reservoir management must be based on an understanding that they are complex and dynamic ecosystems. They are influenced by a set of hydrological conditions, the watershed, the shape of the lake basin, the lake water and the bottom sediments. These physical and chemical components, in turn support a community of organisms that is unique to that water body and make up the ecosystem. All these components are in constant change but the conditions are very much dictated by the origin of the lake or reservoir, the geomorphology and human interference in the water body ecosystem. Although beyond the scope of this chapter, reference to these conditions is made here to underpin the management issues associated with fisheries in lakes and reservoirs.

31.2.1 *Origin, shape and location*

Lakes owe their existence to a variety of causes and can be classified as follows (Wetzel 1983).

- *Glacial lakes*: formed by the scouring action of ice movements in deep valleys and partially dammed by deposition of ice-transported material at the lower end of the valley. They tend to be associated with upland areas in lower latitudes and Alpine regions.
- *Teutonic lakes*: produced by warping and faulting of the earth's crust and are represented by some of the world's largest and deepest lakes. Often they have relatively restricted littoral margins.
- *Volcanic lakes*: formed by volcanic action and prevalent through the major volcanic zones of the world.
- *Deposit lakes*: formed by deposition of sedimentary materials brought down by a river or washed in from landslides, etc. These tend to be shallow and relatively short-lived because the rocky-debris barrier is usually porous and unconsolidated, and is prone to rapid erosion. Floodplain lakes associated with large rivers are important examples of this type of water body.
- *Solution lakes*: formed by solution of limestone and other highly soluble rock formations, especially by groundwater creating caverns, which subsequently collapse. They tend to be shallow with large littoral zones.
- *Man-made lakes*: there are many forms of man-made lakes created for a variety of reasons, but usually for water supply for human consumption, agriculture or industry, for hydropower generation or for flood control. They are created by the construction of a dam, and the purpose and location of an impoundment usually determine its basin size, and the topography of the inundated valley dictates the basin shape. On a smaller scale mineral extraction (e.g. gravel) creates pools

which can become prolific fisheries. Reservoirs have been built throughout the world (Petts 1984) and support important fisheries replacing lost riverine stocks.

31.2.2 *Physicochemical properties*

Although lakes and reservoirs are complex ecosystems, production of the system is driven by a number of key characteristics: water body size and linkages to retention time (time to exchange the entire water mass) and mixing; nutrient fluxes; temperature profiles (stratification); light penetration; and water clarity. Again, although beyond the scope of this chapter to discuss these factors in detail, it should be recognised that these characteristics underpin the fish species assemblages, distribution and abundance of fish species, propensity for eutrophication and extent of aquatic vegetation coverage. These in turn have profound impacts on the fish and fisheries of lakes and reservoirs and provide the bases for understanding the issues concerning the status of the fisheries. For example, eutrophication is linked to increased turbidity and the loss of aquatic vegetation. This leads to a loss of phytophilic fish species or those that use vegetation for refuge or spawning, or because their feeding habits are linked to macrophytic vegetation. Eutrophication also results in reduced oxygenation in deeper waters. Ultimately this may lead to loss of predation pressure on zooplanktivorous and benthivorous fish species which upsets the balance of the ecosystems and exacerbates the problems linked to eutrophication (for full description of problem see Harper 1992).

In the case of reservoirs, additional criteria exist that regulate the structure of the fish communities, especially the indigenous fish fauna in the river prior to inundation need consideration. This often dictates which species are able to survive and thrive in the new environment (Cowx & Kapasa 1995). However, in many cases there are vacant niches, especially in the pelagic zone, which restrict fish production and species introductions are usually considered to exploit this niche (e.g. *Limnothrissa moidon* into Lake Kariba, Marshall 1995). Water level fluctuations and short retention times are also problematic as they restrict spawning opportunities for fish and flush nutrients through the system reducing production potential.

Thus, the size, depth, location and geomorphology and physicochemical characteristics of lakes and reservoirs, and how they are managed for other uses, dictate the structure of the fish assemblages. Wetzel (1983), Harper (1992) and Moss (1998) are recommended for further reading on these issues.

31.3 Stock assessment methods

To manage lake and reservoir fisheries successfully there is a prerequisite for adequate information on the status of the fish stocks on which to formulate the decision-making process or evaluate the impact of a particular management activity. This is problematic because lacustrine fisheries exist in a diverse range of habitats from small ponds to large lakes and reservoirs. Numerous methods are available for sampling fish populations in still waters. These vary from environmental correlation methods, through

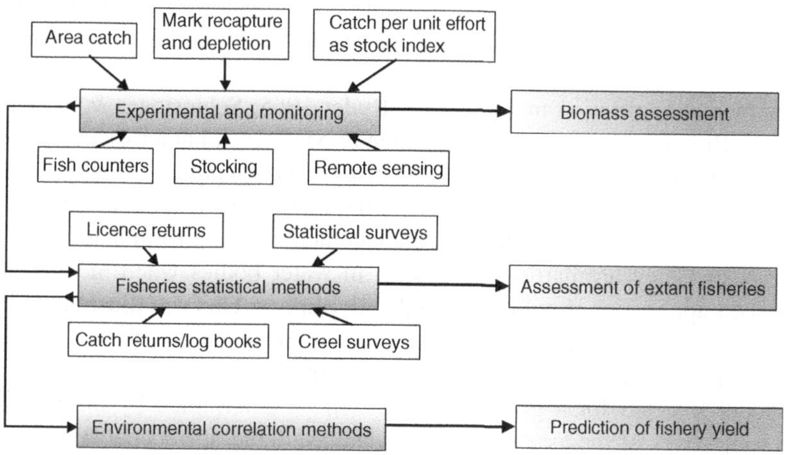

Figure 31.1 Suggested combination of methods and models for fish stock assessment and yield predictions in lakes and reservoirs (from Cowx 1996)

sampling gears and remote methods (e.g. hydroacoustics) to catch assessment surveys (catch effort methods) (Fig. 31.1). However, not all are appropriate for all situations. For example, hydroacoustic surveys are inappropriate for shallow lakes with extensive aquatic vegetation coverage or electric fishing is inefficient except in the shallow margins of lakes. Consequently, there is a need for an array of methods to provide adequate assessment of stocks, and this can only be achieved by combining the appropriate methodology and sampling strategy to optimise the data collection exercise and ensure the information is collected in the most cost-effective manner.

In making an assessment of the most appropriate stock assessment programme, it should be recognised that the various strategies provide output data at different levels of precision and each requires very different levels of input. Experimental and monitoring strategies are typically those used for determining absolute population parameters and use intense, efficient data collection methods to provide input for the models (Cowx 1996; Mkumbo, Ezekiel, Budeba & Cowx, Chapter 8 of this book). It is often this approach which is misused to provide absolute population parameters because the data collected are inadequate to fit the assumptions of the model (particularly mark–recapture and depletion models). The strategy can be, and frequently is, used to provide relative population parameters, which is probably a more appropriate mechanism in large lakes and reservoirs. Greater consideration should be given to using hydroacoustic methods, which have improved considerably over the past decade, particularly through the use of dual beam and side scanning (Hughes & Hateley, Chapter 1; Cadic, Irz & Argillier, Chapter 2; Tumwebaze, Getabu, Bayona, MacLennan & Cowx, Chapter 7).

Fisheries statistical methods are the main methods used where the fisheries are commercially or recreationally exploited (Haldar, Ahmed, Alamgir, Akhter & Rahman, Chapter 12; Petrere Jr, Agostinho, Okada & Julio Jr, Chapter 11; Tito de Morais, Chapter 10). They provide an assessment of the status of the existing fishery but the quality of the output is very dependent on the input information which can be highly variable depending on its source, e.g. scientifically collected or provided by fishermen.

Environmental correlation methods (e.g. Ryder 1965; Schlesinger & Regier 1982; Cryer 1996; Brämick, Chapter 3; Lappalainen & Malinen, Chapter 4) are less precise in their output but do provide a good basis for initiating policy development. They offer a starting point for the development of the management process but need to be supported by more accurate methods if management is to be formulated on a sound scientific basis. They are particularly useful in water bodies where no fishery data exist, but their application should be treated with caution. This was illustrated by Cryer (1996), who showed the potential for gross misrepresentation of the status of the fisheries when only environmental correlation methods were used.

The model selected for a given situation will be determined by the information required (objectives of the assessment), the data available or obtainable, the assumptions which can be made, and the funding, resources and experience available. The choice of model will always be a compromise between these various factors. For example, a simple, cheap model will be easy to apply, but will require many assumptions with unknown risks attached. By contrast, a more detailed model may provide useful insights into the dynamics of the system, but only at the cost of expensive data collection programmes and staff expertise. It should be noted, however, that more complicated models are not necessarily more precise, although it may be easier to remove some of the uncertainties from the assessment. Problems still persist with regards to assessment of juvenile and small-sized fishes, and this is an area that needs attention in the future since an understanding of recruitment dynamics provides a key indicator in managing fisheries (Caddy & Mahon 1995).

31.4 Issues affecting lake and reservoir fisheries

For any management regime to function there is a need to understand the issues and problems relating to the resource that needs to be managed. In the case of lake and reservoir fisheries this can be broken down into four areas: fishery-related issues; environment- or watershed-related issues; the human dimension; and inter-sectoral conflicts.

31.4.1 *Fishery-related issues*

One of the key factors affecting fisheries in lakes and reservoirs is the exploitation pattern. Fisheries in developed countries are exploited by commercial and recreational fisheries, usually for pleasure but also for food (Table 31.1). Commercial fisheries in this region are now declining largely because the fish stocks have been depleted or changed by other human interventions. In developing countries, exploitation is largely by commercial or artisanal fishermen or subsistence fishing by local residents supplementing their food supply (Welcomme 2000), although recreational fishing is developing

Table 31.1 Strategies for management of inland waters for fisheries in developed and developing countries (after Welcomme 2000)

	Conservation (developed)	Production (developing)
Objectives	Conservation Recreation	Provision of food Income
Mechanisms	Sport fisheries Habitat restoration Environmentally sound stocking	Food fisheries Habitat modification Enhancement through intensive stocking and management of ecosystem
	Intensive, discrete, industrialised aquaculture	Extensive, integrated, rural aquaculture
Economic	Capital intensive	Labour intensive

as part of the tourist sector (Cowx 2002). In many cases, fisheries are showing signs of overexploitation typically seen by decline in catch per unit, reduce in size of capture, loss of larger-sized species, change in exploitation to smaller, less valuable fish species (Welcomme 1985; Caddy & Mahon 1995; Mkumbo *et al.*, Chapter 8).

The general effects of overexploitation of the fisheries are thus low abundance of desired species (poor catches) and a poor population or community structure (size and species). Whilst the changes in the stock structure can absorb increased amounts of effort, either as labour or as improved technology, than a fishery targeting only larger-sized species, and, there is a general perception and desire to modify the fish assemblages further through stocking and introductions to support expanding fisheries (Cowx 1998a). Both overfishing and fishery enhancement strategies cause conflict with the general precept to maintain species diversity elaborated in the Code of Conduct for Responsible Fisheries adopted by member countries of the Food and Agriculture Organization (FAO 1995).

From a recreational perspective, many lake and reservoir fisheries are perceived to have declined over the years and again stock enhancement strategies, with their inherent problems, are prevalent to establish a desirable fish assemblage. This can also be compounded by the removal of pest species, often top predators, such as pike, *Esox lucius* L., which often alters the ecosystem dynamics and leads to large numbers of small individuals because predation pressure is reduced (see Ibbotson & Klee, Chapter 16). Such action is common in put-and-take trout and specialist fisheries. Concomitant with these actions has been the development of intensively-stocked fisheries (North, Chapter 23) which target anglers who want to land large catches, at the expense of the other social and environmental attributes of recreational fishing (Cowx 2002).

One of the fundamental constraints on fisheries maintenance and development in lakes and reservoirs is poor recruitment of fish to the size of exploitation. This can arise from lack of optimal spawning habitat, inadequate food resources, especially for larvae after they have absorbed the yolk sac, high mortality of juvenile life stages or poor spawning structures unable to support maintenance of the stock. The typical response of managers is to compensate for the poor production through stocking to bypass the recruitment bottleneck and utilise the potential production of the water body. This action is often in response to pressure from fishermen and requires continuous intervention, often at high cost, with marginal returns (Cowx 1998a). A more appropriate strategy would be to address, if possible, the recruitment bottleneck either through more traditional fishery management regulation or rehabilitation of the water body (see Sections 31.5.2 and 31.5.3)

Finally, an issue that is prevalent in inland fisheries across Europe and North America is the explosion of avian piscivores (Britton, Harvey, Cowx, Holden, Feltham, Wilson & Davies, Chapter 14). This has arisen because many bird species are protected under environmental and species laws and their numbers, especially cormorants (*Phalacocorax* spp.), have expanded in inland regions. Many anglers perceive this expansion of the bird populations as detrimental to the fishery through depredation on fish stocks. However, the general consensus is that the problem is specific to each individual fishery, and that still waters are more prone to damage than rivers (Feltham, Cowx, Davies, Harvey, Wilson, Holden & Britton 1999).

31.4.2 *Environment- or watershed-related issues*

Pollution is probably one of the biggest threats to the freshwater environment (Moss 1998). This is because discharge into fresh waters is a convenient mechanism for disposal, and the capacity of large water bodies to accommodate such wastes is high. However, with the increasing human population, the pressures exerted on lakes and reservoirs have become intolerable and fresh waters have become contaminated, which in turn has resulted in deterioration of fish stocks and fisheries. Pollutants include, organic wastes, nutrients, metals, poisons, suspended solids and cooling water from urban, industrial and agricultural sources. These can act directly on the fish, e.g. toxicity of chemicals, which may have an acute or chronic affect, depending on concentrations of chemicals concerned, or indirectly by changing water quality parameters, and consequently the suitability of the habitat for fish.

Nutrient enrichment, i.e. eutrophication, may in the first instance benefit fisheries through increased levels of production. When the nutrient load is too high, however, excessive primary production and plant growth may lead to reduced production and loss of fish throughout periods of lethally low dissolved oxygen concentrations, especially in the hypolimnion.

Many industries, especially those concerned with the brewing industry, food processing, paper manufacture, produce large quantities of liquid waste, which has a high organic content. The discharge of human excretory wastes also has the immediate effect of raising the concentration of organic material in the water. When these are discharged into natural water bodies at relatively small concentrations the input of nutrient material may raise the level of productivity. However, with excess loads of organic materials, either as solids or in solution, the immediate observable effect is the creation of blooms of green algae or massive growths of rooted or floating higher plants. The water body may suffer further from falls in the levels of dissolved oxygen to the point where the water and its substrate becomes anoxic with the production of hydrogen sulphide and methane. Under these conditions only a few species of organisms are able to survive, although they may do so in populations with very high numbers of individuals. If this species change replaces favoured species with less desirable species the increased levels of production may not be seen to be advantageous.

Acidification of lakes due to acid discharge has the opposite affect of eutrophication because its final stage is an almost dead lake with no decomposition taking place. In part this low production is brought about by the acid environment creating conditions where the nutrients present are not so readily available to the plants. This type of pollution is most notable in North Europe and Canada where large numbers of lakes are affected and they are virtually fishless. Acidification is caused by sulphur and nitrogen compounds being emitted into the atmosphere from burning fossil fuel then being blown hundreds of kilometers before being deposited later as sulphuric and nitric acid. The rain may have a pH of less than 4.

Low levels of dissolved toxic materials may generally lower the productivity of the water body. There is always the risk that constant low concentrations may build up selectively in different organisms to the levels where the concentrations become damaging to other organisms in the food web, including man. High concentrations of toxic materials either as a constant input or a short duration discharge, may create an aquatic environment totally devoid of life.

Many industries directly exploit the water resources as a means of cooling. The biological effects of the discharge of cooling waters is similar to the effects of the discharge of organic materials. In relatively small amounts the raising of the temperature of the receiving waters, and especially the maintenance of higher and constant temperatures throughout the annual seasonal cycle, will raise the level of production. However, as temperatures rise above the optimum for the environment they cause a decrease in the amount of available dissolved oxygen and may create localised, lethal heat stress conditions for fish.

Finally, natural fluctuations in water level are a common feature of most lakes and reservoirs as a result of seasonal and climatic variation in rainfall. The problem is, however, exacerbated in lakes and reservoirs used for hydroelectric power generation, which control the water level in response to power generation requirements and also to accommodate the flood water from their catchments at the start of the rainy season. This drawdown, and the way in which it is achieved, may not be advantageous to the development of fisheries in reservoirs. In all cases the littoral zone becomes barren, with exposed rock, gravels and sands, reducing the potential spawning and nursery areas for many fish species. In particular, the rapid drawdown associated with hydropower generation has an adverse effect on fish that spawn in the littoral zone, such as nest building tilapias, killing eggs and larvae and thus reducing future recruitment to the fishable stocks of these species (Cowx & Kapasa 1995). The re-deposition of sediments below the minimum water level can also lead to a drastic reduction in the availability of benthic animals, the primary food source of many fish species. On the other hand this drawdown may clear trees and brush in the drawdown zone over a period of years (Bernacsek 1984), thus improving the fishing pitches and obviating the need for expensive clearing prior to inundation.

31.4.3 *Human dimension*

Many of the problems pertaining to lake and reservoir fisheries are driven by social and economic demands to provide high quality and quantity of catch. Depending on the objective of the fishery, i.e. to provide food, employment or recreation, the pressures are such that the resource base is inadequate to meet the demands of the users. As indicated in Section 3.4.1, this leads to dissatisfaction in catches in terms of numbers or size of fish caught by fishermen. The classical response is to stock the fishery or introduce a new species to address the problem (e.g. da Costa, Traore & Yte, Chapter 28), but these actions are rarely effective leading to continued dissatisfaction and ironically further pressure to stock (Argillier, Pronier & Changeux, Chapter 25). The underlying issue in these circumstances is that the fishery has often been mismanaged, usually in terms of regulating catch or access. Consequently the stocks are over-exploited or unable to support the fishing pressure imposed. Stocking and introductions are not the solution because the water bodies are rarely able to sustain increased fish biomass or production, and the exercises are fruitless (Cowx 1994). Stock enhancement practices also have deleterious effects on other components of the fishery, including loss of biodiversity and reduced stocks of wild fishes.

31.4.4 *Inter-sectoral conflicts*

One of the major constraints to the development of lake and reservoir fisheries is from cross-sectoral interactions. These issues arise from interactions:

(1) between commercial and recreational fishing;
(2) within recreation fishing groups; and
(3) with other aquatic resource users.

Commercial and recreation fishing interactions
Direct conflicts exist between commercial and recreational angling because they exploit the same resource base. The arguments usually relate to commercial fisheries depleting stocks through overfishing, and demands from the recreation sector to restrict commercial exploitation, although many studies indicate that commercial and recreational fisheries can co-exist (see e.g. Hickley & Tompkins 1998), supplementing each other, creating an overall larger output than would have resulted from, for instance, a sport fishing only scenario. When the commercial and recreational fisheries interfere, the allocation of the harvest generally falls in favour of recreational fishing as it is perceived to have an overall greater benefit to society.

Interactions between recreation fishing groups
There is also increasing conflict between groups of recreational fishermen, for instance between residents and non-residents (e.g. Salmi & Muje 2001). Reasons for this include: loss of available fishing waters has resulted in crowding problems and over-exploitation on the remaining waters; increased specialisation among anglers has put

pressure on attractive waters; and increased focus on tourist fishing in many areas has introduced new angler groups to many districts, which may cause different conflicts related to social and allocation controversies.

Cross-sectoral interactions

Perhaps the greatest threats to fisheries come from outside the fisheries sector. Aquatic resources are subject to numerous anthropogenic perturbations, such as pollution discharge from agricultural, domestic and industrial sources, eutrophication, acidification, afforestation, damming for power generation and water supply (Fig. 31.2; Cowx 1994; Cowx & Welcomme 1998), which have resulted in a shift in the status of the fisheries and a general decline in the yield. In these circumstances fisheries are not considered of sufficiently high priority or value, and thus suffer in the face of economically and socially higher priorities, e.g. agriculture, hydroelectric power production or other water sports.

Problems relating to pollution are in the main being addressed in western industrial countries, and waters that were once fishless or supported poor stocks, are now recovering. Such water quality improvements are, however, rare in the developing world and the scenario remains one moving towards environmental degradation and demise of the fisheries. Notwithstanding, water quality improvements do not always lead to a desirable outcome in terms of fishery status. The cleaning up of water bodies often results in reduced nutrient input and loss of productivity of the water body concerned, which has a knock on effect on the fish community structure and stock size. This results in a change in the catch composition and reduction in catches, which are quickly followed by complaints from anglers (Cowx 1991).

Interactions between other recreational uses of water, as well as recreation in the vicinity of water bodies, are also sources of conflict (Cowx 2002). The following pursuits can impact on recreational fisheries: bird watching, cycling, motor boating, canoeing, cruising, rowing, sailing, diving, swimming and bathing, water-skiing, wind surfing, water bikes and wild fowling. The problems arise from disturbance of fish,

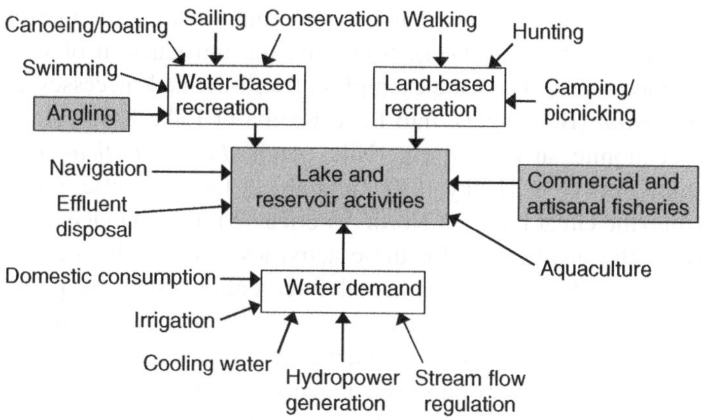

Figure 31.2 Users of lakes and reservoirs that affect fisheries

noise, litter, the loss of access to the water body and general disruption of the angling experience.

31.5 Principles and approaches to management

There are numerous approaches to the management of lake and reservoir fisheries, but these can essentially be broken down into: managing the fish assemblages; managing the fishery; and management of the environment.

31.5.1 *Management of the fish assemblages*

A variety of techniques are used throughout the world to improve production of fish species favoured by commercial or recreation interests, to make up for shortfalls in production arising from overfishing or environmental change, or for conservation initiatives (Cowx 1994; 1998a; Welcomme & Bartley 1998). They include:

- stocking natural waters to improve recruitment, and bias fish assemblage structure (North, Chapter 23);
- stocking to maintain productive species (Argillier *et al.*, Chapter 25);
- introduction of new species to exploit under-utilised parts of the food chain or habitats (da Costa *et al.*, Chapter 28);
- elimination of unwanted species (Ibbotson & Klee, Chapter 16);
- construction of biased and selected faunas.

The enhancement of fisheries through the stocking of individuals or introduction of species is a practice frequently used by fisheries owners, managers and scientists, but the general consensus is that stocking is a much used, but all too often abused, tool in fisheries management (Cowx 1994). This is partly because of the preconception that stocking will improve yields from fisheries. However, there is considerable evidence to suggest this is not the case unless the stocking activity is well managed, and takes into account many of the wider issues that may impinge on the outcome of the stock enhancement exercise (Cowx 1998a). Similarly, the introduction of fish species is shrouded with controversy. There are examples of well-defined successes, e.g. kapenta, *Limnothrissa moidon* (Boulenger), into Lake Kariba or Lake Kivu (Marshall 1995), questionable economic successes, e.g. Nile perch, *Lates niloticus* L., into Lake Victoria (Ogutu-Ohwayo & Hecky 1991), and abject failures, ruffe, *Gymnocephalus cernuus* (L.), into the Great Lakes of North America (Winfield, Dodge & Rösch 1998).

In some cases the justification for these activities is perfectly acceptable, e.g. to compensate for loss due to environmental interventions such as pollution, or to enhance fish yield in depauperate waters. However, despite there being national and international regulations and codes of practice on species that can be stocked or criteria to minimise the transfer of disease, in many cases there appears to be little control over whether the enhancement activity is appropriate or necessary. More recently concerns have also been expressed about the potential risks associated with stocking of fish, particularly with respect to ecological imbalance and change in community structure, and loss of genetic integrity (see Cowx 1994; Carvalho & Cross 1998).

Examples of stocking fish in inland waters in the literature are sparse, mainly because the outcome is rarely documented, but also because many stocking programmes have not proved to be successful (Cowx 1997; Argillier *et al.*, Chapter 25). In developed countries the most successful stock enhancement programmes have been associated with put-and-take and intensive recreational fisheries (e.g. Moehl Jr & Davies 1993). By contrast, successful programmes in developing countries appear to be associated with reservoir fisheries which have been heavily stocked to increase yield, e.g. India (Sunugan 1995), Cuba (Fonticiella, Arboleya & Diaz 1995) and China (Lu 1992). It should be noted that the activities carried out in China are not strictly stocking as they further enhance production through fertilisation of the water and supplementary feeding, so it is more akin to extensive aquaculture. Not all enhancement programmes are successful. For example, stocking of carps in the perennial reservoirs of Sri Lanka have proved to be ineffective and the reservoirs are now being managed mainly for self-reproducing tilapias (Amarasinghe 1998; Amarasinghe, Ajith Kumara & Ariyaratne, Chapter 27).

Although most stocking activities can be categorised into the broad objectives outlined above, the potential for a successful outcome is often limited because the specific objectives of the exercise in relation to perceived problems and available resources are not fully appraised from the onset. Many projects are ill conceived and do not fully address the issues which have led to the requirement to improve the fishery and possible constraints on the enhancement procedures adopted. Furthermore, they often have little consideration for wider cross-sectoral and environmental issues, particularly in relation to long-term impacts. There is one fundamental issue that is often neglected before a stocking programme is undertaken. 'Why does the fish stock need enhancement?' It is a question that is rarely answered before stocking programmes take place because it is often a reflection of poor management of the environment or the fish stocks themselves. Stock enhancement is frequently required because the fishery has been overexploited in the past or suffered some environmental perturbation. In many instances the issue that should be addressed is whether the constraints acting on the fishery can be removed and the fishery can be enhanced based on natural production.

To address the fundamental causes for the poor status of the fish stocks in the first instance has added benefit because a stocking programme is more likely to succeed if bottlenecks to natural recruitment are removed. Without such action, any benefits accrued from the stocking programme are likely to be dissipated quickly and stocking will have to be done on a continuous basis.

31.5.2 *Management of the fishery*

In addition to direct intervention on the fish populations/communities, fisheries are usually controlled by enforcement of various regulatory constraints to prevent the overexploitation of the resources and maintain a suitable stock structure (e.g. Saat, Vaino, Afanasjev & Koncevaja, Chapter 26). The various measures that are commonly operated in inland fisheries, and their expected outcome, are similar to those imposed on major marine commercial fisheries.

Close seasons are imposed to protect the fish mainly during the breeding season or the early development stages. This includes protection of mature fishes as they move towards spawning or feeding grounds, especially at places where they will be particularly vulnerable to heavy exploitation, and provide a respite for fish to spawn unimpeded. In practice this action has been extended to protect stocks which are heavily exploited and thus restrict catch. This restriction has often come under heavy criticism because close seasons are wrongly timed and do not protect the fish when they are most vulnerable, i.e. the season does not always coincide with the reproductive period, as is typically observed in the UK (Hickley, Marsh & North 1995).

Closed areas are designed to protect stocks directly by denying access to the fisherman. These can range from sanctuary areas, where fishing is prohibited to protect vulnerable life stages of fish, to restrictions on fishing in areas where the fish are particularly vulnerable to exploitation, such as the aggregation of spawning fish in the littoral zone.

Where the catch is removed for consumption, limits are frequently placed on total catch to prevent overexploitation and conserve the spawning stock. Such restrictions allow for the sharing of the catch when stocks are low or under intense fishing pressure. Bag limits are commonly applied in migratory game and put-and-take fisheries to provide equity of catch. An alternative to the bag limit is catch and release, where restrictions are placed on catch and all fish must be released back to the water. A combination of both catch restrictions and release is often employed so the angler can still enjoy the experience if fishing is good but has reached the bag limit.

Gear restrictions are used to reduce exploitation of populations by influencing the efficiency of the fishing method, the species caught or the size of fish caught. It is necessary to base the regulations on sound information on the life history cycles and effectiveness of different fishing techniques, but this information is rarely available.

The final regulatory mechanism commonly used is limitation on the size of fish that can be taken. Restrictions of the size of fish that can be removed for consumption are commonplace in commercial fisheries and are equally applicable to recreational fisheries. The restriction is designed to ensure all immature fish are returned to the water to allow a self-sustaining population to persist. For size limitations to work, and they must be based on sound information about the population size structure, size at the sexual maturity and natural mortality rates.

Irrespective of the classical regulation measures, the fundamental problem usually lies with intense fishing pressure brought about by open access to the fishery resources. However, problems with access are further complicated by social issues such as traditional use rights and family obligations. In addition, fishery exploitation cannot be treated as an open access activity because it is usually governed by considerable legislative and administrative protocols. This is often tied up with rights to fish, and regulations governing such issues at close seasons, gear and catch restrictions, and licensing, most of which are set up to protect the fishing, but others to protect the environment.

In many fisheries in the world, management is wholly under the control of a centralised authority. They regulate effort, through access or catch regulations, but it is evident from the previous discussion that this can lead to social inequity by denying access to some existing or potential issues. Centralised authorities are also ineffective because of the inflexibility to responding to the fluctuating nature of the lake or reservoir fishery resources. There is thus a trend world-wide to charge fishing communities with the management and improvement of their resource either directly through assignment of rights by Governments or less directly by extending the period of leases and licences (Welcomme 2000; Crean, Abila, Lwenya, Omwega, Omwega, Atai, Gonga, Nyapendi, Odongkara, Medard, Onyango & Geheb, Chapter 30).

31.5.3 *Management of the environment*

Environmental change produces impacts on fish assemblages that resemble those produced by fishing whereby many of the larger species tend to disappear and smaller more opportunistic species become more abundant. Such pressures usually arise from pressures associated with human population expansion, i.e. urbanisation, pollution from industrial and agricultural development, change in land use patterns, and demands for water resources. Management of the aquatic ecosystems in this multiple resource user environment is thus a prerequisite for maintenance and/or improvement of the fisheries of lakes and reservoirs. A number of responses to environmental degradation are possible.

Do nothing
Sometimes the problem may not be severe enough to warrant intervention measures. It might cure itself, may be intractable or may be so expensive to treat that the resources might be more usefully spent elsewhere. For example, expensive biomanipulation programmes to control eutrophication may be pointless if nutrient inputs from effluents continue at a high level. The only action that might be possible under such circumstances is the 'do nothing approach'. This is not a negative action, but must be considered a viable alternative when other approaches are not possible or socio-economic influences dictate.

Pro-active measures
Environmental impact assessment (EIA) of schemes prior to development may be used to prevent damaging aspects being incorporated. This is a procedure to encourage universally when schemes are being promoted. Perhaps the best mechanism of implementing enforcing recommendations from EIAs is to make negotiation and acceptance of development money conditional on such measures being instituted. Local legislation may be needed to ensure adoption of such assessments, but this must be accompanied by popular support in the locality and thus may be tied in with socio-political factors.

Pollution control and prevention
The many and varied treatment processes that can be applied to potentially polluting wastes before discharge to water are important in the cleaning up or maintenance of

water quality. Important pollution control and prevention measures for rehabilitation of water quality included:

- removal of phosphate from detergents, which is increasingly being adopted in Europe and America to reduce eutrophication;
- phasing out the production and use of persistent pesticides;
- control of acidic emissions;
- diversion of effluents.

Diversion of effluents may be desirable to allow one water to be sacrificed for the sake of another, or may move the problem to allow a greater area of unaffected water. Diversion may not merely transfer the problem elsewhere if the recipient system affords greater dilution or is more robust in other ways. However, this technique needs careful prior assessment in order to avoid unseen pitfalls.

The persistence of pollutants and their transport and cycling mechanisms in the environment are major factors affecting the likely success of such measures once pollution has occurred.

Biomanipulation

Reduction in the nutrient input into the aquatic environment has had little effect in improving water quality in many cases (Walker 1994). Phosphorus and nitrogen locked in the sediments continues to be released over many years, despite concentrations in the water column becoming depleted. An alternative approach, which has received much attention, is that of biomanipulation. In these studies, instead of concentrating solely on the nutrient source, effort is targeted towards the food-chain dynamics as influenced by fish. Several studies have shown that removal of planktivorous fish can lead to clear water, as a result of reduction on predation pressure on the large-bodied zooplankton which graze on the phytoplankton (Carpenter, Kitchell & Hodgson 1985; Meijer, de Haan, Brukelaar & Buitenveld 1990; Perrow, Jowitt, Stansfield & Phillips 1999; Tomlinson, Perrow, Hoare, Pitt, Johnson, Wilson & Alborough, Chapter 15). This approach certainly warrants further consideration, particularly if it is feasible to remove the large numbers of planktivorous fish required to implement the exercise (Backx & Grimm 1994).

Amelioration and treatment

Even though pollution may not be removed, its adverse effects on organisms may be ameliorated by adjusting water quality. Direct intervention can result in dramatic improvements, such as the aeration and destratification which allowed fish to survive in Salford docks (Hendry, Tinsdeall & White 1994) and liming which has improved the survival of fish in the river downstream of Llyn Brianne, Wales (Evans, Wightman, Rogers & Clarke 1994). These must, however, be considered short-term measures whilst more permanent pollution control measures are implemented.

Natural purification processes can often provide longer-term solutions that form part of the pollution control strategy, such as the provision of riparian buffer zones, which help to filter and buffer the effects of pollutants entering lakes and rivers, especially as a result of land use practices (Zalewski, Puchalski, Frankiewicz & Bis 1994).

Some pollutants can be removed by harvesting plants and animals which have absorbed or incorporated them into their tissues. This offers the possibility of using organisms to bioconcentrate pollutants to clean up environments through selective harvesting?

Rehabilitation
Finally, one mechanism that has not received much attention is rehabilitation of the physical habitat. This is only practical when pressures from other users have eased or as a mechanism to ameliorate a bottleneck in the fishery recruitment processes. They include such actions as shoreline development, e.g. reinstatement of riparian vegetation, and creation of artificial or quasi-natural spawning grounds (Zalewski & Frankiewicz, Chapter 17; Winfield, Fletcher & Winfield, Chapter 19). Artificial reefs, made up of, e.g. old tyres, and replanting of submerged and emergent vegetation are also pertinent. However, the scope for physical modification is limited and many of the rehabilitation practices are linked to improvement in water quality discussed earlier.

31.6 Conclusion

Increasing pressures on aquatic resources dictate that fisheries can no longer be treated in isolation and an integrated approach to aquatic resource management is required (Cowx 1998b). As shown in this chapter, the well-being of any lake or reservoir fishery is being constantly eroded, not only by exploitation of the fish directly but also through degradation of their habitat. Inland waters in heavily populated areas have to serve the needs of water supply, waste water and sewerage disposal, and water-based recreation. How the conflicts of interest between these various interests are resolved must depend on involving all stakeholders in the management process. Integration and co-operation will be essential for sustainable development of the fisheries of lakes and reservoirs, and devolvement of management of the fisheries to the communities is considered essential if the fisheries resources are to be preserved for future generations.

References

Amarasinghe U.S. (1998) How effective are the stocking strategies for the management of reservoir fisheries of Sri Lanka? In I.G. Cowx (ed.) *Stocking and Introduction of Fish.* Oxford: Fishing News Books, Blackwell Science, pp. 422–436.

Backx J.J.G.M & Grimm M.P. (1994) Mass-removal of fish from Lake Wolderwijd, The Netherlands. Part II: Implementation phase. In I.G. Cowx (ed.) *Rehabilitation of Freshwater Fisheries.* Oxford: Fishing News Books, Blackwell Science, pp. 401–414.

Bernacsek G.M. (1984) Dam design and operation to optimize fish production in impounded rier basins. *CIFA Technical Paper* 11, 98 pp.

Caddy J.F. & Mahon R. (1995) Reference points for fisheries management. *FAO Fisheries Technical Paper* 347, 83 pp.

Carpenter S.R., Kitchell J.F. & Hodgson J.R. (1985) Cascading trophic interactions and lake productivity. *Bioscience* **35**, 634–638.

Carvalho G.R. & Cross T. (1998) Enhancing fish production through introductions and stocking: genetic perspectives. In I.G. Cowx (ed.) *Stocking and Introduction of Fish*. Oxford: Fishing News Books, Blackwell Science, pp. 329–337.

Cowx I.G. (1991) The use of angler catch data to examine potential fishery management problems in the lower reaches of the River Trent, England. In I.G. Cowx (ed.) *Catch Effort Sampling Strategies: Their Application in Freshwater Fisheries Management*. Oxford: Fishing News Books, Blackwell Science, pp. 154–165.

Cowx I.G. (1994) Stocking strategies. *Fisheries Management and Ecology* **1**, 15–31.

Cowx I.G. (1996) The integration of fish stock assessment into fisheries management. In I.G. Cowx (ed.) *Stock Assessment in Inland Fisheries*. Oxford: Fishing News Books, Blackwell Science, pp. 495–506.

Cowx I.G. (1997) L'introduction d'espèces éstrangères de poissons dans les eaux douces européennes: succès économiques ou désastre écologique? *Bulletin Français de la Pêche et de la Pisciculture* **344/345**, 57–78 (in French).

Cowx I.G. (ed.) (1998a) *Stocking and Introduction of Fish*. Oxford: Fishing News Books, Blackwell Science, 456 pp.

Cowx I.G. (1998b) Aquatic resource management planning for resolution of fisheries management issues. In P. Hickley & H. Tompkins (eds) *Recreational Fisheries: Social, Economic and Management Aspects*. Oxford: Fishing News Books, Blackwell Science, pp. 97–105.

Cowx I.G. (2002) Recreational fisheries In P. Hart & J Reynolds (eds) *The Fisheries Handbook*. Oxford: Blackwell Science (in press).

Cowx I.G. & Kapasa C. (1995) Changes in reservoir fisheries following impoundment, with special reference to Lake Itezhi-tezhi, Zambia. In T.J. Pitcher & P.J.B. Hart (eds) *The Impact of Species Change on African Lakes*. London: Chapman & Hall, pp. 321–332.

Cowx I.G. & Welcomme R.L. (eds) (1998) *Rehabilitation of Rivers for Fish*. Oxford: Fishing News Books, Blackwell Science, 260 pp.

Cryer M. (1996) Trout yield in Lake Taupo, New Zealand: angler's eldorado or just another fishery? In I.G. Cowx (ed.) *Stock Assessment in Inland Fisheries*. Oxford: Fishing News Books, Blackwell Science, pp. 378–385.

Evans D.M., Wightman R.P., Rogers A.P. & Clarke D.R.K. (1994) Preliminary observations on the use of reservoir liming to ameliorate the impact of acidified waters on the receiving watercourse: a case study of the R. Tywi, S. Wales. In I.G. Cowx (ed.) *Rehabilitation of Freshwater Fisheries*. Oxford: Fishing News Books, Blackwell Science, pp. 435–447.

FAO (1995) *Code of Conduct for Responsible Fisheries*. Rome: FAO, 41 pp.

Feltham M.J., Cowx I.G., Davies J.M., Harvey J.P., Wilson B.R., Holden T. & Britton R.J. (1999) Case studies of the impact of fish-eating birds on inland fisheries in England and Wales. *Report to Ministry of Agriculture, Fisheries and Food, London*, 234 pp.

Fonticiella D.W., Arboleya Z. & Diaz G. (1995) La repoblacion como formaa de manejo de pesquerias en la acuicultura de Cuba. *FAO COPESCAL Documento Ocasional* **10**, 45pp.

Harper D. (1992) *Eutrophication of Freshwaters: Principles, Problems and Restoration*. London: Chapman & Hall, 327 pp.

Hendry K., Tinsdeall M. & White K.N. (1994) Restoration of the fishery of a redeveloped freshwater dock. In I.G. Cowx (ed.) *Rehabilitation of Freshwater Fisheries*. Oxford: Fishing News Books, Blackwell Science, pp. 467–479.

Hickley P. & Tompkins H. (eds) (1998) *Recreational Fisheries: Social, Economic and Management Aspects*. Oxford: Fishing News Books, Blackwell Science, 310 pp.

Hickley P., Marsh C. & North R. (1995) Ecological management of angling. In D.M. Harper & A.J.D. Ferguson (eds) *The Ecological Basis for River Management*. Chichester: J. Wiley & Sons, pp. 415–426.

Lu X. (1992) Fishery management approaches in small reservoirs in China. *FAO Fisheries Circular* 854, 69 pp.

Marshall B.E. (1995) Why is *Limnothrissa miodon* such a successful introduced species and is there anywhere else we should put it? In T.J. Pitcher & P.J.B. Hart (eds) *The Impact of Species Change on African Lakes*. London: Chapman & Hall, pp. 527–546.

Meijer M.-L., Haan M.V. de, Brukelaar A.W. & Buitenveld H. (1990) Is reduction of the benthivorous fish an important cause of high transparency following biomanipulation in shallow lakes? *Hydrobiologia* **200/201**, 303–620.

Moehl Jr J.F. & Davies W.D. (1993) Fishery intensification in small water bodies: a review for North America. *FAO Fisheries Technical Paper* 333, 44 pp.

Moss B. (1998) *Ecology of Fresh Waters*, third edition. Oxford: Blackwell Science, 560 pp.

Ogutu-Ohwayo R & Hecky R. (1991) Fish introductions in Africa and some of their consequences. *Canadian Journal of Fisheries and Aquatic Sciences* **48**(Suppl. 1), 8–12.

Perrow M.R., Jowitt A.J.D., Stansfield J.H. & Phillips J.L. (1999) The practical importance of the interaction between fish, zooplankton and microphytes and shallow lakes restoration. *Hydrobiologia* **395/396**, 199–210.

Petts G.E. (1984) *Impounded Rivers*. Chichester: J. Wiley & Sons, 326 pp.

Ryder R.A. (1965) A method for estimating the potential fish production on north-temperate lakes. *Transactions of the American Fisheries Society* **94**, 214–218.

Salmi P. & Muje K. (2001) Local owner-based management of Finnish lake fisheries: social dimensions and power relations. *Fisheries Management and Ecology* **8**, 435–442.

Schlesinger D.A. & Regier H.A. (1982) Climatic and morphoedaphic indices of fish yield in natural lakes. *Transactions of the American Fisheries Society* **111**, 141–150.

Sunugan V.V. (1995) Reservoir fisheries of India. *FAO Fisheries Technical Paper* 345, 423 pp.

Walker P.A. (1994) Development of pike and perch populations after biomanipulation of fish stocks. In I.G. Cowx (ed.) *Rehabilitation of Freshwater Fisheries*. Oxford: Fishing News Books, Blackwell Science, pp. 376–389.

Welcomme R.L. (1985) River Fisheries. *FAO Fisheries Technical Paper* 262, 330 pp.

Welcomme R.L. (2000) Principles and approaches for river fisheries management. In I.G. Cowx (ed.) *Management and Ecology of River Fisheries*. Oxford: Fishing News Books, Blackwell Science, pp. 331–345.

Welcomme R.L. & Bartley D.M. (1998) Current approaches to the enhancement of fisheries. *Fisheries Management and Ecology* **5**, 351–382.

Wetzel R.G. (1983) *Limnology*, second edition. Orlando: Harcourt Brace College Publishers, 767 pp.

Winfield I.J., Dodge D.P. & Rösch R. (1998) Introductions of the ruffe, *Gymnocephalus cernuus*, to new areas of Europe and North America: history, the present situation and management implications. In I.G. Cowx (ed.) *Stocking and Introductions of Fish*. Oxford: Fishing News Books, Blackwell Science, pp. 191–200.

Zalewski M., Puchalski W., Frankiewicz P. & Bis B. (1994) Riparian ecotones and fish communities in rivers -intermediate complexity hypothesis. In I.G. Cowx (ed.) *Rehabilitation of Freshwater Fisheries*. Oxford: Fishing News Books, Blackwell Science, pp. 152–160.

Subject index

Species index

Printed and bound by CPI Group (UK) Ltd, Croydon, CR0 4YY

16/09/2026

14963387-0003